Recent Progress in UAV-AI Remote Sensing

Recent Progress in UAV-AI Remote Sensing

Editors

Yingying Dong
Chenghai Yang
Giovanni Laneve
Wenjiang Huang

Basel • Beijing • Wuhan • Barcelona • Belgrade • Novi Sad • Cluj • Manchester

Editors

Yingying Dong
Aerospace Information
Research Institute, Chinese
Academy of Sciences
Beijing
China

Chenghai Yang
Aerial Application
Technology Research Unit.,
College Station
Texas
United States

Giovanni Laneve
Sapienza University of Rome
Rome
Italy

Wenjiang Huang
Aerospace Information
Research Institute, Chinese
Academy of Sciences
Beijing
China

Editorial Office
MDPI AG
Grosspeteranlage 5
4052 Basel, Switzerland

This is a reprint of articles from the Special Issue published online in the open access journal *Remote Sensing* (ISSN 2072-4292) (available at: https://www.mdpi.com/journal/remotesensing/special_issues/recent_UAV_AI).

For citation purposes, cite each article independently as indicated on the article page online and as indicated below:

Lastname, A.A.; Lastname, B.B. Article Title. *Journal Name* **Year**, *Volume Number*, Page Range.

ISBN 978-3-7258-2083-2 (Hbk)
ISBN 978-3-7258-2084-9 (PDF)
doi.org/10.3390/books978-3-7258-2084-9

Contents

About the Editors

Yingying Dong

Associate Professor, working in the Aerospace Information Research Institute, Chinese Academy of Sciences, Ph.D. The main research areas include remote sensing inversion mechanisms and models of crop physical and chemical parameters, monitoring and early warning mechanisms and methods for the development dynamics of pests and diseases, and the development of an intelligent platform for crop growth and disease monitoring and forecasting.

Chenghai Yang

Research Agricultural Engineer, working in the U.S. Department of Agriculture, Agricultural Research Service's Aerial Application Technology Research Unit of College Station. His research has focused on the development and application of remote sensing technologies for precision agriculture and rangeland assessment.

Giovanni Laneve

Professor, working in Aerospace Engineering School of the Sapienza University of Rome. From 1999 up to 2013, he has been responsible for the satellite remote sensing activities of the San Marco Project ground station in Kenya. Now, he is responsible for the remote sensing activities at the School. Through his laboratory (eosial.psm.uniroma1.it) for Earth Observation satellite imagery acquisition and processing, he is carrying out research concerning the use of remote sensing for environmental applications and the development of automatic procedures for image processing.

Wenjiang Huang

Professor, Director of SINO-UK Crop Pest and Disease Forecasting & Management Joint Laboratory, working in Aerospace Information Research Institute, Chinese Academy of Sciences, Ph.D. His research interests cover quantitative and remote sensing for vegetation, data fusion (multi-scale, multi-sensor, multi-temporal) for global change, monitoring crop, and grass and forest growth using remote sensing technology.

Preface

The rapid advancements in Unmanned Aerial Vehicles (UAV) and Artificial Intelligence (AI) technologies have significantly transformed the landscape of research and application across various domains. These cutting-edge innovations have not only introduced new possibilities, but have also catalyzed a shift towards more automated, efficient, and accurate data processing and modeling. By harnessing the synergies between UAVs and AI, researchers and practitioners are now able to acquire real-time, spatial, and temporal insights into target areas with unprecedented precision, while simultaneously reducing the workload for operators and analysts.

This Special Issue, which is a testament to the growing interest in the integration of UAV and AI technologies, presents a collection of 15 meticulously curated articles. The focus of these contributions lies in the application of UAV imagery for vegetation monitoring and dynamic target tracking, showcasing the remarkable potential of UAV multispectral and hyperspectral imagery in these fields. Through these studies, the authors explore a wide range of topics including the inversion of key physical and chemical parameters of crops, monitoring crop growth, building extraction, target tracking, and the monitoring and forecasting of crop pests and diseases.

Each article in this collection delves into innovative methodologies for extracting valuable information from UAV imagery, providing new insights and advancing our understanding of the application of UAVs in environmental monitoring, agriculture, and beyond. As such, this Special Issue serves as both a reference and a source of inspiration for researchers, professionals, and students who are interested in the intersection of UAV technology and AI.

It is our hope that the knowledge shared within these pages will not only inform but also inspire future research and development in this rapidly evolving field. The integration of UAV and AI technologies holds immense promise, and we are excited to see how these advancements will continue to shape the future of data-driven research and applications.

We extend our deepest gratitude to the authors, reviewers, and everyone who has contributed to the success of this Special Issue. We trust that you will find the articles contained herein both insightful and valuable as you explore the forefront of UAV and AI integration.

Yingying Dong, Chenghai Yang, Giovanni Laneve, and Wenjiang Huang
Editors

 *remote sensing*

Editorial

Editorial for Special Issue: "Recent Progress in UAV-AI Remote Sensing"

Yingying Dong [1,2], Chenghai Yang [3], Giovanni Laneve [4] and Wenjiang Huang [1,2,*]

[1] Key Laboratory of Digital Earth Science, Aerospace Information Research Institute, Chinese Academy of Sciences, Beijing 100094, China; dongyy@aircas.ac.cn
[2] University of Chinese Academy of Sciences, Beijing 100190, China
[3] USDA-Agricultural Research Service, Aerial Application Technology Research Unit, College Station, TX 77845, USA; chenghai.yang@usda.gov
[4] Earth Observation and Satellite Image Applications Laboratory (EOSIAL), School of Aerospace Engineering (SIA), Università di Roma 'La Sapienza', Via Salaria, 851, 00138 Roma, Italy; giovanni.laneve@uniroma1.it
* Correspondence: huangwj@aircas.ac.cn

1. Introduction

The development of unmanned aerial vehicles (UAV) and artificial intelligence (AI) techniques has drawn increasing interest and started a novel area of research applications. Combining the advantages of UAV and AI, automatic and fast processing and modelling can be achieved, instant and spatially and temporally varying knowledge of target areas can be obtained, and the workload of operators and instructors can be greatly reduced. This Special Issue aims at studies covering the uses of AI techniques to interpret data obtained using different UAV sensors, providing references for the application of UAV imagery in various fields.

This Special Issue consists of 15 articles, focusing primarily on vegetation monitoring and target tracking using UAV imagery. These articles analyze the potential applications of UAV multispectral and hyperspectral imagery in vegetation growth monitoring and dynamic target tracking. The topics covered include the inversion of key physical and chemical parameters of crops, crop growth monitoring, building extraction, target tracking, and crop pest and disease monitoring and forecasting, among others. Additionally, these articles present numerous novel methods for extracting valuable information from UAV imagery.

2. Inversion of Crop Physical Parameters and Biomass Estimation

The majority of articles focus on how to use UAV imagery for the accurate estimation and monitoring of crop physical parameters and biomass. In the inversion of crop physical parameters, Ref. [1] focuses on comparing various spectral indices to estimate the nitrogen content of potato plants during different growth periods. This study evaluates the effectiveness of different indices in predicting nitrogen levels, which is crucial for optimizing fertilizer applications and enhancing potato crop yield. Ref. [2] compares the use of UAV RGB imagery and hyperspectral remote-sensing data for monitoring the growth of winter wheat. This study evaluates the capabilities of both data sources in capturing relevant information about wheat growth, such as vegetation health and biomass, to inform agricultural decision-making. Ref. [3] investigates the assessment of maize canopy and leaf chlorophyll content using leaf spectral reflectance data. This study examines the accuracy of estimating chlorophyll content at different growth stages and vertical positions within the maize canopy, providing insights into the crop's physiological state and potential yield.

In the estimation of crop biomass, Ref. [4] uses UAVs equipped with hyperspectral imaging to estimate the above-ground biomass of potato plants. This study employs machine-learning regression techniques to analyze the hyperspectral data and accurately estimate potato biomass, providing valuable information for crop management and yield

Citation: Dong, Y.; Yang, C.; Laneve, G.; Huang, W. Editorial for Special Issue: "Recent Progress in UAV-AI Remote Sensing". *Remote Sens.* **2023**, *15*, 4382. https://doi.org/10.3390/rs15184382

Received: 16 August 2023
Revised: 25 August 2023
Accepted: 4 September 2023
Published: 6 September 2023

estimation. Ref. [5] also utilizes UAVs to estimate potato above-ground biomass. In this case, this study explores the use of vegetation indices and green-edge parameters obtained from UAV data to assess the biomass of potato plants. This information can help farmers and researchers monitor potato growth and health. Ref. [6] focuses on extracting characteristic variables from UAV hyperspectral imagery to estimate the aboveground biomass of potato plants. This study uses specific spectral features and patterns to develop a reliable estimation method for potato biomass, contributing to precision agriculture and crop management.

3. Object Detection and Tracking

Some of the articles focus on research into object detection and tracking technology based on drones. In the field of object detection and extraction, Ref. [7] explores a comprehensive approach for building extraction and floor area estimation in rural Chinese villages using UAV photogrammetry and a novel deep learning network called EDSANet. This study aims to efficiently and accurately analyze building structures and floor areas to support rural development and land-use planning. Ref. [8] focuses on weakly supervised learning to detect transmission lines in UAV imagery. This study introduces a novel algorithm based on unpaired image-to-image translation that requires only image-level labels. The attention module improves detection accuracy, making it effective for detecting transmission lines in UAV images. Ref. [9] discusses the design of a wide-area and real-time object search system for UAVs. The system utilizes high-resolution cameras to collect aerial images with a large field of view, and parallel processing techniques to achieve real-time object search in a wide area.

In the field of object tracking, Ref. [10] addresses challenges in visual tracking for UAVs, particularly dealing with occlusion and deformation. The research proposes an algorithm based on the attention-based mask generative network to enhance the accuracy and robustness of target tracking in UAV imagery. Ref. [11] focuses on using UAVs equipped with AI capabilities and edge computing for remote sensing observation. This study introduces the AERO system, which utilizes deep learning models for object detection and tracking. The edge computing based on UAVs enables real-time data analysis and transmission without relying on cloud computation. Ref. [12] presents a method for multiple object tracking in drone videos using a temporal-association network with a separated-tasks structure. This study aims to improve the efficiency and accuracy of object tracking in drone videos. Ref. [13] proposes a method for designated target anti-interference tracking in UAVs by combining the YOLOv5 and SiamRPN algorithms. This research work focuses on enhancing tracking and landing control performance in UAV operations.

4. Crop Disease Monitoring and Forecasting

In addition, two articles in this Special Issue explore the feasibility of using UAV imagery for crop disease monitoring and forecasting. Ref. [14] proposes a monitoring approach for wheat rust using UAV imagery. This study utilizes the mRMR-XGBoost algorithm, which combines feature selection (mRMR) and the XGBoost machine learning algorithm. The goal is to accurately detect and monitor the occurrence and spread of wheat rust using remote sensing technology. Ref. [15] combines disease mechanism knowledge with machine learning techniques to develop a forecasting model for wheat fusarium head blight. This study aims to improve the accuracy of disease forecasting and enhance early warning systems to mitigate the impact of the disease on wheat.

5. Conclusions

This Special Issue presents innovative applications of UAV and AI techniques in the field of Earth observation. By leveraging the advantages of UAV and AI, rapid and automated data processing and modeling can be achieved, facilitating the acquisition of spatio-temporal information in target areas and reducing information acquisition costs. This Special Issue comprises 15 articles, primarily exploring methods to utilize AI techniques for interpreting

UAV imagery data. The majority of the articles focus on accurately monitoring crop growth parameters and biomass. Additionally, there are studies dedicated to UAV-based object detection and tracking technologies, covering building structures, power transmission lines, etc. Furthermore, this Special Issue discusses methods of using UAV imagery to monitor and forecast crop diseases, including wheat rust and fusarium head blight.

The above-mentioned studies have built UAV-AI systems to address complex problems, promoting the development of UAV-AI and providing references for the application and advancement of UAV technology in various fields.

Author Contributions: All authors contributed to the writing of this Editorial. Y.D. is responsible for summarizing and reviewing the various sections of this Editorial. C.Y. has summarized the content of the "Inversion of Crop Physical Parameters and Biomass Estimation" section. G.L. has summarized the content of the "Object Detection and Tracking" section. W.H. has summarized the content of the "Crop Disease Monitoring and Forecasting" section. All authors have read and agreed to the published version of the manuscript.

Acknowledgments: We would like to acknowledge the authors who contributed to this Special Issue.

Conflicts of Interest: The authors declare no conflict of interest.

References

1. Fan, Y.; Feng, H.; Yue, J.; Liu, Y.; Jin, X.; Xu, X.; Song, X.; Ma, Y.; Yang, G. Comparison of Different Dimensional Spectral Indices for Estimating Nitrogen Content of Potato Plants over Multiple Growth Periods. *Remote Sens.* **2023**, *15*, 602. [CrossRef]
2. Feng, H.; Tao, H.; Li, Z.; Yang, G.; Zhao, C. Comparison of UAV RGB imagery and hyperspectral remote-sensing data for monitoring winter wheat growth. *Remote Sens.* **2022**, *14*, 3811. [CrossRef]
3. Yang, H.; Ming, B.; Nie, C.; Xue, B.; Xin, J.; Lu, X.; Xue, J.; Hou, P.; Xie, R.; Wang, K.; et al. Maize Canopy and Leaf Chlorophyll Content Assessment from Leaf Spectral Reflectance: Estimation and Uncertainty Analysis across Growth Stages and Vertical Distribution. *Remote Sens.* **2022**, *14*, 2115. [CrossRef]
4. Liu, Y.; Feng, H.; Yue, J.; Fan, Y.; Jin, X.; Zhao, Y.; Song, X.; Long, H.; Yang, G. Estimation of Potato Above-Ground Biomass Using UAV-Based Hyperspectral images and Machine-Learning Regression. *Remote Sens.* **2022**, *14*, 5449. [CrossRef]
5. Liu, Y.; Feng, H.; Yue, J.; Fan, Y.; Jin, X.; Song, X.; Yang, H.; Yang, G. Estimation of Potato Above-Ground Biomass Based on Vegetation Indices and Green-Edge Parameters Obtained from UAVs. *Remote Sens.* **2022**, *14*, 5323. [CrossRef]
6. Liu, Y.; Feng, H.; Yue, J.; Li, Z.; Jin, X.; Fan, Y.; Feng, Z.; Yang, G. Estimation of aboveground biomass of potatoes based on characteristic variables extracted from UAV hyperspectral imagery. *Remote Sens.* **2022**, *14*, 5121. [CrossRef]
7. Zhou, J.; Liu, Y.; Nie, G.; Cheng, H.; Yang, X.; Chen, X.; Gross, L. Building extraction and floor area estimation at the village level in rural China via a comprehensive method integrating UAV photogrammetry and the novel EDSANet. *Remote Sens.* **2022**, *14*, 5175. [CrossRef]
8. Choi, J.; Lee, S.J. Weakly Supervised Learning for Transmission Line Detection Using Unpaired Image-to-Image Translation. *Remote Sens.* **2022**, *14*, 3421. [CrossRef]
9. Li, X.; He, B.; Ding, K.; Guo, W.; Huang, B.; Wu, L. Wide-Area and Real-Time Object Search System of UAV. *Remote Sens.* **2022**, *14*, 1234. [CrossRef]
10. Bai, Y.; Song, Y.; Zhao, Y.; Zhou, Y.; Wu, X.; He, Y.; Zhang, Z.; Yang, X.; Hao, Q. Occlusion and Deformation Handling Visual Tracking for UAV via Attention-Based Mask Generative Network. *Remote Sens.* **2022**, *14*, 4756. [CrossRef]
11. Koubaa, A.; Ammar, A.; Abdelkader, M.; Alhabashi, Y.; Ghouti, L. AERO: AI-Enabled Remote Sensing Observation with Onboard Edge Computing in UAVs. *Remote Sens.* **2023**, *15*, 1873. [CrossRef]
12. Lin, Y.; Wang, M.; Chen, W.; Gao, W.; Li, L.; Liu, Y. Multiple Object Tracking of Drone Videos by a Temporal-Association Network with Separated-Tasks Structure. *Remote Sens.* **2022**, *14*, 3862. [CrossRef]
13. Wu, D.; Zhu, H.; Lan, Y. A Method for Designated Target Anti-Interference Tracking Combining YOLOv5 and SiamRPN for UAV Tracking and Landing Control. *Remote Sens.* **2022**, *14*, 2825. [CrossRef]
14. Jing, X.; Zou, Q.; Yan, J.; Gao, W.; Li, L.; Liu, Y. Remote Sensing Monitoring of Winter Wheat Stripe Rust Based on mRMR-XGBoost Algorithm. *Remote Sens.* **2022**, *14*, 756. [CrossRef]
15. Li, L.; Dong, Y.; Xiao, Y.; Liu, L.; Zhao, X.; Huang, W. Combining disease mechanism and machine learning to predict wheat fusarium head blight. *Remote Sens.* **2022**, *14*, 2732. [CrossRef]

remote sensing

Article

Comparison of Different Dimensional Spectral Indices for Estimating Nitrogen Content of Potato Plants over Multiple Growth Periods

Yiguang Fan [1,†], Haikuan Feng [1,2,*,†], Jibo Yue [3], Yang Liu [1,4,5], Xiuliang Jin [6], Xingang Xu [1], Xiaoyu Song [1], Yanpeng Ma [1] and Guijun Yang [1]

[1] Key Laboratory of Quantitative Remote Sensing in Agriculture of Ministry of Agriculture and Rural Affairs, Information Technology Research Center, Beijing Academy of Agriculture and Forestry Sciences, Beijing 100097, China

[2] College of Agriculture, Nanjing Agricultural University, Nanjing 210095, China

[3] College of Information and Management Science, Henan Agricultural University, Zhengzhou 450002, China

[4] Key Lab of Smart Agriculture System, Ministry of Education, China Agricultural University, Beijing 100083, China

[5] Key Laboratory of Agricultural Information Acquisition Technology, Ministry of Agriculture and Rural Affairs, China Agricultural University, Beijing 100083, China

[6] Institute of Crop Sciences, Chinese Academy of Agricultural Sciences/Key Laboratory of Crop Physiology and Ecology, Ministry of Agriculture, Beijing 100081, China

* Correspondence: fenghk@nercita.org.cn

† These authors contributed equally to this work.

Citation: Fan, Y.; Feng, H.; Yue, J.; Liu, Y.; Jin, X.; Xu, X.; Song, X.; Ma, Y.; Yang, G. Comparison of Different Dimensional Spectral Indices for Estimating Nitrogen Content of Potato Plants over Multiple Growth Periods. *Remote Sens.* **2023**, *15*, 602. https://doi.org/10.3390/rs15030602

Academic Editors: Aitazaz A. Farooque and Saeid Homayouni

Received: 22 November 2022
Revised: 6 January 2023
Accepted: 17 January 2023
Published: 19 January 2023

Abstract: The estimation of physicochemical crop parameters based on spectral indices depend strongly on planting year, cultivar, and growing period. Therefore, the efficient monitoring of crop growth and nitrogen (N) fertilizer treatment requires that we develop a generic spectral index that allows the rapid assessment of the plant nitrogen content (PNC) of crops and that is independent of year, cultivar, and growing period. Thus, to obtain the best indicator for estimating potato PNC, herein, we provide an in-depth comparative analysis of the use of hyperspectral single-band reflectance and two- and three-band spectral indices of arbitrary bands for estimating potato PNC over several years and for different cultivars and growth periods. Potato field trials under different N treatments were conducted over the years 2018 and 2019. An unmanned aerial vehicle hyperspectral remote sensing platform was used to acquire canopy reflectance data at several key potato growth periods, and six spectral transformation techniques and 12 arbitrary band combinations were constructed. From these, optimal single-, two-, and three-dimensional spectral indices were selected. Finally, each optimal spectral index was used to estimate potato PNC under different scenarios and the results were systematically evaluated based on a correlation analysis and univariate linear modeling. The results show that, although the spectral transformation technique strengthens the correlation between spectral information and potato PNC, the PNC estimation model constructed based on single-band reflectance is of limited accuracy and stability. In contrast, the optimal three-band spectral index TBI 5 (530,734,514) performs optimally, with coefficients of determination of 0.67 and 0.65, root mean square errors of 0.39 and 0.39, and normalized root mean square errors of 12.64% and 12.17% for the calibration and validation datasets, respectively. The results thus provide a reference for the rapid and efficient monitoring of PNC in large potato fields.

Keywords: potato; plant nitrogen content; unmanned aerial vehicle; hyperspectral; spectral indices

1. Introduction

Studies have shown that yields in agricultural systems must increase 3% annually to meet the global demand for food due to the increasing population and decreasing amount of arable land [1,2]. However, the most widely grown crops globally (rice, wheat, and maize) are constrained by technology and rising costs and have limited room to continue increasing yields. In contrast, potato is a high yield and nutritional food crop rich and is gradually gaining popularity in more than 160 countries and regions due to its cold tolerance, drought resistance, and environmental adaptability. As a result, it has become the fourth largest food crop globally, and its yield and quality have become vital for ensuring world food security [3].

Nitrogen (N) is a critical component of structural molecules, such as chlorophyll, nucleic acids, and proteins, and is essential for potato quality and yield [4]. Plant nitrogen content (PNC) is an important indicator for assessing the N nutritional status of crops, and obtaining PNC information quickly and non-destructively is important for assessing crop N surplus or deficit and growth status, which aids fertilization management and decision-making in precision agriculture [5,6]. Traditional PNC measurement methods tend to combine manual sampling and laboratory chemical analysis, which is accurate and reliable but time-consuming and thus cannot meet the needs of contemporary precision agriculture for high-throughput, real-time monitoring of crop growth conditions [7]. Therefore, it is urgent to develop new platforms and technologies that allow for the low cost, efficient, and robust acquisition of crop phenotype status.

Remote sensing technology can rapidly capture the radiation from crop canopies over large distance scales and without contact. These data can then be analyzed and processed to obtain high-throughput phenotypic information on the crop canopy, which in turn provides technical support for the rapid and non-destructive monitoring of crop physical and chemical parameters [8,9]. Unmanned aerial vehicle (UAV) remote sensing technology provides high spatial and temporal resolution remote sensing images of crops over all growth periods at a lower cost and more conveniently and efficiently than is possible with satellite or ground remote sensing. This technique is thus more conducive to the real-time monitoring of crop growth conditions and provides a reference for field management and scientific decision-making in the field [10,11].

Vegetation indices (VIs) are indicators obtained by the mathematical transformation of two or more specific crop canopy reflectance bands that are simple in form and efficient in the calculation. VIs have thus become an essential tool for monitoring the N nutrient status of crops [12–17]. The results of existing studies have indicated that, although it is feasible to monitor the N nutrient status of crops based on VIs, some limitations remain, mainly in the inconsistency of N monitoring models constructed based on VIs from different periods due to the influence of seasons, growth periods, cultivars, and growth environment, preventing the generalization of the models [18–20]. These factors make it difficult to use optical remote sensing technology to monitor the PNC status of crops in multiple growth periods.

Improved computer performance and image processing technology have provided opportunities to solve these problems. For example, some researchers have introduced new variables, such as morphological parameters (e.g., plant height and cover) [21,22] and texture [23], to monitor the PNC status of crops over multiple growth periods. Simultaneously, sophisticated machine learning methods (e.g., neural networks, random forests, partial least squares support vector machines, etc.) [10,24,25] have also been used to construct PNC estimation models. These measures have achieved some results, but they also introduce problems. On the one hand, the increased technical threshold makes the use of this technology difficult for agricultural workers who are not specialized in remote sensing. On the other hand, the increased model complexity and production operation cost are not conducive to the development and integration of real-time crop PNC-detection devices, which limits the widespread use of these models.

The development of hyperspectral technology has made it possible to obtain high-throughput spectral information on crop canopies [26,27]. Some studies now focus on finding suitable spectral indices rather than using complex methods to improve the accuracy of VIs for estimating PNC in crops with multiple growth periods. For example, the double peak canopy nitrogen (N) index developed by Chen et al. [5] minimizes the interference of the leaf area index to accurately estimate the PNC of winter wheat and maize.

Schlemmer et al. [28] showed that the MERIS terrestrial chlorophyll index provides more accurate estimates of canopy N and chlorophyll content in maize than the conventional VIs. Feng et al. [13] improved leaf N content by developing a water-resistance N index and widened the applicability of the model used to estimate leaf N content under different climatic conditions. Wang et al. [29] showed that combining wavelengths at 924, 703, and 423 nm attenuates the saturation of conventional VIs and enhances the stability and accuracy of the model used to estimate leaf N content. These studies suggest that high dimensional spectral indices, such as three-band spectral indices (TBIs), have a significant potential for estimating PNC over multiple crop growth periods [30]. However, to date, few studies have compared different dimensional spectral indices for estimating N-nutrient status over multiple growth periods. Moreover, unlike crops, such as wheat and maize, potato continuously transfers N to the tuber in all growth periods. Thus, it remains unknown whether existing remote sensing monitoring methods can be applied to potato. Therefore, it is essential to systematically investigate the use of spectral indices of differing dimensions for estimating PNC over multiple potato growth periods.

Thus, this study uses a UAV as a remote sensing platform to acquire hyperspectral images of potato at the bud emergence (S1), tuber formation (S2), tuber growth (S3), and starch accumulation (S4) periods over the years 2018 and 2019. We focus on single-, two-, and three-dimensional spectral indices: hyperspectral single-band reflectance, two- and three-band spectral indices with arbitrary band combinations, respectively, and systematically assess the accuracy of PNC estimates over multiple potato growth periods to determine the optimal indicators for monitoring the N-nutritional status of potato. The first goal of this study is to systematically evaluate the performance of single-, two-, and three-dimensional spectral indices for estimating potato PNC based on spectral transformation techniques and various combinations of optical bands. A second goal is to investigate the optimal spectral indices and models for estimating PNC of potato over different planting years, cultivars, and growth periods. The results of the study should improve the accuracy of PNC estimates of potato over multiple growth periods and provide a reference for developing and integrating PNC-detection devices for the accurate management of N in agricultural fields.

2. Materials and Methods

2.1. Experiment Design

Potato trials were conducted from April to July 2018 and 2019 at the National Precise Agriculture Research and Demonstration Base in Xiaotangshan Town, Changping District, Beijing, China ($40°10'34''$N, $116°26'39''$E). The site has an average altitude of 36 m, an average annual rainfall of 644 mm, and a yearly average temperature of 11.8 °C. The climate is typically warm–temperate, semi-humid, continental monsoon. In the 2018 potato field trial, Zhongshu 5 (Z5) and Zhongshu 3 (Z3) were grown at different densities (T plots), and Zhongshu 195 (Z195) and Z3 were grown at different nitrogen fertilizers (N plots). The T Plots were treated with three densities: 63,000 plants/hm^2 (T0), 72,000 plants/hm^2 (T1), and 81,000 plants/hm^2 (T2), with three replications of each treatment for a total of 18 plots. The N plots were treated with four nitrogen levels: 0 kg/hm^2 urea (N0), 163.05 kg/hm^2 urea (N1), 326.10 kg/hm^2 urea (N2, normal treatment, 152.29 kg pure N), and 489.15 kg/hm^2 urea (N3), with three replications of each treatment for a total of 24 plots. In the 2019 potato field trial, Z5 and Z3 were planted at different densities and with different nitrogen and potassium fertilizers (K plots). The T plots were treated with three density levels: 60,000 plants/hm^2 (T3), 72,000 plants/hm^2 (T4), and 84,000 plants/hm^2 (T5), with three replications of each treatment for a total of 18 plots. The N plots were treated with four nitrogen levels: 0 kg/hm^2 urea (N4), 244.65 kg/hm^2 urea (N5), 489.15 kg/hm^2 urea (N6, normal treatment, 228.43 kg pure N), and 733.50 kg/hm^2 urea (N7), with three replications of each treatment for a total of 24 plots. K plots were treated with three potassium fertilizer treatments: 0 kg/hm^2 potassium fertilizer (K0), 970.50 kg/hm^2 potassium fertilizer (K1, the T plots and N plots received the K1 treatment), and 1941 kg/hm^2 potassium fertilizer (K2), with three replications of each treatment for a total of six plots. In 2018 and 2019, we used

42 and 48 plots, respectively, and the area of each plot was 32.5 m^2. Potato was grown by mulching, and field management included weeding, watering, and spraying. The location of the trial field and the trial profile are shown in Figure 1.

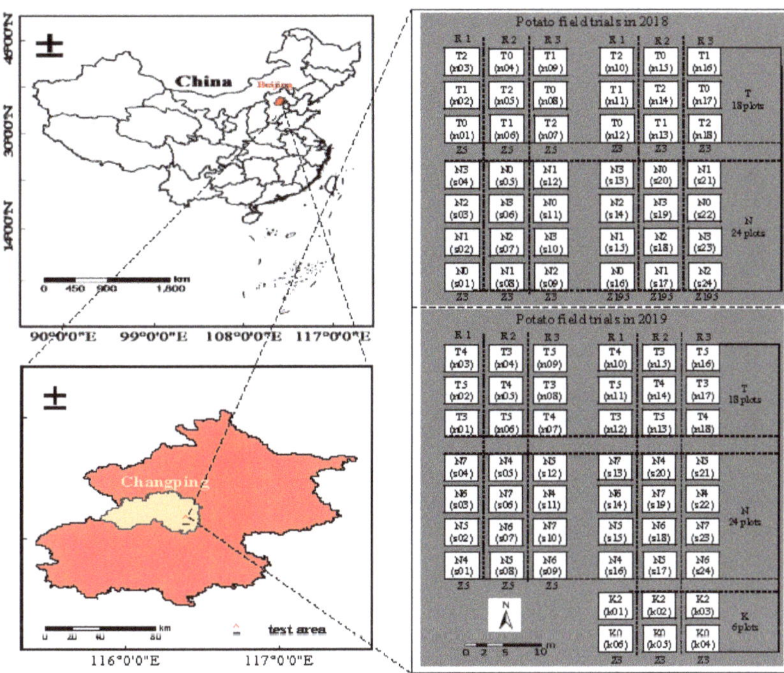

Figure 1. Location of potato field and experimental design. Note: R is an abbreviation for Replication.

2.2. UAV Hyperspectral Data Acquisition and Processing

This study used the DJI M600 UAV with a UHD 185 imaging spectrometer from Cubert, Germany as the remote sensing platform to collect hyperspectral data during the periods of the potato bud emergence (15 May 2018, 13 May 2019), tuber formation (29 May 2018, 28 May 2019), tuber growth (5 June 2018, 10 June 2019), and starch accumulation (19 June 2018, 20 June 2019). The UHD 185 is a full-frame, real-time imaging airborne high speed imaging spectrometer with a spectral range of 450–950 nm, a sampling spectral interval of 4 nm, and a spectral resolution of 8 nm. It measures 195 mm × 67 mm × 60 mm and weighs 470 g. All UAV operations were conducted under windless, cloudless, and light stable conditions from 11:30 am to 1:30 pm local time. The UAV flight altitude was 20 m and its flight speed was about 1.5 m/s. To ensure the quality of the UAV hyperspectral images, the sampling interval of the hyperspectral sensor was set to 1 s, and the overlap of heading and side direction during the flight were set to 85% and 93%, respectively, which translates into a ground resolution of the acquired images of approximately 1.3 cm. Dark-current collection and whiteboard calibration were performed prior to collecting UAV hyperspectral data to ensure accurate reflectance data from the potato canopy at each growth period.

The UAV hyperspectral data were processed by geometric correction, image stitching, image fusion, and spectral data extraction. First, the surface reflectance was calibrated by making radiometric measurements of the digital values of the hyperspectral images based on black-and-white panel data. Second, the grayscale images were stitched together using the motion structure algorithm as implemented in the Agisoft Photoscan software (Agisoft, LLC, St. Petersburg, Russia). The hyperspectral cubes and grayscale images were then fused together by using the Cubert-Pilot software to obtain the complete hyperspectral images. Finally, the average hyperspectral reflectance of the region of interest was extracted from the vector data of each potato test plot using the ENVI 5.3 software.

2.3. Acquisition of Potato PNC Data

After completing the UAV flights, three potato plants representative of the overall plot growth were destructively sampled from each plot and rapidly transported to the laboratory in plastic bags to obtain ground truth PNC data. The samples were washed with running water, and the stems and leaves were separated and killed at 105 °C for 0.5 h and then dried at 80 °C to a constant weight. The dry matter mass of the stems and leaves was measured separately by using a high precision balance (0.0001 g). Finally, the nitrogen content of the stems and leaves was measured independently by using a Kjeldahl nitrogen tester, and the PNC data for each potato plot were calculated as follows:

$$PNC = \frac{LNC \times LDM + SNC \times SDM}{LDM + SDM} \tag{1}$$

where LNC and SNC are the leaf N content and stem N content, respectively, and LDM and SDM are the leaf dry matter and stem dry matter, respectively.

2.4. Calculation of Spectral Index

Currently, most VIs used to monitor crop N status fall into three categories: normalized difference vegetation index, difference vegetation index, and ratio vegetation index. In addition, by adding a constant, soil-adjusted vegetation index is often used to estimate crop N nutrient status. VIs in the form of optimal vegetation indices and chlorophyll indices are also closely related to crop N status. We therefore selected six two-band spectral indices, namely normalized difference spectral index (NDSI), difference spectral index (DSI), ratio spectral index (RSI), soil-adjusted spectral index (SASI, L = 0.5), chlorophyll spectral index (CSI), and optimal spectral index (OSI), and used the "lambda-by-lambda" band-optimization algorithm to determine the optimal band combinations for estimating PNC over multiple potato growth periods. Considering the superior performance of TBIs for monitoring the physicochemical parameters of crops, six TBIs were constructed in this study. The optimal band combinations were also screened using the lambda-by-lambda band-optimization algorithm to investigate the use of TBIs for monitoring PNC in potato over multiple growth periods. This method produced the "optimal spectral indices." Table 1 lists the two- and three-band spectral indices constructed in this study.

Table 1. Formulas for selected two- and three-band spectral indices used in this study.

Type	Spectral Index	Formula	Reference
	NDSI	$(R_{\lambda1} - R_{\lambda2})/(R_{\lambda1} + R_{\lambda2})$	[31]
	RSI	$R_{\lambda1}/R_{\lambda2}$	[32]
Two-band	DSI	$R_{\lambda1} - R_{\lambda2}$	[33]
spectral indices	SASI	$(1 + L)(R_{\lambda1} - R_{\lambda2})/(R_{\lambda1} + R_{\lambda2} + L)$	[34]
	CSI	$(R_{\lambda1} - R_{\lambda2})/R_{\lambda1}$	[35]
	OSI	$(1 + 0.45)(2R_{\lambda2} + 1)/(R_{\lambda1} + 0.45)$	[36]
	TBI 1	$(R_{\lambda1} - R_{\lambda2})/(R_{\lambda2} + R_{\lambda3})$	[37]
	TBI 2	$(R_{\lambda1} - 1.8R_{\lambda2})/(R_{\lambda3} - 1.8R_{\lambda2})$	[38]
Three-band	TBI 3	$R_{\lambda1}/(R_{\lambda2}R_{\lambda3})$	[12]
spectral indices	TBI 4	$R_{\lambda1}/(R_{\lambda2} + R_{\lambda3})$	[30]
	TBI 5	$(R_{\lambda1} - R_{\lambda2})/(R_{\lambda1} + R_{\lambda2} - 2R_{\lambda3})$	[39]
	TBI 6	$(R_{\lambda1} - R_{\lambda2} + 2R_{\lambda3})/(R_{\lambda1} + R_{\lambda2} + 2R_{\lambda3})$	[29]

2.5. Model Construction and Validation

For this study, we collected a total of 360 sets of hyperspectral and PNC data over four potato growth periods in 2018 and 2019. Data from replicates 2 and 3 (240 sets) were used to construct a model to estimate potato PNC. Data from replicate 1 (120 sets) were used to verify the accuracy and stability of the model. Finally, the coefficient of determination (R^2), the root mean square error (RMSE), and the normalized root mean square error (NRMSE) were calculated to quantitatively evaluate the accuracy and stability of the constructed estimation model.

3. Analysis of Results

3.1. PNC of Potato for Different Growth Periods and Years

Table 2 lists the mean, standard deviation (SD), and coefficient of variation (CV) of the PNC used for modeling and validation for different years and different potato growth periods. The PNC data obtained for each potato growth period have a significant coefficient of variation. In most growth periods, the standard deviation and coefficient of variation of the modeling and validation datasets follow similar trends, indicating that the data may be used for further analysis.

Table 2. Descriptive statistics of potato PNC for calibration and validation datasets.

Year	Growth Period	Calibration				Validation			
		Range	Mean	SD	CV (%)	Range	Mean	SD	CV (%)
2018	S1	2.70–4.59	3.70	0.50	13.50	3.06–4.57	3.66	0.47	14.77
	S2	2.32–4.03	3.11	0.50	15.92	2.37–3.46	2.98	0.34	17.80
	S3	2.12–3.98	3.20	0.51	16.01	2.02–3.63	3.00	0.45	13.43
	S4	1.76–3.79	2.62	0.44	16.97	1.90–3.57	2.79	0.44	14.41
2019	S1	2.05–5.15	3.65	0.81	22.18	2.82–5.08	3.91	0.76	19.69
	S2	2.09–4.50	3.15	0.61	19.39	2.47–4.16	3.33	0.59	18.73
	S3	1.61–4.00	2.67	0.60	22.54	1.96–3.54	2.75	0.52	22.47
	S4	1.86–3.74	2.82	0.46	16.41	2.21–3.70	3.11	0.40	16.99

3.2. Correlation Analysis between PNC and Single-, Two-, and Three-Dimensional Spectral Indices

3.2.1. Correlation between Hyperspectral Single-Band Reflectance and PNC

Five spectral transformation techniques, namely the standard normal variate transform (SNVR) [40], the first derivative (FDR), the second derivative (SDR) [41], the logarithm of the reciprocal of the spectra, (Log(1/R)) [42], and continuous removal (CR) [43], were applied to the hyperspectral original reflectance (OR) to explore in depth the connection between hyperspectral single-band reflectance and PNC. The results of the analysis of the correlation between the six forms of single-band reflectance and potato PNC over multiple growth periods (Figure 2) show that FDR, SDR, and CR correlate more with PNC than with OR. In contrast, Log(1/R) and SNVR have the opposite or the same correlation as OR.

Table 3 shows the sensitive band intervals of different types of single-band reflectance with PNC and the locations of the most closely linked wavelengths. The spectral transformation technique can further exploit spectral information to enhance the linkage with PNC, and the wavelengths most closely correlated with PNC for different single-band reflectance are in the red- or green-edge regions.

3.2.2. Association between Two-Band Spectral Indices and PNC

Based on the hyperspectral OR data, the lambda-by-lambda band-optimization algorithm was used to construct the six forms of the two-band spectral indices listed in Table 1. The contour maps of their coefficients of correlation with potato PNC are plotted in Figure 3. The results show that the correlation of OSI with PNC differs from that of other spectral indices. The spectral combinations of the various spectral indices that correlate optimally with PNC are NDSI (618,490), RSI (618,490), DSI (578,494), SASI (586,494), CSI (618,494), and OSI (586,490), corresponding to correlation coefficients of 0.62, 0.61, 0.74, 0.73, 0.62, and −0.72, respectively. The correlation of the six different forms of two-band spectral indices screened by PNC over multiple potato-growth periods is significantly greater than that of the hyperspectral single-band reflectance.

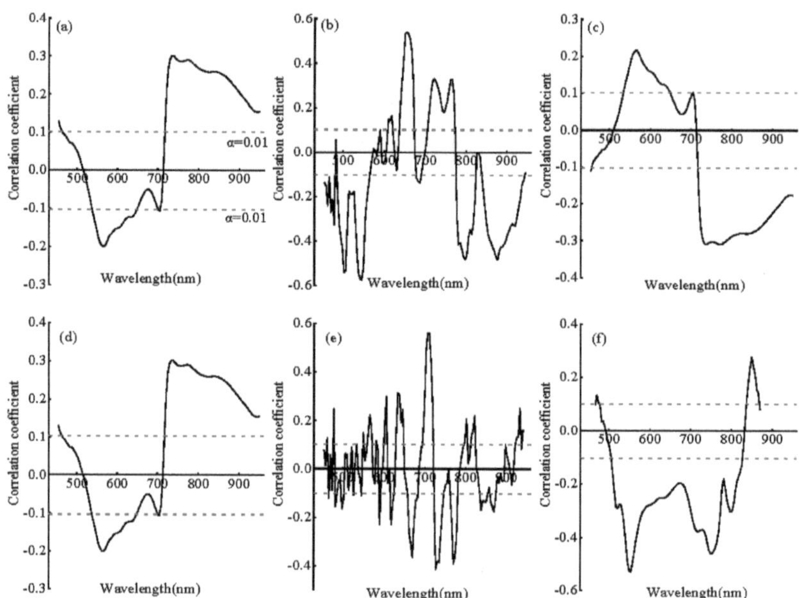

Figure 2. Relationship between PNC and single-band reflectance: (**a**) original reflectance, (**b**) the first derivative, (**c**) the logarithm of the reciprocal of the spectra, (**d**) the standard normal variate transform, (**e**) the second derivative, and (**f**) continuous removal. The dashed lines indicate the critical correlation coefficient at $\alpha = 0.01$.

Table 3. Statistics of bands sensitive to potato PNC and with single-band reflectance.

Reflectance Type	Sensitive Band Range (nm)	Optimal Wavelength (nm)	Optimal Correlation Coefficient
OR	534–646, 718–950	734	0.30
FDR	486–562, 638–670, 706–800,842–942	542	−0.57
Log(1/R)	534–646, 718–950	738	−0.31
SNVR	534–646, 718–950	738	0.30
SDR	630–646, 658–682, 694–742, 766–778	706	0.56
CR	506–826, 838–866	554	−0.53

3.2.3. Correlation between Three-Band Spectral Indices and PNC

Based on the hyperspectral OR data, the six TBIs listed in Table 1 were constructed using the lambda-by-lambda band-optimization algorithm, and the contour maps of their correlation coefficients with potato PNC are plotted in Figure 4. These results show that the six different TBIs most closely correlated with potato PNC over the multiple growth periods are TBI1 (562,494,750), TBI2 (586,666,550), TBI3 (462,574,562), TBI4 (734,514,534), TBI5 (530,734,514), and TBI6 (510,590,514), corresponding to correlation coefficients of −0.76, −0.74, 0.70, −0.81, −0.82, and −0.62, respectively. The TBIs correlate more closely with the PNC of multiple potato-growth periods than the hyperspectral single-band reflectance and two-band spectral indices.

3.3. Estimation of Potato PNC Based on Single-, Two-, and Three-Dimensional Spectral Indices

Table 4 shows the band composition and correlation coefficients of the two- and three-band spectral indices that correlate most strongly with PNC over the multiple potato-growth periods. Tables 3 and 4 show that FDR_{542}, DSI (578,494), and TBI5 (530,734,514) correlate the most strongly with potato PNC.

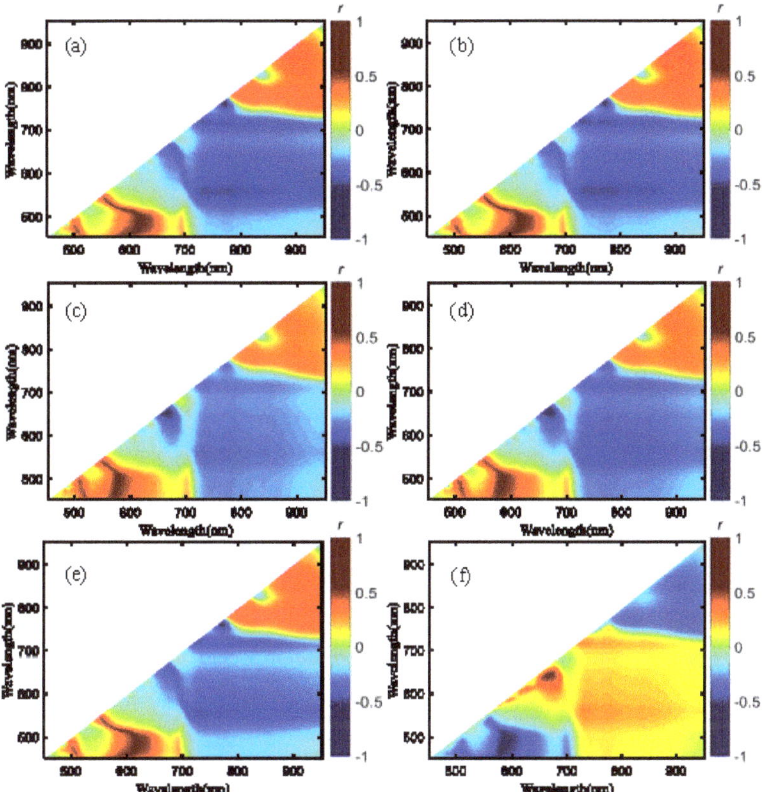

Figure 3. Correlation coefficient between PNC and two-band spectral indices. (**a**) normalized difference spectral index, (**b**) ratio spectral index, (**c**) difference spectral index, (**d**) soil-adjusted spectral index, (**e**) chlorophyll spectral index, and (**f**) optimal spectral index.

Table 4. Sensitive wavelength of the three spectral indices.

Types	Spectral Index	$R_{\lambda 1}$ (nm)	$R_{\lambda 2}$ (nm)	$R_{\lambda 3}$ (nm)	Correlation Coefficient
Two-band spectral indices	NDSI	618	490		0.62
	RSI	618	490		0.61
	DSI	578	494		0.74
	SASI	586	494		0.73
	CSI	618	494		0.62
	OSI	586	490		−0.72
Three-band spectral indices	TBI1	562	494	750	−0.76
	TBI2	586	666	550	−0.74
	TBI3	462	574	562	0.70
	TBI4	734	514	534	−0.81
	TBI5	530	734	514	−0.82
	TBI6	510	590	514	−0.62

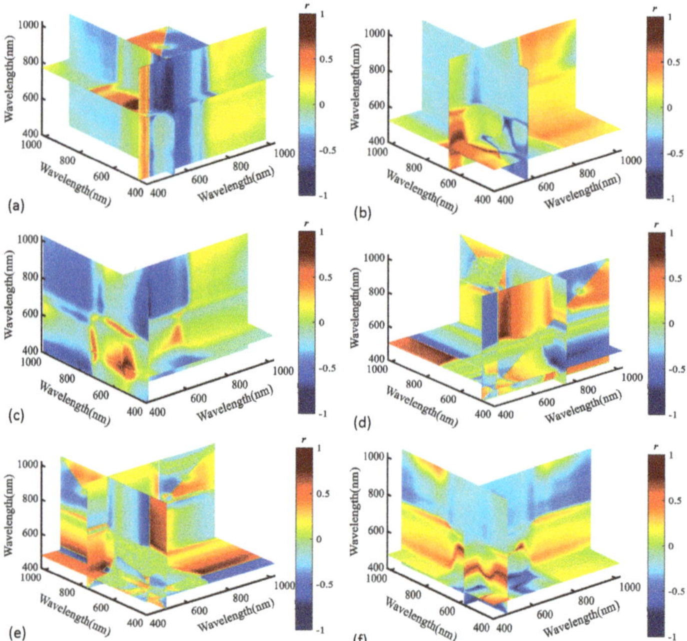

Figure 4. Correlation coefficient between PNC and three-band spectral indices. (a–f) three-band spectral index 1- three-band spectral index 6.

Potato PNC estimation models were constructed using FDR$_{542}$, DSI (578,494), and TBI5 (530,734, 514) as the optimal spectral indices to evaluate their suitability for estimating potato PNC over multiple growth periods. As shown in Figure 5, the PNC estimation model based on hyperspectral optimal single-band reflectance overestimates PNC at low-N levels and underestimates PNC at high-N levels when considering potato PNC data from different years, cultivars, and growth periods. Thus, the estimations are not sufficiently accurate. In contrast, the optimal two- and three-band hyperspectral indices correlate significantly more strongly with potato PNC than the hyperspectral single-band reflectance. In particular, the optimal three-band spectral index TBI5 (530,734,514) is the most strongly correlated with the PNC over multiple potato growth periods ($r = -0.82$). This PNC estimation model offers optimal accuracy and stability, with modeling and validation R^2 values of 0.67 and 0.65, RMSE values of 0.39 and 0.39, and NRMSE values of 12.64% and 12.17%, respectively, and the underestimation and overestimation of potato PNC were significantly improved.

3.4. Using Spectral Indices to Estimate PNC: Effect of Year, Cultivar, and Growth Period

Numerous studies show that planting year, cultivar, and growth period significantly affect the suitability of spectral indices for estimating crop N nutrient status. Therefore, to more comprehensively evaluate the suitability of the three optimal spectral indices for estimating potato PNC, we compare and analyze the use of FDR$_{542}$, DSI (578,494), and TBI5 (530,734,514) for estimating potato PNC over different planting years, cultivars, and growth periods. Figure 6 shows the results of using the three optimal spectral indices to estimate potato PNC over different years. The modeling and validation results show that the model constructed from FDR$_{542}$ significantly overestimates or underestimates in the validation data. Compared with FDR$_{542}$, the model constructed from DSI (578,494) provides better results for the different years, but the accuracy and stability of the model need to be further enhanced. The PNC estimation model constructed from TBI5 (530,734,514) produces stable and accurate results compared with the FDR$_{542}$ and DSI (578,494) models, with $R^2 = 0.40$ and 0.82, RMSE = 0.46 and 0.31, and NRMSE = 14.95% and 9.40% for the 2018 and 2019 validation data, respectively. The measured and predicted values for each sample point are more evenly distributed around the 1:1 line, indicating that the model may be applied to different years. In addition, all three optimal spectral indices constructed for estimating potato PNC produce more accurate estimates in 2019 than in 2018, indicating that the constructed models are more applicable in 2019.

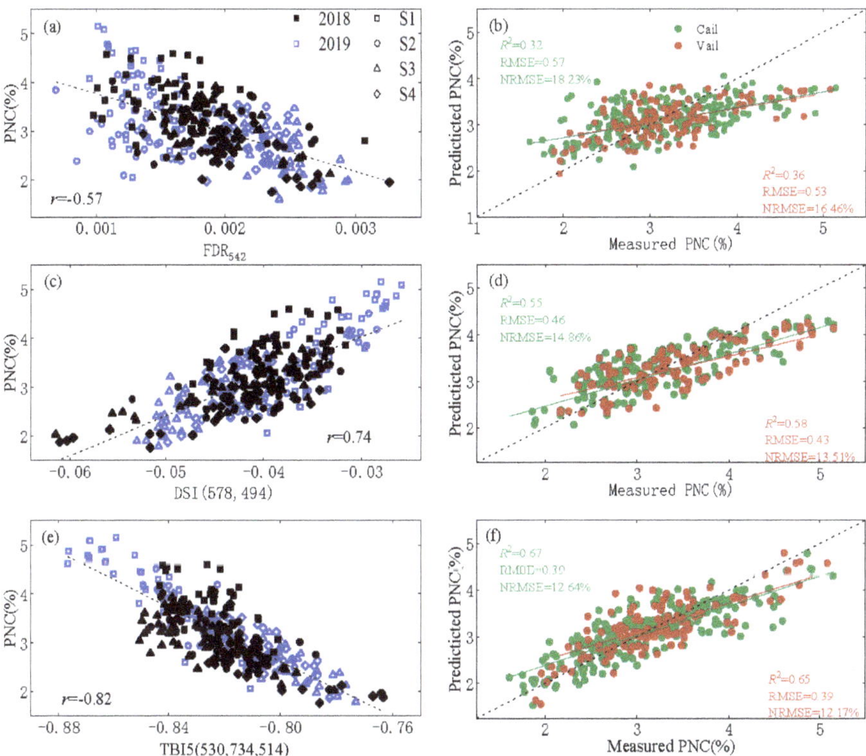

Figure 5. (**a,c,e**) Correlation of three optimal spectral indices of different dimensions with PNC over multiple potato-growth periods. (**b,d,f**) Optimal spectral indices for estimating PNC over multiple potato-growth periods.

Figure 7 shows the accuracy of the three optimal spectral indices for estimating PNC in different potato cultivars. The modeling and validation results show that FDR_{542} does not accurately estimate PNC for any of the three potato cultivars, and the stability of the constructed model needs further improvement. More specifically, the PNC of cultivar Z3 is overestimated, and that of Z5 is underestimated. Compared with FDR_{542}, the PNC estimation model constructed from DSI (578,494) produces more accurate PNC estimates for the different potato cultivars, and the stability of the model is improved. Finally, the potato PNC estimation model based on TBI5 (530,734,514) produces the most accurate PNC estimates for all three different cultivars, and the model is more stable than those based on FDR_{542} and DSI (578,494). For the TBI5 (530,734,514) model, the modeling and validation R^2 values for the Z3 potato cultivar are 0.70 and 0.70, the RMSE values are 0.32 and 0.31, and the NRMSE values are 11.27% and 11.66%, respectively. For the Z5 cultivar, the values are $R^2 = 0.62$ and 0.62, RMSE = 0.42 and 0.36, and NRMSE = 11.91% and 11.24%, respectively, and for the Z195 cultivar, the values are $R^2 = 0.39$ and 0.50, RMSE = 0.48 and 0.61, and NRMSE = 16.30% and 18.79%, respectively. The combined modeling and validation results show that the potato PNC estimation models constructed by the three optimal spectral indices are more applicable for both the Z3 and Z5 cultivars than for the Z195 cultivar.

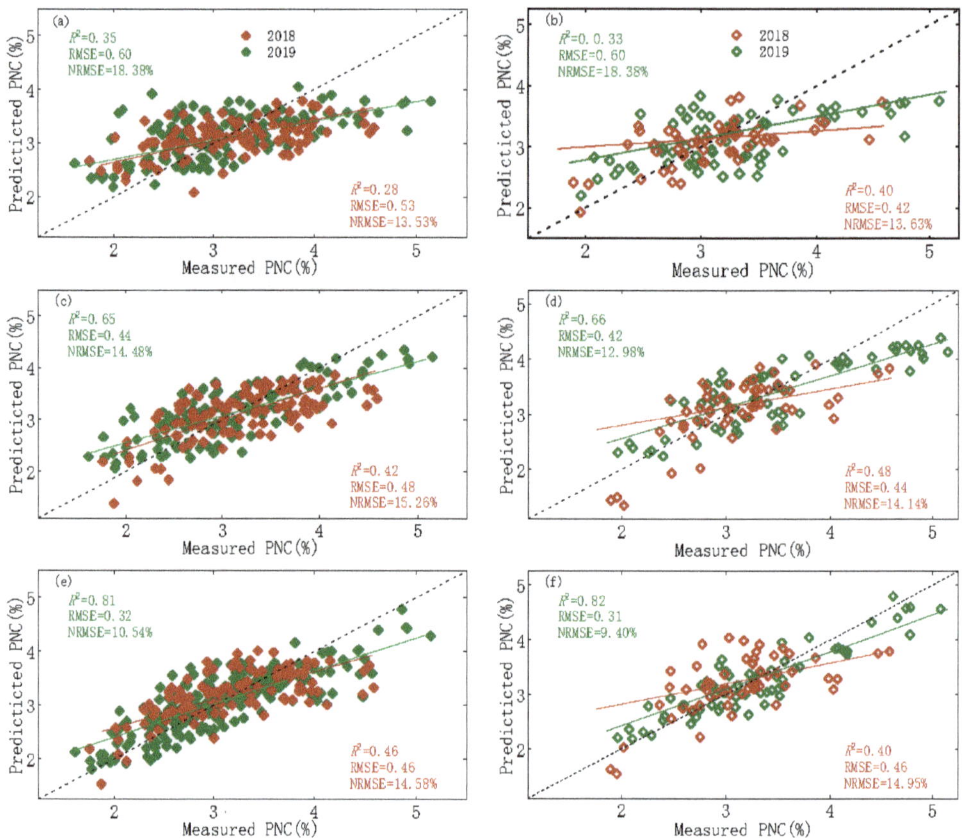

Figure 6. Predicted versus measured PNC. Predicted PNC was estimated by models constructed from the three optimal spectral indices to estimate the effect of potato PNC in different planting years (2018 and 2019). (**a,c,e**) modeling dataset; (**b,d,f**) validation dataset.

Figure 8 shows the accuracy of the three optimal spectral indices for estimating PNC in different potato growth periods. Figure 8a,b show that the PNC estimation model constructed from FDR_{542} for the multiple potato growth periods produces accurate PNC estimates only in S3 but inaccurate PNC estimates in the other three growth periods. Figure 8c,d indicate that, compared with FDR_{542}, the PNC estimation model constructed from DSI (578,494) produces more accurate PNC estimates for all four growth periods. The most accurate PNC estimate is for growth period S3, which has modeling and validation R^2 values of 0.62 and 0.52, RMSE values of 0.40 and 0.36, and NRMSE values of 13.56% and 12.74%, respectively. Figure 8e,f show that the PNC estimation model built from TBI5 (530,734,514) produces the most accurate PNC estimates of the three spectral indices for all potato growth periods. The best estimates are for growth period S3, which has modeling and validation R^2 values of 0.71 and 0.54, RMSE values of 0.35 and 0.45, and NRMSE values of 12.12% and 15.54%, respectively. The poorest estimates are for growth period S1, which has modeling and validation R^2 values of 0.52 and 0.61, RMSE values of 0.51 and 0.45, and NRMSE values of 13.88% and 11.75%, respectively. The combined modeling and validation results show that the PNC estimation models constructed from the three optimal spectral indices all work best in S3 and poorly in S1.

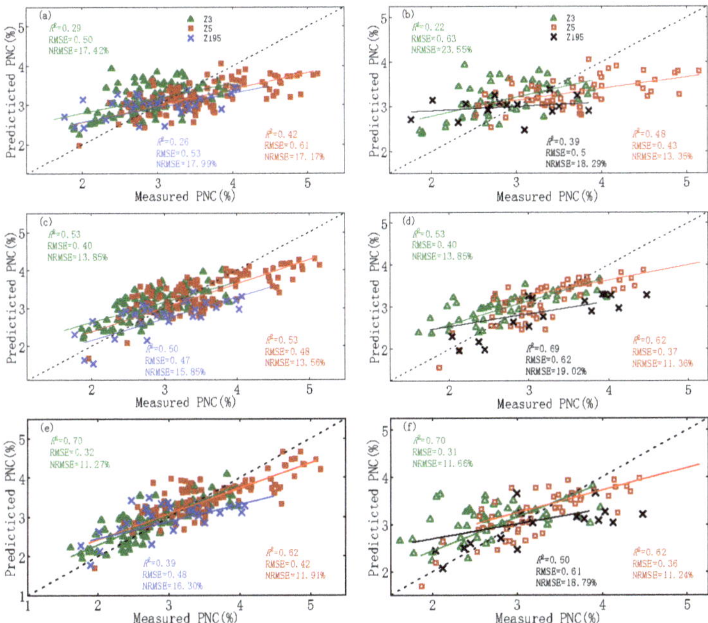

Figure 7. Predicted versus measured PNC. Predicted PNC was estimated by models constructed from the three optimal spectral indices to estimate the effect of potato PNC in different cultivars. (**a**,**c**,**e**) modeling dataset; (**b**,**d**,**f**) validation dataset.

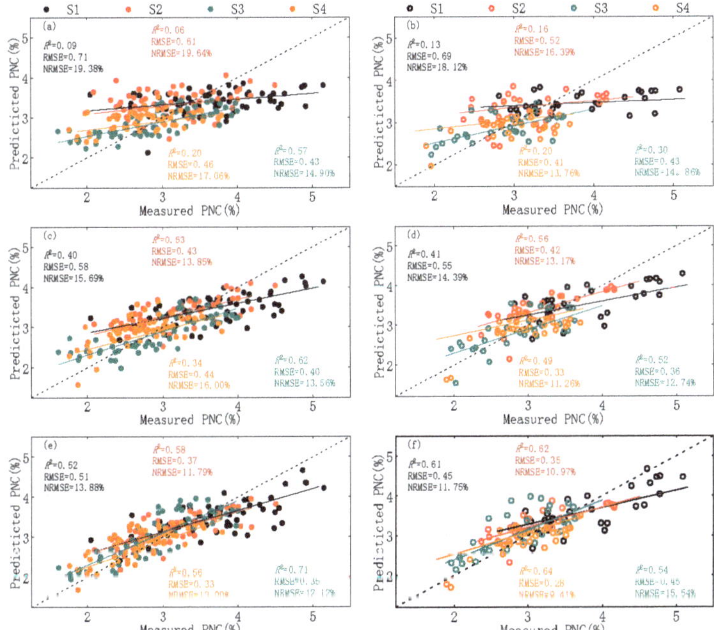

Figure 8. Predicted versus measured PNC. Predicted PNC was estimated by models constructed from the three optimal spectral indices to estimate the effect of potato PNC in different growth periods. (**a**,**c**,**e**) modeling dataset; (**b**,**d**,**f**) validation dataset.

4. Discussion

4.1. Selecting Optimal Spectral Indices in Different Dimensions

The abundance of hyperspectral sensor bands provides a large amount of information on crop canopy conditions, and combining various bands greatly expands their usefulness for monitoring the physical and chemical parameters of crops. Furthermore, the accuracy of estimates of crop physical and chemical parameters can be improved by using multiple linear regression or complex machine learning algorithms based on multiple spectral indices. However, increasing the number of variables and the modeling complexity increases the computational cost and application threshold of the model [13]. Agricultural workers who lack professional mathematical and statistical knowledge and remote sensing skills prefer to use a single spectral index to estimate the N status of crops over multiple growth periods, an approach that is also more conducive to the integration and development of N-monitoring devices [12,32]. Therefore, this study investigates the use of single-, two-, and three-dimensional spectral indices for estimating, via a linear regression method, potato PNC over multiple growth periods.

For hyperspectral single-band reflectance, five spectral transformation techniques, i.e., FDR, Log(1/R), SNVR, SDR, and CR, are used to process the OR data. The reasons for this are as follows: (1) FDR and SDR not only eliminate the effect of background, but also resolve overlapping signals and enhance dark peaks, which improves the sensitivity of spectral features to crop physicochemical parameters [44]. (2) Log(1/R) highlights the spectral differences in the visible and makes spectral reflectance less sensitive to variations in light intensity [42]. (3) SNVR eliminates the dependence of hyperspectral reflectance on solid particles and surface scattering and improves the extraction accuracy of spectral information. (4) CR highlights the location of the feature band, which facilitates its selection.

The results in Figure 2 show that using the spectral transformation technique for OR facilitates the deep mining of practical spectral information, which provides a more comprehensive reference for monitoring potato PNC based on hyperspectral single-band reflectance. Twelve two- and three-band spectral indices, based on spectral indices that have been reported to indicate the crop N status, are constructed by using the lambda-by-lambda band-optimization algorithm. The lambda-by-lambda algorithm considers all possible combinations of bands from 450 to 950 nm, which improves the robustness and accuracy of estimates of vegetation properties [45,46]. To investigate the best indicator of potato PNC, this study investigates the accuracy of estimation models constructed from spectral indices based on hyperspectral reflectance or two- or three-band combinations for estimating potato PNC over different years, cultivars, and growth periods.

4.2. Comparison of Sensitive Wavelengths

Tables 3 and 4 show that the spectral index wavelengths sensitive to potato PNC are all in the visible, which is because N is primarily present in the crop in the form of chlorophyll and proteins [47,48], and chlorophyll reflects mainly in the visible region (400–700 nm). Therefore, the visible region correlates more strongly with crop PNC [4]. More specifically, the wavelengths at which single-band reflectance is sensitive to PNC over multiple potato growth periods are all located in the red-edge (670–740 nm) or green-edge (502–554 nm) and most of the sensitive wavelengths of the three-band spectral indices also include these spectral regions. In contrast, the sensitive wavelengths of the six two-band spectral indices all contain the 490–494 nm band, which is consistent with the results of Hansen and Schjoerring, Pettersson and Eckersten, and Sun et al. [12,49,50] This is because different treatments (density and N and K fertilization) lead to different potato PNC, which is more evident in the green-and red-edge due to the strong absorption and reflection of chlorophyll. Thus, these spectral regions provide a good indication of the PNC. In addition, the 490–494 nm band is in the absorption band of chlorophyll and carotenoids [51] and N is an essential component of chlorophyll and protein, so this band correlates strongly with PNC.

4.3. Comparison of Single-, Two-, and three-Dimensional Spectral Indices for Estimating Potato PNC

To investigate the use of single-, two-, and three-dimensional spectral indices for estimating potato PNC, this study presents an in-depth comparison of hyperspectral single-band reflectance and two- and three-band spectral indices for estimating PNC over different years, cultivars, and growth periods. Consistent with the results of Li and Yu et al. [52,53], this study reveals the significant effects of planting year, cultivar, and growth period on the ability of spectral indices to estimate PNC. In addition, cultivar and growth period more strongly affect PNC estimates than the planting year. This result confirms the importance of exploring generic spectral indices across years, cultivars, and growth periods to ensure accurate estimates of the PNC to monitor the N status of crops. The discussion in Section 3.4 shows that, for potato PNC data from different years, cultivars, and growth periods, the PNC estimation model constructed from single-band reflectance is less accurate and stable. In contrast, the model constructed from the three-band spectral index produces the most accurate and stable PNC estimates over different years, cultivars, and growth periods. Several reasons may be invoked to explain this result: (1) Although spectral transformation techniques such as FDR can improve the correlation between spectra and potato PNC, the information available from single-band reflectance is limited and cannot effectively reflect the dynamic variations in PNC. (2) Compared with the two-band spectral indices, the red-edge and green-edge regions included in the three-band spectral indices carry more information related to PNC [54,55], effectively enhancing the correlation between spectral indices and PNC and improving the accuracy of PNC estimation.

The estimation results for different years (Figure 6) show that the estimation models constructed from the three optimal spectral indices are more accurate in 2019 than in 2018, which may be due to (1) the larger number of samples in 2019 than in 2018, and (2) the greater number of potato cultivars planted in 2018, which requires higher demands on the generality of the models. The PNC estimation results of different cultivars (Figure 7) show that the models constructed by the three optimal spectral indices produce more accurate PNC estimates for Z3 and Z5, whereas the estimates for Z195 are mediocre. This is because Z195 was planted only for a single year, so it has a small sample size and thus contributes less to the modeling. The estimation results of different growth periods (Figure 8) show that the models constructed from the three optimal spectral indices produce the most (least) accurate estimates for growth period S3 (S1). The PNC is underestimated in S1 because the potato plants were still small and the vegetation cover was low, so the difference in the extracted spectral information was slight and perturbed by soil factors. Therefore, the estimation accuracy was poor during this period. In contrast, the potato plants were thriving and the leaves were extended in S3 when the potato were in the closed monopoly state, so the spectral reflectance was less perturbed by soil and other factors and thus better reflected the potato PNC.

4.4. Implications for Future Study

This study investigates the performance of single-, two-, and three-dimensional spectral indices for estimating PNC by using the spectral transformation technique and the lambda-by-lambda band-optimization algorithm based on UAV hyperspectral data of potato over different years and multiple growth periods. The results show that TBI 5 (530,734,514) may be used for different types of potato PNC data and can serve as a reference for the development and integration of rapid N-monitoring devices. In addition, the constituent bands of TBI 5(530,734,514) are also available from existing hyperspectral satellites, such as ZY-1 02D, 02E, and GF5. With the development and maturation of hyperspectral satellite technology, the spatiotemporal resolution is continuously improving, and large area crop N-nutrient monitoring is gradually becoming possible. Therefore, the spectral index TBI 5 (530,734,514) has a strong potential for use in large area PNC monitoring based on hyperspectral satellite platforms.

Although this study used data from two years, three cultivars, and four growth periods to construct the potato PNC estimation models, the effects of different cultivation environments on the spectral indices for estimating PNC should also be considered. Therefore, the reliability and adaptability of the optimal spectral indices should be studied in more regions with differing ecology. In addition, although six forms of hyperspectral single-band reflectance were used in this study and 12 forms of band combinations that are widely used and proven to be effective were screened separately, more spectral transformation methods and band combinations should be investigated to determine their response to potato PNC.

5. Conclusions

This study uses six hyperspectral single-band reflectance and 12 band-combination indices to develop optimal spectral indices of single-, two-, and three-dimensions. Quantitative models to estimate potato PNC over multiple growth periods are then established based on these optimal spectral indices. The results show that (1) the red-edge and green-edge spectral correlate strongly with the N status of potato, which provide relevant information for the rapid and accurate estimation of potato PNC. (2) Although the spectral transformation technique can enhance the correlation between spectral information and PNC, the PNC estimation model based on the optimal single-band reflectance produces less accurate and stable results over different years, cultivars, and growth periods. (3) Compared with the optimal two-band spectral index, the PNC estimation model constructed from the optimal three-band spectral index TBI5 (530,734,514) is more accurate and stable. This index may thus be used to estimate potato PNC over different years, cultivars, and growth periods. The results of this study can thus be used to design fast and efficient N diagnostic systems and enhance the linkage between UAV and satellite sensors to provide technical support for the large area nitrogen monitoring of potato crops.

Author Contributions: H.F., X.X., G.Y. and J.Y. designed the experiments. H.F., Y.L., Y.M. and X.S. collected the AGB, CH, and UAV hyperspectral images. Y.F. and H.F. analyzed the data and wrote the manuscript. X.J. and J.Y. made comments and revised the manuscript. All authors have read and agreed to the published version of the manuscript.

Funding: This study was supported by the Key scientific and technological projects of Heilongjiang province (2021ZXJ05A05), the National Natural Science Foundation of China (41601346), the Platform Construction Funded Program of Beijing Academy of Agriculture and Forestry Sciences (No.PT2022-24), and the Key Field Research and Development Program of Guangdong Province (2019B020216001).

Data Availability Statement: Not applicable.

Acknowledgments: We thanks to Weiguo Li, Hong Chang, Yang Meng, and Yu Zhao for field management and data collection. We thank the National Precision Agriculture Experiment Station for providing the test site and employees.

Conflicts of Interest: The authors declare no conflict of interest.

References

1. Niu, Y.; Zhang, L.; Zhang, H.; Han, W.; Peng, X. Estimating Above-Ground Biomass of Maize Using Features Derived from UAV-Based RGB Imagery. *Remote Sens.* **2019**, *11*, 1261. [CrossRef]
2. Thorp, K.R.; Gore, M.A.; Andrade-Sanchez, P.; Carmo-Silva, A.E.; Welch, S.M.; White, J.W.; French, A.N. Proximal hyperspectral sensing and data analysis approaches for field-based plant phenomics. *Comput. Electron. Agric.* **2015**, *118*, 225–236. [CrossRef]
3. Li, B.; Xu, X.; Zhang, L.; Han, J.; Bian, C.; Li, G.; Liu, J.; Jin, L. Above-ground biomass estimation and yield prediction in potato by using UAV-based RGB and hyperspectral imaging. *ISPRS J. Photogramm. Remote Sens.* **2020**, *162*, 161–172. [CrossRef]
4. Fu, Y.; Yang, G.; Pu, R.; Li, Z.; Li, H.; Xu, X.; Song, X.; Yang, X.; Zhao, C. An overview of crop nitrogen status assessment using hyperspectral remote sensing: Current status and perspectives. *Eur. J. Agron.* **2021**, *124*, 126241. [CrossRef]
5. Chen, P.; Haboudane, D.; Tremblay, N.; Wang, J.; Vigneault, P.; Li, B. New spectral indicator assessing the efficiency of crop nitrogen treatment in corn and wheat. *Remote Sens. Environ.* **2010**, *114*, 1987–1997. [CrossRef]
6. Kefauver, S.C.; Vicente, R.; Vergara-Díaz, O.; Fernandez-Gallego, J.A.; Kerfal, S.; Lopez, A.; Melichar, J.P.E.; Serret Molins, M.D.; Araus, J.L. Comparative UAV and Field Phenotyping to Assess Yield and Nitrogen Use Efficiency in Hybrid and Conventional Barley. *Front. Plant Sci.* **2017**, *8*, 1733. [CrossRef] [PubMed]
7. Zhang, H.; Du, H.; Zhang, C.; Zhang, L. An automated early-season method to map winter wheat using time-series Sentinel-2 data: A case study of Shandong, China. *Comput. Electron. Agric.* **2021**, *182*, 105962. [CrossRef]

8.   Battude, M.; Al Bitar, A.; Morin, D.; Cros, J.; Huc, M.; Sicre, C.M.; Le Dantec, V.; Demarez, V. Estimating maize biomass and yield over large areas using high spatial and temporal resolution Sentinel-2 like remote sensing data. *Remote Sens. Environ.* **2016**, *184*, 668–681. [CrossRef]

9.   Clevers, J.G.P.W.; Kooistra, L. Using Hyperspectral Remote Sensing Data for Retrieving Canopy Chlorophyll and Nitrogen Content. *IEEE J. Sel. Top. Appl. Earth Obs. Remote Sens.* **2012**, *5*, 574–583. [CrossRef]

10.   Yue, J.; Feng, H.; Yang, G.; Li, Z. A Comparison of Regression Techniques for Estimation of Above-Ground Winter Wheat Biomass Using Near-Surface Spectroscopy. *Remote Sens.* **2018**, *10*, 66. [CrossRef]

11.   Bendig, J.; Yu, K.; Aasen, H.; Bolten, A.; Bennertz, S.; Broscheit, J.; Gnyp, M.L.; Bareth, G. Combining UAV-based plant height from crop surface models, visible, and near infrared vegetation indices for biomass monitoring in barley. *Int. J. Appl. Earth Obs. Geoinf.* **2015**, *39*, 79–87. [CrossRef]

12.   Sun, H.; Feng, M.; Yang, W.; Bi, R.; Sun, J.; Zhao, C.; Xiao, L.; Wang, C.; Kubar, M.S. Monitoring Leaf Nitrogen Accumulation With Optimized Spectral Index in Winter Wheat Under Different Irrigation Regimes. *Front. Plant Sci.* **2022**, *13*, 913240. [CrossRef] [PubMed]

13.   Feng, W.; Zhang, H.-Y.; Zhang, Y.-S.; Qi, S.-L.; Heng, Y.-R.; Guo, B.-B.; Ma, D.-Y.; Guo, T.-C. Remote detection of canopy leaf nitrogen concentration in winter wheat by using water resistance vegetation indices from in-situ hyperspectral data. *Field Crop. Res.* **2016**, *198*, 238–246. [CrossRef]

14.   Yao, X.; Ren, H.; Cao, Z.; Tian, Y.; Cao, W.; Zhu, Y.; Cheng, T. Detecting leaf nitrogen content in wheat with canopy hyperspectrum under different soil backgrounds. *Int. J. Appl. Earth Obs. Geoinf.* **2014**, *32*, 114–124. [CrossRef]

15.   Osco, L.P.; Junior, J.M.; Ramos, A.; Furuya, D.; Teodoro, P.E. Leaf Nitrogen Concentration and Plant Height Prediction for Maize Using UAV-Based MultispectralImagery and Machine Learning Techniques. *Remote Sens.* **2020**, *12*, 3237. [CrossRef]

16.   Qiu, Z.; Ma, F.; Li, Z.; Xu, X.; Ge, H.; Du, C. Estimation of nitrogen nutrition index in rice from UAV RGB images coupled with machine learning algorithms. *Comput. Electron. Agric.* **2021**, *189*, 106421. [CrossRef]

17.   Zheng, H.; Cheng, T.; Li, D.; Zhou, X.; Yao, X.; Tian, Y.; Cao, W.; Zhu, Y. Evaluation of RGB, Color-Infrared and Multispectral Images Acquired from Unmanned Aerial Systems for the Estimation of Nitrogen Accumulation in Rice. *Remote Sens.* **2018**, *10*, 824. [CrossRef]

18.   Liu, Y.; Feng, H.; Yue, J.; Li, Z.; Yang, G.; Song, X.; Yang, X.; Zhao, Y. Remote-sensing estimation of potato above-ground biomass based on spectral and spatial features extracted from high-definition digital camera images. *Comput. Electron. Agric.* **2022**, *198*, 107089. [CrossRef]

19.   Yue, J.; Guo, W.; Yang, G.; Zhou, C.; Feng, H.; Qiao, H. Method for accurate multi-growth-stage estimation of fractional vegetation cover using unmanned aerial vehicle remote sensing. *Plant Methods* **2021**, *17*, 1–16. [CrossRef]

20.   Yue, J.; Yang, G.; Tian, Q.; Feng, H.; Xu, K.; Zhou, C. Estimate of winter-wheat above-ground biomass based on UAV ultrahigh-ground-resolution image textures and vegetation indices. *ISPRS J. Photogramm. Remote Sens.* **2019**, *150*, 226–244. [CrossRef]

21.   Lu, J.; Cheng, D.; Geng, C.; Zhang, Z.; Xiang, Y.; Hu, T. Combining plant height, canopy coverage and vegetation index from UAV-based RGB images to estimate leaf nitrogen concentration of summer maize. *Biosyst. Eng.* **2021**, *202*, 42–54. [CrossRef]

22.   Fan, Y.; Feng, H.; Jin, X.; Yue, J.; Liu, Y.; Li, Z.; Feng, Z.; Song, X.; Yang, G. Estimation of the nitrogen content of potato plants based on morphological parameters and visible light vegetation indices. *Front. Plant Sci.* **2022**, *13*, 1012070. [CrossRef] [PubMed]

23.   Fu, Y.; Yang, G.; Li, Z.; Song, X.; Li, Z.; Xu, X.; Wang, P.; Zhao, C. Winter Wheat Nitrogen Status Estimation Using UAV-Based RGB Imagery and Gaussian Processes Regression. *Remote Sens.* **2020**, *12*, 3778. [CrossRef]

24.   Liu, Y.; Feng, H.K.; Yue, J.B.; Jin, X.L.; Li, Z.H.; Yang, G.J. Estimation of potato above-ground biomass based on unmanned aerial vehicle red-green-blue images with different texture features and crop height. *Front. Plant Sci.* **2022**, *13*, 938216. [CrossRef] [PubMed]

25.   Fu, Y.; Yang, G.; Song, X.; Li, Z.; Xu, X.; Feng, H.; Zhao, C. Improved Estimation of Winter Wheat Aboveground Biomass Using Multiscale Textures Extracted from UAV-Based Digital Images and Hyperspectral Feature Analysis. *Remote Sens.* **2021**, *13*, 581. [CrossRef]

26.   Saberioon, M.M.; Amin, M.S.M.; Gholizadeh, A.; Ezri, M.H. A Review of Optical Methods for Assessing Nitrogen Contents During Rice Growth. *Appl. Eng. Agric.* **2014**, *30*, 657–669.

27.   Moharana, S.; Dutta, S. Spatial variability of chlorophyll and nitrogen content of rice from hyperspectral imagery. *ISPRS J. Photogramm. Remote Sens.* **2016**, *122*, 17–29. [CrossRef]

28.   Schlemmer, M.; Gitelson, A.; Schepers, J.; Ferguson, R.; Peng, Y.; Shanahan, J.; Rundquist, D. Remote estimation of nitrogen and chlorophyll contents in maize at leaf and canopy levels. *Int. J. Appl. Earth Obs. Geoinf.* **2013**, *25*, 47–54. [CrossRef]

29.   Wang, W.; Yao, X.; Yao, X.; Tian, Y.; Liu, X.; Ni, J.; Cao, W.; Zhu, Y. Estimating leaf nitrogen concentration with three-band vegetation indices in rice and wheat. *Field Crop. Res.* **2012**, *129*, 90–98. [CrossRef]

30.   Tian, Y.C.; Yao, X.; Yang, J.; Cao, W.X.; Hannaway, D.B.; Zhu, Y. Corrigendum to "Assessing newly developed and published vegetation indices for estimating rice leaf nitrogen concentration with ground- and space-based hyperspectral reflectance" [Field Crops Res. 120 (2011) 299–310]. *Field Crop. Res.* **2011**, *121*, 464. [CrossRef]

31.   Rouse, J.W.; Haas, R.H.; Deering, D.W. *Monitoring the Vernal Advancement and Retrogradation (Green Wave Effect) of Natural Vegetation*; NTRS-NASA Technical Report Server; University of Maryland: College Park, MD, USA, 1973.

32.   Hasituya; Li, F.; Elsayed, S.; Hu, Y.; Schmidhalter, U. Passive reflectance sensing using optimized two- and three-band spectral indices for quantifying the total nitrogen yield of maize. *Comput. Electron. Agric.* **2020**, *173*, 105403. [CrossRef]

33. Li, Y.-J.; Wei, J.; Xu, X.-R. Image Processing Method of Linear Infrared Focal Plane Array. *J. Infrared Millim. Waves* **2010**, *29*, 91–93. [CrossRef]
34. Baghi, N.G.; Oldeland, J. Do soil-adjusted or standard vegetation indices better predict above ground biomass of semi-arid, saline rangelands in North-East Iran? *Int. J. Remote Sens.* **2019**, *40*, 8223–8235. [CrossRef]
35. Zhang, J.; Tian, H.; Wang, D.; Li, H.; Mouazen, A.M. A novel spectral index for estimation of relative chlorophyll content of sugar beet. *Comput. Electron. Agric.* **2021**, *184*, 106088. [CrossRef]
36. Reyniers, M.; Walvoort, D.J.J.; De Baardemaaker, J. A linear model to predict with a multi-spectral radiometer the amount of nitrogen in winter wheat. *Int. J. Remote Sens.* **2006**, *27*, 4159–4179. [CrossRef]
37. Dash, J.; Curran, P.J. The MERIS terrestrial chlorophyll index. *Int. J. Remote Sens.* **2004**, *25*, 5403–5413. [CrossRef]
38. Feng, W.; Guo, B.-B.; Zhang, H.-Y.; He, L.; Zhang, Y.-S.; Wang, Y.-H.; Zhu, Y.-J.; Guo, T.-C. Remote estimation of above ground nitrogen uptake during vegetative growth in winter wheat using hyperspectral red-edge ratio data. *Field Crop. Res.* **2015**, *180*, 197–206. [CrossRef]
39. Sims, D.A.; Gamon, J.A. Relationships between leaf pigment content and spectral reflectance across a wide range of species, leaf structures and developmental stages. *Remote Sens. Environ.* **2002**, *81*, 337–354. [CrossRef]
40. Oliveri, P.; Malegori, C.; Simonetti, R.; Casale, M. The impact of signal pre-processing on the final interpretation of analytical outcomes—A tutorial. *Anal. Chim. Acta* **2019**, *1058*, 9–17. [CrossRef]
41. Meng, X.; Bao, Y.; Liu, J.; Liu, H.; Zhang, X.; Zhang, Y.; Wang, P.; Tang, H.; Kong, F. Regional soil organic carbon prediction model based on a discrete wavelet analysis of hyperspectral satellite data. *Int. J. Appl. Earth Obs. Geoinf.* **2020**, *89*, 102111. [CrossRef]
42. Wang, Y.; Wang, F.; Huang, J.; Wang, X.; Liu, Z. Validation of artificial neural network techniques in the estimation of nitrogen concentration in rape using canopy hyperspectral reflectance data. *Int. J. Remote Sens.* **2009**, *30*, 4493–4505. [CrossRef]
43. Azadi, F.; Saadat, S.; Karimi-Jashni, A. Experimental Investigation and Modeling of Nickel Removal from Wastewater Using Modified Rice Husk in Continuous Reactor by Response Surface Methodology. *Iran. J. Sci. Technol. Trans. Civ. Eng.* **2018**, *42*, 315–323. [CrossRef]
44. Liaghat, S.; Ehsani, R.; Mansor, S.; Shafri, H.Z.M.; Meon, S.; Sankaran, S.; Azam, S.H.M.N. Early detection of basal stem rot disease (*Ganoderma*) in oil palms based on hyperspectral reflectance data using pattern recognition algorithms. *Int. J. Remote Sens.* **2014**, *35*, 3427–3439. [CrossRef]
45. Mariotto, I.; Thenkabail, P.S.; Huete, A.; Slonecker, E.T.; Platonov, A. Hyperspectral versus multispectral crop-productivity modeling and type discrimination for the HyspIRI mission. *Remote Sens. Environ.* **2013**, *139*, 291–305. [CrossRef]
46. Rivera, J.P.; Verrelst, J.; Delegido, J.; Veroustraete, F.; Moreno, J. On the Semi-Automatic Retrieval of Biophysical Parameters Based on Spectral Index Optimization. *Remote Sens.* **2014**, *6*, 4927–4951. [CrossRef]
47. Ollinger, S.V. Sources of variability in canopy reflectance and the convergent properties of plants. *New Phytol.* **2011**, *189*, 375–394. [CrossRef] [PubMed]
48. Kokaly, R.F.; Asner, G.P.; Ollinger, S.V.; Martin, M.E.; Wessman, C.A. Characterizing canopy biochemistry from imaging spectroscopy and its application to ecosystem studies. *Remote Sens. Environ.* **2009**, *113*, S78–S91. [CrossRef]
49. Pettersson, C.G.; Eckersten, H. Prediction of grain protein in spring malting barley grown in northern Europe. *Eur. J. Agron.* **2007**, *27*, 205–214. [CrossRef]
50. Hansen, P.M.; Schjoerring, J.K. Reflectance measurement of canopy biomass and nitrogen status in wheat crops using normalized difference vegetation indices and partial least squares regression. *Remote Sens. Environ.* **2003**, *86*, 542–553. [CrossRef]
51. Clevers, J.G.P.W.; Gitelson, A.A. Remote estimation of crop and grass chlorophyll and nitrogen content using red-edge bands on Sentinel-2 and -3. *Int. J. Appl. Earth Observ. Geoinf.* **2013**, *23*, 344–351. [CrossRef]
52. Yu, K.; Li, F.; Gnyp, M.L.; Miao, Y.; Bareth, G.; Chen, X. Remotely detecting canopy nitrogen concentration and uptake of paddy rice in the Northeast China Plain. *ISPRS J. Photogramm. Remote Sens.* **2013**, *78*, 102–115. [CrossRef]
53. Li, F.; Mistele, B.; Hu, Y.; Chen, X.; Schmidhalter, U. Optimising three-band spectral indices to assess aerial N concentration, N uptake and aboveground biomass of winter wheat remotely in China and Germany. *ISPRS J. Photogramm. Remote Sens.* **2014**, *92*, 112–123. [CrossRef]
54. Shi, T.; Liu, H.; Chen, Y.; Wang, J.; Wu, G. Estimation of arsenic in agricultural soils using hyperspectral vegetation indices of rice. *J. Hazard. Mater.* **2016**, *308*, 243–252. [CrossRef] [PubMed]
55. Wang, J.-J.; Li, Z.; Jin, X.; Liang, G.; Struik, P.C.; Gu, J.; Zhou, Y. Phenotyping flag leaf nitrogen content in rice using a three-band spectral index. *Comput. Electron. Agric.* **2019**, *162*, 475–481. [CrossRef]

remote sensing

MDPI

Article

Comparison of UAV RGB Imagery and Hyperspectral Remote-Sensing Data for Monitoring Winter Wheat Growth

Haikuan Feng [1,2,3,4], Huilin Tao [1,3], Zhenhai Li [1,3,4], Guijun Yang [1,3,4] and Chunjiang Zhao [1,2,*]

1   Key Laboratory of Quantitative Remote Sensing in Agriculture of Ministry of Agriculture and Rural Affairs, Information Technology Research Center, Beijing Academy of Agriculture and Forestry Sciences, Beijing 100097, China
2   National Engineering and Technology Center for Information Agriculture, Nanjing Agricultural University, Nanjing 210095, China
3   National Engineering Research Center for Information Technology in Agriculture, Beijing 100097, China
4   Beijing Engineering Research Center for Agriculture Internet of Things, Beijing 100097, China
*   Correspondence: zhaocj@nercita.org.cn

Abstract: Although crop-growth monitoring is important for agricultural managers, it has always been a difficult research topic. However, unmanned aerial vehicles (UAVs) equipped with RGB and hyperspectral cameras can now acquire high-resolution remote-sensing images, which facilitates and accelerates such monitoring. To explore the effect of monitoring a single crop-growth indicator and multiple indicators, this study combines six growth indicators (plant nitrogen content, above-ground biomass, plant water content, chlorophyll, leaf area index, and plant height) into the new comprehensive growth index (*CGI*). We investigate the performance of RGB imagery and hyperspectral data for monitoring crop growth based on multi-time estimation of the *CGI*. The *CGI* is estimated from the vegetation indices based on UAV hyperspectral data treated by linear, nonlinear, and multiple linear regression (MLR), partial least squares (PLSR), and random forest (RF). The results are as follows: (1) The RGB-imagery indices red reflectance (r), the excess-red index (EXR), the vegetation atmospherically resistant index (VARI), and the modified green-red vegetation index (MGRVI), as well as the spectral indices consisting of the linear combination index (LCI), the modified simple ratio index (MSR), the simple ratio vegetation index (SR), and the normalized difference vegetation index (NDVI), are more strongly correlated with the *CGI* than a single growth-monitoring indicator. (2) The *CGI* estimation model is constructed by comparing a single RGB-imagery index and a spectral index, and the optimal RGB-imagery index corresponding to each of the four growth stages in order is r, r, r, EXR; the optimal spectral index is LCI for all four growth stages. (3) The MLR, PLSR, and RF methods are used to estimate the *CGI*. The MLR method produces the best estimates. (4) Finally, the *CGI* is more accurately estimated using the UAV hyperspectral indices than using the RGB-image indices.

Keywords: comprehensive growth index; vegetation indices; multiple linear regression; partial least squares; wheat; random forest; precision agriculture

Citation: Feng, H.; Tao, H.; Li, Z.; Yang, G.; Zhao, C. Comparison of UAV RGB Imagery and Hyperspectral Remote-Sensing Data for Monitoring Winter Wheat Growth. *Remote Sens.* **2022**, *14*, 3811. https://doi.org/10.3390/rs14153811

Academic Editors: Edoardo Pasolli, Wenjiang Huang, Giovanni Laneve, Yingying Dong and Chenghai Yang

Received: 13 June 2022
Accepted: 3 August 2022
Published: 8 August 2022

1. Introduction

The characteristics of individual plants or groups of plants can be used to evaluate crop growth [1] and can reveal various levels of crop growth within regions [2]. However, in precision agriculture, effective monitoring of growth conditions can provide not only real-time information for field management but also a basis for estimating crop yield [2–4]. Traditionally, field managers determine crop-growth status by visual inspection, which is laborious and time-consuming. However, in precision agriculture, remote sensing technology has become increasingly informative and can now collect spectral information from the crop canopy over a wide range of electromagnetic bands, which may then be translated into physiological and biochemical information about the crop canopy [5]. Thus, field managers may now rely on remote-sensing technology to monitor crop growth.

Crop growth monitoring mainly traces crop growth status and variations in crop growth. Although crop growth is affected by many factors and the growth process is quite complex, it can be estimated using biochemical parameters such as biomass, leaf area index, and chlorophyll content. The crop canopy spectrum obtained by remote sensing technology provides further access to crop canopy biochemical information [6]. The crop canopy spectrum is determined by the leaves, canopy structure, and soil background [7]. Therefore, the relationship between spectral information and crop parameters can be established for estimating crop parameters, such as leaf area index (LAI), above-ground biomass, density, chlorophyll, plant nitrogen content, and photosynthetic pigments [8–12].

Acquiring images from unmanned aerial vehicles (UAV) constitute a remote sensing technology that offers high resolution, high efficiency, rapidity, and low cost, which lead to more timely and accurate crop monitoring [13,14]. Its spatial resolution exceeds that of satellite-based remote sensing, and unlike ground remote sensing, it can generate orthophotos [15–17]. Compared with traditional remote sensing, the flight pattern, time, maneuverability, and the cost of UAV remote sensing is more advantageous [18]. UAV remote sensing technologies are increasingly used in agriculture to monitor crop growth and have achieved good results.

Vegetation indices, which are mathematical constructions involving the reflectance of different spectral bands, lead to more accurate information about vegetation [8]. With the widespread application of remote sensing technology in agriculture, vegetation indices are often used to estimate crop parameters. Chen et al. [19] used vegetation indices and a neural network algorithm to improve the estimation accuracy of maize leaf area index. Han et al. [20] used four machine learning algorithms (multiple linear regression, support vector machine, artificial neural network, and random forest) to invert maize above-ground biomass (AGB) to improve the inversion effect. Swain et al. [21] obtained high-resolution images of rice using an unmanned-helicopter, low-altitude, remote sensing platform to demonstrate that such images work well for estimating rice biomass. Chang et al. [22] constructed a model to estimate corn chlorophyll content using the spectral vegetation indices and difference vegetation index. Schirrmann et al. [23] used low-cost UAV images to monitor the physiological parameters and nitrogen content of wheat. Li et al. [24] used four methods: partial least squares regression (PLSR), support vector machines (SVM), stepwise multiple linear regression (SMLR), and back-propagation neural network (BPN) to estimate the nitrogen content of winter wheat and used partial least squares and support vector machine to improve estimates. These studies all focused on empirical models, whereas others have studied semi-empirical models and physical models. For example, Duan et al. [25] used the PROSAIL model to estimate the LAI of maize, potatoes, and sunflowers and showed that incorporating the direction information improves the estimation accuracy. In addition, Li et al. [26] used the PROSPECT + SAIL model to invert the LAI of multiple crops with high inversion accuracy.

Crop growth is closely related to plant nitrogen content, above-ground biomass, plant water content, chlorophyll, LAI, plant height, and other factors [15]. Therefore, to monitor crop growth, many studies use remote-sensing data to estimate a single parameter (e.g., LAI, AGB) related to crop growth and thereby determine growth status [15]. The estimation of crop growth by a single growth parameter has been extensively studied; however, studies that combine multiple growth parameters to monitor crop growth have not been found. To explore the use of multiple crop-growth indicators to monitor crop growth, the present study uses UAV remote sensing data to monitor surface-scale crop growth. Specifically, we combine plant nitrogen content (*PNC*), above-ground biomass (*AGB*), plant water content (*PWC*), chlorophyll (*CHL*), *LAI*, and plant height (*H*) into a comprehensive growth index (*CGI*). We take into account the different winter wheat growing stages when monitoring single growth phases versus the entire growth phase. More precisely, we evaluate crop-growth monitoring from the vegetation index based on UAV digital images and compare the results with those obtained from the vegetation-index-based UAV hyperspectral multitemporal images. In addition, we combine multiple linear regression (MLR), partial least squares (PLSR), and random forest (RF) with the vegetation index to estimate the *CGI* and map its spatial distribution.

The structure of this paper is as follows:

Section 2 presents the study area, the experimental design, the techniques used for ground sampling, data acquisition, and the processing of digital and hyperspectral remote-sensing data from UAVs. In addition, analytical methods, statistical methods, and vegetation indices are also discussed. Section 3 discusses the selection of indices and how they affect the accuracy of the resulting *CGI*. Specifically, we use the vegetation index based on UAV RGB imagery, the vegetation index based on UAV hyperspectral imagery, and the *PNC, AGB, PWC, CHL, LAI,* and *H*. We also contrast the use of only a single vegetation index with the use of multiple vegetation indices combined with MLR, PLSR, or RF. Section 4 analyzes the advantages and disadvantages of the various methods and of the resulting estimates based on UAV RGB and hyperspectral remote sensing. Finally, Section 5 discusses the potential applications of UAV RGB imagery and hyperspectral imagery in remote monitoring of agriculture.

2. Materials and Methods

2.1. Survey and Test Design of the Research Area

This study was conducted at the National Precision Agriculture Research and Demonstration Base of Xiaotangshan Town, Changping District, Beijing, China. It is located around in the area of the Yanshan branch vein and plain. The north latitude of this region is 40°00′–40°21′, and the east longitude is 116°34′–117°00′. The annual precipitation is approximately 645 mm. The highest and lowest temperatures can reach +40 and $-10\ ^\circ$C, and the average temperature is 11.7 $^\circ$C (from China Meteorological Data Service). The experimental field held a total of 48 plots. There are two winter wheat varieties, J9843 and ZM175, and three water treatments: 200, 100 mm, and rainfall. Water irrigation was applied on 20 October 2014, 9 April 2015, and 30 April 2015. To accentuate the differences in crop nitrogen content between the experimental plots, different amounts of nitrogen were supplied: each plot was provided with either 0(N1), 195(N2), 390(N3), or 585(N4) kg urea/hm^2. Each treatment scheme was repeated three times, and a total of 48 experimental plots were constructed. Figure 1 shows the location of the test area and the experimental design.

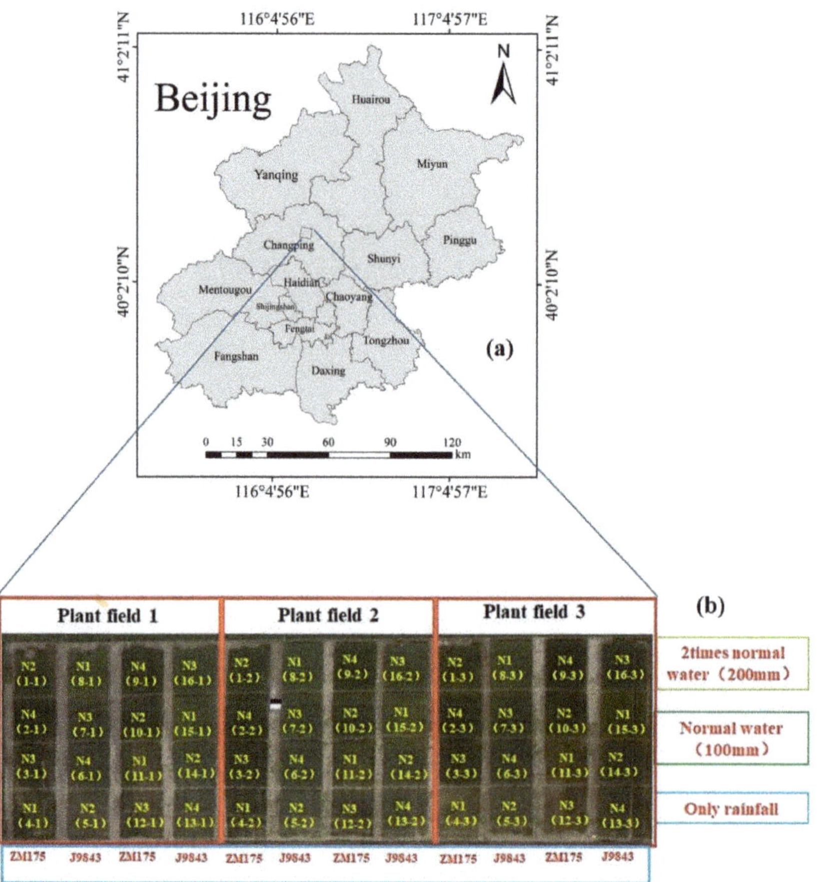

Figure 1. Test area location and experimental design: (**a**) location of Changping District in Beijing City; (**b**) design of experimental and images obtained by UAV.

2.2. Acquisition of Ground Data

We collected the *PNC, AGB, PWC, CHL, LAI*, and *H* data of winter wheat at (GS 31) (21 April 2015), (GS 47) (26 April 2015), and (GS 65) (13 May 2015). For the *PNC*, we used Buchi B-339 (Switzerland) and measured the nitrogen content of each organ (leaves, stems, and spikes) of 20 samples. We selected 20 plants in each growth period that represent the overall growth of the plot and investigated the wheat density of the plot to calculate the biomass of the plot. The dryness and freshness of the spikes in the flowering and filling stages of winter wheat were also considered. For *AGB* acquisition, 20 samples were taken from each growing area (6 × 8 m^2), and stems and leaves were separated. The above-ground parts were heated to 105 °C in an oven for 30 min, then dried at 70 °C for about 24 h (i.e., until achieving a constant weight). The result is the dry mass per sampling area, which is the biomass. We calculated the *PWC* from the fresh and dry mass of sample stems, leaves, and ears. The total fresh quality was subtracted from the total dry quality which was then divided by the total fresh quality. Twenty leaves of different parts of the plant were randomly selected, and the leaf *CHL* content was measured by using a Dualex 4 nitrogen balance. Each leaf was measured five times, and the average value of the 20 leaves was used as the chlorophyll content of the sampling plot. To measure the *LAI*, 20 plant stems and leaves were separated. The leaf area was measured using a CID-203 Laser Leaf Area Meter (CID Company, Bethesda, MD, USA) to obtain the leaf area of a single stem. Next, the number of stems per unit area was determined through field investigation and multiplied by the total number of stems per unit area to calculate the *LAI*. Before the winter wheat began heading, a ruler was used to measure *H* from the stem base to the flag leaf tip. After heading, the distance from the stem base to the topmost end was measured.

2.3. UAV RGB-Data Acquisition and Processing

The experimental UAV remote sensing platform consisted of a DJI S1000 UAV (SZ DJI Technology Co., Ltd., Sham Chun, China) with eight rotors and equipped with two 18,000 mA h (25 V) batteries. It has 30 min of autonomy, its payload capacity is 6 kg, and its flight speed is 8 m/s. It was equipped with an RGB camera (Sony DSC–QX100, Sony, Tokyo, Japan) that weighed 0.179 kg and provided images of 5472 × 3648 pixels. The RGB images obtained were acquired under stable lighting conditions, so the flight time started after 12 a.m. (21 April 2015, 26 April 2015, and 13 May 2015), and the flight height was 80 m. The weather was clear with no wind and few clouds. High-resolution digital images were obtained of winter wheat at (GS 31), (GS 47), (GS 65), with a spatial resolution of 0.013 m. After obtaining the RGB imagery, we used Agisoft PhotoScan software (Agisoft PhotoScan Professional Pro, Version 1.1.6, Agisoft LLC, 11 Degtyarniy per., St. Petersburg, Russia, hereinafter referred to as PhotoScan) to stitch the RGB images. RGB image stitching requires input POS (position point altitude system) data. The POS data contained longitude, latitude, altitude, yaw angle, pitch angle, and rotation angle at the moment of image acquisition. To stitch the UAV RGB imagery, we used the POS data and RGB imagery from the UAV to restore the spatial attitude at the time of image capture and then generated sparse point clouds. A spatial grid was established based on the sparse point cloud, and ground control point information was added to optimize the spatial pose of the image to obtain a sparse point cloud with spatial information. We built a dense point cloud based on the sparse point cloud with spatial information to generate a three-dimensional polygon grid and construct spatial texture. This procedure allowed us to produce a high-definition digital orthophoto mosaic of the UAV flying area.

2.4. UAV Hyperspectral Data Acquisition and Processing

Hyperspectral data acquisition was carried using the FIREFLEYE imaging spectrometer (also known as UHD185, Germany). The UHD185 weighs 0.47 kg and covers a wavelength range from 450 to 950 nm. The hyperspectral data were sampled at 4 nm intervals, producing a 1000 × 1000 pixel image with 125 bands. Before the UAV hyperspectral flight, the UHD185 was calibrated using a black-and-white board. The flight height was 80 m. The acquired UAV hyperspectral images have a spatial resolution of 0.021 m, including the gray pixel with a spatial resolution of 0.01 m. The same flight routes were used to acquire remote-sensing images for the three winter wheat growth stages.

UAV hyperspectral data processing mainly involved image correction, image stitching, and extraction of reflectance. Image correction of hyperspectral images converts the digital number (DN) to the ground surface reflectance [27].

The hyperspectral images have rich spectral information but lack texture information. For the stitching, we used the Cubert Cube-Pilot software (Cube-Pilot, Version 1.4, Cubert GmbH, Ulm, Baden-Württemberg, Germany) to fuse the hyperspectral images with the corresponding full-color image acquired at the same time to generate the fused hyperspectral image. We then used Agisoft PhotoScan software with the point cloud data from the full-color image to complete the stitching. The final spatial resolution of the hyperspectral image was about 5 cm [28].

2.5. Research Methods

Specifically (see Figure 2), we calculated the vegetation indices from both the UAV RGB imagery and from the UAV hyperspectral images and analyzed them both by comparison with *PNC, AGB, PWC, CHL, LAI, H,* and *CGI*. The *CGI* estimation model was constructed using MLR, PLSR, and RF, and the maps of the CGI distribution based on the UAV-based RGB and spectral vegetation indices were generated.

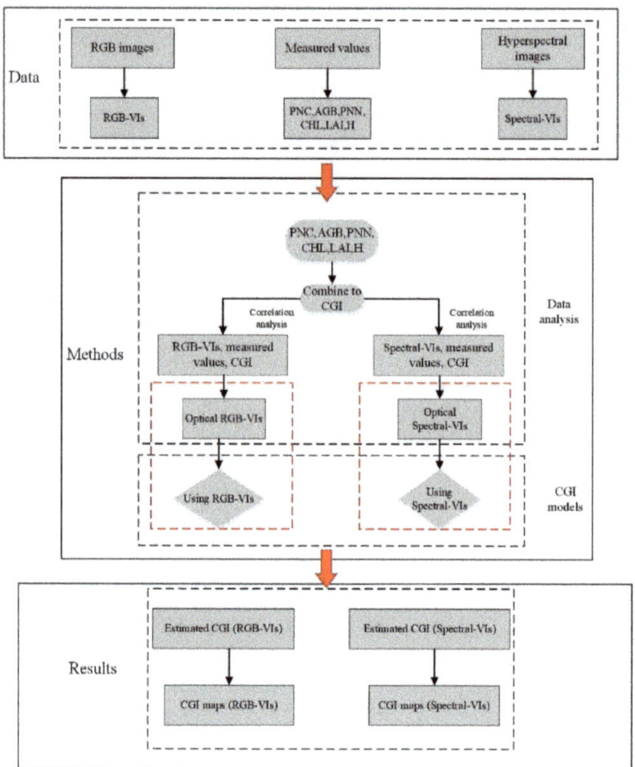

Figure 2. Flowchart showing data processing.

2.6. Analysis Methods

We analyzed the correlation between the vegetation index and *CGI* and used multiple MLR, PLSR, and RF methods to build an estimation model. To estimate the *CGI* from a single vegetation index, to deal with multiple variables, we used MLR, PLSR, and RF. MLR can be used to accurately measure the degree of correlation between various factors and the degree of regression fitting to improve the prediction equation. The larger the absolute value of the standardized regression coefficient, the greater the effect of the corresponding independent variable on the dependent variable.

$$y = a_0 + a_1 x_{1n} + \cdots + a_m x_{mn} + w_m \tag{1}$$

In Equation (1), n is the number of modeling factors and a (= 1, ... , m) is the coefficient.

PLSR can effectively eliminate collinearity among multiple variables, reduce multiple variables to fewer unrelated latent variables, maximize covariance between independent and dependent variables, and then establish regression models [29,30]. RF is based on the bootstrap sampling method whereby multiple samples are extracted from the original sample. The number of random forest algorithm trees we use is set to 1000, and the number of nodes is 50. Each bootstrap sample is modeled using a decision tree, and multiple decision trees are then combined for prediction. Finally, the prediction is determined by voting [31,32].

We used WiMATLAB2018a software (Matrix Laboratory 2014a, MathWorks, Inc., Natick, MA, USA) to calibrate and verify the model with the vegetation index as the input variable and the *CGI* as the output variable.

2.7. Selection of RGB Imagery Indices and Hyperspectral Indices

The vegetation index combines two or more pieces of spectral information, which can simply the measurement of vegetation states. The vegetation index is widely used for monitoring grassland, forest, and drought. To build a model to estimate the *CGI*, 13 RGB-imagery-based vegetation indices and 13 hyperspectral-image-based vegetation indices were selected (Table 1).

Table 1. Indices from RGB imagery and from hyperspectral images.

Type	Index	Formula	References
RGB-imagery-based vegetation indices	R	$R = R$	[15]
	G	$G = G$	[15]
	B	$B = B$	[15]
	r	$r = R/(R + G + B)$	[15]
	g	$g = G/(R + G + B)$	[15]
	b	$b = B/(R + G + B)$	[15]
	EXR	$EXR = 1.4\,r - g$	[33]
	VARI	$VARI = (g - r)/(g + r - b)$	[34]
	GRVI	$GRVI = (g - r)/(g + r)$	[35]
	MGRVI	$MGRVI = (g^2 - r^2)/(g^2 + r^2)$	[35]
	CIVE	$CIVE = 0.441\,r - 0.881\,g + 0.385\,b + 18.78745$	[36]
	EXG	$EXG = 2\,g - b - r$	[37]
	GLA	$GLA = (2G - B - R)/(2G + B + R)$	[38]
Hyperspectral-imagery-based vegetation indices	NDVI	$(R800 - R680)/(R800 + R680)$	[39]
	SR	$R750/R550$	[40]
	MSR	$(R800/R760 - 1)/(R800/R670 + 1)^{1/2}$	[41]
	MCARI	$((R700 - R670) - 0.2 \times (R700 - R550))(R700/R670)$	[42]
	TCARI	$3[(R700 - R670) - 0.2(R700 - R550)(R700/R670)]$	[42]
	MSAVI	$0.5[2R800 + 1 - ((2R800 + 1)^2 - 8(R800 - R670))^{1/2}]$	[43]
	OSAVI	$1.16 \times (R800 - R670)/(R800 + R670 + 0.16)$	[8]
	EVI2	$2.5 \times (R800 - R670)/(R800 + 2.4 \times R670 + 1)$	[9]
	SPVI	$0.4[3.7(R800 - R670) - 1.2\,|R530 - R670|]$	[44]
	LCI	$(R850 - R710)/(R850 + R680)$	[45]
	RDVI	$(R800 - R670)/(R800 + R670)^{1/2}$	[46]
	BGI	$R460/R560$	[47]
	NPCI	$(R670 - R460)/(R670 + R460)$	[48]

Note: R: red reflectance; G: green reflectance; B: blue reflectance; r: normalized red reflectance; g: normalized green reflectance; b: normalized blue reflectance; EXR: excess-red index; VARI: vegetation atmospherically resistant index; GRVI: green-red vegetation index; MGRVI: modified green-red vegetation index; CIVE: color index of vegetation; EXG: excess-green index; GLA: green leaf algorithm index; NDVI: normalized difference vegetation index; SR: simple ratio vegetation index; MSR: modified simple ratio index; MCARI: modified chlorophyll absorption ratio index; TCARI: transformed chlorophyll absorption ratio index; MSAVI: modified soil-adjusted vegetation index; OSAVI: optimized soil-adjusted vegetation index; EVI2: two-band enhanced VI; SPVI: spectral polygon vegetation index; LCI: linear combination index; RDVI: renormalized difference vegetation index; BGI: bare ground index; NPCI: normalized pigment chlorophyll ratio index.

2.8. Construction of Comprehensive Growth Index

In this study, we combine *PNC*, *AGB*, *PWC*, *CHL*, *LAI*, and *H* into a *CGI*. We take into account the different winter wheat growth stages when monitoring single growth phases versus the entire growth phase. In order to combine multiple agronomic parameters into a new index and provide guidance for future remote sensing yield monitoring, the *PNC*, *AGB*, *PWC*, *CHL*, *LAI*, and *H* are combined into *CGI*, which can not only reflect the growth information of crops but can also be correlated with yield. It is also of great significance to the monitoring of crops. Each parameter that constructs the CGI contributes to the construction of the model, so it is calculated in a weighted manner. For the time being, the contribution of each factor is the same; that is, the contribution of each factor to the construction of the model is the same. The *PNC, AGB, PWC, CHL, LAI,* and *H* are normalized separately:

$$W_t = X_t / max(X_t) \tag{2}$$

After normalizing *PNC, AGB, PWC, CHL, LAI,* and *H*, they are weighted by one-sixth and summed to form the CGI [15]:

$$CGI = \frac{1}{6} \sum_{t=1}^{6} W_t \tag{3}$$

$$CGI = a \times W_{PNC} + b \times W_{AGB} + c \times W_{PWC} + d \times W_{CHL} + e \times W_{LAI} + f \times W_H \tag{4}$$

where *t = PNC, AGB, PWC, CHL, LAI,* and *H*; X_t is the value of *PNC, AGB, PWC, CHL, LAI,* and *H* at each the growth stage; $max(X_t)$ is the maximum of *PNC, AGB, PWC, CHL, LAI,* and *H* at each the growth stage; W_t is the normalized value; and *a, b, c, d, e, f* are each one-sixth.

2.9. Verification of Accuracy

We took the 48 data from each of the three winter wheat growth stages, used two sets of data as calibration sets (plant field 1, plant field 2; see Figure 1), and used the remaining set (plant field 3, Figure 1) as the validation set for constructing the model. The correlation coefficient was used to evaluate how the vegetation indices were related to the *CGI*. To evaluate the performance of the proposed model, we use the coefficient of determination R^2, the root-mean-squared error (*RMSE*), and the normalized root-mean-squared error (*NRMSE*). As a model evaluation standard [15], statistically speaking, higher R^2 values and lower *RMSE* and *NRMSE* correspond to a more accurate model. R^2, *RMSE*, and *NRMSE* are calculated as follows:

$$R^2 = \frac{\sum_{i=1}^{n}(x_i - \bar{x})^2 (y_i - \bar{y})^2}{\sum_{i=1}^{n}(x_i - \bar{x})^2 \sum_{i=1}^{n}(y_i - \bar{y})^2} \tag{5}$$

$$RMSE = \sqrt{\frac{\sum_{i=1}^{n}(x_i - y_i)^2}{n}} \tag{6}$$

$$NRMSE = \frac{RMSE}{\overline{X}} \times 100\% \tag{7}$$

where x_i is the measured *CGI* for winter wheat, $\bar{x}$ is the average measured *CGI*, y_i is the predicted *CGI*, $\bar{y}$ is the average predicted *CGI*, and *n* is the number of model samples.

3. Results and Analysis

3.1. Correlation Analysis

To construct the proposed *CGI* and vegetation index, the RGB imagery indices and spectral indices of (GS 31), (GS 47), (GS 65) and for all three stages were combined with *PNC, AGB, PWC, CHL, LAI,* and *H*. Tables 2 and 3 list the correlation between *CGI*, RGB imagery indices, and spectral indices. These results show that the correlation between the RGB imagery indices and *PNC, AGB, PWC, CHL, LAI,* and *H* depends on the growth period, and most of these correlations are very significant (significance level of 0.01).

Table 2. Results of correlation analysis between RGB imagery indices and single indicator and the CGI.

Stages	Index	Correlation Coefficient						
		PNC	AGB	PWC	CHL	LAI	H	CGI
(GS 31)	R	−0.56 **	−0.72 **	−0.66 **	−0.65 **	−0.78 **	−0.50 **	−0.85 **
	r	−0.72 **	−0.61 **	−0.69 **	−0.69 **	−0.65 **	−0.50 **	−0.81 **
	EXR	−0.64 **	−0.63 **	−0.63 **	−0.65 **	−0.69 **	−0.51 **	−0.81 **
	G	−0.47 **	−0.72 **	−0.64 **	−0.59 **	−0.78 **	−0.46 **	−0.80 **
	VARI	0.63 **	0.63 **	0.62 **	0.64 **	0.69 **	0.52 **	0.80 **
	MGRVI	0.63 **	0.63 **	0.61 **	0.64 **	0.69 **	0.52 **	0.80 **
	GRVI	0.63 **	0.63 **	0.61 **	0.64 **	0.69 **	0.52 **	0.80 **
	B	−0.35 *	−0.69 **	−0.52 **	−0.50 **	−0.76 **	−0.45 **	−0.73 **
	CIVE	−0.47 **	−0.61 **	−0.49 **	−0.52 **	−0.69 **	−0.49 **	−0.72 **
	GLA	0.45 **	0.60 **	0.48 **	0.51 **	0.68 **	0.49 **	0.71 **
	EXG	0.45 **	0.60 **	0.48 **	0.51 **	0.68 **	0.49 **	0.71 **
	g	0.45 **	0.60 **	0.48 **	0.51 **	0.68 **	0.49 **	0.71 **
	b	0.66 **	0.23	0.54 **	0.51 **	0.18	0.19	0.43 **
(GS 47)	R	−0.51 **	−0.68 **	−0.66 **	−0.45 **	−0.63 **	−0.48 **	−0.69 **
	r	−0.64 **	−0.73 **	−0.68 **	−0.53 **	−0.74 **	−0.70 **	−0.81 **
	EXR	−0.56 **	−0.71 **	−0.69 **	−0.43 **	−0.72 **	−0.72 **	−0.77 **
	G	−0.41 **	−0.57 **	−0.56 **	−0.41 **	−0.49 **	−0.23	−0.55 **
	VARI	0.55 **	0.70 **	0.69 **	0.42 **	0.72 **	0.72 **	0.76 **
	MGRVI	0.53 **	0.70 **	0.69 **	0.40 **	0.71 **	0.73 **	0.75 **
	GRVI	0.53 **	0.70 **	0.69 **	0.40 **	0.71 **	0.73 **	0.75 **
	B	−0.13	−0.40 **	−0.47 **	−0.11	−0.32	−0.13	−0.31 *
	CIVE	−0.24	−0.53 **	−0.62 **	−0.08	−0.56 **	−0.68 **	−0.52 **
	GLA	0.20	0.50 **	0.60 **	0.03	0.53 **	0.66 **	0.48 **
	EXG	0.20	0.50 **	0.60 **	0.03	0.53 **	0.66 **	0.48 **
	g	0.20	0.50 **	0.60 **	0.03	0.53 **	0.66 **	0.48 **
	b	0.76 **	0.70 **	0.56 **	0.72 **	0.69 **	0.56 **	0.83 **
(GS 65)	R	−0.40 **	−0.63 **	−0.67 **	−0.41 **	−0.63 **	−0.48 **	−0.79 **
	r	−0.49 **	−0.71 **	−0.79 **	−0.45 **	−0.74 **	−0.72 **	−0.81 **
	EXR	−0.37 **	−0.69 **	−0.72 **	−0.35 *	−0.68 **	−0.71 **	−0.79 **
	G	−0.39 **	−0.51 **	−0.55 **	−0.42 **	−0.51 **	−0.20	−0.67 **
	VARI	0.37 **	0.69 **	0.72 **	0.35 *	0.69 **	0.71 **	0.80 **
	MGRVI	0.35 *	0.68 **	0.71 **	0.33 *	0.67 **	0.70 **	0.79 **
	GRVI	0.35 *	0.68 **	0.71 **	0.33 *	0.67 **	0.70 **	0.79 **
	B	−0.14	−0.43 **	−0.39 **	−0.20	−0.39 **	−0.22	−0.59 **
	CIVE	−0.13	−0.59 **	−0.54 **	−0.16	−0.54 **	−0.63 **	−0.69 **
	GLA	0.11	0.57 **	0.52 **	0.15	0.52 **	0.62 **	0.68 **
	EXG	0.11	0.57 **	0.52 **	0.15	0.52 **	0.62 **	0.68 **
	g	0.11	0.57 **	0.52 **	0.15	0.52 **	0.62 **	0.68 **
	b	0.78 **	0.52 **	0.75 **	0.64 **	0.64 **	0.46 **	0.56 **
Total for three stages	R	−0.47 **	−0.64 **	−0.63 **	−0.48 **	−0.51 **	−0.50 **	−0.68 **
	r	−0.64 **	−0.65 **	−0.27 **	−0.45 **	−0.46 **	−0.68 **	−0.71 **
	EXR	−0.55 **	−0.67 **	−0.42 **	−0.44 **	−0.57 **	−0.67 **	−0.74 **
	G	−0.36 **	−0.53 **	−0.65 **	−0.42 **	−0.40 **	−0.32 **	−0.55 **
	VARI	0.54 **	0.68 **	0.42 **	0.43 **	0.57 **	0.68 **	0.74 **
	MGRVI	0.53 **	0.67 **	0.44 **	0.43 **	0.58 **	0.66 **	0.73 **
	GRVI	0.53 **	0.67 **	0.44 **	0.43 **	0.58 **	0.66 **	0.73 **
	B	−0.07	−0.36 **	−0.69 **	−0.24 **	−0.36 *	−0.15	−0.36 **
	CIVE	−0.19 *	−0.50 **	−0.59 **	−0.28 **	−0.57 **	−0.44 **	−0.55 **
	GLA	0.15	0.47 **	0.59 **	0.25 **	0.55 **	0.41 **	0.52 **
	EXG	0.15	0.47 **	0.59 **	0.25 **	0.56 **	0.41 **	0.52 **
	g	0.15	0.47 **	0.59 **	0.25 **	0.56 **	0.41 **	0.52 **
	b	0.61 **	0.41 **	−0.11	0.33 **	0.15	0.48 **	0.44 **

Note: * indicates a significant correlation at 0.05 level; ** indicates a significant correlation at 0.1 level.

Table 3. Results of correlation analysis between the spectral indices, single indicator, and the CGI.

Stages	Index	Correlation Coefficient						
		PNC	AGB	PWC	CHL	LAI	H	CGI
(GS 31)	LCI	0.62 **	0.67 **	0.65 **	0.75 **	0.70 **	0.48 **	0.83 **
	MSR	0.63 **	0.64 **	0.63 **	0.69 **	0.66 **	0.48 **	0.80 **
	SR	0.64 **	0.63 **	0.65 **	0.67 **	0.65 **	0.49 **	0.79 **
	NDVI	0.57 **	0.62 **	0.57 **	0.70 **	0.65 **	0.45 **	0.77 **
	NPCI	−0.68 **	−0.58 **	−0.63 **	−0.67 **	−0.62 **	−0.42 **	−0.77 **
	OSAVI	0.52 **	0.44 **	0.42 **	0.66 **	0.49 **	0.32 *	0.62 **
	BGI	0.57 **	0.45 **	0.70 **	0.49 **	0.45 **	0.26	0.58 **
	RDVI	0.49 **	0.37 **	0.35 *	0.61 **	0.42 **	0.27	0.54 **
	EVI2	0.45 **	0.30 *	0.30 *	0.56 **	0.35 *	0.22	0.47 **
	MSAVI	0.45 **	0.29 *	0.30 *	0.56 **	0.35 *	0.23	0.47 **
	SPVI	0.38 **	0.20	0.21	0.49 **	0.25	0.15	0.36 **
	TCARI	−0.12	−0.31 *	−0.42 **	−0.09	−0.28	−0.27	−0.29 *
	MCARI	0.19	0.01	−0.07	0.22	0.06	0.02	0.11
(GS 47)	LCI	0.68 **	0.78 **	0.64 **	0.60 **	0.74 **	0.69 **	0.85 **
	MSR	0.63 **	0.76 **	0.62 **	0.51 **	0.73 **	0.73 **	0.82 **
	NDVI	0.63 **	0.76 **	0.58 **	0.53 **	0.70 **	0.74 **	0.81 **
	SR	0.62 **	0.75 **	0.63 **	0.49 **	0.73 **	0.72 **	0.81 **
	OSAVI	0.67 **	0.74 **	0.46 **	0.51 **	0.68 **	0.79 **	0.81 **
	NPCI	−0.61 **	−0.75 **	−0.65 **	−0.49 **	−0.71 **	−0.71 **	−0.80 **
	RDVI	0.68 **	0.70 **	0.39 **	0.49 **	0.66 **	0.80 **	0.78 **
	MSAVI	0.67 **	0.70 **	0.38 **	0.48 **	0.65 **	0.80 **	0.78 **
	EVI2	0.67 **	0.69 **	0.36 *	0.47 **	0.64 **	0.80 **	0.77 **
	SPVI	0.66 **	0.63 **	0.28	0.44 **	0.60 **	0.79 **	0.72 **
	BGI	0.58 **	0.65 **	0.67 **	0.54 **	0.61 **	0.44 **	0.71 **
	TCARI	−0.24	−0.37 **	−0.61 **	−0.38 **	−0.37 **	0.01	−0.38 **
	MCARI	0.06	0.05	−0.26	−0.16	0.01	0.45 **	0.05
(GS 65)	LCI	0.55 **	0.77 **	0.79 **	0.53 **	0.78 **	0.66 **	0.84 **
	MSAVI	0.48 **	0.80 **	0.75 **	0.41 **	0.77 **	0.79 **	0.83 **
	SPVI	0.48 **	0.80 **	0.74 **	0.41 **	0.77 **	0.80 **	0.83 **
	EVI2	0.48 **	0.79 **	0.75 **	0.42 **	0.77 **	0.79 **	0.82 **
	RDVI	0.48 **	0.78 **	0.76 **	0.43 **	0.76 **	0.79 **	0.82 **
	SR	0.45 **	0.81 **	0.78 **	0.40 **	0.80 **	0.69 **	0.82 **
	MSR	0.46 **	0.80 **	0.79 **	0.42 **	0.78 **	0.71 **	0.82 **
	OSAVI	0.48 **	0.77 **	0.77 **	0.44 **	0.76 **	0.77 **	0.82 **
	NDVI	0.45 **	0.73 **	0.76 **	0.44 **	0.72 **	0.71 **	0.77 **
	BGI	0.61 **	0.71 **	0.62 **	0.61 **	0.69 **	0.42 **	0.77 **
	NPCI	−0.40 **	−0.75 **	−0.76 **	−0.38 **	−0.72 **	−0.74 **	−0.76 **
	MCARI	−0.05	0.49 **	0.45 **	−0.03	0.40 **	0.81 **	0.40 **
	TCARI	−0.20	0.20	0.19	−0.20	0.12	0.72 **	0.14
Total for three stages	LCI	0.63 **	0.74 **	0.40 **	0.60 **	0.65 **	0.60 **	0.82 **
	MSR	0.58 **	0.74 **	0.52 **	0.52 **	0.63 **	0.65 **	0.80 **
	SR	0.58 **	0.74 **	0.52 **	0.50 **	0.63 **	0.64 **	0.80 **
	NDVI	0.55 **	0.69 **	0.50 **	0.54 **	0.59 **	0.62 **	0.76 **
	OSAVI	0.54 **	0.63 **	0.54 **	0.49 **	0.48 **	0.62 **	0.70 **
	NPCI	−0.55 **	−0.65 **	−0.51 **	−0.41 **	−0.40 **	−0.67 **	−0.67 **
	RDVI	0.52 **	0.58 **	0.54 **	0.46 **	0.42 **	0.61 **	0.65 **
	MSAVI	0.50 **	0.56 **	0.53 **	0.43 **	0.39 **	0.59 **	0.62 **
	EVI2	0.50 **	0.55 **	0.53 **	0.43 **	0.38 **	0.59 **	0.62 **
	SPVI	0.46 **	0.49 **	0.51 **	0.38 **	0.32 **	0.55 **	0.55 **
	BGI	0.50 **	0.49 **	0.43 **	0.33 **	0.15	0.47 **	0.48 **
	TCARI	−0.11	−0.13	0.23 **	−0.16	−0.28 **	0.13	−0.16
	MCARI	0.07	0.13	0.39 **	0.01	−0.05	0.34 **	0.12

Note: * indicates a significant correlation at 0.05 level; ** indicates a significant correlation at 0.1 level.

We find that the RGB imagery indices r, EXR, VARI, and MGRVI correlate with the *CGI* during different growth stages, and the correlation coefficients are all higher than those between (i) r, EXR, VARI, MGRVI and (ii) *PNC, AGB, PWC, CHL, LAI*, and *H*. Thus, the correlation coefficients between RGB imagery indices r, EXR, VARI, MGRVI, and the *CGI* for the four growth stages are greater than the correlation coefficients between the individual crop-growth indicators. For the spectral indices, the correlation between the hyperspectral *CGI* and *PNC, AGB, PWC, CHL, LAI*, and *H* also hovers around 0.01. From (GS 31) to the total for the three stages, the correlation between the spectral index and the six crop-growth indicators varies irregularly. However, the spectral indices LCI, MSR, SR, and NDVI are more strongly correlated with the *CGI* than with the six crop-growth indicators. Thus, the *CGI* provides more accurate estimates of the crop-growth parameters.

3.2. Estimate of CGI Based on RGB Imagery and Spectral Indices

According to the results given in Tables 2 and 3, we select the RGB-imagery indices r, EXR, VARI, and MGRVI and the spectral indices LCI, MSR, SR, and NDVI to estimate CGI.

The RGB-imagery indices r, EXR, VARI, and MGRVI and the spectral indices LCI, MSR, SR, and NDVI are taken as factors in (GS 31), (GS 47), (GS 65) and for all three stages, respectively. We then construct the CGI models based on the RGB-imagery indices r, EXR, VARI, and MGRVI and on the spectral indices LCI, MSR, SR, and NDVI for the different growth stages (see Table 4).

Table 4. CGI estimation model for the single index of winter wheat in different stages.

Growth Stages	Parameters	Equations	Calibration			Verification		
			R^2	RMSE	NRMSE (%)	R^2	RMSE	NRMSE (%)
(GS 31)	r	$y = 8.7269 \times e^{-6.874x}$	0.67	0.05	6.55	0.78	0.04	4.98
	EXR	$y = 1.0729 \times e^{-3.196x}$	0.65	0.05	6.73	0.73	0.04	5.49
	VARI	$y = 0.6664 \times e^{1.863/x}$	0.64	0.05	6.78	0.72	0.04	5.58
	MGRVI	$y = 0.6655 \times e^{1.449x}$	0.64	0.05	6.79	0.71	0.04	5.66
	LCI	$y = 0.3945 \times e^{1.067x}$	0.72	0.04	5.92	0.74	0.04	5.38
	MSR	$y = 0.5264 \times e^{0.1375x}$	0.67	0.05	6.48	0.69	0.05	5.93
	SR	$y = 0.5844 \times e^{0.0259x}$	0.66	0.05	6.65	0.68	0.05	6.03
	NDVI	$y = 0.3222 \times e^{1.0736x}$	0.63	0.05	6.84	0.65	0.05	6.25
(GS 47)	r	$y = -5.5464x + 2.6602$	0.57	0.06	8.14	0.84	0.04	5.09
	EXR	$y = -2.9535x + 1.0445$	0.50	0.07	8.71	0.75	0.05	6.32
	VARI	$y = 1.5784x + 0.615$	0.50	0.07	8.74	0.75	0.05	6.36
	MGRVI	$y = 1.3382x + 0.6099$	0.49	0.07	8.84	0.73	0.05	6.61
	LCI	$y = 0.2967 \times e^{1.4913x}$	0.69	0.05	7.04	0.84	0.04	5.50
	MSR	$y = 0.1306x + 0.3858$	0.62	0.06	7.67	0.81	0.04	5.63
	SR	$y = 0.0223x + 0.5101$	0.60	0.06	7.80	0.80	0.04	5.73
	NDVI	$y = 0.1654 \times e^{1.8498x}$	0.62	0.06	7.71	0.80	0.04	5.80
(GS 65)	r	$y = -5.74x + 2.7027$	0.55	0.06	7.80	0.70	0.05	7.33
	EXR	$y = -2.5008x + 0.9961$	0.45	0.07	8.65	0.59	0.06	8.56
	VARI	$y = 1.3211x + 0.6346$	0.46	0.07	8.62	0.59	0.06	8.50
	MGRVI	$y = 1.0881x + 0.6362$	0.43	0.07	8.78	0.57	0.06	8.76
	LCI	$y = 0.32 \times e^{1.3239x}$	0.67	0.05	6.74	0.75	0.05	6.46
	MSR	$y = 01141x + 0.4495$	0.62	0.06	7.24	0.72	0.05	7.02
	SR	$y = 0.0209x + 0.5405$	0.62	0.06	7.19	0.75	0.05	6.65
	NDVI	$y = 0.2602 \times e^{1.3278x}$	0.58	0.06	7.50	0.64	0.06	7.83
Total for three stages	r	$y = 4.138x + 2.1912$	0.54	0.06	8.04	0.62	0.06	7.67
	EXR	$y = -2.4047x + 1.0012$	0.57	0.06	7.75	0.52	0.07	8.69
	VARI	$y = 1.3196x + 0.6485$	0.57	0.06	7.79	0.52	0.07	8.65
	MGRVI	$y = 1.0988x + 0.6455$	0.57	0.06	7.78	0.50	0.07	8.86
	LCI	$y = 0.893x + 0.2007$	0.69	0.05	6.64	0.70	0.05	6.84
	MSR	$y = 0.4986 \times e^{0.1513x}$	0.63	0.06	7.25	0.66	0.06	7.19
	SR	$y = 0.0203x + 0.5462$	0.62	0.06	7.31	0.67	0.05	7.17
	NDVI	$y = 0.265 \times e^{1.3008x}$	0.59	0.06	7.57	0.61	0.06	7.72

The *CGI* model based on the RGB imagery indices r, EXR, VARI, and MGRVI generally performs well for all three growth stages and for the total of the three stages. The best performance for the RGB imagery index model is for the (GS 31). Overall, R^2 is higher and the *RMSE* and *NMRSE* are lower. At the same time, several pairs of RGB imagery–index calibration sets have the same R^2 and *RMSE*, such as VARI and MGRVI. The effect of the model becomes apparent only upon comparing *NRMSE*. The smaller the *NRMSE*, the higher the prediction accuracy. Of course, we also need to consider R^2, *RMSE*, and *NRMSE* for the validation set. Similarly, the effects of EXR, VARI, and the two RGB imagery indices used to construct the *CGI* model during (GS 47) need to be analyzed by NRMSE. The *CGI* effect of each growth stage is compared. The best-performing RGB indices are r, r, r, and EXR.

In the estimation model based on the spectral indices LCI, MSR, SR, and NDVI, evaluating models based on different spectral indices also requires comparing the magnitudes of R^2, *RMSE*, and *NRMSE*. We find that the difference between the R^2 of different estimation models is relatively clear. Combining these spectral indices with the results of the validation model allows us to evaluate the performance of different spectral index models. A comparison of the results of each model shows that the estimation model based on LCI gives the most accurate results.

3.3. Using RGB VIs and Spectral VIs with Machine Learning to Estimate CGI

Table 5 shows the modeling analysis based on the four RGB imagery indices, the four spectral indices, and the *CGI* estimated by using MLR, PLSR, and RF.

Table 5. Results of using different methods to estimate the *CGI* for different winter wheat growth stages.

Growth Stages	Methods	Data	R^2	RMSE	NRMSE (%)
(GS 31)	MLR	RGB, VIs	0.73	0.04	5.69
		Spectral, VIs	0.77	0.04	5.29
	PLSR	RGB, VIs	0.65	0.05	6.54
		Spectral VIs	0.66	0.05	6.40
	RF	RGB, VIs	0.53	0.06	7.64
		Spectral, VIs	0.64	0.05	6.66
(GS 47)	MLR	RGB, VIs	0.65	0.06	7.37
		Spectral, VIs	0.72	0.05	6.57
	PLSR	RGB, VIs	0.50	0.07	8.73
		Spectral, VIs	0.60	0.06	7.80
	RF	RGB, VIs	0.42	0.08	9.65
		Spectral, VIs	0.50	0.07	8.76
(GS 65)	MLR	RGB, VIs	0.68	0.05	6.63
		Spectral, VIs	0.78	0.04	5.44
	PLSR	RGB, VIs	0.62	0.05	6.93
		Spectral, VIs	0.65	0.05	5.88
	RF	RGB, VIs	0.54	0.06	7.97
		Spectral, VIs	0.60	0.06	7.38
Total for three stages	MLR	RGB, VIs	0.58	0.06	7.71
		Spectral, VIs	0.69	0.05	6.61
	PLSR	RGB, VIs	0.57	0.06	7.77
		Spectral, VIs	0.62	0.06	7.30
	RF	RGB, VIs	0.48	0.07	8.62
		Spectral, VIs	0.57	0.06	7.81

Comparing the estimation of the *CGI* result based on indices built from RGB imagery with that based on the spectral indices from the different growth stages, we find that the latter is superior to the former. In addition, a comparison with different machine learning methods shows that the *CGI* estimation model constructed using MLR is the most accurate of the three methods, followed by the model constructed from PLSR, and finally by the model constructed from RF, which shows that MLR method is the index, followed by PLSR, and finally by RF. We now use the verification set data to verify the RGB-imagery -based index and spectral-based index combined with the MLR, PLSR, and RF methods to estimate the *CGI*. Figures 3 and 4 show the relationship between the measured and predicted values (y = ax + b, R^2, *RMSE*, *NRMSE*). The results show that the verification is consistent with the modeling, and the fit is very good (the fit to the spectral index is better than the fit to the RGB-imagery index). Comprehensive modeling and verification (see Table 5 and Figures 3 and 4) show that the MLR, PLSR, and RF methods all provide a more accurate estimate of the *CGI* than the single RGB-imagery index or the single spectral index (see Tables 4 and 5 and Figures 3 and 4). Thus, these three methods all provide good estimates of the *CGI*.

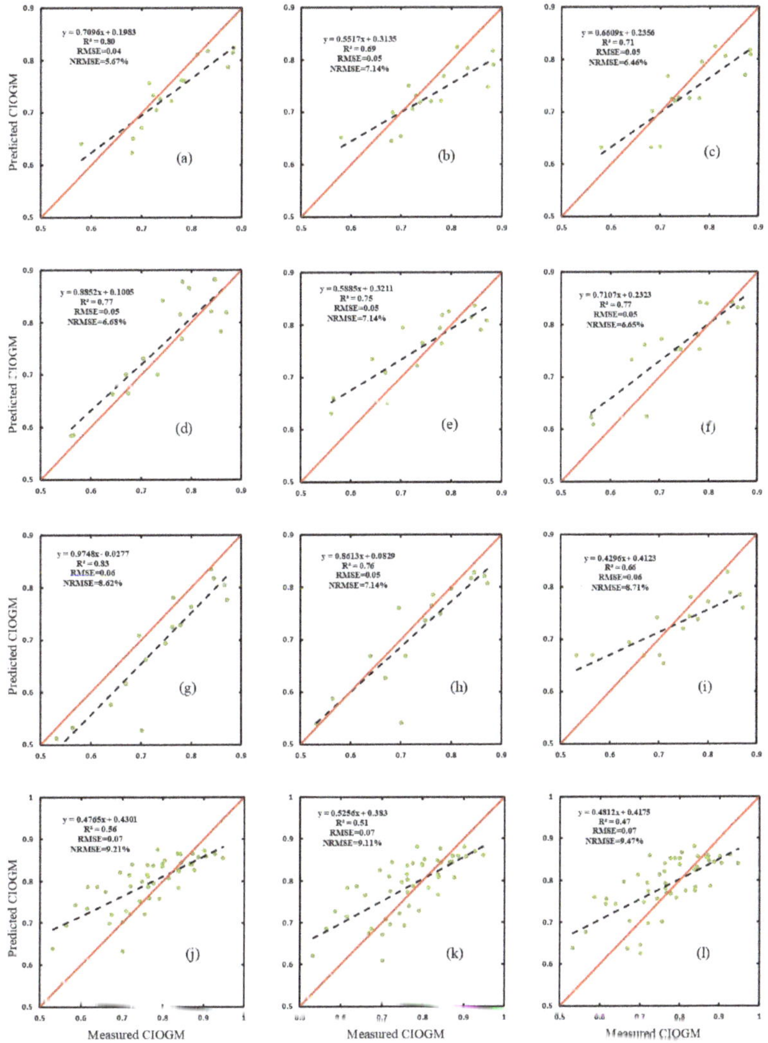

Figure 3. Relationship between the *CGI* predicted vs. the *CGI* measured by RGB VIs using (**a–c**) MLR, PLSR, and RF models at (GS 31), respectively; (**d–f**) MLR PLSR, and RF models at (GS 47), respectively; (**g–i**) MLR, PLSR, and RF models at (GS 65), respectively; and (**j–l**) MLR, PLSR, and RF models over the three stages, respectively.

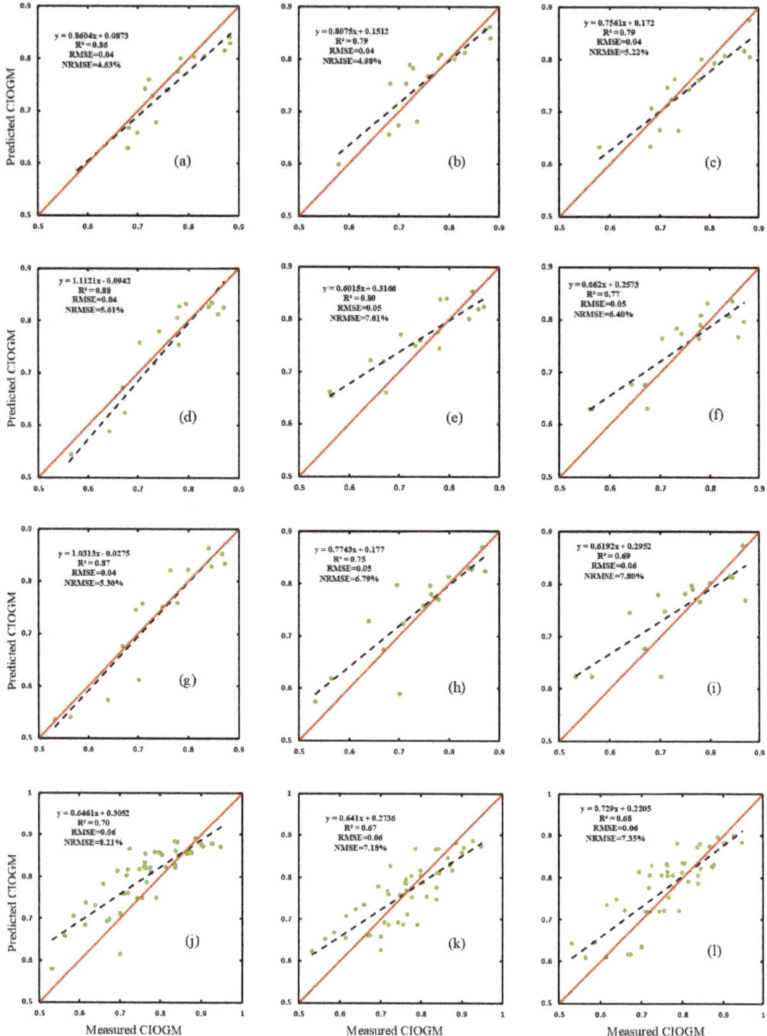

Figure 4. Relationship between the *CGI* predicted vs. the *CGI* measured by Spectral VIs using (**a–c**) MLR, PLSR, and RF models at (GS 31), respectively; (**d–f**) MLR, PLSR, and RF models at (GS 47), respectively; (**g–i**) MLR, PLSR, and RF models at (GS 65), respectively; and (**j–l**) MLR, PLSR, and RF models over the three stages, respectively.

3.4. Map of CGI Distribution

Estimating the *CGI* by using machine learning and based on RGB-imagery-based indices and spectral-based indices allows us to select the best estimation model for the different growth stages. Figures 5 and 6 show the results of applying the MLR estimation model to the UAV RGB and hyperspectral imagery of (GS 31), (GS 47), and (GS 65).

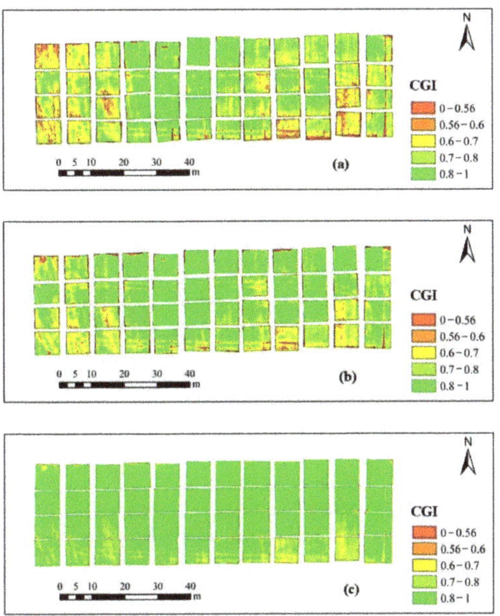

Figure 5. Map of *CGI* distribution based on UAV RGB imagery: (**a**) (GS 31); (**b**) (GS 47); (**c**) (GS 65).

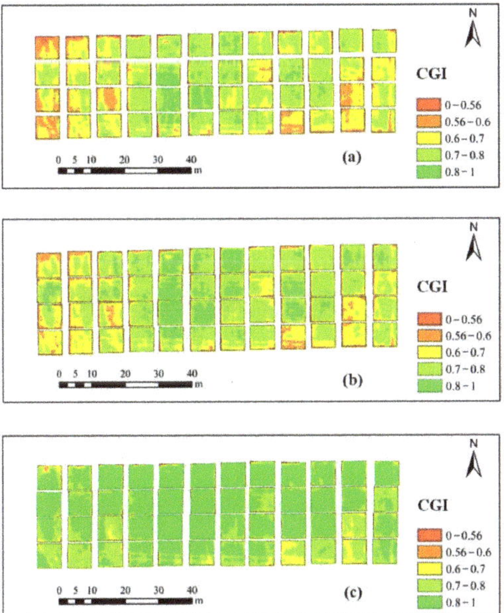

Figure 6. Map of *CGI* distribution based on UAV hyperspectral images: (**a**) (GS 31); (**b**) (GS 47); (**c**) (GS 65).

From jointing to flowering, the average RGB-based *CGI* ranges from 0.74 to 0.79 (Figure 5) and hyperspectral-based CGI averages from 0.74 to 0.78 (Figure 6). In addition, the corresponding *CGI* map is greener, indicating a higher *CGI*. The current map also shows that the *CGI* has increased during the three growing seasons of winter wheat. These results are consistent with the actual observations. The *CGI* distribution based on the RGB and hyperspectral UAV images shows the difference between *CGI* estimation models based on the two types of data (Figure 7). The predicted *CGI* values differ, which leads to different *CGI* distributions for the same growth stage. However, the results for the three growth stages show that winter wheat growth is best in plant field 2 and is relatively stable over the different growth stages. The map of the *CGI* distribution clearly differentiates between the better-growing areas and the poorer-growing areas.

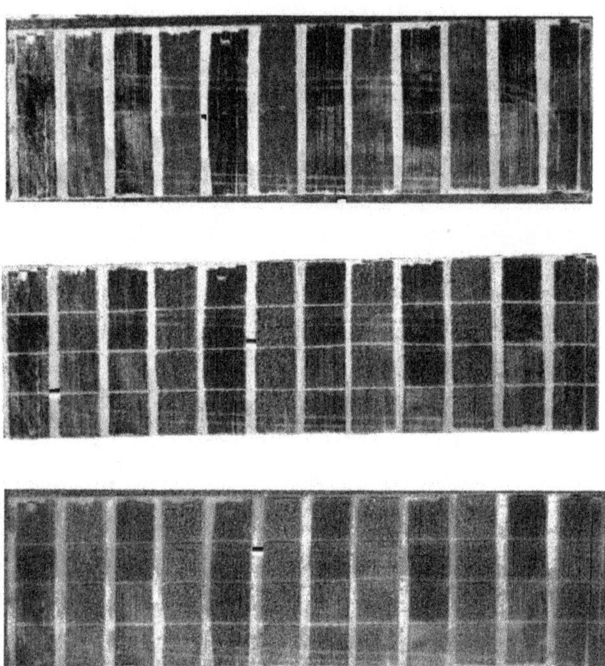

Figure 7. UAV images obtained at different growth stages.

4. Discussion

4.1. Single Growth-Monitoring Indicators and CGI for Winter Wheat

From (GS 31) to the total for the three stages, most of the RGB imagery indices and hyperspectral indices are significantly correlated with *PNC, AGB, PWC, CHL, LAI, H,* and *CGI* (Tables 2 and 3). The RGB imagery indices R, EXR, VARI, MGRVI, and *CGI* are more strongly correlated than r, EXR, VARI, MGRVI and *PNC, AGB, PWC, CHL, LAI,* and *H*, which shows that the four RGB imagery indices for the *CGI* are very sensitive. Zhou et al. [49] used EXR and VARI to monitor rice growth and obtained more accurate predictions. Niu et al. [50] used UAV RGB imagery to show that r and MGRVI can provide good estimates of the *LAI*.

Over the four growth stages, the hyperspectral indices LCI, MSR, SR, and NDVI are more strongly correlated with the *CGI* than LCI, MSR, SR, and NDVI are with *PNC, AGB, PWC, CHL, LAI,* and *H*. These four spectral indices are more sensitive to the *CGI*, mainly because the *CGI* contains more information about crop growth. Yue et al. [15] found that the LCI is useful for monitoring the *AGB* and *LAI* of winter wheat. Wu et al. [42] explored the use of the MSR and NDVI to estimate *CHL*. Liang et al. [51] accurately estimated crop parameters by using the SR. These studies show that these spectral indices are reliable for monitoring crop parameters. At present, the monitoring of crop growth focuses on single indices, which monitor crop growth based on a single growth parameter. Zhu et al. [52] used several spectral LNC indices to confirm their effectiveness for quantitatively inversing the LNC of winter wheat. However, the sensitivities of the RGB imagery, spectral index, and various growth parameters differ. Some RGB imagery and spectral indices may be highly sensitive to one growth parameter and less sensitive to another. The proposed *CGI* combines six growth-monitoring parameters and considers various combinations of growth parameters. Based on the correlation between the RGB-imagery and hyperspectral indices and the *CGI*, the *CGI* can be used as an index to monitor crop growth.

4.2. Estimation of CGI Based on a Single Index

The RGB imagery indices r, EXR, VARI, and MGRVI and the spectral indices LCI, MSR, SR, and NDVI, which are strongly correlated with the *CGI*, are used to construct the *CGI* estimation model. The results show that the models constructed for single growth stages and multiple growth stages both provide good estimates (Table 4). Moreover, the findings indicate that the *CGI* can be accurately estimated from a single RGB imagery index or from a single hyperspectral index. The optimal *CGI* models of the RGB imagery index at (GS 31), (GS 47), (GS 65), and all three stages have different indices that reflect the various sensitivities of the RGB imagery indices in different growth stages. In addition, the indices provide *CGI* predictions of varying accuracy. The performance of the spectral parameters and RGB imagery indices varies over the four stages. In conclusion, the LCI provides the best spectral index for estimating the *CGI* over the different growth stages, and the models constructed for the four growth stages are the most satisfactory. Therefore, the optimal estimation of the *CGI* based on the spectral index is more stable than that based on an RGB imagery index.

Some RGB imagery and spectral indices saturate the growth parameters. Thus, we should further investigate how to estimate the growth parameters [53]. The *PNC, AGB, PWC, CHL, LAI,* and *H* from the (GS 31) to the (GS 65) of winter wheat all change to varying degrees, and these changes modify the *CGI* in the different growth stages. At (GS 31) and (GS 47), the photosynthates are mainly stored in stems and leaves, whereas, at the flowering stage, they are stored in flowers and in the wheat ear. Thus, using a visible-light vegetation index is not recommended for estimating *AGB* [15]. Although the RGB-image and hyperspectral indices at various growth stages are related to six growth parameters, the vegetation index monitoring crop growth effect is related to dry matter and pigment [6], which means that the sensitivity of an index to the *CGI* also depends on the winter wheat's growth stage. For instance, the LCI is highly sensitive to the four growth stages of winter wheat. Therefore, the optimal predictive factor for estimating *CGI* should be determined.

4.3. Estimation of CGI Based on Multiple Indices Combined with Machine Learning

Machine learning methods are increasingly used to estimate parameters to monitor crop growth. The results of this study show that machine-learning estimates of growth parameters based on multiple RGB imagery or spectral indices are more accurate than single-index estimates. Yue et al. [54] used the PLSR to obtain better estimates of the AGB than is possible by using a single vegetation index. Han et al. [20] used multiple linear regression, partial least squares regression, artificial neural networks, and the random forest algorithm to improve the accuracy of estimates of AGB. Better estimates of crop parameters can be obtained by using machine learning methods, which is consistent with the research results presented herein. The *CGI* estimation model constructed by MLR is better than a model based on a single RGB imagery index or a single spectral index. Estimates made using the PLSR or RF models are also better than most (but not all) single RGB imagery index or spectral index models (see Tables 4 and 5 and Figures 3 and 4), which may be related to the number of variables involved in the construction of these models. According to Yue et al. [15], given sufficient input variables, the models constructed by PLSR and RF provide better estimation.

The MLR model based on RGB imagery and spectral indices is superior to the PLSR and RF models over the different growth stages (see Table 5 and Figures 3 and 4), which we attribute to one of the three following reasons: (i) An index that is strongly correlated with the *CGI* is preferable for the construction of the model. (ii) A model made from a linear estimate by a single index provides good results. The sensitivity of multiple indices is combined with the *CGI*, resulting in an abundant amount of information. (iii) RF is suitable for processing large amounts of data but is disadvantaged for processing small data sets. Han et al. [20] and Ozlem et al. [55] used large sample data and the RF algorithm to improve the prediction the model.

In addition, since this paper uses empirical models, although Zhou et al. [49], Niu et al. [50], Yue et al. [15], Wu et al. [42], and Liang et al. [51] used the vegetation index from the estimation model used herein, the crop parameters were estimated by using empirical methods. However, to estimate crop canopy variables, one must consider confounding factors that also affect the estimation, such as leaf or canopy structure and the understory in multiple-scattering processes, soil parameters, and some external parameters [56–61]. To determine how canopy structure affects estimates of crop parameters, Bendig et al. [62] used UAV RGB imagery to obtain crop surface models, extract crop heights, and use the extracted crop heights to estimate biomass, which improved the accuracy of the estimates. In the next step, we further study how to estimate crop parameters based on crop height to explore how various sensors affect estimates.

4.4. CGI Estimation Based on Different Sensors

The results of this study show that using a single index versus multiple indices combined with machine learning to estimate *CGI* produces different results. This research shows that an RGB camera mounted on a UAV can be used to accurately estimate crop-growth parameters, assess the data of cameras mounted on UAVs are reliable [63], but the accuracy of such estimates based on UAV hyperspectral data estimation is higher. Yue et al. [15] monitored crop LAI and AGB and found that the monitoring by using UAV hyperspectral cameras is better than by using UAV RGB cameras. The results are consistent with the performance of the two sensors and with the two estimated *CGIs*. The *CGI* model estimated by the UAV hyperspectral data is superior to that estimated by the UAV RGB imagery. Therefore, hyperspectral sensors are preferred for monitoring crop growth. The results of the RGB camera are close to the results of using a hyperspectral sensor, considering the cost issue, and the simpler data processing of RGB, it is recommended to use an RGB sensor on the UAV.

Note that this study still has several shortcomings. The research area is limited to Changping District, Beijing, China. The growth of winter wheat in other regions, as well as other varieties of wheat grown in different years, should be further investigated to verify the findings of this research. In addition, the *CGI* calculates the average weight of the growth indicators at different growth stages, so the weight of the growth index should be distributed on the basis of growth stages to improve the results. At present, we use UAV to carry RGB cameras and hyperspectral cameras for experiments. In the future, we can use multispectral or lidar sensors to conduct experiments to explore the impact of different sensors on the experimental results. At the same time, in the future, different weight ratios will be assigned to the contribution of different factors in the model construction to explore the estimation effect of the model under different ratios.

5. Conclusions

Six growth indicators were used to construct a new *CGI* index. The *CGI* of winter wheat at different growth stages was estimated by using UAV RGB imagery and hyperspectral data. The linear, nonlinear, MLR, PLSR, and RF models were further developed. The main conclusions are as follows: The RGB imagery indices r, EXR, VARI, and MGRVI are more strongly correlated with the *CGI* for the growth stages than are *PNC, AGB, PWC, CHL, LAI,* and *H*. The spectral indices LCI, MSR, SR, and NDVI are more strongly correlated with the *CGI* than the single growth-monitoring index for the four stages, which indicates that the *CGI* can be used to monitor crop growth. On the one hand, the *CGI* model from a single RGB imagery index has different optimal indices at different growth stages. On the other hand, the model constructed by using a single spectral index has the best index in each growth stage, all of which are in terms of the LCI. By combining multiple RGB imagery indices and spectral indices with MLR, PLSR, and RF to estimate the *CGI*, the MLR model is optimal for different growth stages, the PLSR model is second best, and the RF model is the worst. The best model based on RGB-imagery indices gives R^2, *RMSE*, and *NRMSE* of 0.73, 0.04, 5.69%, and the best model based on hyperspectral data gives an R^2, *RMSE*, and *NRMSE* of 0.78, 0.04, and 5.29%, respectively. The *CGI* can be accurately estimated by using UAV RGB imagery and hyperspectral data. Both methods can estimate the *CGI*. However, the hyperspectral data lead to more accurate estimates of the *CGI* than the RGB imagery, which demonstrates that UAV hyperspectral imaging provides more accurate results. Currently, developments are underway for UAV payloads, including hyperspectral, LIDAR, synthetic aperture radars, and thermal infrared sensors. Fully comparing the advantages of each sensor type requires further study.

Author Contributions: Conceptualization, H.T., H.F. and Z.L.; data curation, G.Y.; formal analysis, H.T.; investigation, H.T., H.F., Z.L. and C.Z.; methodology, H.T.; resources, G.Y.; writing—original draft, H.T. and H.F.; writing—review and editing, C.Z. All authors have read and agreed to the published version of the manuscript.

Funding: The study was funded by the National Key Research and Development Program (2016YFD070030303) and the Nation Science Foundation of China (41601346, 41871333, 41601369, 41501481, 61661136003, 41771370, 41471285, 41471351).

Acknowledgments: We appreciate the help from Hong Chang and Weiguo Li during field data collection. Thanks to all employees of Xiao Tangshan National Precision Agriculture Research Center.

Conflicts of Interest: The authors declare no conflict of interest.

Abbreviations

Parameter	Name
UAVs	unnamed aerial vehicles
CGI	comprehensive growth index
MLR	multiple linear regression
PLSR	partial least squares
RF	random forest
PNC	plant nitrogen content
AGB	above-ground biomass
PWC	plant water content
CHL	chlorophyll
H	plant height

References

1. Li, X.; Zhang, Y.; Luo, J.; Jin, X.; Xu, Y.; Yang, W. Quantification winter wheat LAI with HJ-1CCD image features over multiple growing seasons. *Int. J. Appl. Earth Obs. Geoinf.* **2016**, *44*, 104–112. [CrossRef]
2. Campos, M.; García, F.J.; Camps, G.; Grau, G.; Nutini, F.; Crema, A.; Boschetti, M. Multitemporal and multiresolution leaf area index retrieval for operational local rice crop monitoring. *Remote Sens. Environ.* **2016**, *187*, 102–118. [CrossRef]
3. Jin, X.; Kumar, L.; Li, Z.; Xu, X.; Yang, G.; Wang, J. Estimation of winter wheat biomass and yield by combining the aquacrop model and field hyperspectral data. *Remote Sens.* **2016**, *8*, 972. [CrossRef]
4. Launay, M.; Guerif, M. Assimilating remote sensing data into a crop model to improve predictive performance for spatial applications. *Agric. Ecosyst. Environ.* **2005**, *111*, 321–339. [CrossRef]
5. Di Gennaro, S.F.; Toscano, P.; Gatti, M.; Poni, S.; Berton, A.; Matese, A. Spectral comparison of UAV-Based hyper and multispectral cameras for precision viticulture. *Remote Sens.* **2022**, *14*, 449. [CrossRef]
6. Berger, K.; Atzberger, C.; Danner, M.; D'Urso, G.; Mauser, W.; Vuolo, F.; Hank, T. Evaluation of the PROSAIL model capabilities for future hyperspectral model environments: A review study. *Remote Sens.* **2018**, *10*, 85. [CrossRef]
7. Taniguchi, K.; Obata, K.; Yoshioka, H. Derivation and approximation of soil isoline equations in the red–near-infrared reflflectance subspace. *J. Appl. Remote Sens.* **2014**, *8*, 083621. [CrossRef]
8. Wang, W.; Gao, X.; Cheng, Y.; Ren, Y.; Zhang, Z.; Wang, R.; Cao, J.; Geng, H. QTL mapping of leaf area index and chlorophyll content based on UAV remote sensing in wheat. *EconPapers.* **2022**, *12*, 595. [CrossRef]

9. Zhang, H.; Tang, Z.; Wang, B.; Meng, B.; Qin, Y.; Sun, Y.; Lv, Y.; Zhang, J.; Yi, S. A non-destructive method for rapid acquisition of grassland aboveground biomass for satellite ground verification using UAV RGB images. *Glob. Ecol. Conserv.* **2022**, *33*, e01999. [CrossRef]

10. Xu, L.; Zhou, L.; Meng, R.; Zhang, F.; Lv, Z.; Xu, B.; Zeng, L.; Yu, X.; Peng, S. An improved approach to estimate ratoon rice aboveground biomass by integrating UAV-based spectral, textural and structural features. *Precis. Agric.* **2022**, *23*, 1276–1301. [CrossRef]

11. Jin, X.; Liu, S.; Baret, F.; Hemerlé, M.; Comar, A. Estimates of plant density of wheat crops at emergence from very low altitude UAV imagery. *Remote Sens. Environ.* **2017**, *198*, 105–114. [CrossRef]

12. Yu, N.; Li, L.; Schmitz, N.; Tian, L.F.; Greenberg, J.A.; Diers, B.W. Development of methods to improve soybean yield estimation and predict plant maturity with an unmanned aerial vehicle based platform. *Remote Sens. Environ.* **2016**, *187*, 91–101. [CrossRef]

13. Maimaitijiang, M.; Ghulam, A.; Sidike, P.; Hartling, S.; Maimaitiyiming, M.; Peterson, K.; Shavers, E.; Fishman, J.; Peterson, J.; Kadam, S.; et al. Unmanned Aerial System (UAS)-based phenotyping of soybean using multi-sensor data fusion and extreme learning machine. *ISPRS J. Photogramm. Remote Sens.* **2017**, *134*, 43–58. [CrossRef]

14. Adar, S.; Sternberg, M.; Paz-Kagan, T.; Henkin, Z.; Dovrat, G.; Zaady, E. Argaman E. Estimation of aboveground biomass production using an unmanned aerial vehicle (UAV) and VENμS satellite imagery in Mediterranean and semiarid rangelands. *Remote Sens. Appl. Soc. Environ.* **2022**, *26*, 100753.

15. Yue, J.; Feng, H.; Jin, X.; Yuan, H.; Li, Z.; Zhou, C.; Yang, G.; Tian, Q. A Comparison of Crop Parameters Estimation Using Images from UAV-Mounted Snapshot Hyperspectral Sensor and High-Definition Digital Camera. *Remote Sens.* **2018**, *10*, 1138. [CrossRef]

16. Geipel, J.; Link, J.; Claupein, W. Combined spectral and spatial modeling of corn yield based on aerial images and crop surface models acquired with an unmanned aircraft system. *Remote Sens.* **2014**, *6*, 10335–10355. [CrossRef]

17. Zaman-Allah, M.; Vergara, O.; Araus, J.; Tarekegne, A.; Magorokosho, C.; Zarco-Tejada, P.; Hornero, A.; Albà, A.; Das, B.; Craufurd, P.; et al. Unmanned aerial platform-based multi-spectral imaging for field phenotyping of maize. *Plant Methods* **2015**, *11*, 35. [CrossRef]

18. Kefauver, S.C.; Vicente, R.; Vergara-Diaz, O.; Fernandez-Gallego, J.A.; Serret, M.M.D.; Araus, J.L.; Kerfal, S.; Lopez, A.; Melichar, J.P.E. Comparative UAV and field phenotyping to assess yield and nitrogen use efficiency in hybrid and conventional barley. *Front. Plant Sci.* **2017**, *8*, 1733. [CrossRef]

19. Chen, P.; Li, G.; Shi, Y.; Xu, Z.; Yang, F.; Cao, Q. Validation of an unmanned aerial vehicle hyperspectral sensor and its application in maize leaf area index estimation. *Sci. Agric. Sin.* **2018**, *51*, 1464–1474.

20. Han, L.; Yang, G.; Dai, H.; Xu, B.; Yang, H.; Feng, H.; Li, Z.; Yang, X. Modeling maize above-ground biomass based on machine learning approaches using UAV remote-sensing data. *Plant Methods* **2019**, *15*, 10. [CrossRef]

21. Swain, K.; Thomson, S.; Jayasuriya, H. Adoption of an Unmanned Helicopter for Low-Altitude Remote Sensing to Estimate Yield and Total Biomass of a Rice Crop. *Trans. ASABE* **2010**, *53*, 21–27. [CrossRef]

22. Chang, X.; Chang, Q.; Wang, X.; Chu, D.; Guo, R. Estimation of maize leaf chlorophyll contents based on UAV hyperspectral drone image. *Agric. Res. Arid. Areas* **2019**, *37*, 66–73.

23. Schirrmann, M.; Giebel, A.; Gleiniger, F.; Pflanz, M.; Lentschke, J.; Dammer, K.-H. Monitoring agronomic parameters of winter wheat crops with low-cost UAV imagery. *Remote Sens.* **2016**, *8*, 706. [CrossRef]

24. Li, Z.; Nie, C.; Wei, C.; Xu, X.; Song, X.; Wang, J. Comparison of four chemometric techniques for estimating leaf nitrogen concentrations in winter wheat (*Triticum aestivum*) based on hyperspectral features. *J. Appl. Spectrosc.* **2016**, *83*, 240–247. [CrossRef]

25. Duan, S.; Li, Z.; Wu, H.; Tang, B.; Ma, L.; Zhao, E.; Li, C. Inversion of the PROSAIL model to estimate leaf area index of maize, potato, and sunflower fields from unmanned aerial vehicle hyperspectral data. *Int. J. Appl. Earth Obs. Geoinf.* **2014**, *26*, 12–20. [CrossRef]

26. Li, J.; Zhu, X.; Ma, L.; Zhao, Y.; Qian, Y.; Tang, L. Leaf Area Index Retrieval and Scale Effect Analysis of Multiple Crops from UAV-based Hyperspectral Data. *Remote Sens. Technol. Appl.* **2017**, *32*, 427–434.

27. Lucieer, A.; Malenovsky, Z.; Veness, T.; Wallace, L. HyperUAS-imaging spectroscopy from a multirotor unmanned aircraft system. *J. Field Robot.* **2014**, *31*, 571–590. [CrossRef]

28. Turner, D.; Lucieer, A.; Wallace, L. Direct Georeferencing of Ultrahigh-Resolution UAV Imagery. *IEEE Trans. Geosci. Remote Sens.* **2014**, *52*, 2738–2745. [CrossRef]

29. Darvishzadeh, R.; Skidmore, A.; Schlerf, M.; Atzberger, C.; Corsi, F.; Cho, M. Lai and chlorophyll estimation for a heterogeneous grassland using hyperspectral measurements. *ISPRS J. Photogramm. Remote Sens.* **2008**, *63*, 409–426. [CrossRef]

30. Kuangnan, F.; Jianbin, W.; Jianping, Z.; Bangchang, X. A review of technologies on random forests. *Stat. Inf. Forum* **2011**, *26*, 32–38.

31. Breiman, L. Random Forests. *Mach. Learn.* **2001**, *45*, 5–32. [CrossRef]

32. Yue, J.; Feng, H.; Yang, G.; Li, Z. A comparison of regression techniques for estimation of above-ground winter wheat biomass using near-surface spectroscopy. *Remote Sens.* **2018**, *10*, 66. [CrossRef]

33. Atzberger, C.; Darvishzadeh, R.; Immitzer, M.; Schlerf, M.; Skidmore, A.; le Maire, G. Comparative analysis of different retrieval methods for mapping grassland leaf area index using airborne imaging spectroscopy. *Int. J. Appl. Earth Obs. Geoinf.* **2015**, *43*, 19–31. [CrossRef]

34. Gitelson, A.; Kaufman, Y.; Stark, R.; Rundquist, D. Novel algorithms for remote estimation of vegetation fraction. *Remote Sens. Environ.* **2002**, *80*, 76–87. [CrossRef]

35. Bendig, J.; Yu, K.; Aasen, H.; Bolten, A.; Bennertz, S.; Broscheit, J.; Gnyp, M.; Bareth, G. Combining UAV-based plant height from crop surface models, visible, and near infrared vegetation indices for biomass monitoring in barley. *Int. J. Appl. Earth Obs. Geoinf.* **2015**, *39*, 79–87. [CrossRef]

36. Kataoka, T.; Kaneko, T.; Okamoto, H.; Hata, S. Crop growth estimation system using machine vision. In Proceedings of the IEEE/ASME International Conference on Advanced Intelligent Mechatronics, Kobe, Japan, 20–24 July 2003; pp. 1079–1083.

37. Torressnchez, J. Multi-temporal mapping of the vegetation fraction in early-season wheat fields using images from UAV. *Comput. Electron. Agric.* **2014**, *103*, 104–113. [CrossRef]

38. Chianucci, F.; Disperati, L.; Guzzi, D.; Bianchini, D.; Nardino, V.; Lastri, C.; Rindinella, A.; Corona, P. Estimation of canopy attributes in beech forests using true colour digital images from a small fixed-wing UAV. *Int. J. Appl. Earth Obs. Geoinf.* **2016**, *47*, 60–68. [CrossRef]

39. Penuelas, J.; Isla, R.; Filella, I.; Araus, J. Visible and near-infrared reflectance assessment of salinity effects on barley. *Crop Sci.* **1997**, *37*, 198–202. [CrossRef]

40. Baret, F.; Guyot, G.; Major, D.J. TSAVI: A vegetation index which minimizes soil brightness effects on LAI and APAR estimation. *Symp. Remote Sens. Geosci. Remote Sens. Symp.* **1989**, *3*, 1355–1358.

41. Wu, C.; Niu, Z.; Tang, Q.; Huang, W. Estimating chlorophyll content from hyperspectral vegetation indices: Modeling and validation. *Agric. For. Meteorol.* **2008**, *148*, 1230–1241. [CrossRef]

42. Haboudane, D.; Miller, J.R.; Tremblay, N.; Zarco-Tejada, P.J.; Dextraze, L. Integrated narrow-band vegetation indices for prediction of crop chlorophyll content for application to precision agriculture. *Remote Sens. Environ.* **2002**, *81*, 416–426. [CrossRef]

43. Huete, A. A Modified soil adjusted vegetation index. *Remote Sens. Environ.* **1994**, *48*, 119–126.

44. Vincini, M.; Frazzi, E. Angular dependence of maize and sugar beet VIs from directional CHRIS/Proba data. *Cuore* **2005**, *2*, 5–9.

45. Datt, B. A new reflectance index for remote sensing of chlorophyll content in higher plants: Tests using Eucalyptus leaves. *J. Plant Physiol.* **1999**, *154*, 30–36. [CrossRef]

46. Roujean, J.L.; Breon, F.M. Estimating PAR absorbed by vegetation from bidirectional reflectance measurements. *Remote Sens. Environ.* **1995**, *51*, 375–384. [CrossRef]

47. Zarco-Tejada, P.J.; Berjón, A.; López-Lozano, R.; Miller, J.R.; Martín, P.; Cachorro, V.; González, M.R.; De Frutos, A. Assessing vineyard condition with hyperspectral indices: Leaf and canopy reflflectance simulation in a row-structured discontinuous canopy. *Remote Sens. Environ.* **2005**, *99*, 271–287. [CrossRef]

48. Peñuelas, J.; Gamon, J.A.; Fredeen, A.L., Mcrino, J.; Field, C.B. Reflectance indices associated with physiological changes in nitrogen- and water-limited sunflower leaves. *Remote Sens. Environ.* **1994**, *48*, 135–146. [CrossRef]

49. Zhou, X.; Zheng, H.B.; Xu, X.Q.; He, J.Y.; Ge, X.K.; Yao, X.; Cheng, T.; Zhu, Y.; Cao, Y.X.; Tian, Y.C. Predicting grain yield in rice using multi-temporal vegetation indices from UAV-based multispectral and digital imagery. *ISPRS J. Photogramm. Remote Sens.* **2017**, *130*, 246–255. [CrossRef]

50. Niu, Q.; Feng, H.; Yang, G.; Yang, H.; Xu, B.; Zhao, Y. Monitoring plant height and leaf area index of maize breeding material based on UAV digital images. *Nongye Gongcheng Xuebao/Trans. Chin. Soc. Agric. Eng.* **2018**, *34*, 73–82.

51. Liang, L.; Di, L.; Zhang, L.; Deng, M.; Qin, Z.; Zhao, S.; Lin, H. Estimation of crop LAI using hyperspectral vegetation indices and a hybrid inversion method. *Remote Sens. Environ.* **2015**, *165*, 123–134. [CrossRef]

52. Zhu, H.; Liu, H.; Xu, Y.; Yang, G. UAV-based hyperspectral analysis and spectral indices constructing for quantitatively monitoring leaf nitrogen content of winter wheat. *Appl. Opt.* **2018**, *57*, 7722–7732. [CrossRef] [PubMed]

53. Nguy-Robertson, A.; Gitelson, A.; Peng, Y.; Viña, A.; Arkebauer, T.; Rundquist, D. Green leaf area index estimation in maize and soybean: Combining vegetation indices to achieve maximal sensitivity. *Agron. J.* **2012**, *104*, 1336–1347. [CrossRef]

54. Yue, J.; Yang, G.; Li, C.; Li, Z.; Wang, Y.; Feng, H.; Xu, B. Estimation of Winter Wheat Above-Ground Biomass Using Unmanned Aerial Vehicle-Based Snapshot Hyperspectral Sensor and Crop Height Improved Models. *Remote Sens.* **2017**, *9*, 708. [CrossRef]

55. Ozlem, A. Mapping land use with using Rotation Forest algorithm from UAV images. *Eur. J. Remote Sens.* **2017**, *50*, 269–279.

56. Knyazikhin, Y.; Schull, M.A.; Stenberg, P.; Mõttus, M.; Rautiainen, M.; Yang, Y.; Marshak, A.; Carmona, P.L.; Kaufmann, R.K.; Lewis, P.; et al. Hyperspectral remote sensing of foliar nitrogen content. *Proc. Natl. Acad. Sci. USA* **2013**, *110*, 811–812. [CrossRef]

57. Knyazikhin, Y.; Lewis, P.; Disney, M.I.; Mottus, M.; Rautiainen, M.; Stenberg, P.; Kaufmann, R.K.; Marshak, A.; Schull, M.A.; Carmona, P.L.; et al. Reply to Ollinger et al.: Remote sensing of leaf nitrogen and emergent ecosystem properties. *Proc. Natl. Acad. Sci. USA* **2013**, *110*, 2438. [CrossRef]

58. Ollinger, S.V.; Richardson, A.D.; Martin, M.E.; Hollinger, D.Y.; Frolking, S.E.; Reich, P.B.; Plourde, L.C.; Katul, G.G.; Munger, J.W.; Oren, R.; et al. Canopy nitrogen, carbon assimilation, and albedo in temperate and boreal forests: Functional relations and potential climate feedbacks. *Proc. Natl. Acad. Sci. USA* **2008**, *105*, 19336–19341. [CrossRef]

59. Knyazikhin, Y.; Lewis, P.; Disney, M.I.; Stenberg, P.; Mõttus, M.; Rautiainen, M.; Kaufmann, R.K.; Marshak, A.; Schull, M.A.; Carmona, P.L.; et al. Reply to Townsend et al.: Decoupling contributions from canopy structure and leaf optics is critical for remote sensing leaf biochemistry. *Proc. Natl. Acad. Sci. USA* **2013**, *110*, 1075. [CrossRef]

60. Townsend, P.A.; Serbin, S.P.; Kruger, E.L.; Gamon, J.A. Disentangling the contribution of biological and physical properties of leaves and canopies in imaging spectroscopy data. *Proc. Natl. Acad. Sci. USA* **2013**, *110*, 1074. [CrossRef]

61. Dashti, H.; Glenn, N.F.; Ustin, S.; Mitchell, J.J.; Qi, Y.; Ilangakoon, N.T.; Flores, A.N.; Silván-Cárdenas, J.L.; Zhao, K.; Spaete, L.P.; et al. Empirical Methods for Remote Sensing of Nitrogen in Drylands May Lead to Unreliable Interpretation of Ecosystem Function. *IEEE Trans. Geosci. Remote Sens.* **2019**, *57*, 3993–4004. [CrossRef]

62. Bendig, J.; Bolten, A.; Bennertz, S.; Broscheit, J.; Eichfuss, S.; Bareth, G. Estimating biomass of barley using crop surface models (CSMs) derived from UAV-based RGB imaging. *Remote Sens.* **2014**, *6*, 10395–10412. [CrossRef]
63. Rasmussen, J.; Ntakos, G.; Nielsen, J.; Svensgaard, J.; Christensen, S.; Poulsen, R.N. Are vegetation indices derived from consumer-grade cameras mounted on UAVs sufficiently reliable for assessing experimental plots? *Eur. J. Agron.* **2016**, *74*, 75–92. [CrossRef]

remote sensing

Article

Maize Canopy and Leaf Chlorophyll Content Assessment from Leaf Spectral Reflectance: Estimation and Uncertainty Analysis across Growth Stages and Vertical Distribution

Hongye Yang [1,†], Bo Ming [1,†], Chenwei Nie [1], Beibei Xue [1], Jiangfeng Xin [1,2], Xingli Lu [2], Jun Xue [1], Peng Hou [1], Ruizhi Xie [1], Keru Wang [1,*] and Shaokun Li [1]

[1] Key Laboratory of Crop Physiology and Ecology, Institute of Crop Sciences, Chinese Academy of Agricultural Sciences, Beijing 100081, China; 82101195022@caas.cn (H.Y.); mingbo@caas.cn (B.M.); niechenwei@caas.cn (C.N.); 82101205031@caas.cn (B.X.); xinjiangfeng2020@126.com (J.X.); xuejun@caas.cn (J.X.); houpeng@caas.cn (P.H.); xieruizhi@caas.cn (R.X.); lishaokun@caas.cn (S.L.)
[2] School of Agriculture, Ningxia University, Yinchuan 750021, China; xinglilu@nxu.edu.cn
* Correspondence: wangkeru@caas.cn; Tel.: +86-010-82105791
† These authors contributed equally to this work.

Abstract: Accurate estimation of the canopy chlorophyll content (CCC) plays a key role in quantitative remote sensing. Maize (*Zea mays* L.) is a high-stalk crop with a large leaf area and deep canopy. It has a non-uniform vertical distribution of the leaf chlorophyll content (LCC), which limits remote sensing of CCC. Therefore, it is crucial to understand the vertical heterogeneity of LCC and leaf reflectance spectra to improve the accuracy of CCC monitoring. In this study, CCC, LCC, and leaf spectral reflectance were measured during two consecutive field growing seasons under five nitrogen treatments. The vertical LCC profile showed an asymmetric 'bell-shaped' curve structure and was affected by nitrogen application. The leaf reflectance also varied greatly between spatio–temporal conditions, which could indicate the influence of vertical heterogeneity. In the early growth stage, the spectral differences between leaf positions were mainly concentrated in the red-edge (RE) and near-infrared (NIR) regions, whereas differences were concentrated in the visible region during the mid-late filling stage. LCC had a strong linear correlation with vegetation indices (VIs), such as the modified red-edge ratio (mRER, $R^2 = 0.87$), but the VI–chlorophyll models showed significant inversion errors throughout the growth season, especially at the early vegetative growth stage and the late filling stage (rRMSE values ranged from 36% to 87.4%). The vertical distribution of LCC had a strong correlation with the total chlorophyll in canopy, and sensitive leaf positions were identified with a multiple stepwise regression (MSR) model. The LCC of leaf positions L6 in the vegetative stage (R^2-adj = 0.9) and L11 + L14 in the reproductive stage (R^2-adj = 0.93) could be used to evaluate the canopy chlorophyll status (L12 represents the ear leaf). With a strong relationship between leaf spectral reflectance and LCC, CCC can be estimated directly by leaf spectral reflectance (mRER, rRMSE = 8.97%). Therefore, the spatio–temporal variations of LCC and leaf spectral reflectance were analyzed, and a higher accuracy CCC estimation approach that can avoid the effects of the leaf area was proposed.

Keywords: leaf chlorophyll content; canopy chlorophyll content; vertical distribution; leaf spectral reflectance; maize

Citation: Yang, H.; Ming, B.; Nie, C.; Xue, B.; Xin, J.; Lu, X.; Xue, J.; Hou, P.; Xie, R.; Wang, K.; et al. Maize Canopy and Leaf Chlorophyll Content Assessment from Leaf Spectral Reflectance: Estimation and Uncertainty Analysis across Growth Stages and Vertical Distribution. *Remote Sens.* **2022**, *14*, 2115. https://doi.org/10.3390/rs14092115

Academic Editor: Yingying Dong

Received: 23 February 2022
Accepted: 26 April 2022
Published: 28 April 2022

Publisher's Note: MDPI stays neutral with regard to jurisdictional claims in published maps and institutional affiliations.

1. Introduction

The canopy chlorophyll content (CCC) can reflect the total photosynthetic productivity of a population, and it forms an important basis for judging the growth and nutritional status of individual plants [1–3]. Accurate monitoring of the chlorophyll content at the canopy and leaf scales by remote sensing is key in crop growth status determination and yield prediction [2,4]. However, the chlorophyll content in the maize canopy has a strong non-uniform vertical distribution, which makes it difficult to monitor CCC by standard spectral methods [5,6].

The non-uniform vertical profile of CCC in the canopy is generally abstracted as a slightly skewed "bell-shaped" model or Gaussian curve structure [7,8], with higher levels in the middle and lower levels at the top and bottom of the canopy. This means that the leaf chlorophyll content (LCC) at each leaf position is different. Light and nitrogen availability are the main factors that affect the vertical distribution of LCC in the canopy [7,9,10]. Transference and reuse of nitrogen between leaf layers during the growth season amplifies the spatio–temporal variability of LCC in the canopy [5,11]. Therefore, comprehensive measurements of the temporal and spatial distribution of chlorophyll in the canopy are required for accurate evaluation of the growth and nutritional status of maize in the field.

Vegetation indices (VIs) based on visible and near infrared (NIR) regions are well correlated with LCC, and the sensitive wavebands are primarily between 520 and 585 nm and 695 and 740 nm [12–14]. The empirical VI method, which is based on combinations of wavebands, can be used for accurate quantitative remote sensing of the chlorophyll content; this has been demonstrated in numerous studies. Common and widely used VIs include the normalized difference red edge index (NDRE) [15], MERIS Terrestrial Chlorophyll Index (MTCI) [16], and Vogelman Red Edge Index 2 (VOG2) [17]. In particular, the red edge chlorophyll index ($CI_{red-edge}$) has been found to have a strong correlation with LCC ($R^2 > 0.94$) [18] and can be used to accurately estimate LCC within a range of 10–80 µg/cm^2 [8,18]. Based on the spectral reflectance, VI has strong adaptability and expansibility, and previous studies developed diverse VI models to reduce the influence of factors such as the leaf structure, surface reflectance, and soil background on chlorophyll inversion accuracy [4,13,19]. Other studies developed transferable leaf nitrogen content assessment models based on support vector regression (SVR) and partial least squares regression (PLSR) to account for variations in plant species and growth conditions [20,21]. Physically based radiation transfer models (RTM) such as PROSPECT and PROSAIL have been widely used to simulate leaf hyperspectral reflectance and to optimize chlorophyll inversion methods at the leaf and canopy scales [22–24].

Although the methods described above function well in some applications, the vertical heterogeneity of LCC in the canopy and between growth stages causes significant differences in reflectance spectrum characteristics of leaves in different positions [14,25]. Soil and Plant Analyzer Development (SPAD-502) is a widely used chlorophyll meter, and it has been demonstrated that the relationship between SPAD values and LCC is not consistent between different leaf positions [26,27]. The high-value area of SPAD generally appears in the middle–upper rank of leaves and changes with nitrogen supply [7]. The correlation between leaf reflectance and the chlorophyll content has been shown to be highest in the middle–upper positions [14,27], and LCC-based inversion accuracy of VI is higher in the middle than in upper or lower positions. In studies of canopy vertical distribution, the canopy is often artificially divided into upper, middle, and lower layers, and corresponding spectral index models have been established based on that system [27–29]. There has been no clear conclusion regarding how the relationships between leaf spectral characteristics and LCC change in different growth stages and leaf positions in the canopy.

CCC is a comprehensive evaluation of the chlorophyll content at the canopy scale; estimation of CCC is affected by the vertical distribution of LCC [5,30]. The vertical distribution of LCC in the canopy can be adjusted to increase the photosynthetic rate of leaves in the upper canopy to cope with stress conditions [31]. Symptoms of chlorophyll deficiency often occur in the lower layers, and the mid-upper layers show significant changes in LCC when nitrogen supply is altered [14,28,31]. The collar leaf and the ear leaf have generally been used before and after silking, respectively, as the "sensitive leaf" to estimate the canopy chlorophyll status [30,32]. Previous studies demonstrated a strong relationship between chlorophyll accumulation in the ear leaf and CCC ($R^2 = 0.86$) [30]. It has also been suggested that the mean LCC of upper canopy layers can be used to represent all canopy layers, and therefore CCC can be estimated as the mean $LCC_{upper} \times$ leaf area index (LAI) ($R^2 = 0.97$) [33]. Estimation of CCC based on canopy spectral reflectance is an efficient and convenient method but seldom takes non-uniform chlorophyll distribution into consideration [34,35]. This limits the accuracy of CCC remote sensing ($R^2 = 0.60$) and decreases the practical value [35]. Due to the vertical variability of LCC between growth conditions, it is not known whether CCC can be accurately evaluated by LCC. If it can, LCC measurements could be used as a quick method to determine the canopy chlorophyll status.

In this study, we used a gradient of five nitrogen treatments to establish a range of canopy architectures, sampled all of the leaves in the canopy throughout the growth season, and measured LCC and leaf spectral reflectance. Three objectives were addressed: (1) to understand the effect of nitrogen supply on the vertical distribution of chlorophyll in the maize canopy and dynamic changes in chlorophyll distribution during the growth season; (2) to explore differences in leaf spectral reflectance characteristics in the canopy and verify whether a VI model based on leaf spectral reflectance can accurately invert LCC under variable spatio–temporal conditions; and (3) to identify the sensitive leaf positions (those that can be used to characterize the relationship between LCC and CCC) and evaluate the robustness and accuracy of a VI model based on leaf spectra to assess the canopy chlorophyll status.

2. Materials and Methods

2.1. Study Area and Experimental Design

This experiment was conducted in 2019 and 2020 at the Xinxiang Experimental Station ($35°7'51.57''$N, $113°45'36.63''$E) of the Chinese Academy of Agricultural Sciences, Henan Province, Eastern China, located in the Huang-Huai-Hai maize ecological zone. The experimental site was located in a semi-moist monsoon temperate continental climate region, with annual average temperature, precipitation, and sunlight duration of 14 °C, 573.4 mm, and 1993 h, respectively. The maize hybrid 'Jingnongke 728' (JNK728) was used, and the growing season was from 18 June to 29 September 2019 and from 11 June to 28 September 2020. The planting density was 75,000 plants/ha, and an isometric planting mode was used. The experimental area was ~0.66 hm^2 with a long-term rotation between summer maize (*Zea mays* L.) and winter wheat (*Triticum aestivum* L.). The soil composition was measured prior to nitrogen fertilizer application; organic matter content was 17.78 g/kg, available P was 12.93 mg/kg, available K was 139.09 mg/kg, total N was 1.34 g/kg, and the pH was 7.6.

The experiment in 2019 was a trial, which was followed by the formal experiment in 2020. To obtain a wide range of LCC values and a variety of canopy structures, the experiment was designed with five nitrogen application rate (N rate) treatments: 0, 100, 200, 300, and 400 kg/hm^2 (denoted as N0–N400). The same nitrogen treatment plants and field management practices were followed in 2019 and 2020. In 2019, only the N0 and N400 plants were sampled, whereas plants from all five N treatments were sampled in 2020. Nitrogen fertilizer was applied at the three-leaf (V3) and silking (VT) stages for a total of two applications. Details about the volume and timing of N treatments are shown in Table 1.

Table 1. Nitrogen treatments (kg/hm^2).

Stage	N0	N100	N200	N300	N400
V3	0	60	120	180	240
VT	0	40	80	120	160

A drip irrigation system was used in this study with water and nitrogen fertilizer integrated. Each plot was equipped with a flow gauge to ensure uniform irrigation. Each N treatment had three independent replicate plots, each 9.6 m × 9.6 m (92.16 m^2) in size. The N treatment consisted of urea with a nitrogen content of 46%. Superphosphate was used as phosphate fertilizer (P$_2$O$_5$, 12%, 90 kg/hm^2), and potassium chloride was used as potash fertilizer (K$_2$O, 60%, 120 kg/hm^2), both of which were applied once as basal fertilizer before sowing. The layout and management of the experimental site were shown in Figure 1.

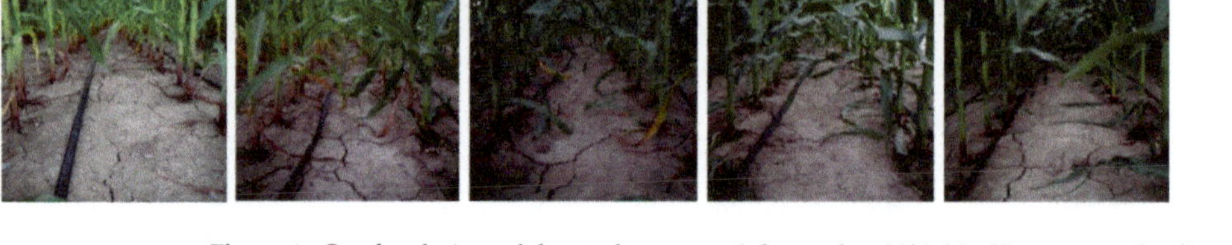

Figure 1. Overhead view of the study area on 2 September 2020 (**a**). Hyperspectral reflectance measurement system (**b**). Benchtop chlorophyll spectrophotometer (**c**). Canopy conditions under five nitrogen treatments (N) on 8 August 2020 (**d**).

2.2. Leaf Sampling and Chlorophyll Measurement

In 2019, samples were collected at three stages during the maize growing season: the sixteen-leaf stage (day 51 after emergence), silking stage (day 64), and filling stage (day 87). In 2020, samples were taken at five stages: the ninth-leaf stage (day 38), silking stage (day 57), blister stage (day 71), filling stage (day 87), and physiological maturity stage (day 100).

The leaf positions L1–L18 denote leaf order from the bottom to the top of the canopy, with the first leaf after emergence designated L1 and subsequent leaves numbered in order. Three maize plants were measured in each plot. For each plot, leaves at the same position were measured as biological replicates of LCC and leaf spectral reflectance. Leaf height was measured as the vertical distance from the leaf collar to the ground. Chlorophyll and spectral reflectance measurements were carried out in the laboratory using ~3800 destructively sampled leaves.

LCC (μg/cm^2) was measured under dark conditions in the laboratory. A circular punch (10 mm in diameter) was used to obtain leaf discs from symmetrical positions on both sides of the leaf vein. LCC was measured in the basal, central, and top portions of each leaf. Approximately 20 leaf discs were obtained for each leaf position, from which six discs were randomly selected, placed in a test tube, soaked in 95% ethanol (10 mL), and incubated in the dark for 48 h until the chlorophyll was completely extracted. The absorbance values of the chlorophyll solution were measured at 645 and 663 nm using a V-1800BPC spectrophotometer (V-1800, Inc., MAPADA, Shanghai, China), with 95% ethanol as the blank sample. The Arnon method was used to calculate the chlorophyll content [36]:

$$Chl_{a+b}(\mathrm{mg/L}) = 8.04 \times A_{663} + 20.29 \times A_{645} \tag{1}$$

$$LCC \left(\mu g/cm^2 \right) = \frac{Chl_{a+b}(mg/L) \times V_T(mL)}{disc\ area(cm^2)} \tag{2}$$

where A_{663} and A_{645} represent the absorbance values of the chlorophyll solution at 663 nm and 645 nm, respectively; V_T (mL) represents the volume of the chlorophyll soaking solution; and *disc area* (cm^2) represents the area of the leaf disc.

The canopy chlorophyll content (CCC, g/m^2) was used to represent the chlorophyll status of the canopy. CCC was calculated as the total chlorophyll accumulation per unit of ground area. The leaf area at each leaf position was calculated by manually measuring the maximum *length* and *width* of the leaf [37]:

$$Leaf\ area \left(cm^2 \right) = \begin{cases} length \times width \times 0.5\ (Unexpanded\ leaf) \\ length \times width \times 0.75\ (expanded\ leaf) \end{cases} \tag{3}$$

$$CCC \left(g/m^2 \right) = \sum_{i=1}^{n}(LCC_i \times LAI_i) \times 100 \tag{4}$$

where n represents the number of leaves in the canopy; LCC_i and LAI_i represent the leaf chlorophyll content ($\mu g/cm^2$) and leaf area index (m^2/m^2) of each leaf position, respectively.

2.3. Leaf Spectral Reflectance Measurements

Leaf spectral reflectance was measured from 350 nm to 2500 nm using an ASD Field-Spec4 spectrometer (Analytical Spectral Devices, Inc., Boulder, CO, USA) attached via a fiber optic cable to a plant probe (Field of View = 25°) equipped with a leaf clip containing a halogen bulb inside. A white reference panel on the leaf clip was used for reflectance correction. The sampling interval was 1.4 nm from 350 to 1000 nm and 2 nm from 1001 nm to 2500 nm. The hyperspectral data were resampled to 1 nm resolution using a self-driven interpolation method in RS3 6.4.0 (ASD Operation Software, Inc., Spectris, Westborough, MA, USA). The wavelength range between 350 nm to 900 nm was used in this experiment because chlorophyll is most strongly associated with that region [12–14]. To avoid potential errors, reflectance spectra were measured at six points symmetrically distributed over the basal, central, and top portions of the leaf. Each point was scanned 10 times, and the average value was recorded. Standard white reference panel calibration was performed before measurements and once every 10–20 min while measurements were taken.

Measurements at several (three to four) leaf positions were halted during the vegetative period because the leaves grew rolled together at the top of the canopy, which hindered chlorophyll extraction and spectral measurements. Leaf hyperspectral reflectance data were acquired in three stages in 2019 and five stages in 2020 as described above for field sampling.

2.4. Vegetation Index Extraction

LCC has a direct correlation with leaf spectral reflectance. Previous research has resulted in the development of various vegetation indices (VIs). We calculated narrow-band vegetation indices from leaf reflectance spectra [38], and 24 classic chlorophyll-related VIs were used to verify their relationship with LCC (Table 2).

2.5. Statistical Analysis

2.5.1. Construction of the VI–Chlorophyll Model

To ensure a sufficient sample size to fully verify the VI–LCC relationship under various spatio–temporal conditions, 170 of the 824 samples from the 2020 dataset were used in model building to determine the relationship between leaf spectral reflectance and LCC. This dataset contained all possible combinations of leaf positions, nitrogen treatments, and sampling time points. A linear model of VI and LCC was constructed using the least squares method.

Table 2. Vegetation indices (VIs) studied in this experiment.

Vegetation Index (Abbr.)	Formula	Reference
Simple ratio (SR)	Rnir/Rred	[39]
Vogelman Red Edge Index 1 (VOG1)	R740/R720	[17]
Normalized difference vegetation index (NDVI)	(Rnir − Rred)/(Rnir + Rred)	[40]
Normalized difference red edge index (NDRE)	(Rnir − Rre)/(Rnir + Rre)	[15]
Green NDVI (GNDVI)	(Rnir − Rgreen)/(Rnir + Rgreen)	[41]
Plant Pigment ratio (PPR)	(Rgreen − Rblue)/(Rgreen + Rblue)	[42]
Canopy chlorophyll content (CCCI)	NDRE/NDVI	[43]
MERIS Terrestrial Chlorophyll Index (MTCI)	(Rnir − Rre)/(Rre − Rred)	[16]
Vogelman Red Edge Index 2 (VOG2)	(R734 − R747)/(R715 + R726)	[17]
Vogelman Red Edge Index 3 (VOG3)	(R734 − R747)/(R715 + R720)	[17]
Red edge chlorophyll index (CIred-edge)	(Rnir/Rre) − 1	[8]
Green chlorophyll index (CIgreen)	(Rnir/Rgreen) − 1	[12]
Transformed Chl absorption in reflectance index (TCARI)	3[(Rre − Rred) − 0.2(Rre − Rgreen)(Rre/Rred)]	[44]
Structure independent pigment index (SIPI)	(R800 − R445)/(R800 − R680)	[45]
Double difference index (DD)	(R750 − R720) − (R700 − R670)	[46]
Modified normalized difference (mND705)	(R750 − R705)/(R750 + R705 − 2 × R445)	[13]
Modified simple ratio (mSR705)	(R750 − R445)/(R705 − R445)	[13]
Triangular vegetation index (TVI)	60 × (Rnir − Rgreen) − 100 × (Rred − Rgreen)	[47]
mTVI (red-edge)	60 × (Rnir − Rgreen) − 100 × (Rre − Rgreen)	[47]
Modified chlorophyll absorption ratio index (MCARI)	(Rre − Rred) − 0.2 × (Rre − Rgreen) × (Rre/Rred)	[3]
mNDblue	(Rblue − Rre)/(Rblue + Rnir)	[48]
Double-peak canopy nitrogen index (DCNII)	(R750 − R700)/(R700 − R670)/(R750 − R670 + 0.09)	[49]
Modified red-edge ratio (mRER)	(R759 − 1.8 × R419)/(R742 − 1.8 × R419)	[50]
Enhanced vegetation index (EVI)	2.5 (Rnir − Rred)/(Rnir + 6 Rred − 7.5 Rblue + 1)	[51]

Rblue, Rgreen, Rred, Rre, and Rnir refer to differences in band reflectance. The characteristic bands were designated as follows: 475 nm (Rblue), 560 nm (Rgreen), 668 nm (Rred), 718 nm (Rre), and 842 nm (Rnir).

2.5.2. Model Testing and Verification

Approximately 13 samples for each growth stage and vertical position were used to test the model. The remaining samples (654 from the 2020 dataset and 427 from the completely independent 2019 dataset) were used as a verification set to determine the correlation between chlorophyll and spectral reflectance. The differences in chlorophyll inversion ability of the unified model based on vertical leaf position were verified at the ninth-leaf stage (day 38), silking stage (day 57), blister stage (day 71), filling stage (day 87), and physiological maturity stage (day 100).

2.5.3. Multivariate Regression Model for LCC and CCC

Multivariable stepwise regression (MSR) is a simple and effective method to avoid multicollinearity among factors and to screen characteristic variables [52]. Each new variable was evaluated with an *F*-test, and the existing variables were evaluated with a *t*-test to ensure that all were significant in MSR. Variance inflation factor (VIF) analysis was used to evaluate the collinearity among variables. When VIF > 10, it is generally considered that there is multicollinearity among variables [53]. The MSR model was evaluated with an adjusted coefficient of determination (R^2-adj), which is used to avoid an increase in R^2 caused by the inclusion of a large number of variables in the model. In this study, it was used to determine the sensitive leaf position that can represent CCC and to construct the multiple linear relationship between LCC and CCC. The stepwise algorithm in SPSS v.22 (SPSS Inc., Chicago, IL, USA) was used for leaf layer filtering and model building. MSR was calculated as follows:

$$Y = \alpha_0 + \alpha_1 X_1 + \alpha_2 X_2 + \cdots + \alpha_n X_n + \varepsilon \tag{5}$$

$$R^2 - adj = 1 - \frac{(1 - R^2)(n - 1)}{n - k - 1} \tag{6}$$

where Y is the value of CCC; a_0 is a constant term; X_1, X_2, ..., X_n are the variables; a_1, a_2, ..., a_n are the coefficients; ε represents a random error term; R^2 represents the coefficient of determination; n represents the number of samples; and k is the number of coefficients in the equation. The dataset from 2020 was used to construct the MSR model of LCC and CCC, and the 2019 dataset was used to verify the robustness and accuracy of the model.

2.5.4. Validation Metrics

ViewSpecPro5.7 spectrum processing software was used for leaf spectral data extraction. Python 3.8 was used for data preprocessing and plotting. The coefficient of determination (R^2), root mean square error (RMSE), and relative root mean square error (rRMSE) were used as indices to evaluate the inversion accuracy of the VI model. Pearson's correlation coefficient (r) was used to express the correlation between spectral wavebands and LCC, and p-values were used to evaluate the degree of significance; $p < 0.05$ was considered a statistically significant difference. The statistical equations were as follows:

$$R^2 = 1 - \frac{\sum_{i=1}^{n}(O_i - P_i)^2}{\sum_{i=1}^{n}(O_i - \overline{O}_i)^2} \tag{7}$$

$$RMSE = \sqrt{\frac{\sum_{i=1}^{n}(O_i - P_i)^2}{n}} \tag{8}$$

$$rRMSE\ (\%) = \frac{RMSE}{\overline{O}} \times 100\% \tag{9}$$

where n is the number of samples; O_i is the observed value; P_i is the estimated value from the regression model; and $\overline{O}_i$ is the average observed value. Higher R^2 and lower RMSE and rRMSE values indicate higher accuracy and precision of the model.

3. Results

3.1. Vertical Profile and Temporal Variation of the Leaf Chlorophyll Content

There were drastic differences in the vertical distribution of LCC at different growth periods (Figure 2). In the early growth stage, LCC showed low levels (2.12 to 76.23 $\mu g/cm^2$). After the vegetative growth stage, LCC increased rapidly and peaked at the early filling stage, ranging from 37.35 to 97.02 $\mu g/cm^2$ (Figure 2a–c). After peaking, LCC slowly decreased until the end of the growing season, at which point it ranged from 1.02 to 59.29 $\mu g/cm^2$. The vertical profile of LCC showed a rapid increase from the top to the middle and a slow decrease from the middle to the bottom. During vegetative stages, the vertical profile of LCC increased toward the bottom layer. During the reproductive period, LCC was obviously skewed toward the higher layers (Figure 2a,b). Maximum LCC values were observed in the two to three leaf positions above the ear leaf (Figure 2b,c). The amount of nitrogen applied had a significant effect on LCC in positions throughout the canopy; those with sufficient nitrogen supply maintained high CCC. At the late growth stage, nitrogen-deficient plants showed a rapid decrease in LCC, which was accompanied by chlorosis and senescence in lower leaf positions (Figure 2c–e).

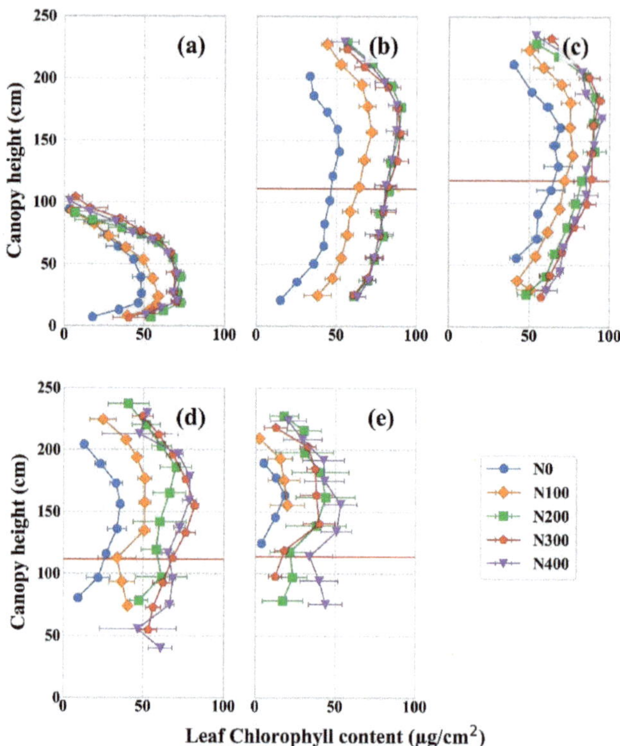

Figure 2. Vertical profile of the leaf chlorophyll content (LCC) within the maize canopy during the 2020 growing season. The y-axis shows the mean height above the ground for each leaf position at five sampling points: (**a**) ninth-leaf stage (day 38), (**b**) silking stage (day 57), (**c**) blister stage (day 71), (**d**) filling stage (day 87), and (**e**) physiological maturity (day 100). The red line denotes the position of the ear leaf (L12). Error bars show standard deviation.

3.2. The Vertical–Temporal Variation of Leaf Reflectance Spectral Characteristics

Samples collected at 38, 57, and 87 days after emergence were used to represent the early (Figure 3a,b), middle (Figure 3c,d), and late (Figure 3e,f) stages of maize growth, respectively. There were significant differences in leaf spectral reflectance curves under different temporal and spatial conditions. In the early growth stage, the main variations in leaf spectral reflectance curves were concentrated in the near-infrared (NIR) band (around 750–900 nm) and the red-edge (RE) band (around 700–721 nm), whereas differences were mostly concentrated in the visible wave band (around 510–680 nm) in the late growth stage. In the middle growth stage, the leaf spectral reflectance curves at different leaf positions were relatively uniform. These obvious fluctuations in reflectance spectra between leaf positions reflect a vertical non-uniformity of leaf properties in the maize canopy.

Leaf position was shown to influence the leaf spectral reflectance. Based on the correlation between wavebands and LCC at different leaf positions, the characteristic of LCC wavebands showed changes in sensitivity at different leaf positions (Figure 3). With the increase in LCC, the RE position moved in the infrared direction, the reflectance spectra in the visible light area showed a descending trend, and the NIR area showed an ascending trend. The sensitive wavebands for LCC were concentrated in the green band (531–567 nm; $r = -0.65$) and the RE band (712–731 nm; $r = -0.77$). There was a weak positive correlation between the NIR band and LCC.

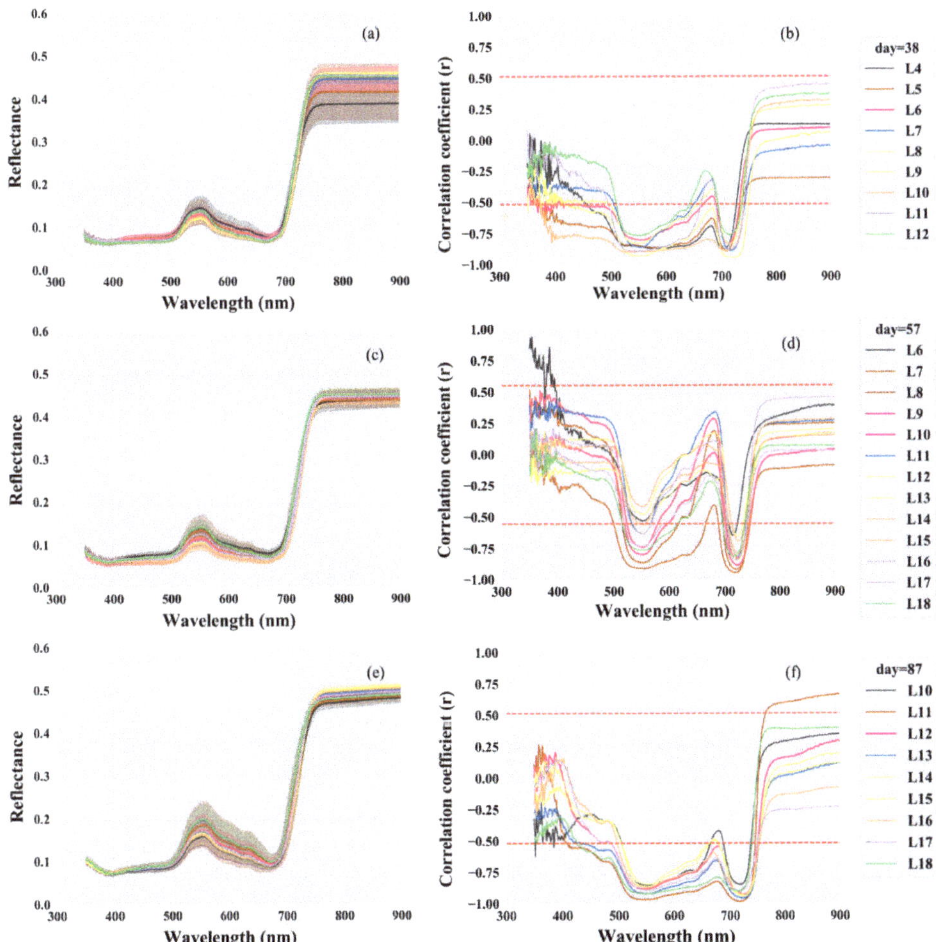

Figure 3. (**a,c,e**) Spectral reflectance curves for leaves at different positions. The standard deviation interval is shaded in gray. (**b,d,f**) Correlation coefficient curve for wavebands and the leaf chlorophyll content (LCC) at different leaf positions. The red dashed line indicates $p = 0.05$. L1–L18 indicate leaf positions from the bottom to the top of the canopy. Chlorophyll-related reflectance in the range of 350 to 900 nm was used.

3.3. Sensitivity of Vegetation Indices

Linear models were established to represent the relationship between vegetation indices (VIs) and LCC using the modeling dataset. After models were built, VI sensitivity to LCC was analyzed by calculating the coefficient of determination (R^2), root mean square error (RMSE), and relative RMSE (rRMSE) (Table 3).

The results showed that several commonly used VIs, such as the normalized difference red edge (NDRE), simple ratio (SR), and $CI_{red-edge}$ had a relatively strong linear correlation with LCC (for example, R^2 for NDRE was 0.85). These VIs are calculated with a combination of near-infrared (NIR) and red-edge band (RE) data, indicating that those two bands had a good inversion ability for LCC under relatively complex conditions. In contrast, green NDVI (GNDVI, NIR and green band; $R^2 = 0.77$) and the normalized difference vegetation index (NDVI, NIR and red band; $R^2 = 0.41$) did not perform as well. These results indicate that the red edge band was key for monitoring LCC through the spectral reflectance.

Table 3. Verification of vegetation index (VI) sensitivity to leaf chlorophyll content (LCC).

NO.	VI	Linear Model	R^2	RMSE	rRMSE (%)
1	mRER	y = 496.61x − 499.81	0.87	7.98	14.42
2	VOG2	y = −441.07x + 6.09	0.85	8.38	15.13
3	SR(Rnir, Rre)	y = 65.38x − 66.96	0.85	8.49	15.34
4	CIred-edge	y = 65.38x − 1.58	0.85	8.49	15.34
5	VOG3	y = −377.01x + 8.38	0.85	8.58	15.49
6	NDRE	y = 250.7x − 18.53	0.85	8.59	15.51
7	mTVI	y = 5.85x + 9.47	0.82	9.21	16.64
8	VOG1	y = 103.37x − 104.75	0.82	9.31	16.82
9	MTCI	y = 43.88x − 1.33	0.81	9.48	17.12
10	DD	y = 316.21x + 11.8	0.81	9.50	17.16
11	mNDblue	y = 321.09x + 161.16	0.80	9.93	17.94
12	mSR705	y = 11.44x − 1.51	0.79	10.03	18.11
13	CIgreen	y = 27.06x − 11.9	0.78	10.30	18.61
14	CI705	y = 28.17x − 0.47	0.78	10.40	18.79
15	GNDVI	y = 236.7x − 72.78	0.77	10.50	18.96
16	CCCI	y = 222.07x − 37.6	0.75	10.96	19.80
17	DCNII1	y = 20.16x + 0.38	0.74	11.27	20.36
18	TCARI	y = −201.47x + 115.49	0.49	15.72	28.38
19	MCARI	y = −604.4x + 115.49	0.49	15.72	28.38
20	PPR	y = −245.18x + 125.12	0.42	16.75	30.25
21	NDVI	y = 150.7x − 49.53	0.41	16.89	30.50
22	EVI	y = 131.18x − 34.95	0.27	18.77	33.90
23	SIPI	y = −79.59x + 137.67	0.22	19.43	35.10
24	TVI	y = 1.14x + 26.76	0.02	21.71	39.21

Note: x and y represent VIs and LCC, respectively. A linear model was used to establish the relationship between VI and LCC. $n = 169$.

Some VIs that are composed of three or four wavebands, including multiple red-edge and near-infrared bands, also showed good inversion ability for LCC, such as VOG2 ($R^2 = 0.85$), mRER ($R^2 = 0.87$), VOG3 ($R^2 = 0.85$), MTCI ($R^2 = 0.81$), and DD ($R^2 = 0.81$). Previous studies have shown that such VIs have increased robustness and sensitivity compared to dual-band VIs and can reduce the influence of saturation effects [54]. Here, the combination of multiple red-edge bands improved the sensitivity and robustness of the correlation between VIs and LCC.

3.4. Establishing an Inversion Model for Chlorophyll Prediction through Vegetation Indices

To comprehensively verify the accuracy of VI–LCC models throughout the whole growth period, six typical vegetation indices (mRER, VOG2, $CI_{red\text{-}edge}$, NDRE, MTCI, and DD) with different formula types and waveband compositions were selected based on R^2 and RMSE values (Table 3). All six VI models constructed had strong linear (or close to linear) relationships with LCC (Figure 4), with the exception of DD, which showed a slight tendency toward exponential function distribution. The model dataset showed a normal distribution (Figure 4), which met the requirements for random sampling and could therefore be used to represent the whole dataset. However, mRER (Figure 4a), $CI_{red\text{-}edge}$ (Figure 4c), and MTCI (Figure 4e) showed a more uniform data distribution, whereas NDRE (Figure 4d) and DD (Figure 4f) showed a slight skewness toward high values, indicating that there was some saturation at higher LCC values. VOG2 (Figure 4b) was the only VI among the six selected that was negatively correlated with LCC.

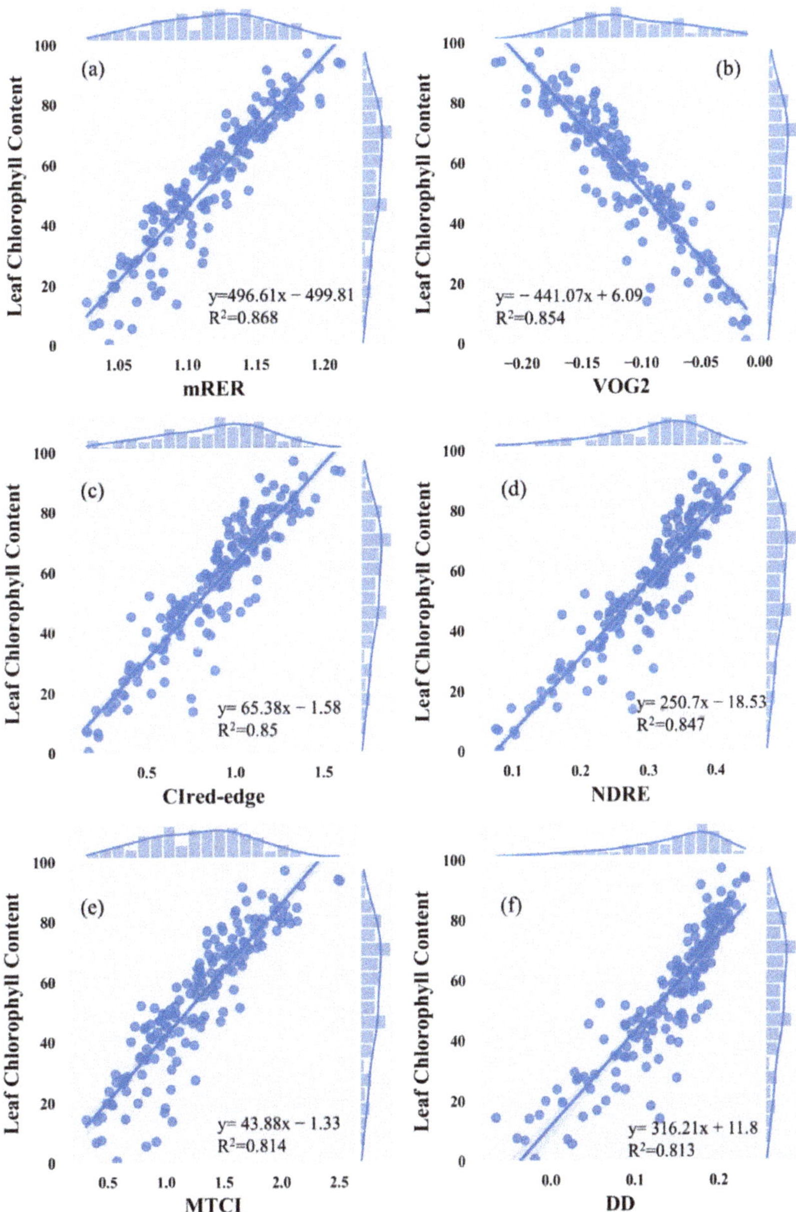

Figure 4. Relationship between LCC and six selected vegetation indices (VIs): (**a**) modified red-edge ratio (mRER), (**b**) Vogelman red edge index 2 (VOG2), (**c**) red edge chlorophyll index (CI$_{\text{red-edge}}$), (**d**) normalized difference red edge index (NDRE), (**e**) MERIS terrestrial chlorophyll index (MTCI), and (**f**) double difference index (DD). Linear models were built using the 2020 modeling dataset. Correlation coefficients and *p*-values are shown above each scatterplot, and the distributions of VI and LCC values are shown as histograms at the top and side of each graph, respectively.

3.5. Validation and Testing under Spatio–Temporal Variation

To verify the relationship between leaf spectral reflectance and the vertical distribution of LCC, VI–chlorophyll models were established to predict LCC under different spatio–temporal conditions (Figure 5). rRMSE was used to express the degree of deviation between the predicted and observed values to measure the robustness and accuracy of each model.

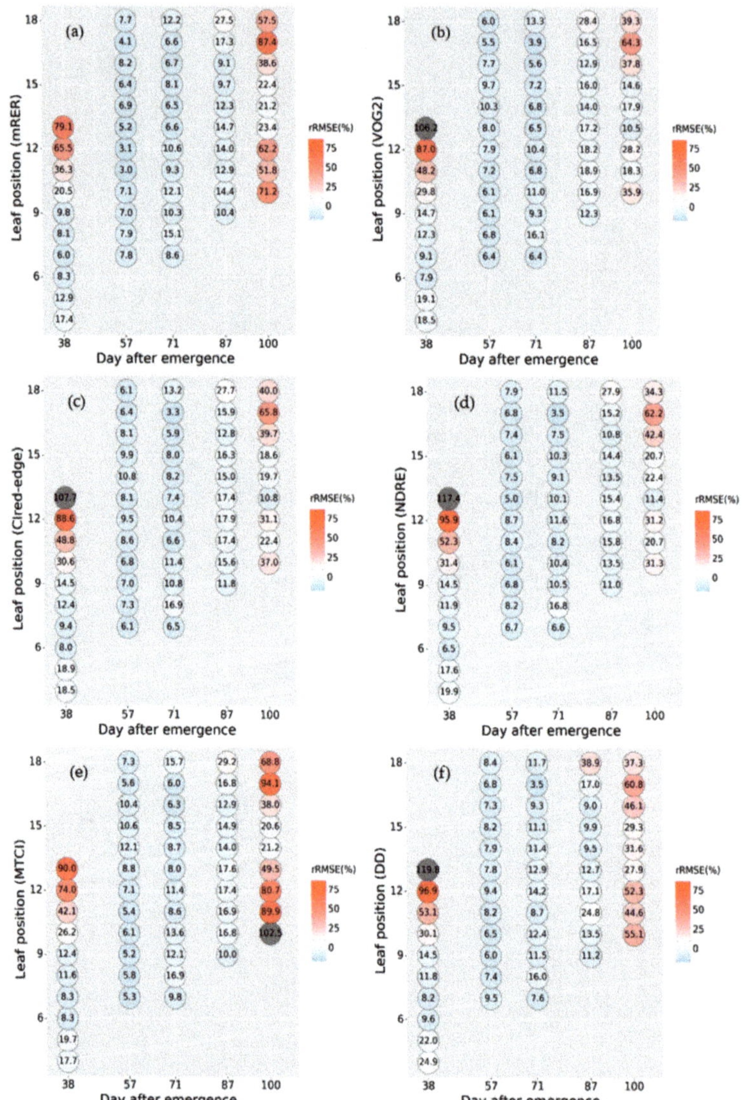

Figure 5. rRMSE (%) for the six LCC-VI models: (**a**) mRER, (**b**) VOG2, (**c**) CIred-edge, (**d**) NDRE, (**e**) MTCI, and (**f**) DD. rRMSE was used to evaluate the model inversion accuracy. Lower values of rRMSE correspond to closer predicted and observed values. The x-axis represents days after first leaf emergence, and the y-axis represents the canopy position where the leaves were located (L1–L18 from the bottom to the top of the plant). Blue indicates a low rRMSE value (low error), and red indicates a high rRMSE value between the observed and predicted LCC. Gray indicates error of more than 100%.

All of the VI models showed a similar pattern in performance throughout the growing season (Figure 5). LCC predictions for the silking stage (day 57) and early filling stage (day 71) had the highest precision, with average rRMSE values of ~6.2% to 9.4%. Precision was lowest in the vegetation stage (day 38) and late growth stage (day 100), with average rRMSE values of ~26.4% to 48.4%. The inversion accuracy of the model also differed by leaf position. In the vegetative growth stage (day 38), the chlorophyll model had a good inversion effect for the lower and middle mature leaves (rRMSE $\approx$ 10.4%) but not for the top leaf positions with tender leaves (rRMSE $\approx$ 20% to 80%). In the early and middle reproductive growth stages (days 57, 71, and 87), all leaves in the maize canopy were in a robust mature functional stage, and the chlorophyll model could be used to conduct high-precision inversion for all leaf positions in the canopy (rRMSE $\approx$ 10.9%). At the end of the growing season when leaves became senescent (day 100), the one to two leaves above the ear leaf (L12), which still had green tissue and maintained basic physiological functions, had the best inversion effect (rRMSE $\approx$ 21.2%).

In the vegetative growth stage before silking and the late reproductive growth stage, the inversion performance differed significantly by leaf position. This indicated that the relationship between LCC and leaf spectral reflectance was variable, and the inversion accuracy of the VI model was affected by leaf properties or other factors that were vertically distributed in the canopy. A completely independent data set (collected in 2019) was used to test the accuracy of the VI–LCC model. The results are consistent with those shown in Figure 5 (Appendix A/Figure A1). The model had the best inversion precision in the middle growth period (days 51 and 64), and the inversion effect was poor for senescent leaves in the mature stage.

3.6. Relationship between LCC and CCC throughout Growth Stages

Different nitrogen treatments and growth stages were associated with changes in the vertical distribution of chlorophyll in the canopy (Figure 2). There were clear relationships between the chlorophyll content in individual leaves (LCC, $\mu g/cm^2$) located at different leaf positions and the total chlorophyll content in the canopy (CCC, g/m^2), and these relationships varied between growth stages (Figure A2). A multiple stepwise regression model (MSR) was used to determine the sensitive leaf position and to construct a multiple linear relationship between LCC and CCC using the 2020 dataset (Table 4). Variance inflation factor (VIF) analysis was used to measure the degree of collinearity among variables. Variables are generally considered to have multicollinearity when VIF > 10. It is important to note that, due to the correlation between LCC and vertical leaf position, the multivariate linear model for LCC and CCC could incorporate data from a maximum of two leaf positions.

Table 4. Relationships between the leaf chlorophyll content (LCC) and canopy chlorophyll content (CCC) in the 2020 dataset.

MSR Model	Parameters	Regression Model	R^2-adj	Beta	VIF
Early model (day 38)	X_1:L6	$Y = 0.042X_1 - 1.6$	0.9	0.95	1
Middle model (day = 57 + 71)	X_1:L9	$Y = 0.052X_1 - 0.334$	0.89	0.95	1
	X_1:L9, X_2:L16	$Y = 0.038X_1 + 0.018X_2 - 0.79$	0.93	(0.7, 0.3)	2.64
Late model (day = 87 + 100)	X_1:L14	$Y = 0.039X_1 - 0.67$	0.92	0.96	1
	X_1:L14, X_2:L9	$Y = 0.032X_1 + 0.009X_2 - 0.362$	0.97	(0.79, 0.27)	1.74
Reproductive model (day 57−100)	X_1:L11	$Y = 0.042X_1 - 0.09$	0.91	0.95	1
	X_1:L11, X_2:L14	$Y = 0.022X_1 + 0.024X_2 - 0.679$	0.93	(0.51, 0.48)	8.428

Notes: Using the 2020 data, the samples were divided into early, middle, late, and reproductive stages based on the number of days after emergence. CCC (g/m^2) was the dependent variable, and LCC ($\mu g/cm^2$) was the independent variable. X_1 and X_2 represent the sensitive leaf positions determined by MSR. Beta represents the contribution of different parameters to the model, and VIF represents the degree of collinearity between parameters. VIF > 10 indicates strong collinearity between parameters. The number of samples at each stage were as follows: n = 15 (early stage), n = 28 (middle stage), n = 34 (late stage), and n = 62 (reproductive stage).

The corn growth season could be divided into the early (day 38), middle (day 57, 71), late (day 87, 100), and reproductive stages (day 57–100) based on the number of days after emergence. The correlation between CCC and vertical LCC varied between growth stages. In the early growth stage, the highest correlation between LCC and CCC was found at the L6 position (R^2-adj = 0.9, Beta = 0.95). In the middle growth stage, L9 and L16 had the best correlation with CCC (R^2-adj = 0.93). In the late growth stage, the sensitive leaf positions were L9 and L14 (R^2-adj = 0.97). The reproductive stage refers to the entire period from silking to maturity (i.e., the middle and late stages), and in this stage, L11 and L14 were the sensitive leaf positions (R^2-adj = 0.93).

According to the relationship between the chlorophyll content in per unit leaf areas and the chlorophyll content per unit ground area, LAI can be ignored in CCC estimation. Based on the relationships described above for each growth stage, CCC can be estimated directly by leaf spectral reflectance.

3.7. Estimation and Validation of CCC by Leaf Spectral Reflectance

Chlorophyll-related vegetation indices were used to estimate the LCC of sensitive leaf positions at each growth stage based on leaf spectral reflectance. The data from 2019 and 2020 verified that mRER had the best inversion precision for LCC (R^2 = 0.87, Section 3.5), and the average rRMSE of LCC inversion at sensitive leaf positions (such as L14 and L11) was 11.28%. The independent data were sampled at the sixteen-leaf stage (day 51, V16), silking stage (day 64, R1), and filling stage (day 87, R3) in 2019, and these data were used to assess the accuracy of CCC estimation using an MSR model (Figure 6). Separate models were built for the middle, late, and reproductive stages to verify the performance at the three growth stages (Table 5).

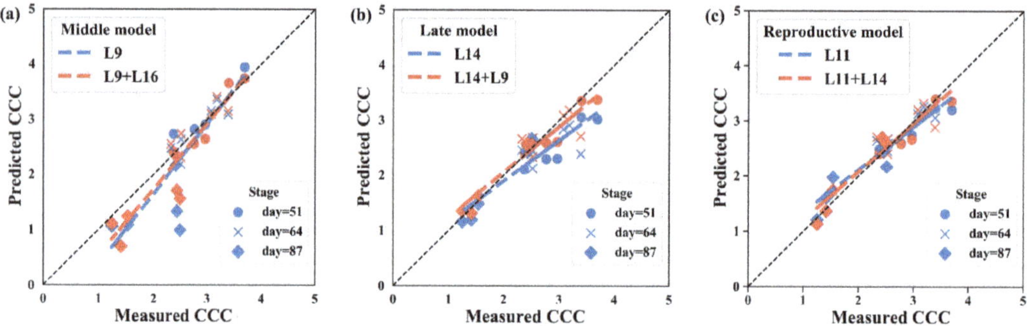

Figure 6. The canopy chlorophyll content (CCC) predicted by the middle model (**a**), late model (**b**), and reproductive model (**c**) compared to measured CCC in 2019. LCC was predicted from mRER, which had the highest accuracy of chlorophyll inversion in this study. The blue dotted line indicates predicted CCC from LCC of one leaf position, and the red line indicates predicted CCC from LCC of two leaf positions. The 2019 dataset included three growth stages: the sixteen-leaf stage (day 51, V16), silking stage (day 64, R1), and filling stage (day 87, R3).

Table 5. Verification of the stepwise regression model in 2019.

Stage	Middle Model		Late Model		Reproductive Model	
	L9	L9 + L16	L14	L14 + L9	L11	L11 + L14
Sixteen-leaf stage (day 51)	7.68%	6.81%	16.60%	7.52%	8.65%	6.95%
Silking stage (day 64)	7.49%	7.26%	16.31%	11.40%	6.96%	9.99%
Filling stage (day 87)	43.95%	30.20%	6.47%	5.46%	13.58%	9.54%
All stages (day 51–87)	20.74%	14.90%	15.35%	9.21%	9.37%	8.97%

Notes: Sample data for the sixteen-leaf stage (day 51, V16), silking stage (day 64, R1), filling stage (day 87, R3), and all stages (days 51–87, n = 18) in 2019 were verified with the three MSR models (middle model, late model, reproductive model). LCC was estimated by mRER (R^2 = 0.87). rRMSE (%) was used to represent error between predicted CCC and measured CCC; smaller rRMSE values correspond to higher model accuracy.

The validation results (Table 5) showed that MSR models could accurately estimate CCC from leaf spectral reflectance. The middle stage model had good estimation accuracy in the V16 and silking stages (with rRMSE values of 6.81% and 7.26%, respectively) but poor estimation at the filling stage (rRMSE = 30.2%). The late model had high precision for CCC estimation at the filling stage (rRMSE = 6.47%), but low precision at the V16 and silking stages (rRMSE values of 16.6% and 16.31%, respectively). The reproductive stage model was constructed by integrating data from silking through maturation stages in 2020, and this model showed high accuracy for CCC estimation throughout all three growth stages in 2019 (rRMSE ranged from 6.95–9.99%).

Compared with establishing relationships between CCC and LCC at a single leaf position, estimating CCC using LCC values from two sensitive leaf positions showed higher accuracy and robustness (Table 5). Thus, CCC (g/m^2) can be accurately estimated from LCC ($\mu g/cm^2$) at the L11 and L14 positions based on leaf spectral reflectance (rRMSE = 8.97%) at the reproductive stage.

4. Discussion

Remote sensing of chlorophyll is a fundamental component of the quantitative remote sensing of crops. Many studies have addressed the vertical distribution of chlorophyll in the canopy because understanding canopy architecture is important for crop growth management and monitoring [5,27]. Solar radiation interception of the canopy structure and nitrogen transference have been identified as the main contributors to vertical heterogeneity of the chlorophyll content in the canopy [5,9,10]. In the present study, vertical distribution of the leaf chlorophyll content (LCC) in the canopy and changes in the correlation between LCC and leaf spectral reflectance were explored. The vertical profile of LCC showed asymmetric distribution, with higher values in the upper–middle layer than in the top or bottom layers [7,8]. In the vegetative stage, the highest values were found in the lower–middle layer (Figure 2). The bottom layers of the canopy were significantly affected by nitrogen supply [14,31]; it appears that a sufficient nitrogen supply can ensure that leaves continue to synthesize chlorophyll and remain vital as the plant matures.

We created a variety of canopy architecture types by applying different amounts of nitrogen fertilizer (Figure 2), as has been done in previous studies [7,8]. The green band at 531–567 nm and the red-edge (RE) band at 712–731 nm were sensitive to chlorophyll (Figure 3) [3]. Vegetation index (VI) models based on leaf spectral reflectance could effectively achieve LCC inversion. mRER was the VI with the best LCC modeling performance in this study (R^2 = 0.87). Red edge (RE) and near-infrared (NIR) were the necessary bands to measure to build the VI–chlorophyll model (Table 3) [3,27]. The RE waveband had a moderate correlation with chlorophyll (r = −0.77), but the correlation with NIR was not significant. Previous studies showed that there are specific absorption bands in the NIR waveband that are associated with nitrogen and proteins, and these are related to the vertical distribution of nitrogen in the canopy [55].

Leaf spectral reflectance has been widely used in previous research for remote sensing, and it has been shown that leaf reflectance varies throughout the canopy [56,57]. However, there are few conclusions about whether there is a unified chlorophyll–spectral relationship for different leaf positions in the canopy. To determine the relationship between LCC and leaf spectral reflectance, a VI–LCC model was established here and verified under several spatio-temporal conditions. The results showed that there was a great deal of variability in the relationship between LCC and leaf spectra at different positions in the canopy, and a single VI–chlorophyll model was limited in its capacity to predict LCC (Figure 5). However, the inversion effect of the chlorophyll model was consistent within each growth stage. Other researchers have published similar results in using SPAD to monitor leaf chlorophyll and the nitrogen content [58–61]. Here, the VI model could invert LCC well in the middle growth stage (day 57–71), and L13–L15 had the best inversion accuracy (rRMSE $\approx$ 6.6%). However, VI and LCC had weak positive correlations in the tender upper leaves at the vegetative growth stage (day 38, L10–L14, rRMSE $\approx$ 70.3%) and in the aged leaves in the upper-lower positions at the late reproductive growth stage (day 100, L10–L12 and L16–L18, rRMSE $\approx$ 61.5%). In the early growth stage, differences in leaf spectra between leaf positions were concentrated in the RE and NIR waveband regions, indicating that the leaf microstructure differed by leaf position. In contrast, in the late stage, differences were concentrated in the visible region, indicating a difference in the chlorophyll content between leaf positions [14,37]. However, there were six VI models for chlorophyll inversion that performed with consistent accuracy, showing that the variation may be caused by changes in leaf structure or chlorophyll spectral properties rather than by VI sensitivity [13,62].

Previous studies reported that crops respond to deficiency or stress by altering the chlorophyll content at different vertical leaf positions to maintain photosynthetic efficiency [56,63,64]. The chlorophyll content in the collar leaf at the vegetation stage and in the ear leaf at the reproductive stage is often used to evaluate the canopy chlorophyll status [30,32], and other research suggested that bottom leaf positions were more suitable as indicators of canopy chlorophyll diagnosis [65,66]. In the present study, data collected in 2019 and 2020 demonstrated a strong correlation between CCC and LCC at various growth stages. Based on a multiple stepwise regression (MSR) model, CCC could be accurately estimated from the chlorophyll content at multiple leaf positions. L6 in the vegetative stage (R^2-adj = 0.9) and L11 + L14 in the reproductive stage (R^2-adj = 0.93) were used as the sensitive leaf positions for CCC estimation. Based on the close correlation between leaf spectral reflectance and the chlorophyll content per unit of leaf area (LCC, $\mu g/cm^2$), the MSR model of CCC estimation was established using a VI model; mRER represents LCC, and the high accuracy was verified at different growth stages (rRMSE = 8.97%). Due to the strong correlation between LCC at vertical leaf positions [18], CCC estimated using LCC from two sensitive leaf positions was a better choice to avoid multilinearity and improve model precision.

The leaves were usually considered to be uniform objects in previous studies, which means the relationship between LCC and leaf spectral reflectance is consistent in the same plant species [33,67,68]. Other studies proposed that the leaf thickness, water content, and leaf structure would lead to inconsistencies between LCC and spectral reflectance [22,69,70]. A common viewpoint is that there are strong correlations between LCC and spectral characteristics in the middle leaf position of a canopy, with a weak relationship in the upper and lower positions of the canopy [56]. In the present study, a more elaborate division of growth stages and vertical distribution was conducted, illustrating the variability of the correlation between LCC and leaf spectral reflectance under various spatial-temporal conditions. The mismatch between the LCC and spectral would be masked by the soil background and canopy architecture that make chlorophyll remote sensing at a canopy level more complicated with a lack of robustness.

Canopy spectral reflectance (measured using a UAV or satellite) is an efficient and non-destructive chlorophyll remote sensing method. However, ignoring the significant vertical gradient of chlorophyll and treating the canopy as a uniform plane field would limit robustness and decrease the practical value of canopy remote sensing [7,71]. On the one hand, previous studies proposed that chlorophyll deficiency is mainly exhibited in the lower positions of the canopy rather than in the upper leaves [14,65]. On the other hand, canopy spectral reflectance is mainly contributed by the upper–middle leaves, and few studies explored CCC estimation from part of LCC in the canopy [72,73]. In this study, a CCC estimation model based on the LCC vertical distribution was proposed, which avoids the influence of LAI and biomass. The LCC of lower leaves (L6) at the vegetative stage and upper–mid leaves (L11, L14) at the reproductive stage was sensitive to CCC changes. This result provides theoretical support and reference sensitive leaf positions for canopy chlorophyll remote sensing, considering the non-uniform vertical distribution [29,34,74].

Some limitations of this study should be mentioned. The vegetative stage is an important period in which crops accumulate material, and due to limitations on the number of sampling time points, the vertical distribution of LCC and verification of sensitive leaf positions in this stage have not been fully demonstrated. Additionally, only the maize variety JNK728 was studied, preventing comparison among different varieties. The factors leading to the observed inconsistent relationship between LCC and VI at the early and late growth stages need further study. Additional time points and maize varieties will be studied in future research to address these issues.

5. Conclusions

Five nitrogen application levels were used to construct diverse maize canopy architectures and reveal the vertical heterogeneity of the leaf chlorophyll content (LCC) and leaf spectral reflectance characteristics in the maize canopy. A multiple stepwise regression (MSR) model was built to accurately monitor the canopy chlorophyll content (CCC) based on the vertical distribution of LCC within the canopy; LCC showed an asymmetric vertical distribution, tending to be lower in the bottom layer, increasing in the middle layer, then decreasing in the upper layer. Nitrogen treatments significantly changed LCC, and the vertical profile of LCC distribution remained similar between treatments. Leaf spectral reflectance characteristics under variable spatio–temporal conditions were analyzed. The green band (531–567 nm) and the red-edge band (712–731 nm) were the sensitive wavebands for monitoring LCC. Six classical VIs were used to construct VI–chlorophyll models, the best of which was the model built with modified red-edge ratio (mRER, $R^2 = 0.87$). The VI model could accurately predict LCC at the middle growth stage (rRMSE = 10.9%), but the correlation between VI and LCC changed in the upper and lower leaf layers during the early vegetative and mature stages (rRMSE ranged from 36% to 87%). Through a combination of the inversion accuracy results and multiple stepwise regression, the leaves at positions L6 in the vegetative stage and L11 and L14 in the reproductive stage (L12 is the ear leaf) were found to be the most sensitive in CCC estimation. In this way, a VI–LCC–CCC model was constructed based on leaf spectral reflectance to estimate the canopy chlorophyll status. The model was evaluated using data from field experiments in 2019 and 2020 and was found to be robust and accurate (rRMSE = 8.97%).

Author Contributions: Conceptualization, H.Y., K.W., B.M. and S.L.; methodology, H.Y., R.X., C.N. and B.M.; investigation, H.Y., B.X., X.L., J.X. (Jiangfeng Xin), P.H. and J.X. (Jun Xue); writing—original draft preparation, H.Y. and B.M. All authors have read and agreed to the published version of the manuscript.

Funding: This research was supported by the National Key Research and Development Program of China (2016YFD0300605), China Agriculture Research System of MOF and MARA, and the Agricultural Science and Technology Innovation Program (CAAS-ZDRW202004).

Data Availability Statement: Not applicable.

Acknowledgments: We gratefully thank the Xinxiang Experimental Station for the experiment support.

Conflicts of Interest: The authors declare no conflict of interest.

Appendix A

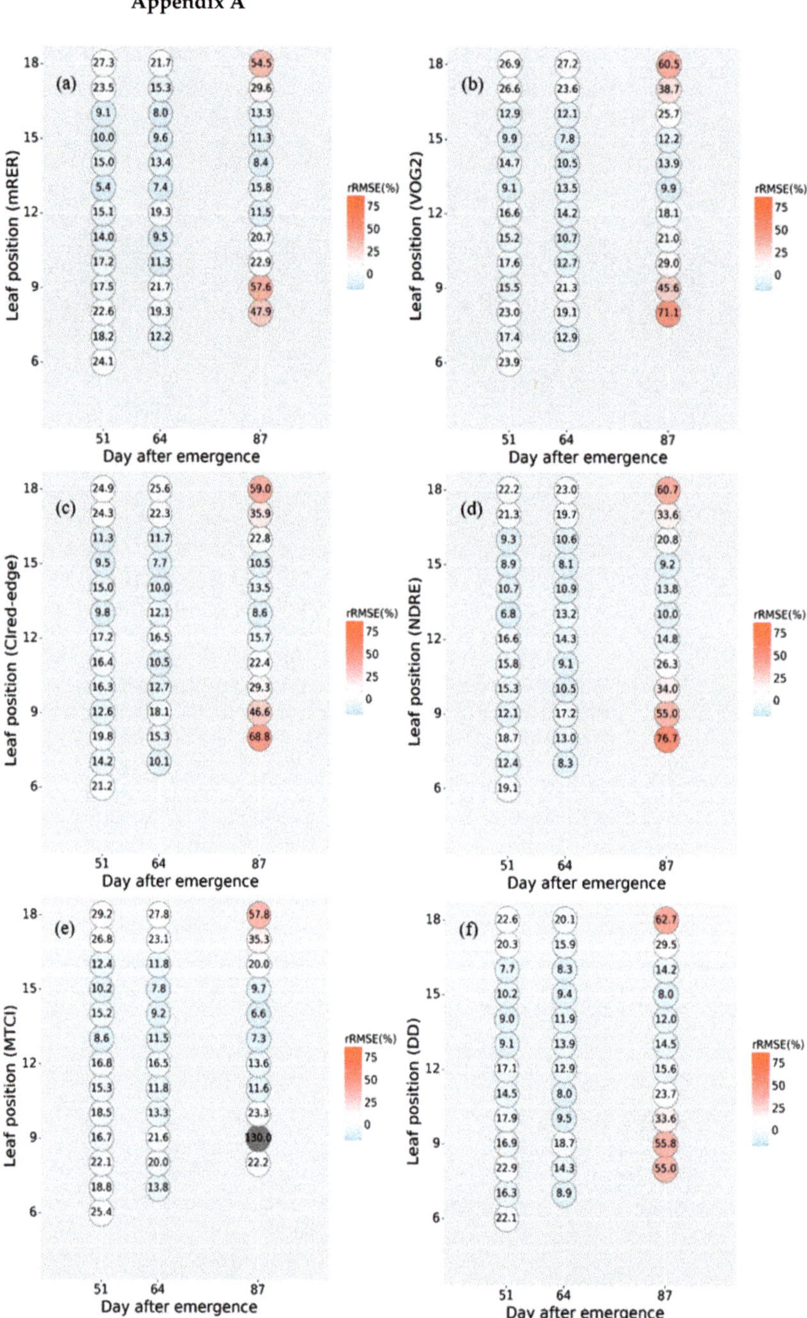

Figure A1. rRMSE (%) of the six LCC–VI models (2019): (**a**) mRER, (**b**) VOG2, (**c**) CIred-edge, (**d**) NDRE, (**e**) MTCI, and (**f**) DD. rRMSE was used to evaluate the consistency of model inversion.

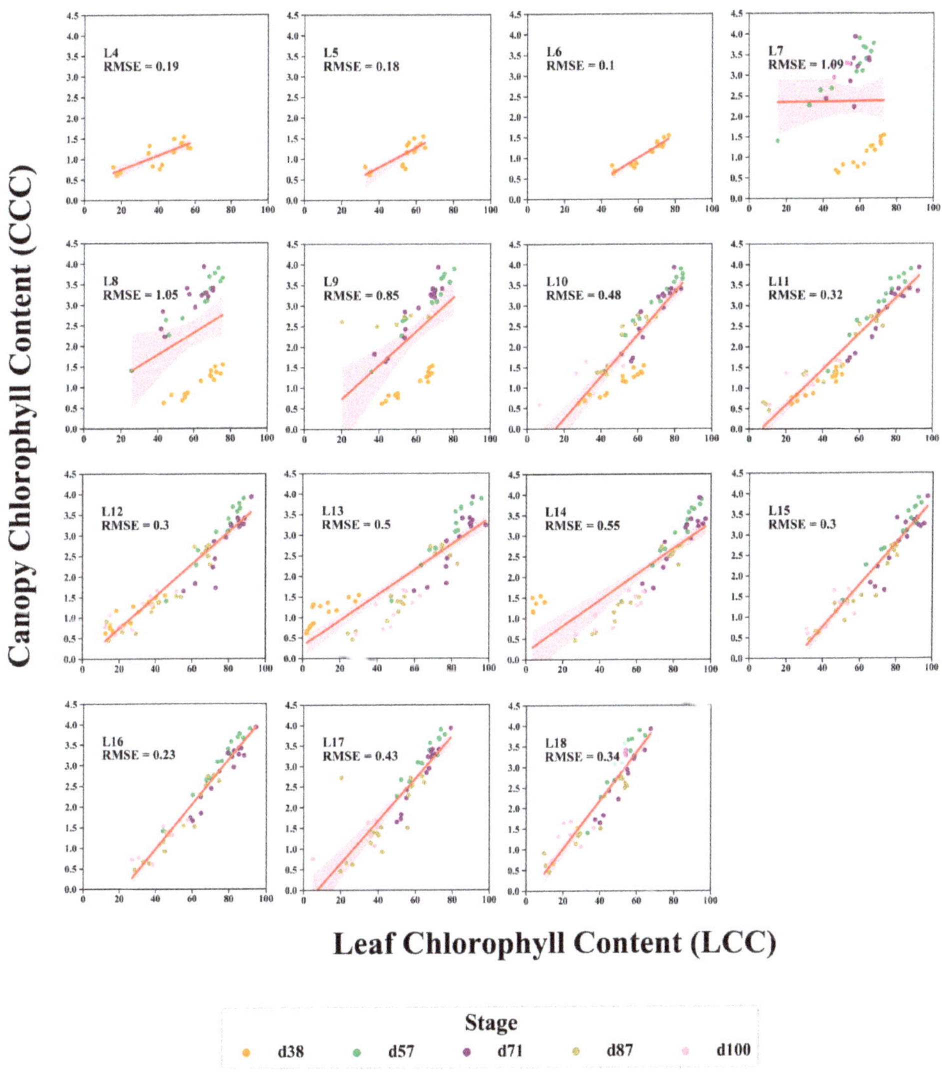

Figure A2. Relationships between the chlorophyll content in the leaf (LCC) and chlorophyll content in the canopy (CCC) at each leaf position and growth stage. Leaf position corresponds to the order of emergence from the bottom to the top of the canopy (L4–L18).

References

1. Xue, L.H.; Cao, W.X.; Yang, L.Z. Predicting Grain Yield and Protein Content in Winter Wheat at Different N Supply Levels Using Canopy Reflectance Spectra. *Pedosphere* **2007**, *17*, 646–653. [CrossRef]
2. Prey, L.; Hu, Y.; Schmidhalter, U. High-Throughput Field Phenotyping Traits of Grain Yield Formation and Nitrogen Use Efficiency: Optimizing the Selection of Vegetation Indices and Growth Stages. *Front. Plant Sci.* **2020**, *10*, 1672. [CrossRef]
3. Daughtry, C.S.T.; Walthall, C.L.; Kim, M.S.; de Colstoun, E.B.; McMurtrey, J.E. Estimating Corn Leaf Chlorophyll Concentration from Leaf and Canopy Reflectance. *Remote Sens. Environ.* **2000**, *74*, 229–239. [CrossRef]
4. Li, D.; Chen, J.M.; Zhang, X.; Yan, Y.; Zhu, J.; Zheng, H.; Zhou, K.; Yao, X.; Tian, Y.; Zhu, Y.; et al. Improved estimation of leaf chlorophyll content of row crops from canopy reflectance spectra through minimizing canopy structural effects and optimizing off-noon observation time. *Remote Sens. Environ.* **2020**, *248*, 111985. [CrossRef]

5.	Li, H.; Zhao, C.; Huang, W.; Yang, G. Non-uniform vertical nitrogen distribution within plant canopy and its estimation by remote sensing: A review. *Field Crops Res.* **2013**, *142*, 75–84. [CrossRef]

6.	Wang, Q.; Li, P. Canopy vertical heterogeneity plays a critical role in reflectance simulation. *Agric. For. Meteorol.* **2013**, *169*, 111–121. [CrossRef]

7.	Winterhalter, L.; Mistele, B.; Schmidhalter, U. Assessing the vertical footprint of reflectance measurements to characterize nitrogen uptake and biomass distribution in maize canopies. *Field Crops Res.* **2012**, *129*, 14–20. [CrossRef]

8.	Ciganda, V.; Gitelson, A.; Schepers, J. Vertical Profile and Temporal Variation of Chlorophyll in Maize Canopy: Quantitative "Crop Vigor" Indicator by Means of Reflectance-Based Techniques. *Agron. J.* **2008**, *100*, 1409–1417. [CrossRef]

9.	Hikosaka, K.; Anten, N.P.; Borjigidai, A.; Kamiyama, C.; Sakai, H.; Hasegawa, T.; Oikawa, S.; Iio, A.; Watanabe, M.; Koike, T.; et al. A meta-analysis of leaf nitrogen distribution within plant canopies. *Ann. Bot.* **2016**, *118*, 239–247. [CrossRef]

10.	Hirose, T.; Werger, M. Maximizing Daily canopy photosynthesis with respect to the leaf nitrogen allocation pattern in the canopy. *Oecologia* **1987**, *72*, 520–526. [CrossRef]

11.	Zhao, C.; Hang, W.; Wang, J.; Liu, L.; Song, X.; Ma, Z.; Li, C. Extracting winter wheat chlorophyll concentration vertical distribution based on bidirectional canopy reflected spectrum. *Trans. Chin. Soc. Agric. Eng.* **2006**, *22*, 104–109. (In Chinese) [CrossRef]

12.	Gitelson, A.; Gritz, Y.; Merzlyak, M. Relationships between leaf chlorophyll content and spectral reflectance and algorithms for non-destructive chlorophyll assessment in higher plant leaves. *J. Plant Physiol.* **2003**, *160*, 271–282. [CrossRef]

13.	Sims, D.; Gamon, J. Relationships between Leaf Pigment Content and Spectral Reflectance across a Wide Range of Species, Leaf Structures and Developmental Stages. *Remote Sens. Environ.* **2002**, *81*, 337–354. [CrossRef]

14.	Lu, Y.-L.; Bai, Y.-L.; Ma, D.-L.; Wang, L.; Yang, L.-P. Nitrogen Vertical Distribution and Status Estimation Using Spectral Data in Maize. *Commun. Soil Sci. Plant Anal.* **2018**, *49*, 526–536. [CrossRef]

15.	Gitelson, A.; Merzlyak, M. Spectral Reflectance Changes Associated with Autumn Senescence of *Aesculus hippocastanum* L. and *Acer platanoides* L. Leaves. Spectral Features and Relation to Chlorophyll Estimation. *J. Plant Physiol.* **1994**, *143*, 286–292. [CrossRef]

16.	Dash, J.; Curran, P. The MERIS terrestrial chlorophyll index. *Int. J. Remote Sens.* **2004**, *25*, 5003–5013. [CrossRef]

17.	Vogelmann, J.E.; Rock, B.N.; Moss, D.M. Red edge spectral measurements from sugar maple leaves. *Int. J. Remote Sens.* **1993**, *14*, 1563–1575. [CrossRef]

18.	Ciganda, V.; Gitelson, A.; Schepers, J.S. How deep does a remote sensor sense? Expression of chlorophyll content in a maize canopy. *Remote Sens. Environ.* **2012**, *126*, 240–247. [CrossRef]

19.	Lee, Y.; Yang, C.-M.; Chang, K.W.; Shen, Y. Effects of nitrogen status on leaf anatomy, chlorophyll content and canopy reflectance of paddy rice. *Bot. Stud.* **2011**, *52*, 295–303.

20.	Wan, L.; Zhou, W.; He, Y.; Wanger, T.C.; Cen, H. Combining transfer learning and hyperspectral reflectance analysis to assess leaf nitrogen concentration across different plant species datasets. *Remote Sens. Environ.* **2022**, *269*, 112826. [CrossRef]

21.	Li, L.T.; Jakli, B.; Lu, P.P.; Ren, T.; Ming, J.; Liu, S.S.; Wang, S.Q.; Lu, J.W. Assessing leaf nitrogen concentration of winter oilseed rape with canopy hyperspectral technique considering a non-uniform vertical nitrogen distribution. *Ind. Crops Prod.* **2018**, *116*, 1–14. [CrossRef]

22.	Wang, Z.H.; Skidmore, A.K.; Darvishzadeh, R.; Heiden, U.; Heurich, M.; Wang, T.J. Leaf Nitrogen Content Indirectly Estimated by Leaf Traits Derived From the PROSPECT Model. *IEEE J. Sel. Top. Appl. Earth Obs. Remote Sens.* **2015**, *8*, 3172–3182. [CrossRef]

23.	Jacquemoud, S.; Verhoef, W.; Baret, F.; Bacour, C.; Zarco-Tejada, P.J.; Asner, G.P.; François, C.; Ustin, S.L. PROSPECT+SAIL models: A review of use for vegetation characterization. *Remote Sens. Environ.* **2009**, *113*, S56–S66. [CrossRef]

24.	Chakhvashvili, E.; Siegmann, B.; Muller, O.; Verrelst, J.; Bendig, J.; Kraska, T.; Rascher, U. Retrieval of Crop Variables from Proximal Multispectral UAV Image Data Using PROSAIL in Maize Canopy. *Remote Sens.* **2022**, *14*, 1247. [CrossRef]

25.	Zhang, D.; Wang, X.; Ma, W.; Zhao, C. Research Vertical Distribution of Chlorophyll Content of Wheat Leaves Using Imaging Hyperspectra. *Intell. Autom. Soft Comput.* **2012**, *18*, 1111–1120. [CrossRef]

26.	Yang, H.; Li, J.; Yang, J.; Wang, H.; Zou, J.; He, J. Effects of nitrogen application rate and leaf age on the distribution pattern of leaf SPAD readings in the rice canopy. *PLoS ONE* **2014**, *9*, e88421. [CrossRef]

27.	Wu, B.; Ye, H.; Huang, W.; Wang, H.; Luo, P.; Ren, Y.; Kong, W. Monitoring the Vertical Distribution of Maize Canopy Chlorophyll Content Based on Multi-Angular Spectral Data. *Remote Sens.* **2021**, *13*, 987. [CrossRef]

28.	Huang, W.J.; Yang, Q.Y.; Peng, D.L.; Huang, L.S.; Zhang, D.Y.; Yang, G.J. Nitrogen vertical distribution by canopy reflectance spectrum in winter wheat. In *IOP Conference Series: Earth and Environmental Science*; IOP Publishing: Bristol, UK, 2014; Volume 17. [CrossRef]

29.	Wu, B.; Huang, W.; Ye, H.; Luo, P.; Ren, Y.; Kong, W. Using Multi-Angular Hyperspectral Data to Estimate the Vertical Distribution of Leaf Chlorophyll Content in Wheat. *Remote Sens.* **2021**, *13*, 1501. [CrossRef]

30.	Ciganda, V.; Gitelson, A.; Schepers, J. Non-destructive determination of maize leaf and canopy chlorophyll content. *J. Plant Physiol.* **2009**, *166*, 157–167. [CrossRef]

31.	Klem, K.; Rajsnerová, P.; Novotná, K.; Míša, P.; Křen, J. Changes in Vertical Distribution of Spectral Reflectance within Spring Barley Canopy as an Indicator of Nitrogen Nutrition, Canopy Structure and Yield Parameters. *Agriculture* **2014**, *60*, 50–59. [CrossRef]

32.	Li, Z.; Zhang, Y.; Liu, H.; Zhang, F. Application of chlorophyll meter on N nutritional diagnosis for summer corn. *Plant Nutr. Fertitizer Sci.* **2005**, *11*, 764–768.

33. Gitelson, A.A.; Vina, A.; Ciganda, V.; Rundquist, D.C.; Arkebauer, T.J. Remote estimation of canopy chlorophyll content in crops. *Geophys. Res. Lett.* **2005**, *32*, L08403. [CrossRef]
34. Duan, D.D.; Zhao, C.J.; Li, Z.H.; Yang, G.J.; Zhao, Y.; Qiao, X.J.; Zhang, Y.H.; Zhang, L.X.; Yang, W.D. Estimating total leaf nitrogen concentration in winter wheat by canopy hyperspectral data and nitrogen vertical distribution. *J. Integr. Agric.* **2019**, *18*, 1562–1570. [CrossRef]
35. Ye, H.; Huang, W.; Huang, S.; Wu, B.; Dong, Y.; Cui, B. Remote Estimation of Nitrogen Vertical Distribution by Consideration of Maize Geometry Characteristics. *Remote Sens.* **2018**, *10*, 1995. [CrossRef]
36. Li, D.; Guo, Y.; Yun, H.; Zhang, M.; Gong, X.; Fang, M. Dtermined Methods of Chlorophyll from Maize. *Chin. Agric. Sci. Bull.* **2005**, *21*, 153. (In Chinese)
37. Wen, P.F.; Shi, Z.J.; Li, A.; Ning, F.; Zhang, Y.H.; Wang, R.; Li, J. Estimation of the vertically integrated leaf nitrogen content in maize using canopy hyperspectral red edge parameters. *Precis. Agric.* **2021**, *22*, 984–1005. [CrossRef]
38. Cheng, M.; Jiao, X.; Liu, Y.; Shao, M.; Yu, X.; Bai, Y.; Wang, Z.; Wang, S.; Tuohuti, N.; Liu, S.; et al. Estimation of soil moisture content under high maize canopy coverage from UAV multimodal data and machine learning. *Agric. Water Manag.* **2022**, *264*, 107530. [CrossRef]
39. Jordan, C.F. Derivation of Leaf-Area Index from Quality of Light on the Forest Floor. *Ecology* **1969**, *50*, 663–666. [CrossRef]
40. Rouse, J.; Haas, R.; Schell, J.; Deering, D.; Harlan, J. *Monitoring the Vernal Advancement and Retrogradation (Green Wave Effect) of Natural Vegetation*; NASA/GSFC Type III, Final Report; NASA: Greenbelt, MD, USA, 1974.
41. Gitelson, A.; Merzlyak, M. Remote sensing of chlorophyll concentration in higher plant leaves. *Adv. Space Res.* **1998**, *22*, 689–692. [CrossRef]
42. Metternicht, G. Vegetation indices derived from high-resolution airborne videography for precision crop management. *Int. J. Remote Sens.* **2003**, *24*, 2855–2877. [CrossRef]
43. Fitzgerald, G.; Rodriguez, D.; O'Leary, G. Measuring and predicting canopy nitrogen nutrition in wheat using a spectral index—The canopy chlorophyll content index (CCCI). *Field Crops Res.* **2010**, *116*, 318–324. [CrossRef]
44. Haboudane, D.; Miller, J.; Tremblay, N.; Zarco-Tejada, P.; Dextraze, L. Integrated narrow-band vegetation indices for prediction of crop chlorophyll content for application to precision agriculture. *Remote Sens. Environ.* **2002**, *81*, 416–426. [CrossRef]
45. Strachan, I.; Pattey, E.; Boisvert, J. Impact of nitrogen and environmental conditions on corn as detected by hyperspectral reflectance. *Remote Sens. Environ.* **2002**, *80*, 213–224. [CrossRef]
46. le Maire, G.; Francois, C.; Dufrêne, E. Towards universal broad leaf chlorophyll indices using PROSPECT simulated database and hyperspectral reflectance measurements. *Remote Sens. Environ.* **2004**, *89*, 1–28. [CrossRef]
47. Broge, N.; Leblanc, E. Comparing prediction power and stability of broadband and hyperspectral vegetation indices for estimation of green leaf area index and canopy chlorophyll density. *Remote Sens. Environ.* **2001**, *76*, 156–172. [CrossRef]
48. Jay, S.; Gorretta, N.; Morel, J.; Maupas, F.; Bendoula, R.; Rabatel, G.; Dutartre, D.; Comar, A.; Frederic, B. Estimating leaf chlorophyll content in sugar beet canopies using millimeter- to centimeter-scale reflectance imagery. *Remote Sens. Environ.* **2017**, *198*, 173–186. [CrossRef]
49. Chen, P.; Haboudane, D.; Tremblay, N.; Wang, J.; Baoguo, L. New spectral indicator assessing the efficiency of crop nitrogen treatment in corn and wheat. *Remote Sens. Environ.* **2010**, *114*, 1987–1997. [CrossRef]
50. Feng, W.; Guo, B.-B.; Zhang, H.-Y.; He, L.; Zhang, Y.-S.; Wang, Y.-H.; Zhu, Y.-J.; Guo, T.-C. Remote estimation of above ground nitrogen uptake during vegetative growth in winter wheat using hyperspectral red-edge ratio data. *Field Crops Res.* **2015**, *180*, 197–206. [CrossRef]
51. Peng, Y.; Nguy-Robertson, A.; Arkebauer, T.; Gitelson, A. Assessment of Canopy Chlorophyll Content Retrieval in Maize and Soybean: Implications of Hysteresis on the Development of Generic Algorithms. *Remote Sens.* **2017**, *9*, 226. [CrossRef]
52. Huang, Z.; Turner, B.J.; Dury, S.J.; Wallis, I.R.; Foley, W.J. Estimating foliage nitrogen concentration from HYMAP data using continuum removal analysis. *Remote Sens. Environ.* **2004**, *93*, 18–29. [CrossRef]
53. Feng, A.; Zhou, J.; Vories, E.D.; Sudduth, K.A.; Zhang, M. Yield estimation in cotton using UAV-based multi-sensor imagery. *Biosyst. Eng.* **2020**, *193*, 101–114. [CrossRef]
54. Zhang, H.Y.; Ren, X.X.; Yi, Z.; Wu, Y.P.; Li, H.; Ya-Rong, H.; Wei, F.; Wang, C.Y. Remotely assessing photosynthetic nitrogen use efficiency with in situ hyperspectral remote sensing in winter wheat. *Eur. J. Agron.* **2018**, *101*, 90–100. [CrossRef]
55. Berger, K.; Verrelst, J.; Féret, J.-B.; Wang, Z.; Wocher, M.; Strathmann, M.; Danner, M.; Mauser, W.; Hank, T. Crop nitrogen monitoring: Recent progress and principal developments in the context of imaging spectroscopy missions. *Remote Sens. Environ.* **2020**, *242*, 111758. [CrossRef]
56. Zhang, Y.-J.; Wang, L.; Bai, Y.-L. Nitrogen Nutrition Diagnostic Based on Hyperspectral Analysis about Different Layers Leaves in Maize. *Spectrosc. Spectr. Anal.* **2019**, *39*, 2829. (In Chinese)
57. Xu, X.; Li, Z.; Yang, X.; Yang, G.; Teng, C.; Zhu, H.; Liu, S. Predicting leaf chlorophyll content and its nonuniform vertical distribution of summer maize by using a radiation transfer model. *J. Appl. Remote Sens.* **2019**, *13*, 034505. [CrossRef]
58. Zhang, Y.-J.; Wang, L.; Bai, Y.; Lu, Y.-L.; Zhang, J.-J.; Ge, L. Relationship of physiological and biochemical indicators with SPAD values in maize leaves at different layers. *J. Plant Nutr. Fertil.* **2020**, *26*, 61–73. (In Chinese)
59. Zhao, B.; Tahir, A.; Liu, Z.; Zhang, J.; Xiao, J.; Liu, Z.; Qin, A.; Ning, D.; Yang, Q.; Zhang, Y. Simple Assessment of Nitrogen Nutrition Index in Summer Maize by Using Chlorophyll Meter Readings. *Front. Plant Ence* **2018**, *9*, 11. [CrossRef]

60. Dang, R.J.; Li, S.-Q.; Mu, X.-H.; Li, S.-X. Effect of nitrogen on vertical distribution of canopy nitrogen and chlorophyll relative value (SPAD value) of summer maize in sub-humid areas. *Chin. J. Eco-Agric.* **2009**, *17*, 54–59. [CrossRef]
61. Peng, S.; Laza, M.; Garcia, F.V.; Cassman, K.G. Chlorophyll meter estimates leaf area-based nitrogen concentration of rice. *Commun. Soil Sci. Plant Anal.* **1995**, *26*, 927–935. [CrossRef]
62. Féret, J.-B.; Francois, C.; Asner, G.; Gitelson, A.; Martin, R.; Bidel, L.; Ustin, S.; le Maire, G.; Jacquemoud, S. PROSPECT-4 and 5: Advances in the leaf optical properties model separating photosynthetic pigments. *Remote Sens. Environ.* **2008**, *112*, 3030–3043. [CrossRef]
63. Li, Z.; Jin, X.; Yang, G.; Drummond, J.; Yang, H.; Clark, B.; Li, Z.; Zhao, C. Remote Sensing of Leaf and Canopy Nitrogen Status in Winter Wheat (*Triticum aestivum* L.) Based on N-PROSAIL Model. *Remote Sens.* **2018**, *10*, 1463. [CrossRef]
64. Li, L.; Sheng, K.; Yin, H.; Guo, Y.; Wang, D.; Wang, Y. Selecting the sensitive position of maize leaves for nitrogen status diagnosis of summer maize by considering vertical nitrogen distribution in plant. *Trans. Chin. Soc. Agric. Eng.* **2020**, *36*, 64–73. [CrossRef]
65. Huang, W.; Yang, Q.; Pu, R.; Yang, S. Estimation of Nitrogen Vertical Distribution by Bi-Directional Canopy Reflectance in Winter Wheat. *Sensors* **2014**, *14*, 20347–20359. [CrossRef]
66. Wang, Z.; Wang, J.; Zhao, C.; Zhao, M.; Huang, W.; Wang, C. Vertical Distribution of Nitrogen in Different Layers of Leaf and Stem and Their Relationship with Grain Quality of Winter Wheat. *J. Plant Nutr.* **2005**, *28*, 73–91. [CrossRef]
67. Wan, L.; Zhang, J.; Xu, Y.; Huang, Y.; Zhou, W.; Jiang, L.; He, Y.; Cen, H. PROSDM: Applicability of PROSPECT model coupled with spectral derivatives and similarity metrics to retrieve leaf biochemical traits from bidirectional reflectance. *Remote Sens. Environ.* **2021**, *267*, 112761. [CrossRef]
68. Wang, H.F.; Huo, Z.G.; Zhou, G.S.; Liao, Q.H.; Feng, H.K.; Wu, L. Estimating leaf SPAD values of freeze-damaged winter wheat using continuous wavelet analysis. *Plant Physiol Biochem.* **2016**, *98*, 39–45. [CrossRef]
69. Liu, L.; Song, B.; Zhang, S.; Liu, X. A Novel Principal Component Analysis Method for the Reconstruction of Leaf Reflectance Spectra and Retrieval of Leaf Biochemical Contents. *Remote Sens.* **2017**, *9*, 1113. [CrossRef]
70. Jacquemoud, S.; Ustin, S.L.; Verdebout, J.; Schmuck, G.; Andreoli, G.; Hosgood, B. Estimating leaf biochemistry using the PROSPECT leaf optical properties model. *Remote Sens. Environ.* **1996**, *56*, 194–202. [CrossRef]
71. Li, H.; Zhao, C.; Yang, G.; Feng, H. Variations in crop variables within wheat canopies and responses of canopy spectral characteristics and derived vegetation indices to different vertical leaf layers and spikes. *Remote Sens. Environ.* **2015**, *169*, 358–374. [CrossRef]
72. Xiao, C.H.; Li, S.K.; Wang, K.R.; Lu, Y.L.; Bai, J.H.; Xie, R.Z.; Gao, S.J.; Li, X.J.; Tan, H.Z. The Response of Canopy Direction Reflectance Spectrum for the Wheat Vertical Leaf Distributing. *Sens. Lett.* **2009**, *9*, 77–85.
73. Liu, S.; Peng, Y.; Du, W.; Le, Y.; Li, L. Remote Estimation of Leaf and Canopy Water Content in Winter Wheat with Different Vertical Distribution of Water-Related Properties. *Remote Sens.* **2015**, *7*, 4626–4650. [CrossRef]
74. Luo, J.; Ma, R.; Feng, H.; Li, X. Estimating the Total Nitrogen Concentration of Reed Canopy with Hyperspectral Measurements Considering a Non-Uniform Vertical Nitrogen Distribution. *Remote Sens.* **2016**, *8*, 789. [CrossRef]

remote sensing

MDPI

Article

Estimation of Potato Above-Ground Biomass Using UAV-Based Hyperspectral images and Machine-Learning Regression

Yang Liu [1,2,3,†], Haikuan Feng [1,4,*,†], Jibo Yue [5], Yiguang Fan [1], Xiuliang Jin [6], Yu Zhao [1], Xiaoyu Song [1], Huiling Long [1] and Guijun Yang [1]

[1] Key Laboratory of Quantitative Remote Sensing in Agriculture of Ministry of Agriculture and Rural Affairs, Information Technology Research Center, Beijing Academy of Agriculture and Forestry Sciences, Beijing 100097, China
[2] Key Lab of Smart Agriculture System, Ministry of Education, China Agricultural University, Beijing 100083, China
[3] Key Laboratory of Agricultural Information Acquisition Technology, Ministry of Agriculture and Rural Affairs, China Agricultural University, Beijing 100083, China
[4] College of Agriculture, Nanjing Agricultural University, Nanjing 210095, China
[5] College of Information and Management Science, Henan Agricultural University, Zhengzhou 450002, China
[6] Institute of Crop Sciences, Chinese Academy of Agricultural Sciences/Key Laboratory of Crop Physiology and Ecology, Ministry of Agriculture, Beijing 100081, China
* Correspondence: fenghk@nercita.org.cn
† These authors contributed equally to this work.

Citation: Liu, Y.; Feng, H.; Yue, J.; Fan, Y.; Jin, X.; Zhao, Y.; Song, X.; Long, H.; Yang, G. Estimation of Potato Above-Ground Biomass Using UAV-Based Hyperspectral images and Machine-Learning Regression. *Remote Sens.* **2022**, *14*, 5449. https://doi.org/10.3390/rs14215449

Academic Editors: Giovanni Laneve, Chenghai Yang, Wenjiang Huang and Yingying Dong

Received: 8 October 2022
Accepted: 27 October 2022
Published: 29 October 2022

Publisher's Note: MDPI stays neutral with regard to jurisdictional claims in published maps and institutional affiliations.

Abstract: Above-ground biomass (AGB) is an important indicator for monitoring crop growth and plays a vital role in guiding agricultural management, so it must be determined rapidly and nondestructively. The present study investigated the extraction from UAV hyperspectral images of multiple variables, including canopy original spectra (COS), first-derivative spectra (FDS), vegetation indices (VIs), and crop height (CH) to estimate the potato AGB via the machine-learning methods of support vector machine (SVM), random forest (RF), and Gaussian process regression (GPR). High-density point clouds were combined with three-dimensional spatial information from ground control points by using structures from motion technology to generate a digital surface model (DSM) of the test field, following which CH was extracted based on the DSM. Feature bands in sensitive spectral regions of COS and FDS were automatically identified by using a Gaussian process regression-band analysis tool that analyzed the correlation of the COS and FDS with the AGB in each growth period. In addition, the 16 Vis were separately analyzed for correlation with the AGB of each growth period to identify highly correlated Vis and excluded highly autocorrelated variables. The three machine-learning methods were used to estimate the potato AGB at each growth period and their results were compared separately based on the COS, FDS, VIs, and combinations thereof with CH. The results showed that (i) the correlations of COS, FDS, and VIs with AGB all gradually improved when going from the tuber-formation stage to the tuber-growth stage and thereafter deteriorated. The VIs were most strongly correlated with the AGB, followed by FDS, and then by COS. (ii) The CH extracted from the DSM was consistent with the measured CH. (iii) For each growth stage, the accuracy of the AGB estimates produced by a given machine-learning method depended on the combination of model variables used (VIs, FDS, COS, and CH). (iv) For any given set of model variables, GPR produced the best AGB estimates in each growth period, followed by RF, and finally by SVM. (v) The most accurate AGB estimate was achieved in the tuber-growth stage and was produced by combining spectral information and CH and applying the GPR method. The results of this study thus reveal that UAV hyperspectral images can be used to extract CH and crop-canopy spectral information, which can be used with GPR to accurately estimate potato AGB and thereby accurately monitor crop growth.

Keywords: potato; canopy original spectra; first-derivative spectra; vegetation indices; plant height; support vector machine; random forest; Gaussian process regression

---

1. Introduction

Potatoes are the fourth largest food crop in the world after wheat, corn, and rice. It is an important crop for national food security given its ability to adapt to its environment, its ease of cultivation, its short production cycle, and its high yield, so research on potatoes is particularly important [1–3]. Timely and accurate information on potato crop growth is vital for supporting agricultural production management and for macroregulation of cropping patterns and excavating production potential [4–7].

Above-ground biomass (AGB) is an important physiological indicator and is closely related to the nutritional status of crops and the ability of stems and leaves to accumulate organic matter, so it is often used to monitor crop growth [8–11]. Traditionally, AGB is measured by field-sampling surveys, which not only consume significant time and energy but also damage crops. In addition, the method is inefficient due to the limitation of sampling points because it can only be applied to small areas and thus cannot be scaled up to monitor crops grown in large areas [12,13].

Ground objects reflect and absorb electromagnetic waves, which sensors can remotely detect. Through data processing and comprehensive analysis, the characteristics of ground objects can thus be monitored remotely [14]. In recent years, remote-sensing technology has been used to dynamically estimate physiological and biochemical crop indicators to monitor crop growth. This continuous observation allows for nondestructive acquisition of crop-canopy spectral information [15–17]. In particular, remote sensing by using unmanned aerial vehicles (UAVs) has proven mobile, flexible, easy to implement, and produces high-resolution images. This approach has thus rapidly developed for acquiring small-scale regional crop phenotype information to monitor growth [18–21]. In contrast with the digital and multispectral sensors often carried by UAV platforms, hyperspectral sensors detect in narrow spectral bands that cover a broad spectrum. Subtle differences in ground objects may thus be distinguished by deep mining of the hyperspectral data. Based on the multilevel and multi-angle analysis of the hyperspectral characteristics of a crop canopy, an AGB estimation model may be developed to study the real-time dynamic change in AGB, which allows crop growth to be monitored [22–25]. Therefore, UAV hyperspectral remote-sensing technology is gaining attention in precision agriculture as a way to rapidly obtain accurate crop information.

Currently, the estimation of crop AGB using visible–near-infrared remote-sensing techniques is mainly done by using physical and statistical models. The physical models are based on the radiative transfer mechanism and require numerous parameters (meteorological data, soil data, crop variety information, etc.) as input to simulate crop growth and finally estimate the AGB. For example, physiological development time, soil data, temperature data, and vegetation indices (VIs) have been used to estimate maize AGB [26,27]. Although the physical model has a strong potential for generalization, it is mechanistically complex and requires numerous experimental supports in practical applications. It is also limited by its input parameters, which makes it difficult to apply on a large scale [28–32]. The statistical models are based on the relationship between crop-canopy spectral information and AGB. An AGB estimation model is constructed from spectral feature parameters, VIs, or regression techniques, which are simple in form, efficient in operation, and widely adaptable for monitoring crop growth. For example, a continuous projection algorithm was used to extract eight sensitive spectral bands (706, 724, 734, 806, 808, 810, 812, and 816 nm) to estimate winter wheat AGB [33]. Based on the optimal VIs, AGB estimation models were constructed to monitor the growth of rice [8], maize [11], and winter wheat [34]. Winter wheat AGB was accurately estimated by combining different VIs with partial-least-squares regression (PLSR) [35], stepwise regression (SWR) [36], and multiple linear regression (MLR) [23], respectively. Other studies estimate AGB by applying several regression techniques simultaneously, using correlation analysis to confirm sensitive spectral parameters, and then building an AGB estimation model through PLSR and MLR. They report that the PLSR-AGB model is more stable and consistent than the MLR-AGB model [37].

Most reports establish the correlation between crop AGB and VIs by traditional regression techniques. In fact, the estimations of crop AGB over multiple growth periods using VIs formed from visible–near-infrared spectra are limited due to the canopy spectral saturation caused by the growth stages effect [38]. However, although crop height (CH) that has been extracted based on the structure-from-motion technique combined with canopy spectral information can inhibit canopy saturation and allow for more comprehensive AGB estimation in multiple growth periods [39–41], few studies have been used to monitor variations in potato AGB. In recent years, machine-learning methods combined with remote-sensing data have been more commonly used to monitor AGB [42,43]. The machine-learning methods include random forest (RF) and support-vector machine (SVM) and are increasingly used to estimate crop AGB. In addition, Gaussian process regression (GPR) is gradually gaining importance for monitoring crop AGB because it has fewer input parameters, is better able to deal with nonlinearities using kernel functions, and solves the black-box problem in the machine-learning regression process [44,45]. Heretofore, no studies have extracted CH information and canopy spectral information from UAV hyperspectral images and fed these data to machine-learning methods to explore the accuracy of AGB estimation models over the various potato growth stages.

The main objectives of this study thus include (1) verifying the reliability of potato CH data that have been extracted by using a digital surface model (DSM) generated by the structure-from-motion technique; (2) comparing the accuracy of AGB estimates for different potato growth stages made by the GPR, RF, and SVM methods fed with (a) spectral data and (b) spectral data and CH; and (3) exploring the use of machine-learning regression to quantify the AGB based on UAV remote-sensing data.

2. Materials and Methods

2.1. Study Area Location and Experimental Design

The study area was in the National Precision Agriculture Research and Demonstration Base in Xiaotangshan Town, Changping District, Beijing, China, which is located at 40°10′34″ N, 116°26′39″ E.

A planting-density test area, a nitrogen test area, and a potassium-fertilizer test area were set up in the trial field, and two early-maturing potato varieties, Zhongshu 5 (Z5) and Zhongshu 3 (Z3), were selected to increase the spatial variability of crop growth in the field. The planting-density test area was planted in three densities: 60,000(P1), 72,000(P2), and 84,000 (P3) tubers/ha, and 18 test plots were treated six times with three replications. The nitrogen test area was treated at four levels: 0 (N0), 112.5 (N1), 225 (N2), and 337.5 (N3) kg/ha pure N, and 24 test plots were treated eight times with three replications. The potassium-fertilizer test area was treated at three levels: 0 (K0), 495 (the planting-density and nitrogen test areas were treated this way), and 990 (K2) kg/ha KO_2, and six test plots were treated twice with three replications. All plots were treated with 90 kg/ha P_2O_5. The plots totaled 48, and each had a sample area of 6.5 m × 5 m. To accurately locate the test plots of the split-plot experiment, 11 ground control points were evenly distributed around the test plots for terrain correction. The specific experimental design is shown in Figure 1.

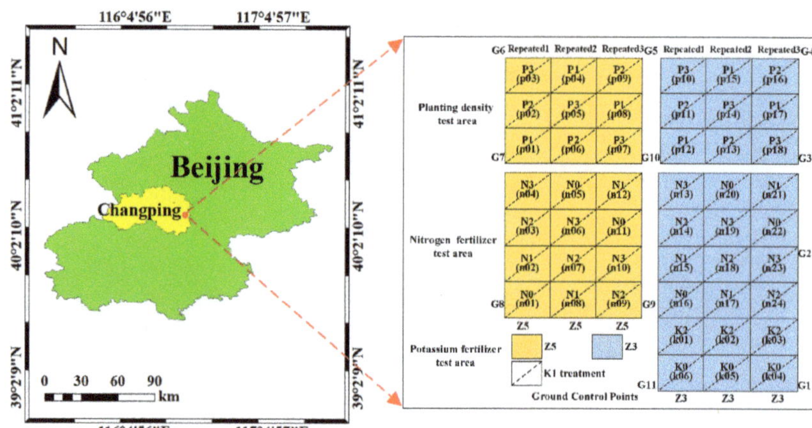

Figure 1. Location and design of test area.

2.2. Ground Data Acquisition and Processing

We obtained the AGB and CH for the potato tuber-formation stage (28 May 2019, BBCH-41), tuber-growth stage (10 June 2019, BBCH-44), and starch-storage stage (20 June 2019, BBCH-47). (1) AGB data acquisition: To ensure that reasonable samples were acquired, three plants representative of the overall growth level were destructively sampled from each plot and immediately sealed in plastic bags and transported to the laboratory. After separating stem from leaf, they were washed with running water, put into an oven at 105 °C for 1 h, then held at 80 °C to dry for over 48 h until the mass was constant. The dry mass of the stems and leaves was determined by using a high-precision balance (accuracy 0.001 g) and summed to obtain the dry mass of the sample. Finally, the potato AGB of each plot was obtained based on the population density and sample dry mass. (2) CH data acquisition: Four representative plants were selected from each plot and the distance from the bottom of the stem to the tip of the top leaf was measured with a straightedge. The average of four such measurements was taken as the potato CH of the plot.

2.3. Acquisition and Processing of UAV Hyperspectral Remote-Sensing Data

This experiment used a six-rotor electric UAV (M600, SZ DJI Technology Co., Ltd., Shenzhen, Guangdong, China) equipped with a UHD185 Firefly imaging spectrometer (Cubert, Germany), with dimensions of 195 mm × 67 mm × 60 mm, a mass of 470 g, a spectral range of 450–950 nm, and a spectral resolution of 4 nm with 125 spectral channels. The imaging speed was one hyperspectral image cubes per second. To ensure the accuracy of the resulting DSM, the takeoff position was the same each time and the flight was undertaken at 12:00 PM when the sky was clear and without wind or clouds. In addition, the same flight route was used for each growth stage. The flight altitude was set to 20 m (the transect width was 6 m) and had a speed of about 1.5 m/s. The overlap of heading and side direction was set to 85%. Before each flight, black and white plate data were collected on the ground for radiometric calibrations. The UAV carried sensors to acquire hyperspectral images of the bare-soil stage (April 20), tuber-formation stage, tuber-growth stage, and starch-storage stage. The spatial resolution of the resulting images was 1.3 cm.

Once the images were acquired, they were preprocessed in two steps: (1) The first step was stitching and geometric correction of the UAV hyperspectral images. The grayscale images within the selected aerial zone were stitched together with Agisoft PhotoScan software (version 1.1.6; Agisoft LLC, St. Petersburg, Russia) by using the position information of the ground control points (the correction error in each period was less than 2 cm). Next, using Cuber-Pilot software (Germany), the hyperspectral images were fused with the grayscale images to form new hyperspectral images. Finally, the digital orthophoto map and DSM of the test area were generated by using Agisoft PhotoScan software. (2) The second step involved extracting the canopy reflectance of the plot. The maximum-area vectors of different plots were plotted by using ArcGIS software (version 10.2; Esri, Redlands, CA, USA), the vector data were numbered by plot, and the average spectral reflectance of each area of interest was extracted by using the interactive data language implemented in ENVI software (version 5.0; Boulder, CO, USA). The result was used as the spectral reflectance of the potato canopy of the different plots.

2.4. Selection of Vegetation Indices

Vis are closely related to the physiological and biochemical parameters of crops and are often used to monitor crop growth. Therefore, we selected 16 commonly used Vis for estimating potato AGB. Table 1 lists the parameters and their mathematical expressions.

Table 1. Vis used in this study.

Vegetation indices	Equation	Reference
OSAVI (optimizing soil-adjusted vegetation index)	$1.16 \times (R800 - R670)/(R800 + R670 + 0.16)$	[46]
MTVI2 (modified triangular vegetation index 2)	$1.5 \times (1.2 \times (R800 - R500) - 2.5 \times (R670 - R550))$ $/(2 \times (R800 + 1)2 - (6 \times R800 - 5 \times (R670)1/2) - 0.5)^{1/2}$	[10]
SAVI (soil-adjusted vegetation index)	$(1 + 0.5) \times (R800 - R670)/(R800 + R670 + 0.5)$	[8]
RVI (ratio vegetation index)	$R810/R660$	[10]
NDVI (normalized-difference vegetation index)	$(R800 - R680)/(R800 + R680)$	[10]
EVI (enhanced vegetation index)	$2.5 \times (R800 - R670)/(R800 + 6 \times R670 - 7.5 \times R450 + 1)$	[47]
MCARI (modified chlorophyll-absorption ratio index)	$((R700 - R670) - 0.2 \times (R700 - R550))(R700/R670)$	[36]
RDVI (renormalized-difference vegetation index)	$(R800 - R670)/(R800 + R670)^{1/2}$	[36]
SPVI (spectral-polygon vegetation index)	$0.4 \times [3.7 \times (R800 - R670) - 1.2 \times \,|R550 - R670\,|]$	[36]
GNDVI (green normalized-difference vegetation index)	$(R750 - R550)/(R750 + R550)$	[48]
CI1 (red-edge chlorophyll index 1)	$R800/R740 - 1$	[49]
MSR (modified simple ratio index)	$(R800/R670 - 1)/(R800/R670 + 1)^{1/2}$	[48]
SIPI (structure-insensitive pigment index)	$(R800 - R450)/(R800 + R680)$	[50]
VARI (visible atmospherically resistance index)	$(R555 - R680)/(R555 + R680 - R480)$	[51]
NGRDI (normalized green–red difference index)	$(R560 - R680)/(R560 + R680)$	[42]
TVI (triangular vegetation index)	$0.5 \times [120 \times (R750 - R550) - 200 \times (R670 - R550)]$	[52]

2.5. Analysis Methods

RF is a method of data classification and statistical regression first proposed by Breiman and Cutler [53]. The method forms a training dataset by bootstrap sampling, generating random decision trees based on the classifier integrated into the system, then combining multiple decision trees to predict the dependent variable, and, finally, deciding the prediction results by voting. SVM is a class of generalized linear classifiers that perform binary classification of data by supervised learning. SVM are mainly classified as linear or nonlinear vector machines [42]. The training dataset is binary-classified by a kernel function to minimize the distance of all samples from a hyperplane and then the sample data are fit for prediction purposes [54]. GPR is a nonparametric probabilistic statistical model based on Bayes' theorem that learns the relationship between independent variables (e.g., spectral features) and dependent variables (e.g., AGB) by using mean and covariance functions based on maximum-likelihood estimation method. Compared with conventional machine-learning methods, parameter optimization is simpler and more suitable for training with small sample data, and the greatest advantage is the automatic identification of the best spectral features through the Gaussian process regression-band analysis tool (GPR-DAT) [45,55]. For sample training and prediction, this study uses the machine-learning regression algorithm of the ARTMO software (version 3.3.0; The University of Valencia, Spain, Europe).

2.6. Statistical Analysis

Forty-eight datasets were obtained for each fertility period. The data of the repeat 2 and repeat 3 plots were selected as the modeling set, and the data of the repeat 1 plots were used as the validation set. The detailed statistics are shown in Table 2. To reduce overfitting and underfitting, a tenfold cross-validation approach was used to construct the AGB estimation model for potatoes at different growth periods. The coefficient of determination (R^2), root–mean–square error of prediction (RMSE), and normalized root–mean–squared error (NRMSE) were used to assess the model fit and stability.

Table 2. Descriptive statistics for AGB (kg/hm^2) and CH (cm) of calibration and validation datasets.

Dataset	Crop Parameters	Min	Mean	Max	Standard Deviation	Coefficient of Variation (%)
Calibration	AGB	307	1144	2897	493	43.17
	CH	15.12	28.68	40.87	5.69	20.66
Validation	AGB	608	1281	2268	405	31.67
	CH	15.75	27.55	37.25	4.56	15.92

3. Results and Analysis

3.1. Extraction of Potato Crop Height

By using the raster calculation tool in the ArcGIS software, (i) the DSM of the potato tuber-formation, the tuber-growth, and the starch-storage stages and (ii) the DSM of bare-soil stage were calculated by difference, respectively. This gave the spatial distribution of CH for each growth period. Finally, the average CH of potatoes in each plot was extracted by using the area-of-interest tool. A total of 144 sets of average CH data were extracted for the three growth periods. To verify the reliability of based DSM extraction CH, the extracted CH and the measured CH were linearly fit (Figure 2). Figure 2 shows that the extracted CH with the measured CH fits better (R^2 = 0.84, RMSE = 2.52 cm, NRMSE = 9.05%), indicating that the extracted CH is reliable.

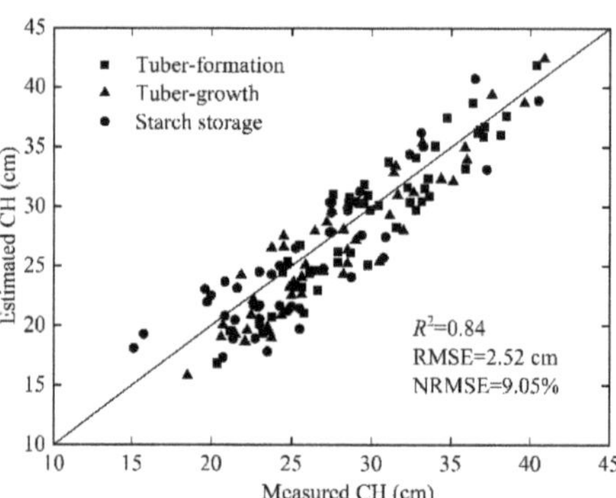

Figure 2. Measured and estimated potato CH.

3.2. Potato AGB Estimates Based on Canopy Spectra

In addition to the Vis formed by the common combination of bands, spectral feature parameters can also be formed from transformations such as spectral differentiation. First-order differentiation gives the change in reflectance, that is, the slope of spectral intensity with respect to wavelength. Hyperspectral data are suitable for first-order differentiation because of the high number of bands they contain. Figure 3 shows the correlation between (i) the canopy original spectra (COS) and first-derivative spectra (FDS) and (ii) the potato AGB at different growth stages. During the tuber-formation stage, the COS correlates significantly with the AGB in the near-infrared region (754–950 nm; $p < 0.01$), as does the FDS in both the red-edge (682–750 nm) and near-infrared regions. During the tuber-growth stage, the COS correlates significantly with the AGB from the visible to the near-infrared ($r > 0.53$, $p < 0.01$), whereas the FDS correlates significantly with the AGB mainly in the green (502–598 nm), red-edge, and near-infrared regions. During the starch-storage stage, the COS correlates significantly with the AGB over the whole band ($r > 0.53$, $p < 0.01$), whereas the FDS correlates significantly with the AGB in the red (602–678 nm), red-edge, and near-infrared regions.

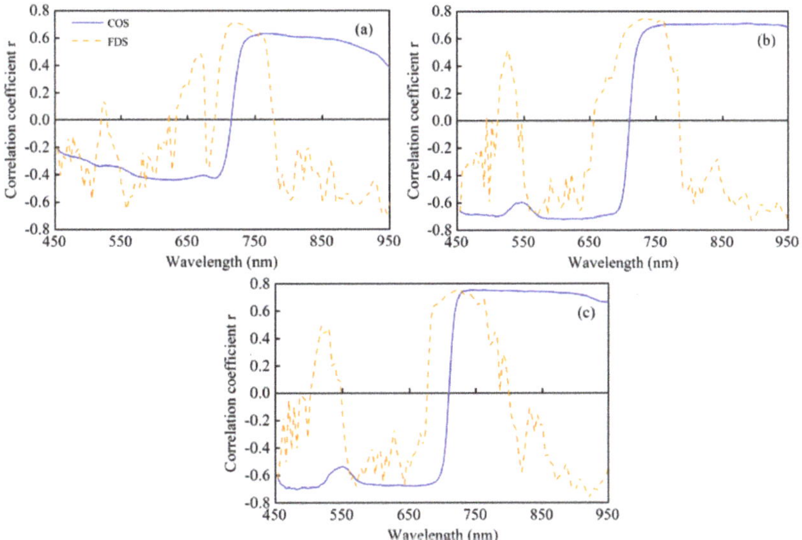

Figure 3. Correlation of canopy original spectra (COS) and first-derivative spectra (FDS) with above-ground biomass (AGB) of potatoes at different growth stages: (**a**) tuber-formation stage, (**b**) tuber-growth stage, (**c**) starch-storage stage.

The GPR-BAT was used to automatically identify spectral features in sensitive regions of the COS and FDS of the potato tuber-formation stage, tuber-growth stage, and starch-storage stage. A tenfold cross-validation was used in the GPR-BAT program. Figure 4 shows the mean RMSEcv, standard deviation, and min–max extreme value range of AGB estimated for each growth period by using the GPR method. Based on these results, the GPR-BAT determined that the optimal number of COS features for the tuber-formation, tuber-growth, and starch-storage stages were three (Figure 4a), six (Figure 4b), and four (Figure 4c), respectively. Likewise, the optimal number of FDS features were six (Figure 4d), seven (Figure 4e), and 11 (Figure 4f), respectively. The optimal COS and FDS features selected for each growth period were used to estimate potato AGB (Table 3).

Table 3. COS and FDS features selected by GPR-BAT for the tuber-formation, tuber-growth, and starch-storage periods.

Growth Stages	Feature Types	Selected Spectra Features (nm)
BBCH-41	COS	778, 802, 950
	FDS	682, 718, 754, 762, 946, 950
BBCH-44	COS	742, 746, 750, 934, 938, 942
	FDS	558, 774, 798, 806, 862, 898, 950
BBCH-47	COS	570, 698, 702, 850
	FDS	610, 618, 642, 710, 722, 730, 758, 766, 850, 870, 922

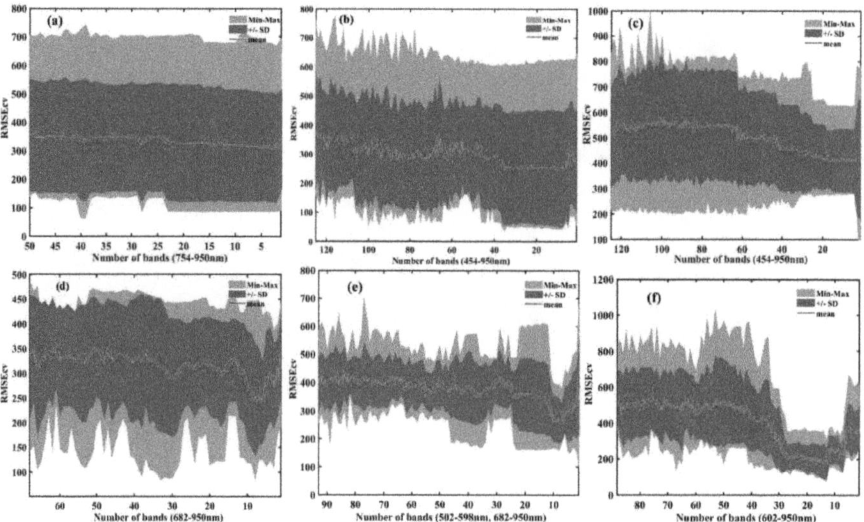

Figure 4. Determination of COS and FDS spectral features for use in estimating potato AGB. Panels (**a**) and (**d**) show the cross-validation RMSEcv statistics (mean, standard deviation, and min–max ranges) obtained by applying the GPR-BAT to the COS and FDS, respectively, for the tuber-formation stage. Panels (**b,e**) show the same for the tuber-growth stage, and panels (**c,f**) show the same for the starch-storage stage.

To evaluate the reliability of potato AGB estimates based on canopy spectra, three machine-learning methods, SVM, RF, and GPR, were used to form potato AGB estimation models for the tuber-formation, tuber-growth, and starch-storage stages with the selected COS and FDS features. The values of R^2, RMSE, and NRMSE were used to evaluate the fit and stability of the models. The estimates obtained for each growth period are listed in Tables 4 and 5. The results show that the potato AGB of multiple growth periods cannot be accurately estimated based on the canopy spectral information. In addition, under the same conditions, AGB estimates made directly from the COS features (Table 4) are significantly less accurate than those made directly from the FDS features (Table 5). AGB estimates for the three potato growth stages made by using SVM, RF, and GPR, with the help of the canopy spectra COS and FDS, are most accurate for the tuber-growth period and degrade thereafter (Tables 4 and 5).

With COS as input, AGB estimates by the three machine-learning methods, SVM, RF, and GPR, for the tuber-growth period produced R^2 = 0.49, 0.55, and 0.58, respectively (SVM: RMSE = 299.97 kg/hm^2, NRMSE = 23.27%; RF: RMSE = 271.48 kg/hm^2, NRMSE = 21.06%; GPR: RMSE = 238.22 kg/hm^2, NRMSE = 18.48%). The AGB estimates with FDS as input produced R^2 = 0.56, 0.61, and 0.65, respectively (SVM: RMSE = 278.57 kg/hm^2, NRMSE = 21.61%; RF: RMSE = 242.09 kg/hm^2, NRMSE = 18.78%; GPR: RMSE = 223.92 kg/hm^2, NRMSE = 17.37%). The GPR method produced the highest R^2 for all three growth stages (COS: 0.42, 0.58, and 0.35; FDS: 0.58, 0.65, and 0.43) and the lowest RMSE (COS: 250.69, 238.22, and 341.41 kg/hm^2; FDS: 226.27, 223.92, and 326.32 kg/hm^2) and NRMSE (COS: 20.53%, 18.48%, and 23.53%; FDS: 18.53%, 17.37%, and 22.49%), indicating that the GPR method produces more accurate potato AGB estimates than the RF or SVM methods.

Table 4. Potato AGB estimates based on selected COS features using machine-learning methods for different growth stages.

Growth Stages	Methods	Modeling			Validation		
		R^2	RMSE (kg/hm^2)	NRMSE (%)	R^2	RMSE (kg/hm^2)	NRMSE (%)
BBCH-41	SVM	0.31	308.33	25.25	0.41	218.17	21.43
	RF	0.38	276.21	22.62	0.45	205.14	20.15
	GPR	0.42	250.69	20.53	0.56	187.52	18.42
BBCH-44	SVM	0.49	299.97	23.27	0.53	201.50	22.19
	RF	0.55	271.48	21.06	0.59	166.36	18.32
	GPR	0.58	238.22	18.48	0.61	148.38	16.34
BBCH-47	SVM	0.29	377.69	26.03	0.38	221.89	25.72
	RF	0.33	357.23	24.62	0.42	207.82	24.09
	GPR	0.35	341.41	23.53	0.53	193.51	22.43

Table 5. Potato AGB estimates based on selected FDS features using machine-learning methods for different growth periods.

Growth Stages	Methods	Modeling			Validation		
		R^2	RMSE (kg/hm^2)	NRMSE (%)	R^2	RMSE (kg/hm^2)	NRMSE (%)
BBCH-41	SVM	0.37	284.88	23.33	0.54	210.23	20.65
	RF	0.47	247.64	20.28	0.57	196.18	19.27
	GPR	0.58	226.27	18.53	0.61	175.61	17.25
BBCH-44	SVM	0.56	278.57	21.61	0.58	168.17	18.52
	RF	0.61	242.09	18.78	0.62	161.45	17.78
	GPR	0.65	223.92	17.37	0.68	143.65	15.82
BBCH-47	SVM	0.32	369.27	25.45	0.41	212.05	24.58
	RF	0.34	335.61	23.13	0.51	197.91	22.94
	GPR	0.43	326.32	22.49	0.58	185.92	21.55

3.3. Potato AGB Estimates Based on Vegetation Indices

Figure 5a shows the results of a Pearson correlation analysis between the selected VIs and potato AGB for multiple growth stages. The selected VIs correlate significantly with the AGB over the three growth stages ($p < 0.01$), with the correlation first increasing and then decreasing. The VI most strongly correlated with AGB during the tuber-formation stage is SPVI ($r = 0.672$, $p < 0.01$), and the VI most strongly correlated with AGB during both the tuber-growth and starch-storage stages is GNDVI ($r = 0.759$ and 0.756, respectively, $p < 0.01$). The scatter plots in Figure 5b,c reveal a linear correlation between the optimal VIs and AGB for each growth stage, although the sensitivity of the VIs to AGB decreases as the growth stage advances. The VIs saturate when the AGB is large. These results indicate that potato AGB estimates that use only a single VI are not reliable.

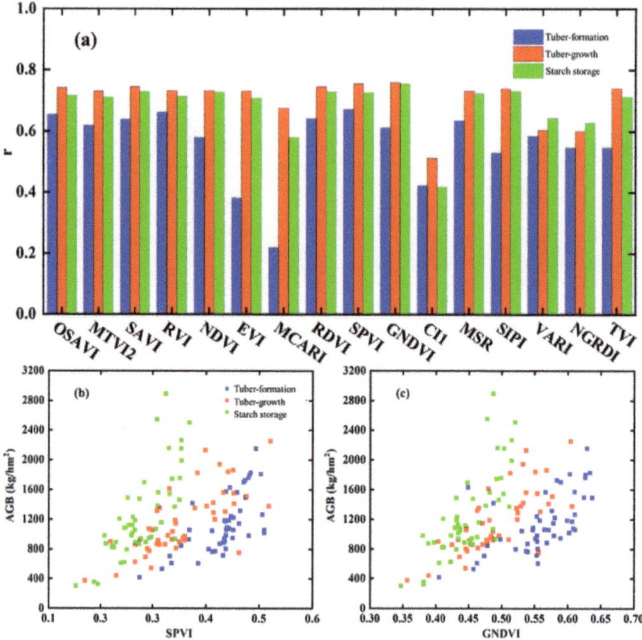

Figure 5. Pearson correlation analyses of above-ground biomass (AGB) of potatoes with VIs for different growth stages. (**a**) Absolute correlation coefficient for VIs versus AGB for tuber-formation, tuber-growth, and starch-storage periods. (**b,c**) Scatter plots of the best performing VIs versus AGB for each growth period.

Before constructing the potato AGB estimation model based on multiple VIs, we considered the autocorrelation of the independent input variables and used the "findcorrelation" function in the "caret" package of the R language to remove variables with high redundancy. The cutoff value of this function was set to 0.99. With the highly correlated independent variables removed, the remaining VIs used to estimate the AGB for each growth period are shown in Table 6. These are SAVI, NDVI, MSR, and TVI for all three potato growth stages, indicating that these Vis are essential for estimating the AGB.

Table 6. Remaining VIs after removing highly correlated variables for each growth stage.

Growth Stages	Selected Vegetation Indices
BBCH-41	OSAVI, SAVI, NDVI, MSR, NGRDI, TVI
BBCH-44	OSAVI, MTVI2, SAVI, NDVI, RDVI, SPVI, MSR, VARI, TVI
BBCH-47	MTVI2, SAVI, NDVI, RDVI, SPVI, MSR, NGRDI, TVI

The potato AGB during tuber-formation, tuber-growth, and starch-storage stages were estimated by using SVM, RF, and GPR with the selected VIs. The values of R^2, RMSE, and NRMSE for the regression results are given in Table 7. The results show that the AGB estimates become more accurate upon using the same machine-learning technique and VIs for each growth stage (Table 4, Table 5, and Table 7). The estimation accuracy peaks in the tuber-growth stage. Using the three machine-learning methods to estimate AGB for the tuber-growth stage produces R^2 = 0.59, 0.68, and 0.71 (SVM: RMSE = 266.90 kg/hm^2, NRMSE = 20.71%; RF: RMSE = 231.13 kg/hm^2, NRMSE = 17.93%; GPR: RMSE = 218.11 kg/hm^2, NRMSE = 16.92%). The largest R^2 value and the smallest RMSE and NRMSE values occur when using the GPR machine-learning method based on VIs, so this combination produced the best modeling results. However, the scatter plot (Figure 6) shows that, during crop growth, potato AGB is underestimated when using VIs, regardless of the machine-learning method.

Table 7. Potato AGB estimates based on selected VIs for each growth stage.

Growth Stages	Methods	Modeling			Validation		
		R^2	RMSE (kg/hm^2)	NRMSE (%)	R^2	RMSE (kg/hm^2)	NRMSE (%)
BBCH-41	SVM	0.45	270.47	22.15	0.61	205.44	20.18
	RF	0.62	243.61	19.95	0.65	174.70	17.16
	GPR	0.64	213.45	17.48	0.68	165.33	16.24
BBCH-44	SVM	0.59	266.90	20.71	0.63	163.63	18.02
	RF	0.68	231.13	17.93	0.72	139.57	15.37
	GPR	0.71	218.11	16.92	0.75	135.66	14.94
BBCH-47	SVM	0.43	356.21	24.55	0.58	205.93	23.87
	RF	0.53	329.66	22.72	0.62	184.79	21.42
	GPR	0.60	318.20	21.93	0.64	170.47	19.76

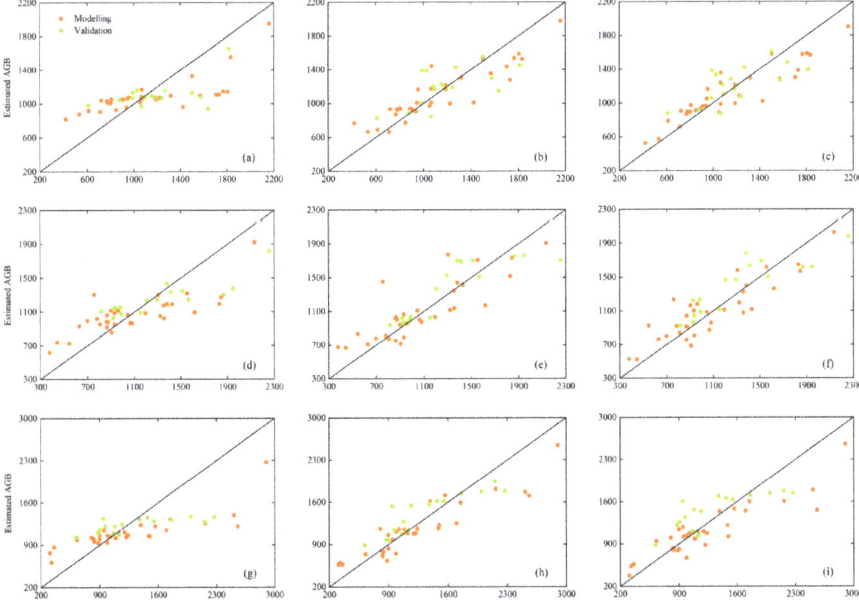

Figure 6. Scatter plots of measured versus estimated potato AGB (kg/hm^2) for modeling and validation datasets and for different growth stages. SVM, RF, and GPR methods are used with VIs as input. (**a–c**) The relationship between measured and estimated potato AGB for the tuber-formation stage. Estimates were made by SVM, RF, and GPR, respectively. (**d–f**) Same as panels (**a–c**) but for the tuber-growth stage. (**g–i**) Same as panels (**a–c**) but for the starch-storage stage.

The validation dataset was used to verify the reliability of AGB estimates based on VIs combined with RF, SVM, or GPR for the three fertility stages of potato growth (Table 7 and Figure 6). The results show that the RF, SVM, and GPR methods were consistent across all growth stages. In other words, the fact that R^2, RMSE, and NRMSE follow the same trends for both the modeling and validation sets means that the models constructed by the three machine-learning methods fit the data well and are stable. Meanwhile, estimates based on VIs using the same machine-learning method were also consistent with the modeling set and gradually improved when going from the tuber-formation stage to the tuber-growth stage, followed by deterioration. A comprehensive analysis, for each growth stage, of the accuracy metrics of the AGB estimation model (Table 7) revealed that the GPR method not only improves the accuracy but also enhances the stability, thereby providing the optimum AGB estimates and solving the saturation problem.

3.4. Estimation of Potato AGB Using Canopy Spectra and Vegetation Indices Combined with Crop Height

To evaluate how CH affects potato AGB estimates, the SVM, RF, and GPR machine-learning methods were used to develop AGB estimation models for the three growth stages by combining crop-canopy spectral information (COS, FDS), Vis, and CH. Before constructing the models, excessive redundancy between input parameters was avoided by using the "findCorrelation" function to exclude highly correlated independent variables. The final model factors for estimating potato AGB are listed in Table 8.

Table 8. Model factors involved after removing highly relevant variables for each growth stage.

Growth Stages	Selected Vegetation Indices
BBCH-41	718(FDS), NDVI, RDVI, SPVI, MSR, VARI, CH
BBCH-44	746(COS), 762(FDS), SAVI, EVI, RDVI, SPVI, MSR, SIPI, VARI, TVI, CH
BBCH-47	722(FDS), SAVI, NDVI, RDVI, SPVI, MSR, VARI, TVI, CH

AGB was estimated for each growth period by the three machine-learning methods (SVM, RF, and GPR) using the modeling dataset based on the model variables listed in Table 8. The regression results for R^2, RMSE, and NRMSE are listed in Table 9, and the fitted scatter plots are shown in Figure 7. The results show that the best estimates are obtained for each growth stage when estimating AGB by using the same machine-learning method with spectral information combined with CH (Table 4, Table 5, Table 7, and Table 9). The accuracy of the estimates produced by each machine-learning method vary from good to poor, peaking at the tuber-growth stage (Table 9). AGB estimates for the tuber-growth period using the three machine-learning methods produced R^2 = 0.64, 0.74, and 0.76, respectively (SVM: RMSE = 245.19 kg/hm^2, NRMSE = 19.02%; RF: RMSE = 223.79 kg/hm^2, NRMSE = 17.36%; GPR: RMSE = 199.68 kg/hm^2, NRMSE=15.49%).

Table 9. Potato AGB estimates based on spectral information combined with CH for each growth stage.

Growth Stages	Methods	Modeling			Validation		
		R^2	RMSE (kg/hm^2)	NRMSE (%)	R^2	RMSE (kg/hm^2)	NRMSE (%)
BBCH-41	SVM	0.58	263.39	21.57	0.66	196.38	19.29
	RF	0.69	228.71	18.73	0.75	172.35	16.93
	GPR	0.72	201.24	16.48	0.78	158.30	15.55
BBCH-44	SVM	0.64	245.19	19.02	0.69	168.17	18.52
	RF	0.74	223.79	17.36	0.79	141.84	15.62
	GPR	0.76	199.68	15.49	0.82	130.31	14.35
BBCH-47	SVM	0.56	334.59	23.06	0.62	198.25	22.98
	RF	0.62	310.80	21.42	0.65	176.94	20.51
	GPR	0.68	291.35	20.08	0.72	161.93	18.77

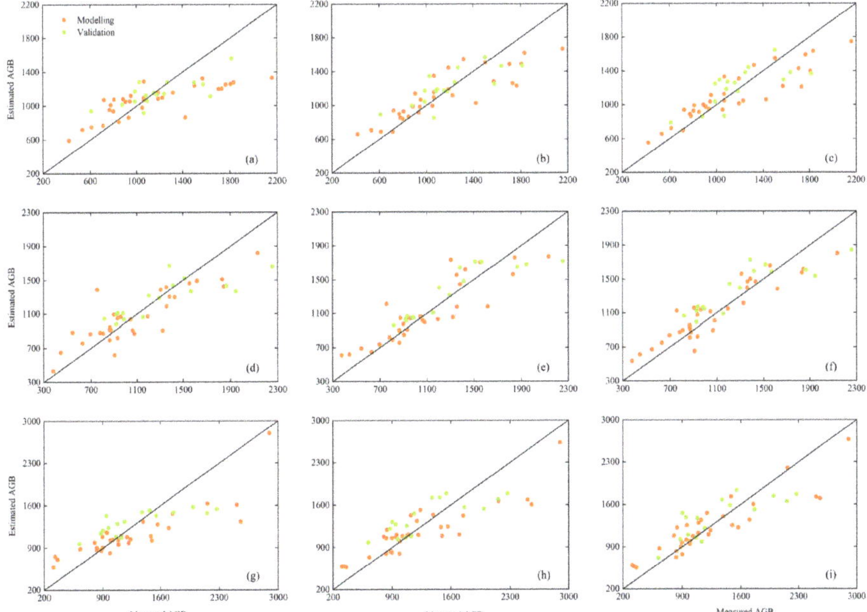

Figure 7. Scatter plots of measured versus estimated potato AGB (kg/hm^2) for modeling and validation datasets at different growth stages based on spectral information combined with CH and using SVM, RF, and GPR methods. (**a–c**) Relationship between measured and estimated potato AGB for the tuber-formation stage, as per SVM, RF, and GPR methods, respectively. (**d–f**) Same as panels (**a–c**) but for tuber-growth stage. (**g–i**) Same as panels (**a–c**) but for starch-storage stage.

The use of GPR fed with spectral information combined with CH improved the R^2 value and reduced the RMSE and NRMSE values, indicating that this method provides the most accurate estimate of potato AGB. At the same time, using any of the three machine-learning methods with the combination of CH and spectral information reduced the occurrence of AGB underestimation in each growth stage (Figures 6 and 7). Finally, the GPR method based on spectral information combined with CH handled the saturation problem of AGB estimation model better than the SVM and RF methods (Figure 7).

The stability of AGB estimation models constructed using SVM, RF, and GPR based on spectral information combined with CH was then verified by using the validation dataset. Figure 7 shows the relationship between the measured and estimated values for each potato growth stage. The results show that the SVM, RF, and GPR validation results are consistent with the modeling results for each fertility stage, and the higher the modeling accuracy, the higher the validation accuracy (Table 9), indicating that the models constructed by the three machine-learning methods are stable and that the prediction results are reliable. The prediction results obtained using the same machine-learning methods at each fertility stage were also consistent with the modeling dataset: a gradual improvement occurs upon going from tuber formation to tuber growth, following which deterioration sets in. In terms of model fit and stability, estimating potato AGB using the GPR method produced a higher R^2 and lower NRMSE, indicating that the GPR-AGB estimation method is more accurate than the SVM and RF methods.

4. Discussion

4.1. Monitoring Potato Crop Height

Variation in potato CH can provide researchers with important information on crop health, growth, and response to environmental impacts. Therefore, CH plays a decisive role in monitoring crop growth [56–59], so CH data can not only guide agricultural production management but also provide a reliable theoretical basis for plant morphology research and crop selection and breeding. In this study, the CH information is extracted without generating a digital elevation model of bare-soil points by kriging interpolation because potatoes are planted in monopolies that reach a certain height, so ignoring their height and taking the difference would cause large extraction errors [60].

On the contrary, this study generates a high-density point cloud of UAV hyperspectral remote-sensing data by using structure-from-motion technology and then combining the location information of ground control points to generate a DSM for each fertility stage. Thus, the DSM for tuber-formation, tuber-growth, and starch-storage stages are differentially calculated from those of the bare-soil stage to obtain the CH of the corresponding fertility stage [61,62]. The relationship between extracted and measured CH is analyzed by using data from 144 samples from three fertility stages (Figure 2). The results give $R^2 = 0.84$ and NRMSE = 9.05%, indicating an accurate CH, which is consistent with the findings from Tao et al. [47], Bendig et al. [56,63], and Li et al. [49].

A comparative analysis reveals that the overall CH extracted in this study is small compared with the measured value, which is mainly because the UAV acquires hyperspectral data from the potato canopy, which has a tight spatial structure, so the spatial information from the top leaves may be removed as noise when performing three-dimensional point-cloud reconstruction. However, the measured CH is the height of the highest point of the leaves in the natural growth, which leads to an overall lower CH than the measured CH. At the same time, the images acquired by the UAV contain elements of potato-plant leaves and bare soil, so part of the soil affects the calculation of CH, which causes significant deviations. Therefore, the accuracy of three-dimensional point-cloud reconstruction of smaller structures in canopy space must be improved to accurately extract CH based on DSMs.

4.2. Estimation of Potato AGB Based on Canopy Spectra

How COS and FDS correlate with AGB must be analyzed before estimating potato AGB for the three reproductive stages. The results show that the correlations of both variables gradually increase upon going from the tuber-formation stage to the tuber-growth stage, following which they start to decrease. This result may be related to the potato crop growth cycle. Starting in the tuber-formation stage, the nutritional reproductive organs such as stems and leaves gradually develop, and the potato crop grows gradually and increases the vegetation cover. The spectral data are thus less affected by the ground soil when they are extracted. From the end of the tuber-growth stage to the beginning of the starch-storage stage, the organic matter created by above-ground photosynthesis is continuously transported to the underground tubers, robbing nutrients from the above-ground stems and leaves. At the same time, due to rainy weather, the plant leaves rapidly wither and fall off, causing the ground soil to seriously interfere with the extraction of spectral information from the crop canopy. The correlation between the FDS and AGB at this point is stronger than that for COS (Figure 3), which is consistent with the findings of Fan et al. [64] and Masjedi et al. [65] and is due to the ability to remove background noise and thereby enhance the spectral sensitivity by spectral differentiation [37]. Due to the numerous bands in the hyperspectral data, a serious collinearity problem exists between different spectral data. The GPR-BAT automatically identifies the characteristic bands in the sensitive spectral areas of the canopy (Figure 4) and determines the COS and FDS for estimating AGB at each growth stage (Table 3). More FDS features are screened in each growth period than are COS features, which indicates that the original spectrum contains more redundant information. The first-order derivative refines the spectral information and enhances the spectral sensitivity [63].

The SVM, RF, and GPR machine-learning methods are used to estimate AGB at each fertility stage based on the two spectral variables, COS and FDS, and the accuracy of each model is analyzed (Tables 4 and 5). The results show that, with the same machine-learning methods, the AGB estimates based on the two spectral variables progressively get more accurate when going from the tuber-formation stage to the tuber-growth stage and then start to deteriorate, which may be closely related to potato growth. In addition, at each reproductive stage, AGB estimates based on FDS are better than those based on COS, which is consistent with the results of Feng et al. [65], indicating that the FDS better reflects the real AGB situation. However, comparing the estimation accuracy throughout the reproductive period shows that, although AGB may be estimated by using only crop-canopy spectral information (COS, FDS), the estimation accuracy is unsatisfactory (Tables 4 and 5).

4.3. Estimation of Potato AGB Based on Vegetation Indices

Based on previous studies, we selected 16 VIs that produce accurate AGB estimates (Table 1). These VIs correlated significantly with AGB ($p < 0.01$) across all fertility stages, with the correlations progressively increasing when going from the tuber-formation stage to the tuber-growth stage and then deteriorating. This result is consistent with the correlation of canopy spectral information with AGB (Figures 3 and 5). VIs are determined by the reflectance intensity in two or more spectral bands; however, the intensity of spectral reflectance has a limited response to variations in AGB [66]. As the growth stages advance, VIs tend to saturate (Figure 5), so estimating AGB by using a single VI is unreliable [37,44]. To reduce the autocorrelation between variables and improve efficiency, the model variables used to estimate AGB were obtained during each growth stage after excluding variables with high redundancy (Table 6). SAVI, NDVI, MSR, and TVI correlated strongly with AGB in each growth period, indicating that these indices contain important AGB information. This result is consistent with the findings of Gnyp et al. [8].

The SVM, RF, and GPR methods based on VIs were used to estimate the AGB at each growth stage (Table 5). The results show that AGB estimates using the same machine-learning method at each fertility stage are consistent with the variations based on canopy spectral information, which gradually improve upon going from tuber formation to tuber growth and then start to deteriorate. The models based on Vis (Table 7, Figure 6) were more accurate for each fertility period than those based on canopy spectral information (Tables 4 and 5), which is consistent with the results of Tao et al. [36]. In addition, differences appeared in the Vis used to estimate the three growth stages of potatoes, which is due to the different magnitude of variation in the AGB, leaf-area index, chlorophyll, and water content during crop growth. The VIs are related to these physicochemical parameters, making them sensitive to the AGB depending on the crop growth stage. Thus, the optimization of model-independent variables is vital for estimating the AGB based on UAV hyperspectral remote-sensing data.

4.4. Estimation of Potato AGB Based on Canopy Spectra, Vegetation Indices, and Crop Height

Estimating potato AGB using different machine-learning methods combined with crop-canopy spectral information (COS, FDS) and VIs is inaccurate because of crop spectral-response mechanisms (Table 4, Table 5, and Table 7 and Figure 6). Therefore, to overcome the saturation problem of the model and to estimate the AGB more accurately and across multiple fertility stages, the crop-canopy spectral information, Vis, and structural information (i.e., CH) are integrated to estimate potato AGB during the tuber-formation, tuber-growth, and starch-storage stages by using SVM, RF, and GPR. Before constructing the AGB estimation model for each fertility period, the final model variables used to estimate the AGB for each fertility period are obtained after excluding the variables with high autocorrelation (Table 8).

Model effects obtained using the same machine-learning method are consistent with the estimation effects based on canopy spectral information and Vis, which also gradually improve upon going from the tuber formation to the tuber-growth stage, and then start to deteriorate. Comparing the model effects obtained when using the three variables at each fertility stage (Tables 4, 5, 7 and 8) shows that combining the extracted CH with canopy spectral information and Vis significantly improves the accuracy of AGB estimates and the stability of the model. This is consistent with the results of Banerjee et al. [23],

Zarco et al. [61], and Bendig et al. [62], who reported that the inclusion of CH information significantly improves the accuracy of AGB estimates. Using the three machine-learning methods based on spectral information and CH tends to underestimate the AGB less as the fertility stage progresses (Figures 6 and 7), indicating that CH is particularly important for monitoring potato AGB.

The R^2 values for AGB estimates produced by the SVM, RF, and GPR methods applied to the tuber-growth stage are 0.64, 0.74, and 0.76, respectively, and NRMSE = 19.02%, 17.36%, and 15.49%, respectively. Compared with the canopy-based original spectra, R^2 increases by 23%, 26%, and 24%, respectively, and NRMSE decreases by 22%, 21%, and 19%, respectively. Compared with the canopy-based first-derivative spectra, R^2 improves by 13%, 18%, and 14%, and NRMSE decreases by 14%, 8%, and 12%, respectively. Compared with VI-based estimates, R^2 improves by 7%, 8%, and 7%, and NRMSE decreases by 9%, 3%, and 9%, respectively. These results indicate that AGB estimates improve when using spectral information combined with CH.

4.5. Estimation of Potato AGB Using SVM, RF, and GPR Methods

SVM, RF, and GPR machine-learning methods were used to estimate AGB for each fertility stage based on different variables, with the results assessed by modeling and validating the dataset based on the magnitude of R^2, RMSE, and NRMSE. The results showed that the GPR method significantly improves the accuracy of potato AGB based on different types of variables (Tables 4, 5, 7 and 9, Figures 6 and 7). This indicates that the model constructed using the GPR method is superior, which is consistent with the results of Fu et al. [44], who estimated nitrogen indicators for winter wheat. This is mainly because, in the process of building the AGB estimation model with the GPR method, GPR-BAT automatically identifies important feature subsets related to potato AGB through the embedded feature-selection function, which not only reduces the complexity of the model but also improves its interpretability. At the same time, small datasets with nonlinear characteristics can be well-processed through the built-in kernel function, which makes the data fit better [44].

The AGB estimate at each fertility stage, produced by using the RF method based on different variables, is less accurate than that produced by the GPR method, mainly because the RF method is suitable for large datasets. The tuber-formation, tuber-growth, and starch-storage stages belong to the single-growth stage of potatoes with a small sample size (32 modeling data and 16 validation data qualify as small datasets), so AGB estimates produced by the RF method are not accurate.

However, using SVM to estimate AGB across all fertility stages and based on different variables is the least accurate, mainly because constructing the model with the SVM method is limited by penalty factors and kernel functions, which reduce the accuracy of AGB estimates. In addition, SVM depends on the collinearity of input variables. There may be some collinearity with the parameters involved in the final model construction in this study, resulting in the poor generalization of SVM to the data.

5. Conclusions

AGB plays an important role in monitoring crop growth, so the rapid and nondestructive acquisition of AGB data is vital for precision agriculture. Unlike satellite and aerial remote sensing, UAVs equipped with a UHD185 sensor can acquire potato crop-canopy images under cloud cover and can gather large amounts of spectral data with high spectral resolution, thereby providing a reliable technical support for crop AGB estimations. This study investigates the feasibility of using SVM, RF, and GPR methods based on COS, FDS, CH, Vis, and combinations thereof to estimate the potato AGB during tuber formation, tuber growth, and starch storage. The models constructed, based on different variables but using the same machine-learning method, all produce progressively more accurate AGB estimates when going from the tuber-formation stage to the tuber-growth stage, following which the accuracy deteriorates. The most accurate AGB estimates for each fertility period are obtained by using the GPR method (independent of variables). Based on the combination of spectral information and CH, the SVM, RF, and GPR methods significantly improve the AGB estimation accuracy compared with the use of single-spectral features or VIs. However, the combination of spectral information and CH produces the most accurate AGB estimates and the most stable model, which is an important guideline for monitoring crop growth.

Author Contributions: H.F., Y.L., H.L., G.Y. and J.Y. designed the experiments. H.F., Y.F., Y.Z. and X.S. collected the AGB, CH, and UAV hyperspectral images. Y.L. and H.F. analyzed the data and wrote the manuscript. X.J. and J.Y. made comments and revised the manuscript. All authors have read and agreed to the published version of the manuscript.

Funding: This study was supported by the Key scientific and technological projects of Heilongjiang province (2021ZXJ05A05), the National Natural Science Foundation of China (41601346), the Platform Construction Funded Program of Beijing Academy of Agriculture and Forestry Sciences (No.PT2022-24), and the Key Field Research and Development Program of Guangdong Province (2019B020216001).

Acknowledgments: We thanks to Bo Xu, Hong Chang, Weiguo Li, Yang Meng, and Yu Zhao for field management and data collection. We thank the National Precision Agriculture Experiment Station for providing the test site and employees.

Conflicts of Interest: The authors declare no conflict of interest.

References

1. Swain, K.C.; Thomson, S.J.; Jayasuriya, H.P. Adoption of an unmanned helicopter for low-altitude remote sensing to estimate yield and total biomass of a rice crop. *Trans. Asabe* **2010**, *53*, 21–27. [CrossRef]
2. Mueller, N.D.; Gerber, J.S.; Johnston, M.; Ray, D.K. Closing yield gaps through nutrient and water management. *Nature* **2013**, *494*, 254–261. [CrossRef]
3. Morier, T.; Cambouris, A.N.; Chokmani, K. In-season nitrogen status assessment and yield estimation using hyperspectral vegetation indices in a potato crop. *Agron. J.* **2015**, *107*, 1295–1309. [CrossRef]
4. Franceschini, M.D.; Harm, B.; Apeldoorn, D.; Suomalainen, J.; Kooistra, L. Intercomparison of unmanned aerial vehicle and ground-based narrow band spectrometers applied to crop trait monitoring in organic potato production. *Sensors* **2017**, *17*, 1428. [CrossRef]
5. Mahlein, A.K.; Rumpf, T.; Welke, P.; Dehne, H.W.; Plumer, L.; Steiner, U.; Oerke, E.C. Development of spectral indices for detecting and identifying plant diseases. *Remote Sens. Environ.* **2013**, *128*, 21–30. [CrossRef]
6. Greaves, H.E.; Vierling, L.A.; Eitel, J.H.; Boelman, N.T.; Magney, T.S.; Prager, C.M. Estimating aboveground biomass and leaf area of low stature arctic shrubs with terrestrial LiDAR. *Remote Sens. Environ.* **2015**, *164*, 26–35. [CrossRef]
7. Fu, Y.Y.; Yang, G.J.; Wang, J.H.; Song, X.Y.; Feng, H.K. Winter wheat biomass estimation based on spectral indices, band depth analysis and partial least squares regression using hyperspectral measurements. *Comput. Electron. Agric.* **2014**, *100*, 51–59. [CrossRef]
8. Gnyp, M.L.; Miao, Y.X.; Yuan, F.; Ustin, S.L.; Yu, K.; Yao, Y.K.; Huang, S.Y.; Bareth, G. Hyperspectral canopy sensing of paddy rice aboveground biomass at different growth stages. *Field Crop. Res.* **2014**, *155*, 42–55. [CrossRef]
9. Thenkabail, P.S.; Smith, R.B.; Pauw, E.D. Hyperspectral vegetation indices and their relationships with agricultural crop characteristics. *Remote Sens. Environ.* **2000**, *71*, 158–182. [CrossRef]
10. Angela, K.; Heather, M.; David, L.; Mark, S.; Catherine, C. Assessment of RapidEye vegetation indices for estimation of leaf area index and biomass in corn and soybean crops. *Int. J. Appl. Earth Obs.* **2015**, *34*, 235–248.
11. Asari, N.; Suratman, M.N.; Jaafar, J. Modelling and mapping of above ground biomass (AGB) of oil palm plantations in Malaysia using remotely-sensed data. *Int. J. Remote Sens.* **2017**, *38*, 4741–4764. [CrossRef]
12. Melian, J.M.; Jimenez, A.; Diaz, M.; Morales, A.; Horstrand, P.; Guerra, R.; Lopez, S. Real-time hyperspectral data transmission for UAV-based acquisition platform. *Remote Sens.* **2021**, *13*, 850. [CrossRef]
13. Yu, N.; Li, L.; Schmitz, N.; Tian, L.F.; Greenberg, J.A.; Diers, B.W. Development of methods to improve soybean yield estimation and predict plant maturity with an unmanned aerial vehicle based platform. *Remote Sens. Environ.* **2016**, *187*, 91–101. [CrossRef]
14. Lydia, S.; Iolanda, F.; Josep, P. Remote sensing of biomass and yield of winter wheat under different nitrogen supplies. *Crop. Sci.* **2000**, *40*, 723–731.
15. David, A.J.; Hernan, D.B.; Jocelyn, C. Graph-based data fusion applied to: Change detection and biomass estimation in rice crops. *Remote Sens.* **2020**, *12*, 2683–2705.
16. Han, L.; Yang, G.; Dai, H.Y.; Xu, B.; Yang, H.; Feng, H.K.; Li, Z.H.; Yang, X.D. Modeling maize above-ground biomass based on machine learning approaches using UAV remote-sensing data. *Plant. Methods* **2019**, *15*, 10–24. [CrossRef]

17. Ma, B.L.; Dwyer, L.M.; Costa, C.; Cober, E.R.; Morrison, M.J. Early prediction of soybean yield from canopy reflectance measurements. *Agron. J.* **2001**, *93*, 1227–1234. [CrossRef]
18. Jan, C.; Lammert, K.; Marnix, V.B. Using sentinel-2 data for retrieving LAI and Leaf and canopy chlorophyll content of a potato crop. *Remote Sens.* **2017**, *9*, 405–427.
19. Gitelson, A.A.; Vian, A.; Arkebauer, T.J.; Rundquist, D.C.; Keydan, G.; Leavitt, B. Remote estimation of leaf area index and green leaf biomass in maize canopies. *Geophys. Res. Lett.* **2003**, *30*, 1248–1270. [CrossRef]
20. Kanemasu, E.T. Seasonal canopy reflectance patterns of wheat, sorghum, and soybean. *Remote Sens. Environ.* **1974**, *3*, 43–47. [CrossRef]
21. Kooistra, L.; Clevers, J.W. Estimating potato leaf chlorophyll content using ratio vegetation indices. *Remote Sens. Lett.* **2016**, *7*, 611–620. [CrossRef]
22. Daughtry, C.T.; Walthall, C.L.; Kim, M.S.; Colstoun, E.B.; Mcmurtrey, J.E. Estimating corn leaf chlorophyll concentration from leaf and canopy reflectance. *Remote Sens. Environ.* **2000**, *74*, 229–239. [CrossRef]
23. Banerjee, B.P.; Spangenberg, G.; Kant, S. Fusion of spectral and structural information from aerial images for improved biomass estimation. *Remote Sens.* **2020**, *12*, 3164. [CrossRef]
24. Monica, H.; Pablo, R.G.; Katy, M.R. Yield prediction by machine learning from UAS-based multi-sensor data fusion in soybean. *Plant. Methods* **2020**, *16*, 78–91.
25. Guo, A.T.; Huang, W.J.; Dong, Y.Y.; Ye, H.C.; Ma, H.Q.; Liu, B. Wheat yellow rust detection using UAV-based hyperspectral technology. *Remote Sens.* **2021**, *13*, 123. [CrossRef]
26. Eweys, O.A.; Elwan, A.A.; Borham, T.I. Integrating WOFOST and Noah LSM for modeling maize production and soil moisture with sensitivity analysis, in the east of The Netherlands. *Field Crop. Res.* **2017**, *210*, 147–161. [CrossRef]
27. Zhou, G.X.; Liu, X.N.; Liu, M. Assimilating remote sensing phenological information into the WOFOST model for rice growth simulation. *Remote Sens.* **2019**, *11*, 268. [CrossRef]
28. Onisimo, M.; Elhadi, A.; Moses, A.C. High density biomass estimation for wetland vegetation using WorldView-2 imagery and random forest regression algorithm. *Int. J. Appl. Earth Obs.* **2012**, *18*, 399–406.
29. Hansen, P.M.; Schjoerring, J.K. Reflectance measurement of canopy biomass and nitrogen status in wheat crops using normalized difference vegetation indices and partial least squares regression. *Remote Sens. Environ.* **2003**, *86*, 542–553. [CrossRef]
30. Behmann, J.; Mahlein, A.K.; Rumpf, T.; Romer, C.; Plumer, L. A review of advanced machine learning methods for the detection of biotic stress in precision crop protection. *Precis. Agric.* **2015**, *16*, 239–260. [CrossRef]
31. Andres, V.; Gitelson, A.A.; Nguy-Robertson, A.L.; Peng, Y. Comparison of different vegetation indices for the remote assessment of green leaf area index of crops. *Remote Sens. Environ.* **2011**, *115*, 3468–3478.
32. Yue, J.B.; Yang, G.J.; Tian, Q.J.; Feng, H.K.; Xu, K.J.; Zhou, C.Q. Estimate of winter-wheat above-ground biomass based on UAV ultrahigh-ground-resolution image textures and vegetation indices. *ISPRS J. Photogramm.* **2019**, *150*, 226–244. [CrossRef]
33. Jia, M.; Li, W.; Wang, K.; Zhou, C.; Cheng, T.; Tian, Y.; Zhu, Y.; Cao, W. A newly developed method to extract the optimal hyperspectral feature for monitoring leaf biomass in wheat. *Comput. Electron. Agric.* **2019**, *165*, 104942. [CrossRef]
34. Jin, X.L.; Yang, G.J.; Xu, X.G.; Yang, H.; Feng, H.K.; Li, Z.K.; Zhao, C.J. Combined multi-temporal optical and radar parameters for estimating LAI and biomass in winter wheat Using HJ and RADARSAR-2 Data. *Remote Sens.* **2015**, *7*, 13251. [CrossRef]
35. Liu, Y.; Feng, H.K.; Yue, J.B.; Li, Z.H.; Yang, G.J.; Song, X.Y.; Yang, X.D. Remote-sensing estimation of potato above-ground biomass based on spectral and spatial features extracted from high-definition digital camera images. *Comput. Electron. Agric.* **2022**, *198*, 107089. [CrossRef]
36. Tao, H.L.; Feng, H.K.; Xu, L.J.; Miao, M.K.; Long, H.L.; Yue, J.B. Estimation of crop growth parameters using UAV-based hyperspectral remote sensing data. *Sensors* **2020**, *20*, 1296. [CrossRef] [PubMed]
37. Yue, J.B.; Feng, H.K.; Yang, G.J.; Li, Z.H. A Comparison of regression techniques for estimation of above-ground winter wheat biomass using near-surface spectroscopy. *Remote Sens.* **2018**, *10*, 66. [CrossRef]
38. Pugh, N.A.; Horne, D.W.; Murray, S.C.; Carvalho, G.; Malambo, L.; Jung, J.H. A Temporal estimates of crop growth in sorghum and maize breeding enabled by unmanned aerial systems. *Plant. Phenome J.* **2017**, *28*, 170006. [CrossRef]
39. Varela, S.; Assefa, Y.; Prasad, P.V.; Peralta, N.R.; Griffin, T.W.; Sharda, A.; Ferguson, A. Spatio-temporal evaluation of plant height in corn via unmanned aerial systems. *J. Appl. Remote Sens.* **2017**, *11*, 12–32. [CrossRef]
40. Muharam, F.M.; Bronson, K.F.; Maas, S.J.; Ritchie, G.L. Inter-relationships of cotton plant height, canopy width, ground cover and plant nitrogen status indicators. *Field Crop. Res.* **2014**, *169*, 58–69. [CrossRef]
41. Fenner, H.; Andrew, R.; Adam, M.; Wooster, M.J.; Hawkesford, M.J. High throughput field phenotyping of wheat plant height and growth rate in field plot trials using UAV based remote sensing. *Remote Sens.* **2016**, *8*, 1031.
42. Zheng, B.H.; Zhang, L.; Xie, D.; Yin, X.L.; Liu, C.J.; Liu, G. Application of synthetic NDVI time series blended from Landsat and MODIS data for grassland biomass estimation. *Remote Sens.* **2016**, *8*, 10. [CrossRef]
43. Wang, L.A.; Zhou, X.D.; Zhu, X.K.; Dong, Z.D.; Guo, W.S. Estimation of biomass in wheat using random forest regression algorithm and remote sensing data. *Crop. J.* **2016**, *4*, 212–219. [CrossRef]
44. Fu, Y.Y.; Yang, G.J.; Li, Z.H.; Song, X.Y.; Li, Z.H.; Xu, X.G.; Wang, P.; Zhao, C.J. Winter wheat nitrogen status estimation using UAVbased RGB imagery and gaussian processes regression. *Remote Sens.* **2020**, *12*, 3778. [CrossRef]
45. Verrelst, J.; Alonso, L.; Camps-Valls, G.; Delegido, J.; Moreno, J. Retrieval of vegetation biophysical parameters using gaussian process techniques. *IEEE Trans. Geosci. Remote. Sens.* **2012**, *50*, 1832–1843. [CrossRef]

46. Rondeaux, G.; Steven, M.; Baret, F. Optimization of soil-adjusted vegetation indices. *Remote Sens. Environ.* **1996**, *55*, 95–107. [CrossRef]

47. Tao, H.L.; Feng, H.K.; Xu, L.J.; Miao, M.K.; Yang, G.J. Estimation of the yield and plant height of winter wheat using uav-based hyperspectral images. *Sensors* **2020**, *20*, 1231. [CrossRef]

48. Zheng, H.; Cheng, T.; Zhou, M.; Li, D.; Yao, X. Improved estimation of rice aboveground biomass combining textural and spectral analysis of UAV imagery. *Precis. Agric.* **2018**, *20*, 611–629. [CrossRef]

49. Wu, C.; Niu, Z.; Tang, Q.; Huang, W. Estimating chlorophyll content from hyperspectral vegetation indices: Modeling and validation. *Agric. For. Meteorol.* **2008**, *148*, 1230–1241. [CrossRef]

50. Li, B.; Xu, X.; Zhang, L.; Han, J.; Bian, C. Above-ground biomass estimation and yield prediction in potato by using UAV-based RGB and hyperspectral imaging. *ISPRS J. Photogramm.* **2020**, *162*, 161–172. [CrossRef]

51. Broge, N.H.; Leblanc, E. Comparing prediction power and stability of broadband and hyperspectral vegetation indices for estimation of green leaf area index and canopy chlorophyll density. *Remote Sens. Environ.* **2001**, *76*, 156–172. [CrossRef]

52. Gitelson, A.A.; Kaufman, Y.J.; Stark, R.; Rundquist, D. Novel algorithms for remote estimation of vegetation fraction. *Remote Sens. Environ.* **2002**, *80*, 76–87. [CrossRef]

53. Prasad, A.M.; Iverson, L.R.; Liaw, A. Newer classification and regression tree techniques: Bagging and random forests for ecological prediction. *Ecosystems* **2006**, *9*, 181–199. [CrossRef]

54. Gleason, C.J.; Junho, I. Forest biomass estimation from airborne LiDAR data using machine learning approaches. *Remote Sens. Environ.* **2012**, *125*, 80–91. [CrossRef]

55. Verrelst, J.; Rivera, J.P.; Gitelson, A.; Delegido, J.; Moreno, J.; Camps-Valls, G. Spectral band selection for vegetation properties retrieval using Gaussian processes regression. *Int. J. Appl. Earth Obs.* **2016**, *52*, 554–567. [CrossRef]

56. Bendig, J.; Bolten, A.; Bennertz, S.; Broscheit, J.; Bareth, G. Estimating biomass of barley using crop surface models (CSMs) derived from UAV-based RGB imaging. *Remote Sens.* **2014**, *6*, 10395. [CrossRef]

57. Brocks, S.; Bareth, G. Estimating barley biomass with crop surface models from oblique RGB imagery. *Remote Sens.* **2018**, *10*, 268. [CrossRef]

58. Souza, C.D.; Lamparelli, R.C.; Rocha, J.V. Height estimation of sugarcane using an unmanned aerial system (UAS) based on structure from motion (SfM) point clouds. *Int. J. Remote Sens.* **2017**, *38*, 2218–2273. [CrossRef]

59. Ballesteros, R.; Ortega, J.F.; Hernandez, D.; Moreno, M.A. Onion biomass monitoring using UAV-based RGB imaging. *Precis. Agric.* **2018**, *19*, 840–857. [CrossRef]

60. Liu, Y.; Feng, H.K.; Huang, J.; Sun, Q.; Yang, F.Q. Estimation of potato biomass based on UAV digital images. *Trans. Chin. Soc. Agric. Eng.* **2020**, *36*, 182–192.

61. Zarco, P.J.; Diza, R.; Angileri, V.; Loudjani, P. Tree height quantification using very high resolution imagery acquired from an unmanned aerial vehicle (UAV) and automatic 3D photo-reconstruction methods. *Eur. J. Agron.* **2014**, *55*, 89–99. [CrossRef]

62. Bendig, J.; Bolten, A.; Bareth, G. UAV-based imaging for multi-temporal, very high resolution crop surface models to monitor crop growth variability. *Photogramm. Fernerkund. Geoinf.* **2013**, *6*, 551–562. [CrossRef]

63. Fan, L.L.; Zhao, J.L.; Xu, X.G.; Liang, D.; Yang, G.J.; Feng, H.K. Hyperspectral-based estimation of leaf nitrogen content in corn using optimal selection of multiple spectral variables. *Sensors* **2019**, *19*, 2898. [CrossRef] [PubMed]

64. Masjedi, A.; Crawford, M.M.; Carpenter, N.R.; Tuinstra, M.R. Multi-Temporal predictive modelling of sorghum biomass using UAV-based hyperspectral and LiDAR data. *Remote Sens.* **2020**, *12*, 3587. [CrossRef]

65. Feng, W.; Guo, B.B.; Zhang, H.Y.; He, L.; Zhang, Y.S.; Wang, Y.H.; Zhu, Y.J.; Guo, T.C. Remote estimation of above ground nitrogen uptake during vegetative growth in winter wheat using hyperspectral red-edge ratio data. *Field Crop. Res.* **2015**, *180*, 197–206. [CrossRef]

66. Oppelt, N.; Mauser, W. Hyperspectral monitoring of physiological parameters of wheat during a vegetation period using AVIS data. *Int. J. Remote Sens.* **2004**, *25*, 145–159. [CrossRef]

Article

Estimation of Potato Above-Ground Biomass Based on Vegetation Indices and Green-Edge Parameters Obtained from UAVs

Yang Liu [1,2,3,†], Haikuan Feng [1,4,*,†], Jibo Yue [5], Yiguang Fan [1], Xiuliang Jin [6], Xiaoyu Song [1], Hao Yang [1] and Guijun Yang [1]

[1] Key Laboratory of Quantitative Remote Sensing in Agriculture of Ministry of Agriculture and Rural Affairs, Information Technology Research Center, Beijing Academy of Agriculture and Forestry Sciences, Beijing 100097, China

[2] Key Lab of Smart Agriculture System, Ministry of Education, China Agricultural University, Beijing 100083, China

[3] Key Laboratory of Agricultural Information Acquisition Technology, Ministry of Agriculture and Rural Affairs, China Agricultural University, Beijing 100083, China

[4] College of Agriculture, Nanjing Agricultural University, Nanjing 210095, China

[5] College of Information and Management Science, Henan Agricultural University, Zhengzhou 450002, China

[6] Institute of Crop Sciences, Chinese Academy of Agricultural Sciences/Key Laboratory of Crop Physiology and Ecology, Ministry of Agriculture, Beijing 100081, China

[*] Correspondence: fenghk@nercita.org.cn

[†] These authors contributed equally to this work.

Abstract: Aboveground biomass (AGB) is an important indicator to evaluate crop growth, which is closely related to yield and plays an important role in guiding fine agricultural management. Compared with traditional AGB measurements, unmanned aerial vehicle (UAV) hyperspectral remote sensing technology has the advantages of being non-destructive, highly mobile, and highly efficient in precision agriculture. Therefore, this study uses a hyperspectral sensor carried by a UAV to obtain hyperspectral images of potatoes in stages of tuber formation, tuber growth, starch storage, and maturity. Linear regression, partial least squares regression (PLSR), and random forest (RF) based on vegetation indices (Vis), green-edge parameters (GEPs), and combinations thereof are used to evaluate the accuracy of potato AGB estimates in the four growth stages. The results show that (i) the selected VIs and optimal GEPs correlate significantly with AGB. Overall, VIs correlate more strongly with AGB than do GEPs. (ii) AGB estimates made by linear regression based on the optimal VIs, optimal GEPs, and combinations thereof gradually improve in going from the tuber-formation to the tuber-growth stage and then gradually worsen in going from the starch-storage to the maturity stage. Combining the optimal GEPs with the optimal VIs produces the best estimates, followed by using the optimal VIs alone, and using the optimal GEPs produces the worst estimates. (iii) Compared with the single-parameter model, which uses the PLSR and RF methods based on VIs, the combination of VIs with the optimal GEPs significantly improves the estimation accuracy, which gradually improves in going from the tuber-formation to the tuber-growth stage, and then gradually deteriorates in going from the starch-storage to the maturity stage. The combination of VIs with the optimal GEPs produces the most accurate estimates. (iv) The PLSR method is better than the RF method for estimating AGB in each growth period. Therefore, combining the optimal GEPs and VIs and using the PLSR method improves the accuracy of AGB estimates, thereby allowing for non-destructive dynamic monitoring of potato growth.

Keywords: potato; above-ground biomass; vegetation indices; green-edge parameters; partial least squares regression; random forest

Citation: Liu, Y.; Feng, H.; Yue, J.; Fan, Y.; Jin, X.; Song, X.; Yang, H.; Yang, G. Estimation of Potato Above-Ground Biomass Based on Vegetation Indices and Green-Edge Parameters Obtained from UAVs. *Remote Sens.* **2022**, *14*, 5323. https://doi.org/10.3390/rs14215323

Academic Editor: Yingying Dong

Received: 25 September 2022
Accepted: 22 October 2022
Published: 24 October 2022

Publisher's Note: MDPI stays neutral with regard to jurisdictional claims in published maps and institutional affiliations.

1. Introduction

As the basis of organic matter accumulation, aboveground biomass (AGB) is often used to evaluate crop growth and farmland productivity [1,2]. As a result, knowing the spatial-temporal dynamic variations in AGB is critical for field decision-making and yield prediction [3]. Fast and accurate access to AGB information not only helps to tap into production potential and boost crop yield [4,5] but also helps to formulate macro agricultural trade policies and implement digital agriculture [6]. AGB measurement is traditionally done mostly by manual destructive sampling, which requires significant time and effort, is only applicable to small areas, and is difficult to scale up [7,8].

Because of its high throughput, rapidity, and wide sensing range, remote-sensing technology offers a new technical method for nondestructive dynamic monitoring of crop growth over large areas. It can establish a relationship between the acquired crop canopy spectral information and growth parameters, which allows for the analysis of periodic quantitative variations in crop parameters, which has become a hot spot in agricultural quantitative remote-sensing research [9,10]. In recent years, remote-sensing platforms based on the ultralow-altitude unmanned aerial vehicles (UAVs) carrying various microsensors have been widely implemented in precision agricultural management because of their high mobility, low cost, high operational efficiency, and capacity to acquire crop canopy images with higher temporal and spatial resolution beneath cloud cover [11–13]. Currently, hyperspectral imaging spectrometers carried by UAVs can capture copious detailed spectral information of crop canopies due to their vast number of wavebands, allowing them to better monitor crop-growth metrics [14,15].

The crop growth model, the semi-mechanical model, and the empirical statistical model are the three main models for estimating AGB [16–18]. Crop-growth models, such as WOFOST [19], CROPSYST [20], and CERES [21], are extremely mechanistic. However, these models require as input a large number of parameters that affect crop growth, such as soil moisture, meteorological data, physiological and biochemical parameters, and other indicators, which are often difficult to obtain, limiting the wide application of the model in crop AGB monitoring [22,23]. The semi-mechanical model (also known as the parametric model) involves the light-energy utilization model [24]. Variations in AGB are obtained by simulating crop growth based on photosynthesis in the canopy leaves, but this model is affected by geographical restrictions and some parameters are difficult to simulate quantitatively [25]. The empirical statistical model directly establishes the regression relationship between remote sensing factors and crop parameters. Because of its simple form, easy operation, and high efficiency, it is widely applicable in crop growth monitoring. For example, Yuri et al. [26] and Mansaray et al. [27] used vegetation indices (VIs) to estimate the AGB of sugarcane and rice.

VIs are usually combined mathematically with two or more spectral bands to reduce or eliminate the impact of background noise on crop canopy spectral information, thereby enhancing estimates of vegetation structure [28,29]. For example, Zhang et al. [30], Devia et al. [31], and Han et al. [32] estimated the AGB of winter wheat, maize, and rice at different growth stages using five, six, and seven vegetation indices, respectively. Although these studies use UAV remote-sensing technology to obtain accurate AGB estimates, there are some shortcomings. The model easily saturates when only the original spectral information of the crop canopy is used to build the VIs to estimate AGB in multiple growth periods of crops [33,34]. To solve this problem, researchers use three main methods to improve the accuracy of AGB estimates: (i) Various regression techniques are used, such as multiple linear regression [35], partial least squares regression (PLSR) [36], random forests (RF) [28], support vector machines [37], and artificial neural networks [38]. (ii) Spectral transformation techniques are used, such as spectral differentiation [39], wavelet analysis [40], band-depth analysis [41], and Fourier transforms [42]. (iii) Finally, to improve AGB estimates, new model factors are introduced such as crop height [43,44], texture features [45–47], crop coverage [48,49], and red-edge parameters [15].

Previous research has thus used UAV remote-sensing technology to estimate the growth parameters of wheat, rice, corn, soybean, and other crops in different phases. However, the structure and dry-matter accumulation process of potatoes differs significantly from that of other common crops, so additional research is required to accurately estimate potato AGB. Multiband crop canopy spectral information can be obtained through hyperspectral sensors with many narrow bands carried by UAVs, and the sensitivity of reflectance information and growth parameters varies across the different spectral bands. Therefore, AGB can be estimated by extracting characteristic spectral parameters. Due to the influence of the characteristics and the internal structure of crop leaves, the spectral reflectance of the crop canopy slowly increases and peaks within the 502–554 nm green-edge region, which differs significantly from the background features. Therefore, the sensitivity of characteristic spectral parameters in this area should be thoroughly examined to improve the accuracy of growth-parameter estimates.

No research has yet been conducted to extract the green-edge parameters (GEPs) from potato canopy hyperspectral images, nor to combine them with traditional VIs and apply regression techniques to evaluate the accuracy of AGB estimates for multiple growth periods, as is required for monitoring potato growth. The main purpose of this study is thus to evaluate the applicability of VIs, GEPs, and combinations thereof to estimate potato AGB over multiple growth stages. Specifically, we evaluate the use of PLSR and RF techniques to estimate AGB based on VIs, GEPs, and combinations thereof and determine the best model for AGB estimation in each growth period.

2. Experiment and Methods

2.1. Overview of Study Area and Experimental Design

From March to July 2019, experimentation was carried out in the experimental potato field of the National Precision Agriculture Research Demonstration Base in Xiaotangshan Town, Changping District, Beijing, China. This area is located in the east of Xiaotangshan Town (40°10′34″N, 116°26′39″E), with an average altitude of about 36 m, an average annual precipitation of about 600 mm, an average annual temperature of about 11 °C, and an average annual frost-free period of 200 days. The climate of the study area is warm temperate semi-humid continental monsoon. The location of the test area is shown in Figure 1.

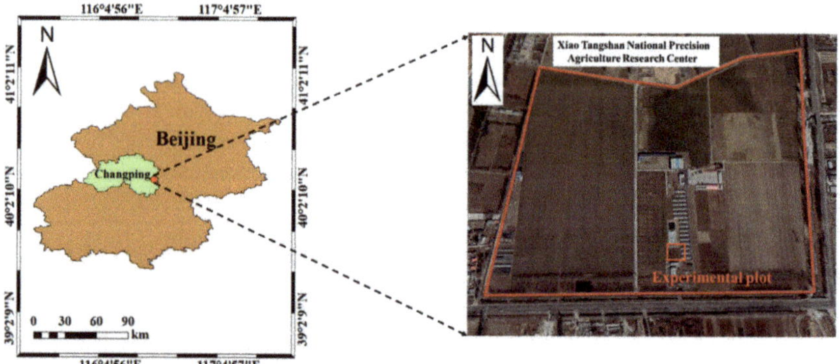

Figure 1. Location of experimental area.

Two early maturing potato varieties, Zhongshu 5 (Z5) and Zhongshu 3 (Z3), were planted in the experimental field at different planting densities and with different treatments of nitrogen and potassium fertilizers. The planting density test area had 18 plots, which were planted at three densities: 60,000 plants/hm^2 (T1), 72,000 plants/hm^2 (T2), and 84,000 plants/hm^2 (T3). There were 24 plots in the nitrogen test area, which underwent four fertilization treatments: 0 kg/hm^2 urea (N0), 244.65 kg/hm^2 urea (N1), 489.15 kg/hm^2 urea

(N2, normal treatment, 15 kg pure nitrogen), and 733.5 kg/hm^2 urea (N3). There were six plots in the potassium fertilizer test area, which underwent two potassium fertilization treatments: 0 kg/hm^2 potassium fertilizer (K0) and 1941 kg/hm^2 potassium fertilizer (K2). A treatment of 970.5 kg/hm^2 potassium fertilizer (K1) was applied to the planting density and nitrogen test area. There was a total of 48 test plots, each with an area of 32.5 m^2. Eleven ground control points were uniformly arranged around the test area to calibrate the hyperspectral images acquired by UAVs. The specific test scheme appears in Figure 2.

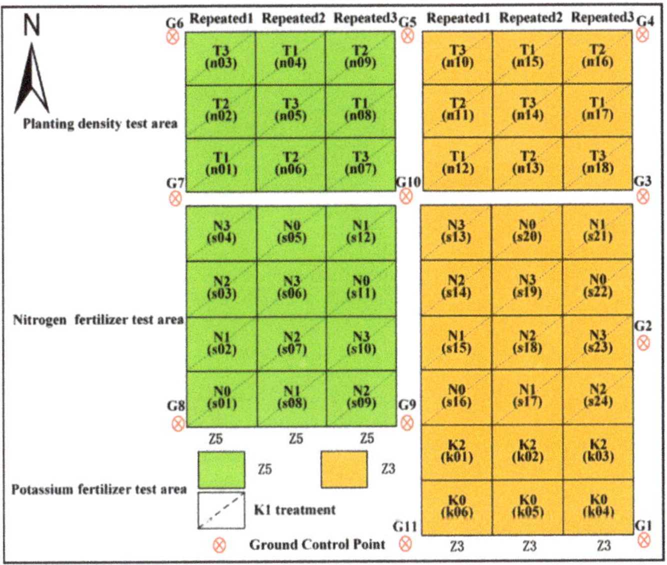

Figure 2. Experimental design.

2.2. Ground Data Acquisition and Processing

AGB data were obtained during the four key stages of potato growth: tuber formation (28 May 2019), tuber growth (10 June 2019), starch storage (20 June 2019), and maturity (3 July 2019). The specific process of AGB acquisition in each growth period was to select three plant samples in each plot that represent the growth level of the entire plot. After artificial destructive sampling, they were packaged and then quickly taken to the laboratory. After separating the stems and leaves, they were washed in running water, then placed in an oven for one hour at 105 °C, then at 80 °C to dry for over 48 h until obtaining constant mass. The dry mass of stems and leaves was weighed by using a high-precision balance (accuracy of 0.001 g). The dry mass of the stems was added to that of the leaves to obtain the dry mass of the sample. Finally, the potato AGB in each plot was obtained based on the population density and the dry mass of the sample.

2.3. UAV Hyperspectral Data Acquisition and Processing

UAV hyperspectral remote-sensing operation and ground data acquisition were carried out simultaneously. A six-rotor electric UAV equipped with a UHD185 Firefly imaging spectrometer was used to obtain canopy hyperspectral data during the four potato growth periods (Figure 3). The UHD185 sensor is a non-scanning, full frame, real-time-imaging airborne spectrometer that measures 195 mm × 67 mm × 60 mm, weighs 470 g, and detects over a spectral range of 450–950 nm with 125 spectral channels, 4 nm per spectral sampling interval, 8 nm spectral resolution, and 12-bit digital resolution. All UAV flights were conducted under windless and cloudless conditions from 11:30 to 12:30 because the sun shines directly on the ground at this time and the light intensity is stable. The UAV flew at 20 m (the transect width was 6 m) and had a speed of about 1.5 m/s. To obtain high-quality hyperspectral images, the UAV's heading and lateral overlap rates were 85% and 93%, respectively. The image capture interval was set to one second. The ground resolution of the resulting images was about 1.3 cm. Before each UAV took off, a black and white panel was used to radiometrically correct the canopy spectral reflectance.

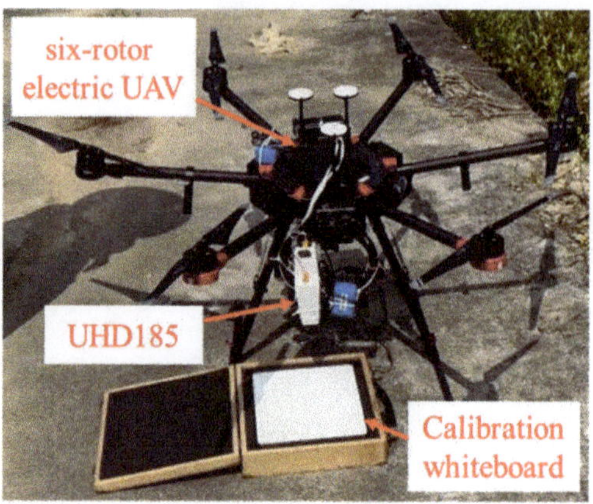

Figure 3. Components of the UHD 185 hyperspectral system.

UAV hyperspectral image data processing mainly includes four parts: geometric correction, image stitching, image fusion, and spectral extraction. Before stitching, we radiometrically calibrated the image digital values to surface reflectance. First, the selected grayscale images and 11 ground control points in the flight zone were imported into the PhotoScan software, and the three-dimensional coordinate information of the ground control points was used to correct the images for the terrain. The correction error in each period was less than 2 cm. Second, the grayscale images were stitched through the dense point cloud generated by the motion structure algorithm. Hyperspectral cube and grayscale images were then fused by using Cubert-Pilot software to obtain complete hyperspectral images. Finally, vector data for each plot were plotted based on the maximum boundaries by using ArcGIS 10.2 software, and ENVI 5.3 software was used to extract the average hyperspectral reflectance from the region of interest (Figure 4). To evaluate the performance of different spectra to estimate AGB, we analyzed the correlation between potato canopy spectra and AGB (Figure 5). Finally, we found that the spectral regions sensitive to AGB were mainly located in the red edge and near-infrared range, which provided a basis for a reasonable selection of VIs.

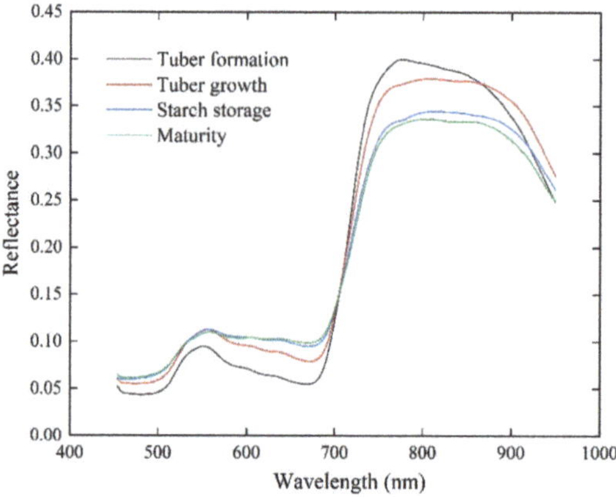

Figure 4. Canopy spectral reflectance of potato in four key growth stages.

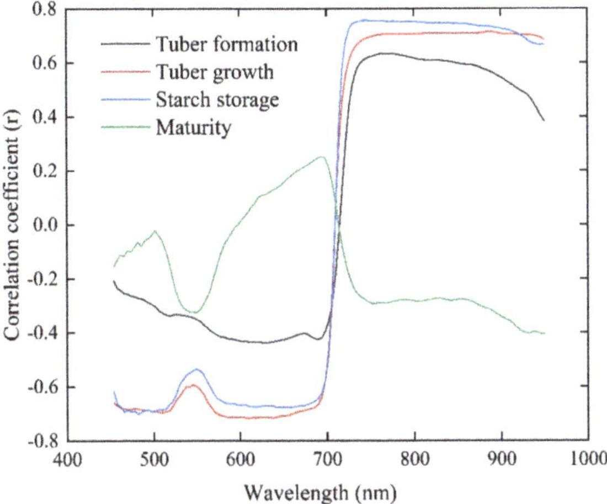

Figure 5. Correlation coefficient between potato canopy spectra and AGB.

2.4. Selection of Vegetation Indices

According to previous research results, fifteen VIs were selected for this study. The specific names and mathematical expressions are shown in Table 1.

Table 1. VIs used in this study.

Vegetation Indices	Definition	References
MSR (modified simple ratio index)	$(R800/R670 - 1)/(R800/R670 + 1)^{1/2}$	[15]
MSAVI (modified soil adjusted vegetation index)	$0.5 \times [2 \times R800 + 1 - ((2 \times R800 + 1)^2 - 8 \times (R800 - R670))^{1/2}]$	[34]
OSAVI (optimizing soil adjusted vegetation index)	$1.16 \times (R800 - R670)/(R800 + R670 + 0.16)$	[15]
PSRI (plant senescence reflectance index)	$(R680 - R500)/R750$	[46]
RDVI (renormalized difference vegetation index)	$(R800 - R670)/(R800 + R670)^{1/2}$	[15]
TCARI (transformed chlorophyll absorption ratio index)	$3 \times [(R710 - R680) - 0.2 \times (R700 - R560)(R710/R680)]$	[15]
TVI (triangular vegetation index)	$0.5 \times [120 \times (R750\text{-}R550) - 200 \times (R670 - R550)]$	[15]
SPVI (spectral polygon vegetation index)	$0.4 \times [3.7 \times (R800\text{-}R670) - 1.2 \times |R550 - R670|]$	[34]
DVI (difference vegetation index)	$R890\text{-}R670$	[15]
MCARI (modified chlorophyll absorption ratio index)	$((R700 - R670) - 0.2 \times (R700\text{-}R550))(R700/R670)$	[34]
NDVI (normalized difference vegetation index)	$(R800 - R680)/(R800 + R680)$	[29]
PBI (plant biochemical index)	$R810/R560$	[38]
LCI (linear combination index)	$(R850 - R710)/(R850 + R670)^{1/2}$	[38]
SRI (simple ratio vegetation index)	$R800/R680$	[45]
SAVI (soil adjusted vegetation index)	$(1 + 0.5) \times (R800 - R670)/(R800 + R670 + 0.5)$	[45]

2.5. Selection of Green-Edge Parameters

This work uses the maximum reflectivity of the green edge (Rmax), the sum of the reflectance of the green edge (Rsum), the area of the green edge (SDr), the amplitude of the green edge (Dr), the minimum amplitude (Drmin), and the ratio Dr/Drmin. Rmax and Rsum are the maximum reflectance and total reflectance of the original spectrum of the green edge area, respectively. SDr is the sum of the first-order differential spectral reflectance of the green edge area, and Dr is the maximum of the first-order differential spectrum of the green-edge region. Drmin is the minimum of the first-order differential spectrum of the green-edge region.

2.6. Analysis Methods

PLSR is a combination of multiple linear regression and principal component analysis. It can provide a many-to-many regression model on the basis of fully considering the multicollinearity between independent variables. Especially when the number of independent variables is large, and there is a certain degree of autocorrelation with a small number of dependent variables, the model built by PLSR is more accurate than the traditional regression analysis, which allows us to make estimates based on a small number of factors. RF was first proposed by Breiman as a machine-learning algorithm [15]. Through bootstrap sampling, some of the original samples are recycled for multiple sampling to form a training set, then multiple decision trees are combined to predict the dependent variables, and the prediction results are determined by voting. In this study, PLSR and RF were implemented by using MATLAB2018a software.

To estimate the potato AGB for each growth stage, this study used repeat 2 and repeat 3 data (32 groups) as the calibration set, and repeat 1 data (16 groups) as the validation set. The coefficient of determination (R^2), root mean square error (RMSE), and normalized root mean square error (NRMSE) were used to evaluate the accuracy of the model.

3. Results and Analysis

3.1. Correlation between VIs, GEPs, and AGB

Table 2 shows the correlation coefficients between (i) fifteen VIs and six GEPs and (ii) the AGB for each potato growth period. According to Table 2, most of the correlations between twenty-one spectral parameters and AGB reach a significance level of 0.01. The correlation coefficient between spectral parameters and AGB in each growth stage fluctuates, and the difference is obvious. The correlation increases in going from the tuber-formation stage to the tuber-growth stage and decreases in going from the starch-storage stage to the mature stage. Overall, the VIs are more strongly correlated with AGB than are the GEPs.

Table 2. Correlation coefficients between (i) VIs and GEPs and (ii) AGB for the four potato growth stages.

Spectral Parameters		AGB			
		Tuber Form Stage	**Tuber Growth Stage**	**Starch Store Stage**	**Maturity Stage**
VIs	MSR	0.636	0.733	0.724	−0.237
	MSAVI	0.651	0.748	0.729	−0.266
	OSAVI	0.655	0.742	0.717	−0.291
	PSRI	−0.727	−0.682	−0.696	0.396
	RDVI	0.642	0.746	0.731	−0.270
	TCARI	0.410	0.686	0.641	−0.304
	TVI	0.659	0.741	0.714	−0.314
	SPVI	0.673	0.756	0.728	−0.256
	DVI	0.650	0.742	0.733	−0.302
	MCARI	0.220	0.676	0.581	−0.197
	NDVI	0.580	0.733	0.729	−0.255
	PBI	0.654	0.757	0.745	−0.047
	LCI	0.661	0.725	0.751	−0.284
	SRI	0.662	0.726	0.715	−0.238
	SAVI	0.639	0.746	0.731	−0.270
GEPs	Rmax	−0.367	−0.613	−0.546	−0.311
	Rsum	−0.342	−0.668	−0.644	−0.236
	SDr	0.023	0.471	0.462	−0.447
	Dr	−0.284	0.321	0.401	−0.415
	Drmin	−0.545	−0.509	−0.382	0.196
	Dr/Drmin	0.122	0.581	0.491	−0.260

Note: A correlation coefficient with an absolute value greater than 0.361 indicates a significance level of 0.01.

During tuber formation, the most highly correlated VIs and GEPs are PSRI and Drmin, respectively. In the tuber-growth stage, VIs and GEPs are more strongly correlated with AGB than in the other three growth stages, and the most strongly correlated VIs and GEPs are PBI and Rsum, respectively. During the starch-storage period, all spectral parameters and the AGB reach a significance level of 0.01, but the correlation weakens. The VIs and GEPs most strongly correlated with AGB are LCI and Rsum, respectively. During the mature stage, the correlation between spectral parameters and AGB is the weakest. Only six VIs (PSRI, TVI, TCARI, DVI, OSAVI, and LCI) and two GEPs (SDr and Dr) reach a significance level of 0.01 with AGB, with the best VIs and GEPs being PSRI and SDr, respectively.

3.2. Relationship of Optimal VIs and GEPs with AGB

3.2.1. Relationship between Optimal GEPs and AGB

Based on the optimal GEPs selected in each growth period (Table 2), the AGB estimation models for the four growth periods were established by using linear regression, and the R^2, RMSE, and NRMSE of the models in each growth period were obtained through modeling and validation (Table 3). From the tuber-formation stage to the tuber-growth stage, the modeling set R^2 increases from 0.31 to 0.37, the RMSE decreases from 325.37 to 332.83 kg/hm^2, and the NRMSE decreases from 25.24% to 22.94%. Similarly, the validation set R^2 increases continuously from 0.40 to 0.52, and the RMSE and NRMSE decreases continuously. The accuracy of the AGB estimation model gradually increases in these two growth periods. From the starch-storage stage to the mature stage, the R^2 range of the modeling set ranges from 0.23 to 0.21, and the RMSE and NRMSE range from 404.96 to 428.87 kg/hm^2 and 29.11% to 35.12%, respectively. The variation in the precision obtained with the validation set is consistent with that of the modeling set. R^2 continues to decrease (from 0.38 to 0.27), and RMSE and NRMSE continue to increase (from 259.51 to 279.39 kg/hm^2 and 28.57% to 32.39%, respectively). The precision obtained from the models for these two growth periods gradually decreases, and the estimates worsen.

Table 3. Linear regression between optimal GEPs and potato AGB for the four growth stages.

Growth Stages	Optimal GEPs	Modeling			Validation		
		R^2	RMSE (kg/hm^2)	NRMSE (%)	R^2	RMSE (kg/hm^2)	NRMSE (%)
Tuber form	Drmin	0.31	325.37	25.24	0.40	256.15	25.16
Tuber growth	Rsum	0.37	332.83	22.94	0.52	229.96	21.39
Starch store	Rsum	0.23	404.96	29.11	0.38	259.51	28.57
Maturity	SDr	0.21	428.87	35.12	0.27	279.39	32.39

3.2.2. Relationship between Optimal VIs and AGB

The model for estimating potato AGB at different growth stages was established by linear regression, and the R^2, RMSE, and NRMSE of each model are shown in Table 4. AGB estimates gradually improve in going from the tuber-formation to the tuber-growth stage. R^2 for the modeling set ranges from 0.43 to 0.48, and the RMSE and NRMSE range from 301.88 to 300.51 kg/hm^2 and 23.42% to 20.71%, respectively. R^2 for the validation set increases from 0.59 to 0.62, and the RMSE and NRMSE decrease gradually from 226.59 to 207.16 kg/hm^2 and 22.24% to 19.27%, respectively. In going from the starch-storage stage to the mature stage, AGB estimates gradually deteriorate. The modeling set R^2 decreases from 0.38 to 0.26, and the RMSE and NRMSE increase from 349.76 to 376.33 kg/hm^2 and 25.14% to 30.82%, respectively. The validation set R^2 also decreases continuously from 0.46 to 0.36, and the RMSE and NRMSE increase from 208.98 to 240.49 kg/hm^2 and from 23.01% to 27.88%, respectively. Analysis of the four growth-stage model accuracy-evaluation indicators (Tables 3 and 4) shows that the optimal-VIs estimate of potato AGB is more accurate than the optimal-GEPs estimate.

Table 4. Linear regression relationship between optimal VIs and AGB of potato for the four growth stages.

Growth Stages	Optimal VI	Modeling			Validation		
		R^2	RMSE (kg/hm^2)	NRMSE (%)	R^2	RMSE (kg/hm^2)	NRMSE (%)
Tuber form	PSRI	0.43	301.88	23.42	0.59	226.59	22.24
Tuber growth	PBI	0.48	300.51	20.71	0.62	207.16	19.27
Starch store	LCI	0.38	349.76	25.14	0.46	208.98	23.01
Maturity	PSRI	0.26	376.33	30.82	0.36	240.49	27.88

3.2.3. Relationship of Optimal VIs and Optimal GEPs with AGB

Table 5 shows the accuracy of AGB estimates made by using linear regression combined with the optimal VIs and the optimal GEPs for each growth period. The results show that the optimal VIs combined with the optimal GEPs significantly improves the accuracy of AGB estimates in each growth period (Tables 3–5). The increase during the tuber-formation stage and the tuber-growth stage is small: R^2 increases by 16% and 20% at the highest, and NRMSE decreases by 35% and 31% at the highest. During the starch accumulation and maturity stages, R^2 increases by 25% and 29% at the highest, and NRMSE decreases by 43% and 33% at the highest.

Table 5. Linear regression between (i) optimal VIs and optimal GEPs and (ii) potato AGB over the four growth stages.

Growth Stages	Optimal VIs, GEPs	Modeling			Validation		
		R^2	RMSE (kg/hm^2)	NRMSE (%)	R^2	RMSE (kg/hm^2)	NRMSE (%)
Tuber form	PSRI, Drmin	0.48	279.46	21.68	0.62	197.43	19.39
Tuber growth	PBI, Rsum	0.54	277.03	19.09	0.65	187.70	17.46
Starch store	LCI, Rsum	0.40	321.64	23.12	0.55	191.33	21.07
Maturity	PSRI, SDr	0.39	331.29	27.13	0.48	221.88	25.71

It can be seen from the comprehensive modeling set and verification set (Tables 3–5) that the variation in the accuracy of AGB estimates throughout the growth stage is consistent with the optimal VIs and the optimal GEPs. The estimation effect gradually improves from the tuber-formation stage to the tuber-growth stage. The modeling-set R^2 increases from 0.48 to 0.54, and the NRMSE decreases from 21.68% to 19.09%. R^2 for the verification set increases from 0.62 to 0.65, and the NRMSE decreases from 19.39% to 17.46%. The estimation effect thus worsens in going from the starch-storage stage to the maturity stage. R^2 for the modeling set ranges from 0.40 to 0.39, and the NRMSE ranges from 23.12% to 27.13%. The accuracy of AGB estimates based on the validation set is like that of the modeling set. R^2 continues to decrease, RMSE and NRMSE continue to increase, and the model accuracy worsens.

3.3. Estimation of AGB Using VIs, Optimal GEPs Combined with PLSR and RF Methods

To improve the accuracy and stability of the AGB estimation model, this study selects the top six VIs and optimal GEPs based on the correlation ranking in Table 2. For example, ranking correlation from best to worst in the tuber-formation stage gives PSRI, SPVI, SRI, LCI, TVI, OSAVI, and Drmin. Doing the same in the tuber-growth stage gives PBI, SPVI, MSAVI, RDVI, SAVI, OSAVI, and Rsum. Doing the same in the starch-storage stage gives LCI, PBI, DVI, RDVI, SAVI, MSAVI, and Rsum. Finally, doing the same in the mature stage gives PSRI, TVI, TCARI, DVI, OSAVI, LCI, and SDr. Based on the spectral parameters obtained at different growth stages, the PLSR and RF methods are used to estimate the potato AGB over the different growth stages.

Tables 6 and 7 show the VIs combined with the optimal GEPs using PLSR and RF regression to estimate the potato AGB over different growth stages, respectively. These results show that, based on the VIs, the PLSR and RF methods produce the most accurate potato AGB estimates for the tuber growth stage. For PLSR modeling, $R^2 = 0.72$, RMSE = 254.63 kg/hm^2, and NRMSE = 17.55%. For RF modeling, $R^2 = 0.69$, RMSE = 265.22 kg/hm^2, and NRMSE = 18.28%. However, the use of PLSR or RF regression techniques combined with VIs and optimal GEPs produce more accurate estimates of potato AGB at each growth stage, with the best estimate being for the tuber growth stage. The best R^2 for PLSR is 0.74 (RMSE = 210.55 kg/hm^2, NRMSE = 14.51%), and the best R^2 for RF is 0.71 (RMSE = 234.92 kg/hm^2, NRMSE = 16.19%). The accuracy of the AGB estimation model constructed by the PLSR or RF method based on VIs and VIs combined with the optimal GEPs as the variable first increase and then decrease with progressing growth periods. The estimates gradually improve in going from the tuber-formation stage to the tuber-growth stage and then deteriorate in going from the starch-storage stage to the maturity stage.

Table 6. PLSR-based estimates of potato AGB at different growth stages.

Growth Stages	Data	Modeling			Validation		
		R^2	RMSE (kg/hm^2)	NRMSE (%)	R^2	RMSE (kg/hm^2)	NRMSE (%)
Tuber form	VIs	0.65	275.74	21.39	0.71	187.02	18.37
	VIs, Drmin	0.68	244.01	18.93	0.75	169.93	16.68
Tuber growth	VIs	0.72	254.63	17.55	0.74	162.23	15.09
	VIs, Rsum	0.74	210.55	14.51	0.78	131.91	12.27
Starch store	VIs	0.62	312.03	22.43	0.69	190.38	20.96
	VIs, Rsum	0.66	268.65	19.31	0.72	171.37	18.87
Maturity	VIs	0.58	315.55	25.84	0.67	205.12	23.78
	VIs, SDr	0.62	290.98	23.83	0.7	186.49	21.62

Table 7. RF-based estimates of potato AGB at different growth stages.

Growth Stages	Data	Modeling			Validation		
		R^2	RMSE (kg/hm^2)	NRMSE (%)	R^2	RMSE (kg/hm^2)	NRMSE (%)
Tuber form	VIs	0.63	289.4	22.45	0.69	194.35	19.09
	VIs, Drmin	0.66	263.86	20.47	0.72	175.95	17.27
Tuber growth	VIs	0.69	265.22	18.28	0.71	192.012	17.86
	VIs, Rsum	0.71	234.92	16.19	0.75	152.122	14.15
Starch store	VIs	0.59	324.27	23.31	0.65	196.105	21.59
	VIs, Rsum	0.64	285.9	20.55	0.68	180.37	19.86
Maturity	VIs	0.55	327.15	26.79	0.62	216.853	25.14
	VIs, SDr	0.61	300.87	24.64	0.64	197.363	22.88

The validation dataset was used to verify the AGB estimate based on VIs and VIs combined with optimal GEPs using PLSR and RF. Figures 6 and 7 show the relationship between measured and estimated AGB at different potato growth stages. The results show that the accuracy of the model constructed from the PLSR and RF methods is consistent with the accuracy of the modeling set. The larger the R^2 value for modeling and verification, the smaller the NRMSE, indicating that the model gains in stability and fits the data better, and the estimation accuracy is reasonable. Compared with estimating AGB based only on VIs (Tables 6 and 7 and Figures 6 and 7), the estimation accuracy obtained by combining VIs and the optimal GEPs is greater, and the model is more stable. Compared with estimating AGB by using only the optimal GEPs (Table 3), optimal VIs (Table 4), or VIs combined with the optimal GEPs (Table 5), the estimation accuracy obtained by using the PLSR and RF methods is significantly improved. In addition, the comprehensive analysis and comparison of the evaluation indexes of model accuracy for the four potato growth stages show that using the PLSR method produces more accurate AGB estimates than using the RF method.

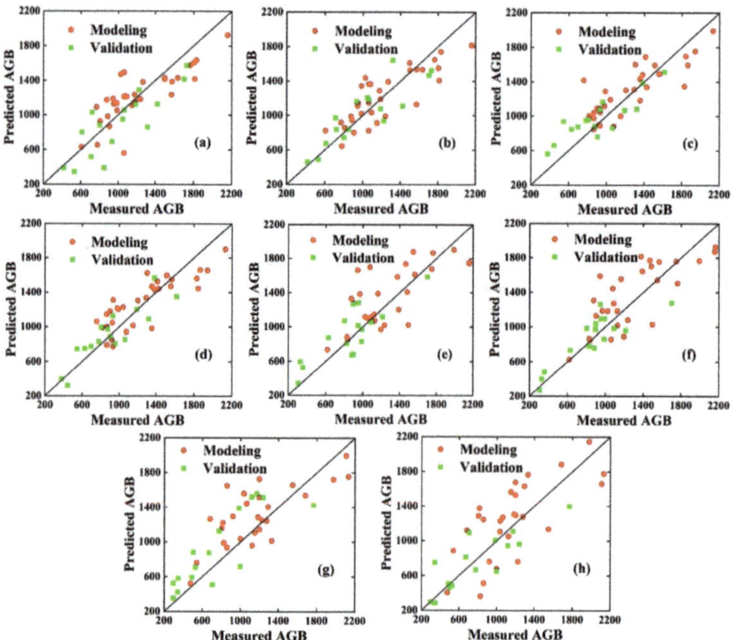

Figure 6. Relationship between measured and PLSR-estimated potato AGB (kg/hm^2) at different growth stages: (**a**) VIs, (**b**) VIs and optimal GEPs (tuber form stage); (**c**) VIs, (**d**) VIs and GEPs (tuber growth stage); (**e**) VIs, (**f**) VIs and GEPs (starch storage stage); (**g**) VIs, (**h**) VIs and GEPs (maturity stage).

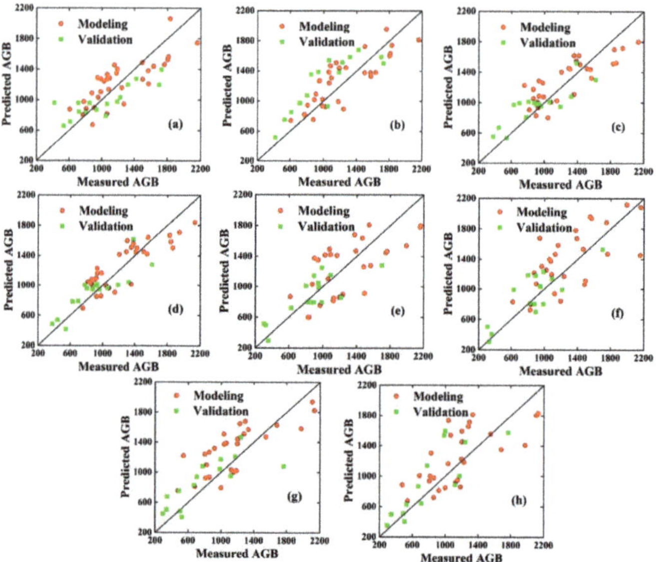

Figure 7. Relationship between measured and RF-estimated potato AGB (kg/hm^2) at different growth stages: (**a**) VIs, (**b**) VIs and optimal GEPs (tuber form stage); (**c**) VIs, (**d**) VIs and optimal GEPs

(tuber growth stage); (**e**) VIs, (**f**) VIs and optimal GEPs (starch storage stage); (**g**) VIs, (**h**) VIs and optimal GEPs (maturity stage).

3.4. Spatial Distribution of AGB

Comparing the accuracy indicators of each model (Tables 3–7) allows us to make the final selection based on VIs combined with the optimal GEPs and using the PLSR method to obtain the spatial distribution of potato AGB during tuber formation, tuber growth, starch storage, and maturity (Figure 8). During tuber formation, the AGBs of the plots differ significantly, with the overall distribution varying from 0 to 2400 kg/hm^2. However, the AGB of most plots falls between 0 and 1600 kg/hm^2. Repeat 1 and 2 plots have higher AGBs than repeat 3 plots, with repeat 2 plots having the highest AGB. During tuber growth, potato stems and leaves grow wildly, so this stage has the highest AGB of all stages. At this time, the overall AGB of each plot exceeds 1600 kg/hm^2, and the AGB of repeat 1 and 2 plots is even as high as 2400 kg/hm^2. Repeat 2 plots have the highest AGB and repeat 3 plots have the lowest. During starch storage, the AGB of each plot begins to decrease, falling mainly between 800 and 1600 kg/hm^2. Similarly, the AGB of the repeat 1 and 2 plots exceeds that of the repeat 3 plot, with the repeat 2 plot having the highest AGB. During the mature stage, the edge of the experimental area saw relatively good growth, but the AGB of the three replicate plots is the lowest of all the growth stages (mainly below 1600 kg/hm^2). In general, ranking the potato growth over the four growth periods from strong to weak gives repeat 2, repeat 1, and repeat 3. The spatial distribution of AGB in Figure 8 is consistent with the measured results, indicating that the selected optimal model produces good inversion.

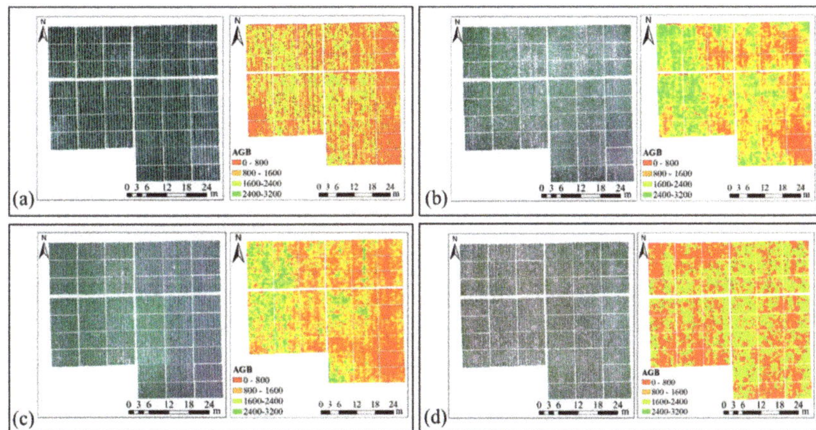

Figure 8. RGB images of potato canopy (left) and distribution (right) of the estimated AGB (kg/hm^2) in the 48 study plots obtained by combining VIs and the optimal GEPs and using the PLSR method: (**a**) tuber formation, (**b**) tuber growth, (**c**) starch storage, (**d**) maturity.

4. Discussion

4.1. AGB Estimation Based on Spectral Parameters

Before building the AGB-estimation model for each potato growth stage, we analyze the correlation between VIs, GEPs, and AGB. The correlation between these two spectral parameters (VIs and GEPs) and AGB first increases and then decreases over the growth period. This is because, from the tuber formation stage to the tuber growth stage, to maintain their own life activities, the vegetative organs such as stem nodes and leaves grow faster, the potato gradually grows better, and the vegetation coverage of the crop canopy increases, thus making the extracted canopy spectral information less vulnerable to the effect of ground soil and more closely linked to AGB. From the starch-storage stage to the maturity stage, the assimilates accumulated by photosynthesis must be continuously transported to the underground tubers. Because of the rainy season, the above-ground stems and leaves continuously wither and yellow, and the canopy vegetation coverage gradually decreases. Spectral information is interfered by the soil background, which worsens the relationship between spectral parameters and AGB. The results also show that the correlation between GEPs and AGB is weaker than that of the VIs, which is consistent with the results of Kasim et al. [50]. This result may be related to the range of the bands and the spectral resolution of the sensor.

Based on the optimal VIs and the optimal GEPs selected at each growth stage, the optimal VIs at the tuber-formation stage and the maturity stage is PSRI, the optimal GEPs is Drmin and SDR, respectively, and the optimal VIs at the tuber-growth and starch-storage stages is PBI and LCI, respectively. The optimal GEPs is Rsum, indicating that the spectral parameters are closely related to the crop growth stage. However, some spectral parameters cannot accurately estimate AGB due to saturation [15,34], so it is important to determine the spectral parameters and AGB sensitivity at different growth stages for monitoring potato growth.

Estimating AGB in each growth period based on the optimal GEPs is not as accurate as doing it based on the optimal VIs (Tables 3 and 4), but combining the VIs can improve the estimation accuracy (Table 5), which is consistent with the results of Zhang et al. [37] and Zheng et al. [47], who reached the same conclusion from estimating the AGB of sugar beet and rice, indicating that the GEPs more strongly affects AGB estimates during the crop growth stage. Therefore, deep mining of the sensitive spectral parameters in the green-edge region helps improve the accuracy of AGB estimates. The study found that AGB estimates based on the optimal VIs, the optimal GEPs, and combinations thereof gradually improve in going from the tuber-formation stage to the tuber-growth stage, and gradually deteriorate in going from the starch-storage stage to the maturity stage, which is related to the early spectral parameters being more correlated with AGB than the later spectral parameters (Table 2). Based on the single-variable estimation model, the maximum R^2 in each growth period is less than 0.55, which indicates that AGB estimates based only on the optimal spectral parameters are insufficient, so multiple sensitive spectral variables must be combined to improve the accuracy of AGB estimates and thereby monitor the growth of potato crops.

A sensitivity analysis of the spectral parameters and AGB indicates that the first six VIs and the optimal GEPs are strongly correlated in each growth period, so these are selected as model factors to estimate AGB. The selected VIs and potato AGB reach a significance level of 0.01 in the first three growth stages and a significance level of 0.05 in the mature stage. However, the optimal GEPs selected at each growth stage reaches a significance level of 0.01 for AGB, but the correlation weakens in the later stage, which is attributed to the potato crop growth cycle and dry matter distribution [51]. We used PLSR and RF based on VIs alone and VIs combined with the optimal GEPs to estimate AGB at each growth stage and analyzed the accuracy of each model. The results show that the accuracy of AGB estimates obtained from the same model with different variables gradually increases in going from the tuber-formation to the tuber-growth stage, and then gradually decreases in going from the starch-storage to the mature stage, which is consistent with the AGB estimates based on the optimal VIs, the optimal GEPs, and the combination thereof (Tables 3–7). Estimating AGB based on VIs alone or on VIs combined with the optimal GEPs is more accurate than doing so based on a single model factor, which is consistent with the results obtained by Han et al. [32] and Zhu et al. [35] for estimating corn AGB, indicating that it is particularly important to use multiple VIs with high sensitivity to improve the accuracy of AGB estimates. Estimating the AGB of each growth period based on VIs and the optimal GEPs is the most accurate approach, which shows that adding GEPs can improve the accuracy of the model. The green edge contains special band information for hyperspectral remote sensing and is highly sensitive to crop parameters, so further research is needed to understand its relationship with growth parameters.

4.2. AGB Estimation Based on Regression Technique

We use PLSR and RF based on VIs alone and on VIs combined with the optimal GEPs to construct AGB estimation models of potatoes at different growth stages, which produces more accurate AGB estimates (Tables 6 and 7), indicating that it is feasible to estimate crop parameters by using models based on PLSR and RF. In addition, the PLSR technique based on VIs combined with the optimal GEPs produces the most accurate AGB estimates in each growth period, whereas estimating AGB with the RF technique is relatively inaccurate. Comparing the precision of the validation set and modeling set (Tables 6 and 7; Figures 6 and 7) shows that the validation results are consistent with the modeling results, indicating that the AGB estimation model built from the PLSR technique is more stable and more accurate than the RF model, which is consistent with the results obtained by Tao et al. [15] and Yue et al. [46] for estimating winter wheat AGB. This is because there is a certain degree of autocorrelation between input variables. The PLSR technique uses principal component analysis to better solve the problem of collinearity between variables. It thus makes full use of spectral information to estimate AGB [15,46,52]. However, using RF techniques produces inaccurate AGB estimates, mainly because machine learning methods are designed for larger datasets and perform poorly with smaller data sets. This study involves 32 modeling-set data and 16 validation-set data, which are small data sets. Second, the spectral variables selected in this study are multicollinear, and the RF algorithm is not sensitive to multicollinearity, so models built on the RF method produce less accurate AGB estimates [11,29].

4.3. Spatial Distribution of AGB

The optimal model constructed by the PLSR method is used to draw the spatial distribution of AGB for each potato growth period. The AGB over the entire growth period first increases and then decreases (Figure 8), which is consistent with the measured results and with the potato growth law. From the tuber-formation to the tuber-growth stage, the vegetative organs such as stems and leaves develop gradually under their own metabolic activity, whereas potato AGB is mainly contained in the stems and leaves, so the AGB of a single potato plant continues to increase in the early stage. From the starch-storage stage to the maturity stage, the organic matter accumulated on the ground must be continuously transported to the underground tubers, which limits plant growth. At the same time, the late-growth stage is subject to continuous rainy weather, resulting in rapid withering and yellowing of the stems and leaves above the ground, which gradually reduces the AGB. In the spatial distribution of potato AGB in the four growth stages (Figure 8), the AGB values of different growth stages differ significantly, as do the AGB values of different experimental plots in the same growth stage. This result may be related to the growth of potato plants themselves, differences in the field, and the influence of external environmental factors. Although this study uses UAV hyperspectral sensors to obtain crop canopy spectral information which is expensive, and the data structure is complex compared with multispectral sensors, it can provide more crop canopy spectral information with the high spectral resolution (narrowband) and multiple bands, which can provide an opportunity to mine and extract spectral parameters (such as green edges) related to crop growth information. In this study, the advantages of UAV hyperspectral remote sensing were used to extract the VIs and GEPs used to characterize crop growth, and potato AGB monitoring was successfully realized.

5. Conclusions

This study uses linear regression, PLSR, and RF methods to estimate the AGB of potatoes during tuber-formation, tuber-growth, starch-storage, and maturity stages based on VIs, GEPs, and combinations thereof. The results show that the correlation between (i) VIs and GEPs and (ii) AGB gradually increase in going from the tuber-formation stage to the tuber-growth stage and decrease in going from the starch-storage stage to the mature stage. In addition, the VIs correlate more strongly with AGB than do the GEPs. AGB estimates based on the optimal VIs, the optimal GEPs, and combinations thereof improve in going from the tuber-formation stage to the tuber-growth stage and worsen in going from the starch-storage stage to the maturity stage. AGB estimates based on the combination of the optimal VIs and the optimal GEPs are the most accurate, and estimates based on the optimal VI and the optimal GEP are the worst. VIs combined with the optimal GEPs using PLSR and RF methods produce more accurate AGB estimates at each growth stage than do methods based on a single optimal VIs, a single optimal GEPs, or combinations thereof. Using the same method with different variables produces AGB estimates that gradually improve in going from the tuber-formation stage to the tuber-growth stage and then gradually worsen in going from the starch-storage stage to the mature stage. The model constructed by the PLSR method with the combination of VIs and the optimal GEPs produces the most accurate and stable AGB estimates for each growth period.

Author Contributions: H.F., Y.L., G.Y. and J.Y. designed the experiments. H.F., Y.L., Y.F. and X.S. collected the AGB and UAV hyperspectral images. Y.L. and H.F. analyzed the data and wrote the manuscript. X.J. and H.Y. made comments and revised the manuscript. All authors have read and agreed to the published version of the manuscript.

Funding: This study was supported by the Key scientific and technological projects of Heilongjiang province (2021ZXJ05A05), the National Natural Science Foundation of China (41601346), the Platform Construction Funded Program of Beijing Academy of Agriculture and Forestry Sciences (No.PT2022-24), and the Key Field Research and Development Program of Guangdong Province (2019B020216001).

Data Availability Statement: Not applicable.

Acknowledgments: The authors thank Bo Xu, Hong Chang, Weiguo Li, Yang Meng, and Yu Zhao for field management and data collection. The authors also thank the National Precision Agriculture Experiment Station for providing the test site and employees.

Conflicts of Interest: The authors declare no conflict of interest.

References

1. Liu, Y.; Feng, H.; Yue, J.; Li, Z.; Yang, G.; Song, X.; Yang, X.; Zhao, Y. Remote-sensing estimation of potato above-ground biomass based on spectral and spatial features extracted from high-definition digital camera images. *Comput. Electron. Agric.* **2022**, *198*, 107089. [CrossRef]
2. Brocks, S.; Bareth, G. Estimating barley biomass with crop surface models from Oblique RGB imagery. *Remote Sens.* **2018**, *10*, 268. [CrossRef]
3. Bendig, J.; Yu, K.; Aasen, H.; Bolten, A.; Bennertz, S.; Broscheit, J.; Gnyp, M.L.; Bareth, G. Combining UAV-based plant height from crop surface models, visible, and near infrared vegetation indices for biomass monitoring in barley. *Int. J. Appl. Earth Obs. Geoinf.* **2015**, *39*, 79–87. [CrossRef]
4. Gnyp, M.L.; Miao, Y.; Yuan, F.; Ustin, S.L.; Yu, K.; Yao, Y.; Huang, S.; Bareth, G. Hyperspectral canopy sensing of paddy rice aboveground biomass at different growth stages. *Field Crops Res.* **2014**, *155*, 42–55. [CrossRef]
5. Li, W.; Niu, Z.; Chen, H.; Li, D.; Wu, M.; Zhao, W. Remote estimation of canopy height and aboveground biomass of maize using high-resolution stereo images from a low-cost unmanned aerial vehicle system. *Ecol. Indic.* **2016**, *67*, 637–648. [CrossRef]
6. Zhu, W.; Sun, Z.; Peng, J.; Huang, Y.; Li, J.; Zhang, J.; Yang, B.; Liao, X. Estimating maize above-ground biomass using 3D point clouds of multi-source unmanned aerial vehicle data at multi-spatial scales. *Remote Sens.* **2019**, *11*, 2678. [CrossRef]
7. Walter, J.D.C.; Edwards, J.; McDonald, G.; Kuchel, H. Estimating biomass and canopy height with LiDAR for field crop breeding. *Front. Plant Sci.* **2019**, *10*, 1145–1157. [CrossRef] [PubMed]
8. Fu, Y.; Yang, G.; Wang, J.; Song, X.; Feng, H. Winter wheat biomass estimation based on spectral indices, band depth analysis and partial least squares regression using hyperspectral measurements. *Comput. Electron. Agric.* **2014**, *100*, 51–59. [CrossRef]
9. Yue, J.; Feng, H.; Jin, X.; Yuan, H.; Li, Z.; Zhou, C.; Yang, G.; Tian, Q. A comparison of crop parameters estimation using images from UAV-mounted snapshot hyperspectral sensor and high-definition digital camera. *Remote Sens.* **2018**, *10*, 1138. [CrossRef]
10. Greaves, H.E.; Vierling, L.A.; Eitel, J.U.H.; Boelman, N.T.; Magney, T.S.; Prager, C.M.; Griffin, K.L. Estimating aboveground biomass and leaf area of low-stature Arctic shrubs with terrestrial LiDAR. *Remote Sens. Environ.* **2015**, *164*, 26–35. [CrossRef]
11. Jin, X.; Liu, S.; Baret, F.; Hemerlé, M.; Comar, A. Estimates of plant density of wheat crops at emergence from very low altitude UAV imagery. *Remote Sens. Environ.* **2017**, *198*, 105–114. [CrossRef]
12. Schirrmann, M.; Hamdorf, A.; Garz, A.; Ustyuzhanin, A.; Dammer, K.-H. Estimating wheat biomass by combining image clustering with crop height. *Comput. Electron. Agric.* **2016**, *121*, 374–384. [CrossRef]

13. Chen, Y.; Li, L.; Lu, D.; Li, D. Exploring bamboo forest above ground biomass estimation using sentinel-2 Data. *Remote Sens.* **2019**, *11*, 7. [CrossRef]
14. Duan, B.; Fang, S.; Gong, Y.; Peng, Y.; Wu, X.; Zhu, R. Remote estimation of grain yield based on UAV data in different rice cultivars under contrasting climatic zone. *Field Crop. Res.* **2021**, *267*, 108148–108166. [CrossRef]
15. Tao, H.; Feng, H.; Xu, L.; Miao, M.; Long, H.; Yue, J.; Li, Z.; Yang, G.; Yang, X.; Fan, L. Estimation of crop growth parameters using UAV-based hyperspectral remote sensing data. *Sensors* **2020**, *20*, 1296–1317. [CrossRef] [PubMed]
16. Kross, A.; McNairn, H.; Lapen, D.; Sunohara, M.; Champagne, C. Assessment of RapidEye vegetation indices for estimation of leaf area index and biomass in corn and soybean crops. *Int. J. Appl. earth Obs.* **2015**, *34*, 235–248. [CrossRef]
17. Ma, G.N.; Huang, J.X.; Wu, W.B.; Fan, J.L.; Zou, J.Q.; Wu, S.J. Assimilation of MODIS-LAI into the WOFOST model for forecasting regional winter wheat yield. *Math. Comput. Model.* **2013**, *58*, 634–643. [CrossRef]
18. Zhao, Y.; Chen, S.; Shen, S. Assimilating remote sensing information with crop model using Ensemble Kalman Filter for improving LAI monitoring and yield estimation. *Ecol. Model.* **2013**, *270*, 30–42. [CrossRef]
19. Zhou, G.; Liu, X.; Liu, M. Assimilating remote sensing phenological information into the WOFOST model for rice growth simulation. *Remote Sens.* **2019**, *11*, 268. [CrossRef]
20. Montoya, F.; Camargo, D.; Domínguez, A.; Ortega, J.; Córcoles, J. Parametrization of Cropsyst model for the simulation of a potato crop in a Mediterranean environment. *Agric. Water Manag.* **2018**, *203*, 297–310. [CrossRef]
21. Lobell, D.B.; Ortiz-Monasterio, J.I. Regional importance of crop yield constraints: Linking simulation models and geostatistics to interpret spatial patterns. *Ecol. Model.* **2005**, *196*, 173–182. [CrossRef]
22. Chahbi, A.; Zribi, M.; Lili-Chabaane, Z.; Duchemin, B.; Shabou, M.; Mougenot, B.; Boulet, G. Estimation of the dynamics and yields of cereals in a semi-arid area using remote sensing and the SAFY growth model. *Int. J. Remote Sens.* **2014**, *35*, 1004–1028. [CrossRef]
23. Choudhury, M.R.; Das, S.; Christopher, J.; Apan, A.; Chapman, S.; Menzies, N.W.; Dang, Y.P. Improving biomass and grain yield prediction of wheat genotypes on sodic soil using integrated high-resolution multispectral, hyperspectral, 3D point cloud, and machine learning techniques. *Remote Sens.* **2021**, *13*, 3482. [CrossRef]
24. Berger, K.; Atzberger, C.; Danner, M.; D'Urso, G.; Mauser, W.; Vuolo, F.; Hank, T.; Berger, K.; Atzberger, C.; Danner, M.; et al. Evaluation of the PROSAIL model capabilities for future hyperspectral model environments: A review study. *Remote Sens.* **2018**, *10*, 85. [CrossRef]
25. Xiao, X.; Zhang, Q.; Braswell, B.; Urbanski, S.; Boles, S.; Wofsy, S.; Moore, B.; Ojima, D. Modeling gross primary production of temperate deciduous broadleaf forest using satellite images and climate data. *Remote Sens. Environ.* **2004**, *91*, 256–270. [CrossRef]
26. Shendryk, Y.; Sofonia, J.; Garrard, R.; Rist, Y.; Skocaj, D.; Thorburn, P. Fine-scale prediction of biomass and leaf nitrogen content in sugarcane using UAV LiDAR and multispectral imaging. *Int. J. Appl. Earth Obs.* **2020**, *92*, 102177–102191. [CrossRef]
27. Mansaray, L.R.; Kanu, A.S.; Yang, L.; Huang, J.; Wang, F. Evaluation of machine learning models for rice dry biomass estimation and mapping using quad-source optical imagery. *GISci. Remote Sens.* **2020**, *57*, 785–796. [CrossRef]
28. Suarez, L.A.; Robson, A.; McPhee, J.; O'Halloran, J.; Van Sprang, C. Accuracy of carrot yield forecasting using proximal hyperspectral and satellite multispectral data. *Precis. Agric.* **2020**, *21*, 1304–1326. [CrossRef]
29. Jin, X.; Li, Z.; Feng, H.; Ren, Z.; Li, S. Deep neural network algorithm for estimating maize biomass based on simulated Sentinel 2A vegetation indices and leaf area index. *Crop J.* **2020**, *8*, 87–97. [CrossRef]
30. Zhang, L.X.; Chen, Y.Q.; Li, Y.X.; Ma, J.C.; Zheng, F.X.; Sun, Z.F. Estimating above ground biomass of winter wheat at early growth stages based on visual spectral. *Spectrosc. Spect. Anal.* **2018**, *39*, 2501–2506.
31. Devia, C.A.; Rojas, J.P.; Petro, E.; Martinez, C.; Mondragon, I.F.; Patino, D.; Rebolledo, M.C.; Colorado, J. High-throughput biomass estimation in rice crops using UAV multispectral imagery. *J. Intell. Robot. Syst.* **2019**, *96*, 573–589. [CrossRef]
32. Han, L.; Yang, G.; Dai, H.; Xu, B.; Yang, H.; Feng, H.; Li, Z.; Yang, X. Modeling maize above-ground biomass based on machine learning approaches using UAV remote-sensing data. *Plant Methods* **2019**, *15*, 10–24. [CrossRef] [PubMed]
33. Asari, N.; Suratman, M.N.; Jaafar, J. Modelling and mapping of above ground biomass (AGB) of oil palm plantations in Malaysia using remotely-sensed data. *Int. J. Remote Sens.* **2017**, *38*, 4741–4764. [CrossRef]
34. Yue, J.; Yang, G.; Li, C.; Li, Z.; Wang, Y.; Feng, H.; Xu, B. Estimation of winter wheat above-ground biomass using unmanned aerial vehicle-based snapshot hyperspectral sensor and crop height improved models. *Remote Sens.* **2017**, *9*, 708. [CrossRef]
35. Zhu, Y.; Zhao, C.; Yang, H.; Yang, G.; Han, L.; Li, Z.; Feng, H.; Xu, B.; Wu, J.; Lei, L. Estimation of maize above-ground biomass based on stem-leaf separation strategy integrated with LiDAR and optical remote sensing data. *PeerJ* **2019**, *7*, 7593–7614. [CrossRef] [PubMed]
36. Nguyen, H.T.; Lee, B.-W. Assessment of rice leaf growth and nitrogen status by hyperspectral canopy reflectance and partial least square regression. *Eur. J. Agron.* **2006**, *24*, 349–356. [CrossRef]
37. Zhang, J.; Tian, H.; Wang, D.; Li, H.; Mouazen, A.M. A novel approach for estimation of above ground biomass of sugar beet based on wavelength selection and optimized support vector machine. *Remote Sens.* **2020**, *12*, 620–638. [CrossRef]
38. Yue, J.; Feng, H.; Yang, G.; Li, Z. A Comparison of regression techniques for estimation of above-ground winter wheat biomass using near-surface spectroscopy. *Remote Sens.* **2018**, *10*, 66. [CrossRef]
39. Sun, Q.; Gu, X.; Chen, L.; Xu, X.; Wei, Z.; Pan, Y.; Gao, Y. Monitoring maize canopy chlorophyll density under lodging stress based on UAV hyperspectral imagery. *Comput. Electron. Agric.* **2022**, *193*, 106671–106685. [CrossRef]

40. Yao, X.; Si, H.; Cheng, T.; Jia, M.; Chen, Q.; Tian, Y.; Zhu, Y.; Cao, W.; Chen, C.; Cai, J.; et al. Hyperspectral estimation of canopy leaf biomass phenotype per ground area using a continuous wavelet analysis in wheat. *Front. Plant Sci.* **2018**, *9*, 1360–1374. [CrossRef]

41. Tian, H.; Shi, S.; Wang, H.; Li, F.; Li, Z.; Alva, A.; Zhang, Z. Estimation of sugar beet aboveground biomass by band depth optimization of hyperspectral canopy reflectance. *J. Indian Soc. Remote* **2017**, *45*, 795–803. [CrossRef]

42. Chen, H.Y.H.; Luo, Y.; Reich, P.B.; Searle, E.B.; Biswas, S.R. Climate change—Associated trends in net biomass change are age dependent in western boreal forests of Canada. *Ecol. Lett.* **2016**, *19*, 1150–1158. [CrossRef] [PubMed]

43. Bendig, J.; Bolten, A.; Bennertz, S.; Broscheit, J.; Eichfuss, S.; Bareth, G. Estimating biomass of barley using crop surface models (CSMs) derived from UAV-based RGB imaging. *Remote Sens.* **2014**, *6*, 10395–10412. [CrossRef]

44. Banerjee, B.; Spangenberg, G.; Kant, S. Fusion of spectral and structural information from aerial images for improved biomass estimation. *Remote Sens.* **2020**, *12*, 3164. [CrossRef]

45. Liu, Y.; Liu, S.; Li, J.; Guo, X.; Wang, S.; Lu, J. Estimating biomass of winter oilseed rape using vegetation indices and texture metrics derived from UAV multispectral images. *Comput. Electron. Agric.* **2019**, *166*, 105026–105036. [CrossRef]

46. Yue, J.; Yang, G.; Tian, Q.; Feng, H.; Xu, K.; Zhou, C. Estimate of winter-wheat above-ground biomass based on UAV ultrahigh-ground-resolution image textures and vegetation indices. *ISPRS J. Photogramm. Remote Sens.* **2019**, *150*, 226–244. [CrossRef]

47. Zheng, H.B.; Cheng, T.; Zhou, M.; Li, D.; Yao, X.; Tian, Y.C.; Cao, W.X.; Zhu, Y. Improved estimation of rice aboveground biomass combining textural and spectral analysis of UAV imagery. *Precis. Agric.* **2019**, *20*, 611–629. [CrossRef]

48. Quirós Vargas, J.J.; Zhang, C.; Smitchger, J.A.; McGee, R.J.; Sankaran, S. Phenotyping of plant biomass and performance traits using remote sensing techniques in pea. *Sensors* **2019**, *19*, 2031. [CrossRef]

49. Sankaran, S.; Zhou, J.; Khot, L.R.; Trapp, J.J.; Mndolwa, E.; Miklas, P.N. High-throughput field phenotyping in dry bean using small unmanned aerial vehicle based multispectral imagery. *Comput. Electron. Agric.* **2018**, *151*, 84–92. [CrossRef]

50. Kasim, N.; Shi, Q.D.; Wang, J.Z.; Sawut, R.; Nurmemet, I.; Isak, G. Estimation of spring wheat chlorophyll content based on hyperspectral features and PLSR model. *Trans. Chin. Soc. Agric. Eng.* **2017**, *33*, 208–215.

51. Li, B.; Xu, X.; Zhang, L.; Han, J.; Bian, C.; Li, G.; Liu, J.; Jin, L. Above-ground biomass estimation and yield prediction in potato by using UAV-based RGB and hyperspectral imaging. *ISPRS J. Photogramm. Remote Sens.* **2020**, *162*, 161–172. [CrossRef]

52. Li, X.; Zhang, Y.; Bao, Y.; Luo, J.; Jin, X.; Xu, X.; Song, X.; Yang, G. Exploring the best hyperspectral features for LAI estimation using partial least squares regression. *Remote Sens.* **2014**, *6*, 6221–6241. [CrossRef]

remote sensing

MDPI

Article

Estimation of Aboveground Biomass of Potatoes Based on Characteristic Variables Extracted from UAV Hyperspectral Imagery

Yang Liu [1,2,3,†], Haikuan Feng [1,4,*,†], Jibo Yue [5], Zhenhai Li [6], Xiuliang Jin [7], Yiguang Fan [1], Zhihang Feng [8] and Guijun Yang [1]

1   Key Laboratory of Quantitative Remote Sensing in Agriculture, Ministry of Agriculture and Rural Affairs, Information Technology Research Center, Beijing Academy of Agriculture and Forestry Sciences, Beijing 100097, China
2   Key Laboratory of Smart Agriculture System, Ministry of Education, China Agricultural University, Beijing 100083, China
3   Key Laboratory of Agricultural Information Acquisition Technology, Ministry of Agriculture and Rural Affairs, China Agricultural University, Beijing 100083, China
4   College of Agriculture, Nanjing Agricultural University, Nanjing 210095, China
5   College of Information and Management Science, Henan Agricultural University, Zhengzhou 450002, China
6   College of Geomatics, Shandong University of Science and Technology, Qingdao 266590, China
7   Institute of Crop Sciences, Chinese Academy of Agricultural Sciences/Key Laboratory of Crop Physiology and Ecology, Ministry of Agriculture, Beijing 100081, China
8   School of Computer and Information Technology, Beijing Jiaotong University, Beijing 100091, China
*   Correspondence: fenghk@nercita.org.cn
†   These authors contributed equally to this work.

Citation: Liu, Y.; Feng, H.; Yue, J.; Li, Z.; Jin, X.; Fan, Y.; Feng, Z.; Yang, G. Estimation of Aboveground Biomass of Potatoes Based on Characteristic Variables Extracted from UAV Hyperspectral Imagery. *Remote Sens.* 2022, 14, 5121. https://doi.org/10.3390/rs14205121

Academic Editors: Giovanni Laneve, Chenghai Yang, Wenjiang Huang and Yingying Dong

Received: 6 September 2022
Accepted: 12 October 2022
Published: 13 October 2022

Publisher's Note: MDPI stays neutral with regard to jurisdictional claims in published maps and institutional affiliations.

Abstract: Aboveground biomass (AGB) is an important indicator for crop-growth monitoring and yield prediction, and accurate monitoring of AGB is beneficial to agricultural fertilization management and optimization of planting patterns. Imaging spectrometer sensors mounted on unmanned aerial vehicle (UAV) remote-sensing platforms have become an important technical method for monitoring AGB because the method is convenient, rapidly collects data and provides image data with high spatial and spectral resolution. To confirm the feasibility of UAV hyperspectral remote-sensing technology to estimate AGB, this study acquired hyperspectral images and measured AGB data over the potato bud, tuber formation, tuber growth, and starch-storage periods. The canopy spectrum obtained in each growth period was smoothed by using the Savitzky–Golay filtering method, and the spectral-reflection feature parameters, spectral-location feature parameters, and vegetation indexes were extracted. First, a Pearson correlation analysis was performed between the three types of characteristic spectral parameters and AGB, and the spectral parameters that reached a significant level of 0.01 in each growth period were selected. Next, the spectral parameters reaching a significance of 0.01 were optimized and screened by moving window partial least squares (MWPLS), Monte Carlo uninformative variable elimination (MC-UVE), and random frog (RF) methods, and the final model parameters were determined according to the thresholds of the root mean square error of cross-validation (RMSEcv), the reliability index, and the selected probability. Finally, the three optimal characteristic spectral parameters and their combinations were used to estimate the potato AGB in each growth period by combining the partial least squares regression (PLSR) and Gaussian process regression (GPR) methods. The results show that, (i) ranked from high to low, vegetation indexes, spectral-location feature parameters, and spectral-reflection feature parameters in each growth period are correlated with the AGB, and these correlations all first improve and then degrade in going from the budding period to the starch storage period. (ii) The AGB estimation model based on the characteristic variables screened by the three methods in each growth period is most accurate with RF, less so with MC-UVE, and least accurate with MWPLS. (iii) Estimating the AGB with the same variables combined with the PLSR method in each growth period is more accurate than the corresponding GPR method, but the estimations produced by the two methods both show a trend of first improving and then worsening from the budding period to the starch-accumulation period.

The accuracy of the estimation models constructed by PLSR and GPR from high to low is based on comprehensive variables, vegetation indexes, spectral-location feature parameters and spectral-reflection feature parameters. (iv) When combined with the RF-PLSR method to estimate AGB in each growth period, the best R^2 values are 0.65, 0.68, 0.72, and 0.67, the corresponding RMSE values are 167.76, 162.98, 160.77, and 169.24 kg/hm^2, and the corresponding NRMSE values are 19.76%, 16.01%, 15.04%, and 16.84%. The results of this study show that a variety of characteristic spectral parameters may be extracted from UAV hyperspectral images, that the RF method may be used for optimizing and screening, and that PLSR regression provides accurate estimates of the potato AGB. The proposed approach thus provides a rapid, accurate, and nondestructive way to monitor the growth status of potatoes.

Keywords: UAV; hyperspectral; spectral feature; location feature; vegetation indexes; potato; aboveground biomass

---

1. Introduction

Aboveground biomass (AGB) is an important agronomic parameter that characterizes the life activities of crops. It is closely related to the nutritional status and growth status of crops and is commonly used to monitor crop seedlings and evaluate farmland productivity [1–3]. Field management and yield prediction therefore rely on the rapid, non-destructive, and accurate determination of the temporal and spatial dynamics of AGB [4–6]. Traditionally, AGB is measured by a destructive sampling method that requires manual crop harvesting, weighing, and recording. This method is not only time-consuming, laborious, and destructive but is also limited to small areas due to the limited number of sampling points and monitoring range. It therefore does not suit the needs of quantitative monitoring of a large-scale crop AGB [7–10].

Because ground objects can reflect and absorb electromagnetic radiation, the spectral information of different crop canopies can be obtained by remote-sensing technology in a long-distance, high-throughput, and nondestructive manner. The application of mathematical analyses to interpret the spectral characteristics from multiple perspectives allows for nondestructive quantitative monitoring of crop physiological and biochemical indicators [11–13]. Currently, the main categories of sensor platforms for earth observation based on remote-sensing technology are ground, spaceborne, and airborne. Using ground sensors to obtain canopy spectral data in the field can sometimes damage crops. At the same time, it is challenging to monitor physical and chemical parameters in large areas for a long time due to physical constraints [14–16]. Although spaceborne remote-sensing technology can be used for nondestructive monitoring, it is expensive, imposes long transit periods, and yields images of coarse spatial resolution, which limits its application in precision agriculture [17–19]. Compared with other remote-sensing technologies, remote sensing from unmanned aerial vehicles (UAVs) has become an essential means to monitor crop growth due to its advantages of solid controllability, simple operation, economic suitability, and the ability to obtain high-resolution crop canopy orthophoto images under clouds [20–22].

At present, UAV platforms are typically equipped with digital, multispectral, and hyperspectral sensors to obtain spectral information on crop canopies and thereby monitor crop growth [23]. UAV hyperspectral remote sensing offers continuous narrow-band spectral information on the crop canopy, which allows the hidden details of the crop canopy spectrum to be mined [24,25]. Seeking the characteristic spectrum establishes the AGB monitoring model, which can effectively evaluate crop growth and predict yield [26]. In addition, UAV hyperspectral remote sensing also provides orthophotos of fields, which facilitates the mapping and display of AGB spatial variations [27]. Therefore, UAV hyperspectral remote-sensing technology with simultaneous acquisition of image and spectral information is vital for the accurate monitoring of crop AGB and to support real-time field management strategies [28,29].

Currently, the methods to estimate AGB from spectral information obtained by hyperspectral sensors fall into three main categories: (i) those based on spectral-reflectance features, (ii) those based on spectral-location features, and (iii) those based on vegetation indexes. The original canopy spectrum characterizes the ability of the crop canopy to radiate energy, which is related to variations in AGB content, so it can be directly used to estimate AGB. For example, Wang et al. [30], Jia et al. [31], and Kong et al. [32] used the successive projection algorithm to screen the characteristic wavelengths of the original crop canopy spectrum and combined different regression methods to estimate the AGB of winter wheat. Spectral-location features reflect the absorption and reflection of biochemical components in crops, and these feature locations contain rich crop-growth information that can be extracted to monitor crop AGB. For example, Tao et al. [33], Fu et al. [34], and Gnyp et al. [35] analyzed the position of the red edge and found that the red-edge parameters correlate strongly with wheat AGB, which means that the position of spectral features can be used to estimate AGB. The vegetation index enhances vegetation information and is usually composed of two or more spectral bands in a specific mathematical way to reduce or eliminate the effect of background noise on the crop canopy spectral information. For example, Jin et al. [36] found that the enhanced vegetation index and the three-band water index can be used to estimate maize AGB. Hansen et al. [37] showed that band combinations of the normalized vegetation index within the central wavelength band 680–750 nm correlate strongly with maize AGB and that an accurate estimation could be achieved by using the partial least squares regression (PLSR) method. Finally, Liu et al. [38] used the chlorophyll content index to estimate rice AGB.

Although these studies show how to effectively monitor AGB using different spectral features, most of them use only one type of spectral variable as an input parameter for their model. No studies have compared the performance of different forms of spectral parameters for estimating AGB. We do not yet know which spectral information is beneficial for potato AGB estimation and what is the accuracy of the model constructed. As a result, they fail to comprehensively evaluate how multiple types of spectral variables affect AGB estimation results, which may prevent hyperspectral information from being fully utilized, thus restricting the accuracy of estimation models. In addition, due to the high information redundancy between adjacent bands of hyperspectral data, most studies estimate AGB by directly inputting spectral characteristic variables into the model without optimizing the model's input data, which reduces the predictive power and robustness of the model. Although research has shown that the use of UAV hyperspectral data can lead to reasonable estimates of the AGB of winter wheat, corn, rice, and other crops, the morphological structure of potato plants differs significantly from those crops, and the crop nutrient absorption, transportation, and transfer also differ considerably. Therefore, to determine which spectral information is most beneficial for potato AGB estimation and to validate whether methods used for AGB estimation in other crops are applicable to potato crops, we evaluated the performance of various spectral variables to estimate AGB in potato multiple growth stages. Briefly, the main objective of this study was to determine the best spectral information and estimation method for estimating potato AGB.

In order to obtain the best spectral information for estimating potato AGB, this study used moving window partial least squares (MWPLS), Monte Carlo uninformative-variable elimination (MC-UVE), and random leapfrog (RF) to optimize spectral-reflection features, spectral-location features, vegetation indexes, and their combinations. Finally, model parameters with low redundancy were selected to establish AGB estimation models combined with PLSR and Gaussian process regression (GPR). We selected the optimal estimation model and regression technique. The resulting model provides rapid and nondestructive monitoring of potato-crop growth.

2. Materials and Methods

2.1. Experimental Design

The potato cultivation experiment took place at the Xiao Tangshan National Precision Agriculture Research Center (40°10′N, 116°26′E), Changping District, Beijing, China. In the field, we planted two precocious potato varieties (Zhongshu 5, (Z5) and Zhongshu 3, (Z3)) with different planting densities (P plots), nitrogen treatments (N plots), and potassium fertilizer treatments (K plots). The experimental area contained 48 plots and each plot covered 32.5 m². Please see Ref. [39] for the specific experimental design. To accurately correct UAV hyperspectral images, the three-dimensional information of eleven ground control points was obtained by using the differential global positioning system. The specific plan of this study is shown in Figure 1.

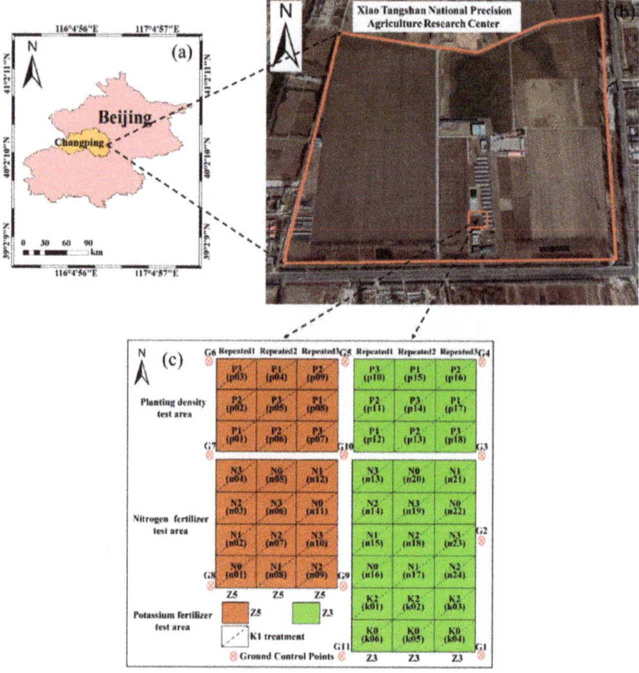

Figure 1. Location and design of experimental area: (**a**) location of Changping District in Beijing, (**b**) Xiao Tangshan National Precision Agriculture Research Center, (**c**) P1, P2, and P3 represent planting density treatments of 60,000, 72,000, and 84,000 plants/hm², respectively. N0, N1, N2, and N3 represent nitrogen fertilization treatments of 0, 244.65, 489.15, and 733.50 kg/hm², respectively. K0, K1, and K2 represent potassium fertilization treatments of 0, 970.50, and 1941 kg/hm², respectively.

2.2. Ground-Data Collection and Processing

AGB data were collected during four potato-growth periods including the budding period (13 May 2019), the tuber-formation period (28 May 2019), the tuber growth period (10 June 2019), and the starch-accumulation period (20 June 2019). To obtain the ground-truth AGB data, three plants representative of the overall growth level were randomly selected from each plot. After random field sampling, we quickly took it back to the laboratory. We used an oven to dry the samples to a constant mass before weighing. For each growth period, the plant density and dry mass of the stems and leaves of each plant sample as measured by a high-precision balance were used to calculate the potato AGB of each plot (in kg/hm²). The specific steps of collection are available in the literature [39].

2.3. UAV Hyperspectral Data Acquisition and Processing

A six-rotor electric UAV (SZ DJI Technology Co., Ltd., Shenzhen, Guangdong, China) was used as the ultralow-altitude remote-sensing platform. The UAV was equipped with a GPS module to record the attitude and spatial position during image shooting and two 1800 mAH batteries (25 V, which can maintain autonomous flight for 25 min on the specified route). The UAV had a maximum takeoff mass of 6 kg and a flight speed of about 8 m/s. The UAV platform carried a UHD 185 firefly imaging spectrometer, referred to as "UHD 185" for short. The flying height of the UAV is 20 m, and the obtained image resolution is 1.3 cm. Before each operation of the UAV, black and white panel data were collected on the ground using UHD185 for radiometric calibration.

UAV hyperspectral image processing includes two main parts:

(i) *Image mosaic and terrain correction.* First, with the help of the Agisoft Photoscan software, which is based on a motion structure algorithm, image mosaic and terrain correction were carried out in combination with the position of ground control points (the correction error of each growth period is less than 2 cm). Second, hyperspectral and grayscale images were fused by using the Cubert Cube-Pilot software to form new hyperspectral digital orthophotos [40].

(ii) *Extraction of canopy spectral reflectance.* According to the ratio method, the digital number value of the hyperspectral image is converted into the surface reflectance based on the black and white panel data collected on the ground. To eliminate the boundary effect, ArcGIS 10.2 software was used to draw the maximum region of interest (a total of 48 plots) for each plot. Based on ENVI 5.3 software, the average spectral reflectance of all pixels in each region of interest was calculated, and the average spectrum obtained served as the spectral reflectance of the potato canopy for each experimental plot [41].

2.4. Selection of Model Parameters

Due to the sensitivity of the imaging spectrometer itself and to the interference of the external environment, the acquired crop canopy spectrum is easily affected by noise in the process of image acquisition. To improve the signal-to-noise ratio of the spectral data and further improve the accuracy of AGB estimates, the spectral data must be smoothed. The Savitzky–Golay (SG) filtering method was proposed by Savitzky and Golay in 1964 and is a low-pass filtering method that smooths time-series data by applying a local polynomial regression model. Its biggest advantage is to remove noise while maintaining the shape and width of the signal [42]. The SG filter used in this study has a width of 5, a smoothing order of 0, and a smoothing degree of 2. Figure 2 shows the canopy spectral-reflectance features of potatoes (Z5 and Z3) before and after SG filtering at each growth stage. Comparative analysis showed that the canopy spectral curve retained the original shape and became smoother after SG filtering. Before constructing the AGB estimation model for each growth period, the spectral-reflectance features, spectral-position features, and vegetation indexes were extracted from the canopy spectral data after SG filtering.

2.4.1. Extraction of Spectral-Reflection Features

The characteristic spectral-reflectance parameters extracted in this study include mainly canopy original reflectance spectra (CRS) and first-order differential spectra (FDS). CRS is the most direct expression of the response of the internal cell structure of the plant leaf to the incident optical spectrum. The depth analysis of the changes in spectral reflection and absorption characteristics in the visible and near-infrared bands provides the basis for the establishment of the AGB estimation model. FDS represents the change of reflectance, that is, the slope of wavelength, which can remove the influence of partial linear background noise on the spectral information of crop canopy, refine the differences between spectra, and enhance the spectral sensitivity. The UHD 185 sensor covers numerous bands, which is suitable for first-order differential processing. The FDS formula is

$$\text{FDS}_{\lambda(i)} = \frac{R_{\lambda(i-1)} - R_{\lambda(i+1)}}{8} \tag{1}$$

where $\text{FDS}_{\lambda(i)}$ is the first-order differential spectral reflectance at a central wavelength i between waveband $i-1$ and $i+1$. $R_{\lambda(i-1)}$ and $R_{\lambda(i+1)}$ are the reflectances in the waveband $i-1$ and $i+1$, respectively.

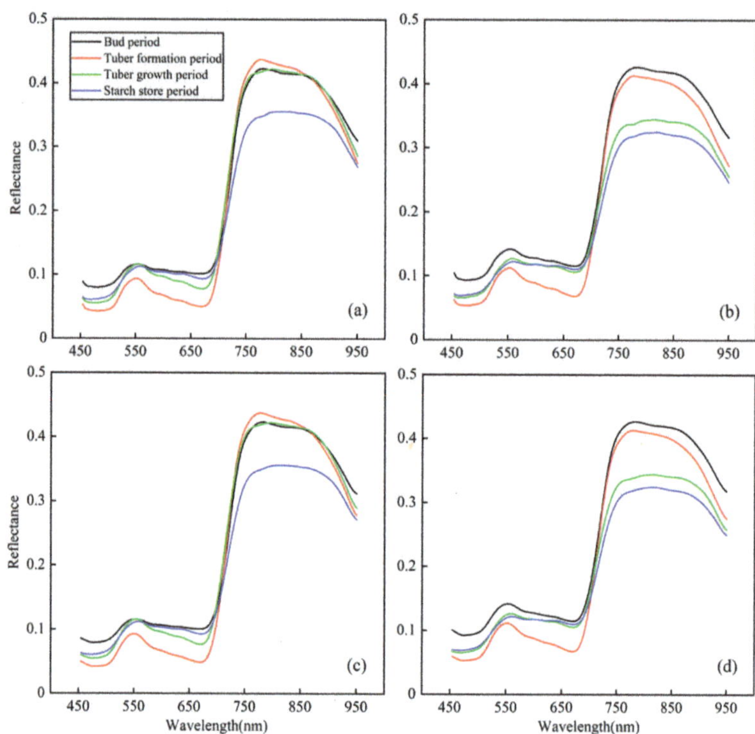

Figure 2. Canopy spectral-reflectance curves of potatoes in each growth period before and after SG filtering: (**a,c**) Z5 potatoes, (**b,d**) Z3 potatoes.

2.4.2. Extraction of Spectral-Position Features

De-enveloping line processing, also known as the continuous removal method, is a processing method that normalizes the canopy spectral data to unity so that the spectral data have the same background, retain most of the information, and effectively highlight the difference between the spectral information and the variation in AGB, which produces accurate AGB estimates [43–45]. The potato canopy spectral data range obtained by the UHD185 sensor ranges from 454 to 950 nm, but as of 750 nm, the shape of the absorption valley of the potato canopy de-envelope spectrum is slightly smaller, and the difference between the spectra does not suffice to express the AGB variations. Therefore, in this study, only the absorption valley between 454 and 750 nm is studied in-depth. As shown in Figure 3, after the spectral data for this study are processed by applying the continuous removal method, the characteristic parameters of the two absorption valleys (V1 and V2) are extracted and mainly include the absorption valley depth (DP1 and DP2), the absorption valley area (A1 and A2), the absorption valley width (W1 and W2), the absorption valley left slope (SL1 and SL2), and the absorption valley right slope (SR1 and SR2). See Table 1 for details.

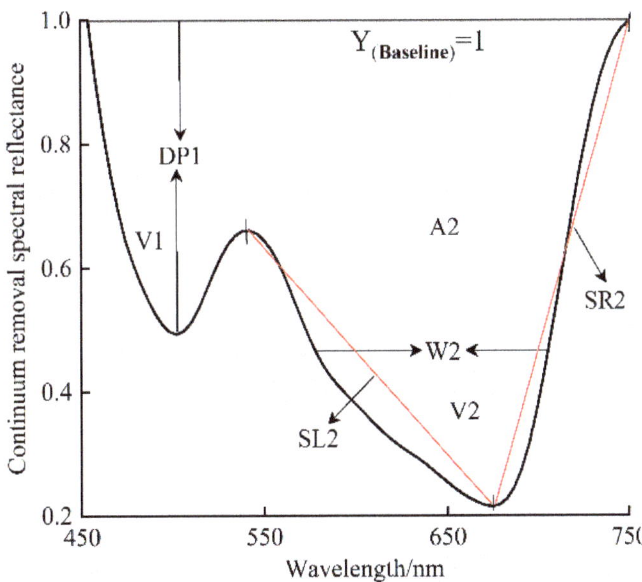

Figure 3. Potato canopy continuous removal spectra. V1 and V2 are two absorption valleys, DP1 is the first absorption depth, W2 is the second absorption width, A2 is the second absorption area, and SL2 and SR2 are the left and right slopes of the second absorption valley, respectively.

Table 1. Spectral position characteristic parameters.

Parameter Names	Variables	Definition and Description
Absorption valley depth	DP	The distance from the lowest point of the absorption valley to the baseline
Absorption valley area	A	Absorption valley integration of continuous removal spectra
Absorption valley width	W	Distance on either side of the absorption valley at half the depth
Slope of the left side of the absorption valley	SL	Slope of the line connecting the starting point on the left side of the absorption valley with the bottom point of the absorption valley
Slope of the right side of the absorption valley	SR	The slope of the line connecting the starting point on the right side of the absorption valley with the bottom point of the absorption valley
Green edge amplitude	Dg	Maximum value of the 1st derivative with a green edge (502–554 nm)
Green edge position	λg	Wavelength at Dg
Green edge area	SDg	Sum of the 1st derivative values within the green edge
Blue edge amplitude	Db	Maximum value of the 1st derivative with a blue edge (490–530 nm)
Blue edge position	λb	Wavelength at Db
Blue edge area	SDb	Sum of the 1st derivative values within the blue edge
Yellow edge amplitude	Dy	Maximum value of the 1st derivative with a yellow edge (562–638 nm)
Yellow edge position	λy	Wavelength at Dy
Yellow edge area	SDy	Sum of the 1st derivative values within the yellow edge
Red edge amplitude	Dre	Maximum value of the 1st derivative with a red edge (682–758 nm)
Red edge position	λre	Wavelength at Dre
Red edge area	SDre	Sum of the 1st derivative values within the red edge
Red valley amplitude	Drv	Maximum value of the 1st derivative with a red valley (650–690 nm)
Red valley position	λrv	Wavelength at Drv
Red valley area	SDrv	Sum of the 1st derivative values within the red valley
Green peak amplitude	Dgp	Maximum value of the 1st derivative with a green peak (510–558 nm)
Green peak position	λgp	Wavelength at Dgp
Green peak area	SDgp	Sum of the 1st derivative values within the green peak

At 502–554, 490–530, 562–638, 682–758, 650–690, and 510–558 nm, the absorption of chlorophyll varies with the scattering degree of leaves and canopy in different directions,

defining the green edge, blue edge, yellow edge, red edge, red valley, and green peak positions, respectively. These bands contain a large amount of sensitive crop canopy spectral information. The features extracted by the first-order differential spectrum are collectively referred to as "four sides, one valley, and one peak" parameters and reflect the growth status of potatoes in different periods. Therefore, the variations in potato AGB in different growth periods can be explored by studying the typical spectral position of characteristic parameters. This study extracts three types of parameters from the "four sides, one valley, and one peak" group: amplitude, position, and area [46].

2.4.3. Selection of Vegetation Indexes

Previous studies confirm that the vegetation index is closely related to physiological and biochemical parameters of crops, so it is often used to monitor crop growth. Therefore, we selected 20 commonly used vegetation indexes to estimate potato AGB. The specific names and mathematical expressions are listed in Table 2.

Table 2. Vegetation indexes used in the study.

Vegetation Indices	Equation	Reference
MCARI (modified chlorophyll absorption ratio index)	$[(R700 - R670) - 0.2 \times (R700R550)](R700/R670)$	[33]
TCARI (transformed chlorophyll absorption reflectance index)	$3 \times [(R700 - R670) - 0.2(R700 - R550)(R700/R670)]$	[33]
TVI (triangular vegetation index)	$0.5 \times [120(R750 - R550) - 200 \times (R670 - R550)]$	[24]
NDVI (normalized difference index)	$(R800 - R680)/(R800 + R680)$	[35]
SIPI (structure-insensitive pigment index)	$(R800 - R445)/(R800 + R680)$	[11]
GNDVI (green normalized difference vegetation index)	$(R800 - R570)/(R800 + R570)$	[12]
RDVI (re-normalized difference vegetation index)	$(R800 - R670)/(R800 + R670)^{1/2}$	[12]
OSAVI (optimized soil adjusted vegetation index)	$1.16 \times (R800 - R670)/(R800 + R670 + 0.16)$	[33]
MSR (modified simple ratio index)	$(R800/R670 - 1)/(R800/R670 + 1)^{1/2}$	[33]
NDRE (normalized difference red edge)	$(R790 - R720)/(R790 + R720)$	[11]
EVI (enhanced vegetation index)	$2.5 \times (R800 - R670)/(1 + R800 + 6 \times R670 - 7.5 \times R500)$	[12]
PSND (pigment specific normalized difference)	$(R800 - R470)/(R800 + R470)$	[33]
SPVI (spectral polygon vegetation index)	$0.4 \times [3.7(R800 - R670) - 1.2 \times |R530 - R670|]$	[33]
PSRI (plant senescence reflectance index)	$(R680 - R500)/R750$	[11]
SAVI (soil adjusted vegetation index)	$1.5 \times (R800 - R670)/(R800 + R670 + 0.5)$	[12]
NRI (nitrogen reflectance index)	$(R570 - R670)/(R570 + R670)$	[12]
CRI (carotenoid reflectance index)	$1/R570 + 1/R800$	[12]
NDI (normalized difference index)	$(R850 - R710)/(R850 + R680)$	[12]
WDRVI (modified wide dynamic range vegetation index)	$(0.1 \times R800 - R670)/(0.1 \times R800 + R670)$	[12]
LCI (linear combination index)	$(R850 - R710)/(R850 + R670)^{1/2}$	[33]

2.5. Analysis Method

This study selects three spectral feature screening methods: MWPLS, MC-UVE, and RF. MWPLS is based on interval partial least squares. The moving window technology is used to scan all wavebands, move one wavelength point backward each time with a specific window width, establish a partial least squares model, and repeat the operations many times. The root mean square error of cross-validation (RMSEcv) is used as the evaluation standard for wavelength screening intervals. This method avoids information redundancy [47]. MC-UVE is a widely used and effective band-selection method in the field of chemometrics. By randomly selecting the spectral variable matrix as noise, adding it to the original matrix, and then establishing the partial least squares model by the cross-validation method to eliminate one by one, the regression coefficient matrix is obtained, and the stability or reliability of the quotient of the mean and standard deviation of the regression coefficient is analyzed. Finally, the threshold is determined based on the importance of the noise variables, and the variables with importance below this threshold are deleted to obtain the optimal model variables [48]. The RF algorithm is a heuristic swarm evolution algorithm that computes efficiently and offers excellent global search capability, similar to the reversible jump Markov chain Monte Carlo algorithm, which calculates the selected probability of each variable by simulating a Markov chain obeying the steady-state distribution in the model space to evaluate the importance of the variable [49].

PLSR and GPR are used to build a model to estimate AGB. PLSR is a combination of multiple linear regression and principal component analysis. It provides a many-to-many regression model by fully considering the multicollinearity between independent variables. Especially when the number of independent variables is large and there is a certain autocorrelation, and the number of dependent variables is small, the model constructed by PLSR is more accurate than the traditional regression analysis and allows predictions to be made from a small number of factors [50]. Gaussian process regression is a nonparametric probabilistic statistical model that learns the relationship between independent variables (e.g., spectral features) and dependent variables (e.g., AGB) based on the Bayesian theorem and uses mean and covariance functions to train samples according to the maximum likelihood estimation method. Compared with conventional machine learning, parameter optimization is simpler and more suitable for training small-sample data. At the same time, the prediction and its related confidence interval are provided, which allows the reliability of the prediction results to be evaluated [51]. The characteristic variable selection and model construction for this study were carried out using MATLAB 2020b software.

2.6. Accuracy Evaluation

To estimate the AGB, this study selects the repeat 2 and repeat 3 data (32 groups total) as the calibration set, and the repeat 1 data (16 groups total) as the validation set to check the reliability and stability of the model. The coefficient of determination (R^2), root mean square error (RMSE), and normalized root mean square error (NRMSE) are used to evaluate the model accuracy.

3. Results and Analysis
3.1. Estimation of AGB Using Spectral-Reflectance Features

Figure 4 shows the correlation coefficient changes of CRS and FDS with AGB at different potato-growth periods. Figure 4a–d shows that the correlation coefficients for CRS, FDS, and AGB in each growth period differ significantly. Overall, the correlation of the two reflection spectra increases from the budding period to the tuber growth period but decreases slightly in the starch-accumulation period. There are also significant differences in the wavelength range where CRS, FDS, and AGB reach a very significant correlation level. At the budding period, CRS correlates significantly with AGB at 454–714 and 742–914 nm, whereas the wavelength range where FDS correlates significantly with AGB is scattered and mainly spread over five ranges: 470–498, 546–602, 702–774, 790–834, and 854–946 nm. During the tuber-formation period, the number of wavelengths in the bands 554–698 and 726–950 nm with a significant correlation between CRS and AGB increases, and the correlation coefficient also increases somewhat. Similarly, the number of wavelength bands (454–506, 542–598, 654–666, 698–950 nm) with significant correlation between FDS and AGB increases, and the correlation coefficient increases. During the tuber-growth period, the correlation between CRS, FDS, and AGB is the highest over all the growth periods, and the wavelength ranges with extremely significant correlations are 454–702, 718–950 nm, and 454–498, 514–534, 546–646, 682–774, and 786–950 nm, respectively. During the starch-accumulation period, the correlation between CRS, FDS and AGB is slightly less than in the previous period. The wavelength ranges that produced an extremely significant correlation were 454–702, 714–950 nm and 454–490, 510–534, 550–666, 678–786, 806–830, and 846–950, respectively.

MWPLS, MC UVE, and RF methods were used to screen the spectral-reflectance features of CRS and FDS that reached a very significant correlation in each potato-growth period. Figure 5 shows the variations in RMSEcv, reliability index, and selected probability obtained by using different wavelength-screening methods in each growth period. If the RMSEcv of the partial least squares model established in the selected wavelength interval is lower or the reliability index and the selected probability of a single wavelength are higher, it is more likely to become the parameter of the AGB estimation model. The combined results of many tests indicate that the threshold for extracting spectral features by using three sensitive wavelength screening methods at each potato-growth period was finally determined. The budding period based on CRS and FDS is 105 kg/hm^2, 3.0, 0.2 and 113 kg/hm^2, 2.0, 0.1, respectively. The tuber-formation period based on CRS and FDS is 240 kg/hm^2, 1.8, 0.1 and 225 kg/hm^2, 2.0, 0.1, respectively. The tuber-growth period based on CRS and FDS is 250 kg/hm^2, 1.0, 0.1 and 250 kg/hm^2, 1.8, 0.1, respectively. Finally, the starch-accumulation period based on CRS and FDS gives 320 kg/hm^2, 1.0, 0.1.

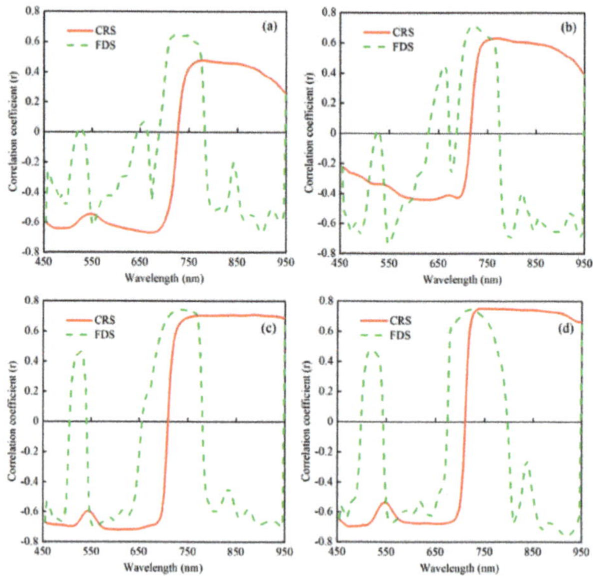

Figure 4. Correlation coefficients between CRS, FDS and AGB at various potato-growth periods: (**a**) bud period, (**b**) tuber-formation period, (**c**) tuber-growth period, (**d**) starch-storage period.

According to the threshold results set in Figure 4, the characteristic wavelengths extracted by MWPLS, MC-UVE, and RF methods based on CRS and FDS at each potato-growth period are shown in Figure 6. From the results of the distribution map, the specific number of characteristic wavelengths obtained by different methods based on the two spectral variables in each growth period is different, but the overall position is roughly the same, mainly located in "four sides, one valley and one peak", which confirms the importance of the special spectral position to the estimation of potato AGB.

To evaluate the performance of estimating AGB based on spectral-reflectance characteristics, PLSR and GPR were used to build AGB estimation models for potatoes at each growth period based on the characteristic spectra of CRS and FDS (Figure 6), and R^2, RMSE, and NRMSE were used to evaluate the fitting and stability of the models. The estimations obtained for each growth period are evaluated in Figures 7 and 8. The results show that AGB estimates based on the two types of canopy spectral information of crops are not good, but, under the same conditions, the effect of estimating AGB directly using CRS features (Figure 7) is significantly weaker than when using FDS features (Figure 8). Based on a comprehensive analysis of the quality indicators (R^2, RMSE, and NRMSE) in Figures 7 and 8, the same modeling method was used to estimate AGB with the characteristic wavelengths selected by MWPLS, MC-UVE, and RF over the whole growth period, and they all showed that the AGB estimates gradually improved from the budding period to the tuber-growth period, and then deteriorated. The model variables selected by RF produce the best results, followed by MC-UVE, and the variables selected by MWPLS produce the worst results. Comparing the AGB estimation ability of PLSR and GPR shows that the former produces a greater R^2 and lower RMSE and NRMSE for four growth periods, indicating that PLSR significantly improves the accuracy of potato AGB estimates.

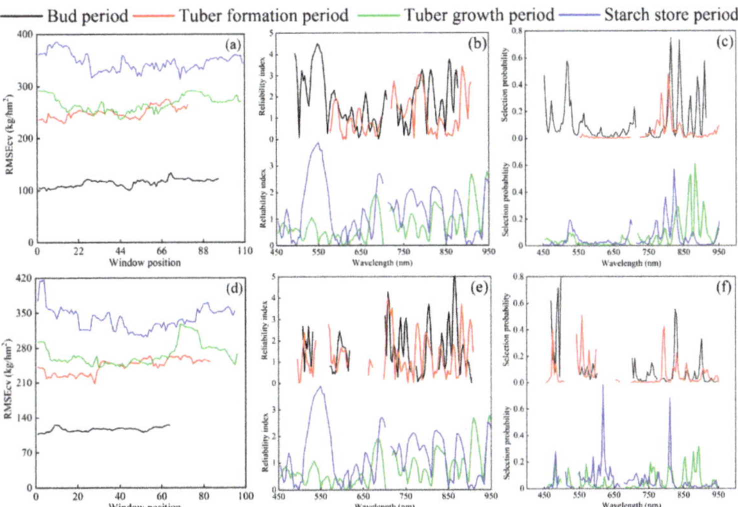

Figure 5. Determination of the influential spectra features in the AGB estimation of potatoes based on the CRS and FDS at the bud period, tuber-formation period, tuber-growth period, starch-storage period. (**a**,**d**) Cross-validation RMSEcv statistics using MWPLS based on CRS and FDS, respectively. (**b**,**e**) Reliability index statistics using MC-UVE based on CRS and FDS, respectively. (**c**,**f**) Selection probability statistics using RF based on CRS and FDS, respectively

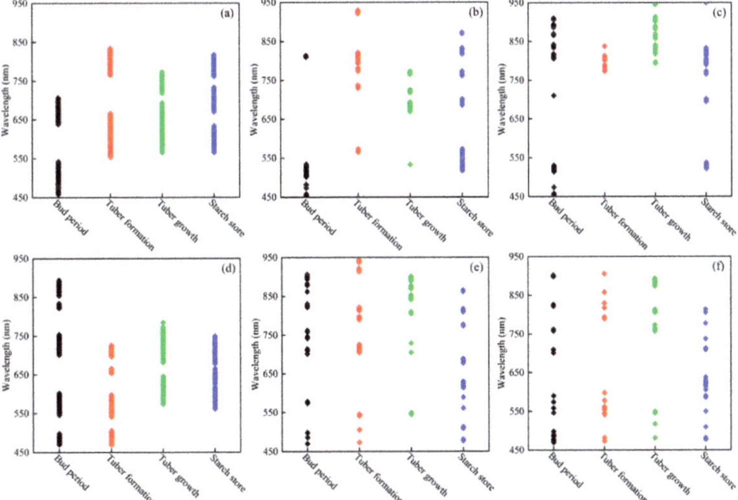

Figure 6. Distribution of wavelengths based on CRS and FDS using three wavelength-screening methods for potatoes for each growth period. (**a**,**d**) Characteristic wavelength distribution using MWPLS based on CRS and FDS, respectively. (**b**,**e**) Characteristic wavelength distribution using MC-UVE based on CRS and FDS, respectively. (**c**,**f**) Characteristic wavelength distribution using RF based on CRS and FDS, respectively.

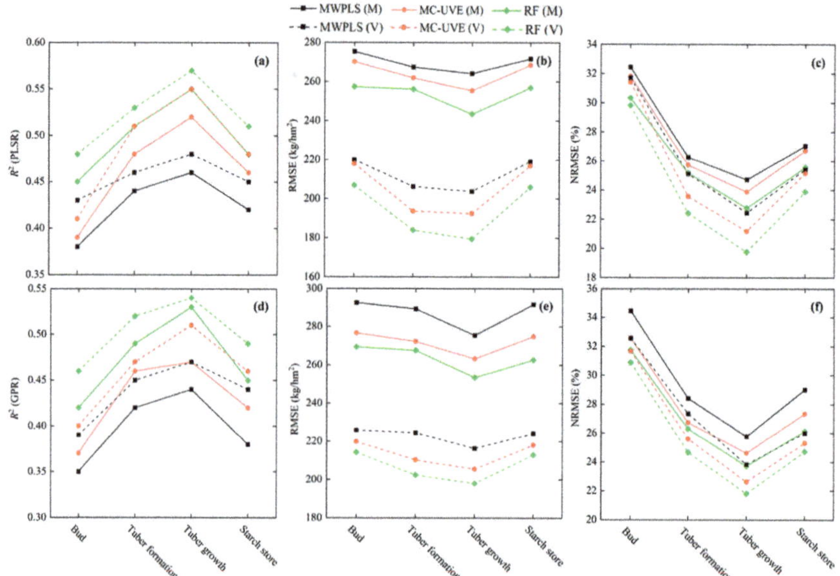

Figure 7. Accuracy of estimating AGB based on the feature variables screened from CRS by MWPLS, MC-UVE and RF. (**a–c**) The R^2, RMSE (kg/hm^2) and NRMSE (%) of AGB was estimated using the PLSR method. (**d–f**) The R^2, RMSE (kg/hm^2) and NRMSE (%) of AGB was estimated using the GPR method.

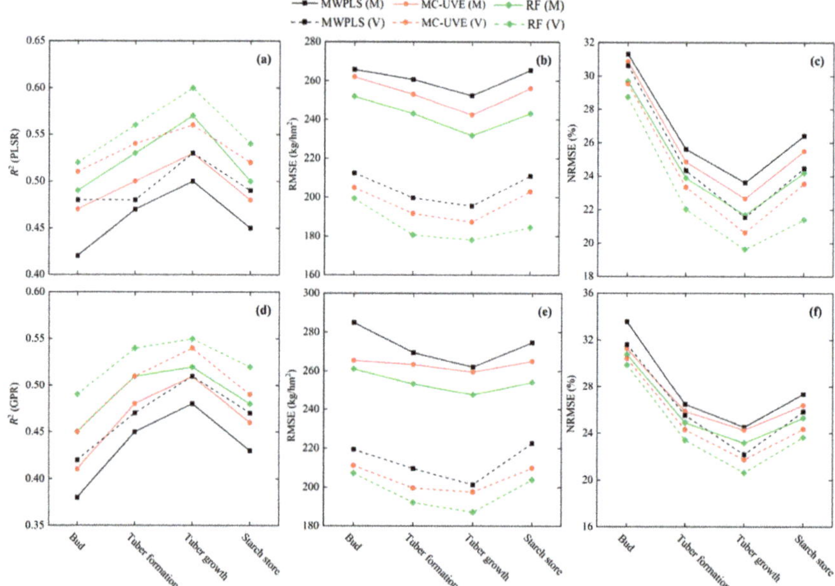

Figure 8. Accuracy of estimating AGB based on the feature variables screened from FDS by MWPLS, MC-UVE and RF. (**a–c**) The R^2, RMSE (kg/hm^2) and NRMSE (%) of AGB was estimated using the PLSR method. (**d–f**) The R^2, RMSE (kg/hm^2) and NRMSE (%) of AGB was estimated using the GPR method.

The AGB estimation model was established by RF-PLSR in each growth period, with the highest fitting accuracy and the strongest model stability. From the budding period to the tuber-formation period, based on CRS and FDS modeling, R^2 varies over the range 0.45–0.55 and 0.49–0.57, RMSE varies over the range 257.50–243.41 and 252.06–231.75 kg/hm^2, and NRMSE varies over the range 30.33–22.77% and 29.69–21.68%, respectively. These results are consistent with the validation results, which show that R^2 gradually increases, whereas RMSE and NRMSE gradually decrease. From the tuber-formation period to the starch-accumulation period, based on CRS and FDS modeling, R^2 varies over the range 0.55–0.48 and 0.57–0.50, RMSE varies over the range 243.41–256.87 kg/hm^2 and 231.75–243.11 kg/hm^2, and NRMSE varies over the range 22.77–25.56% and 21.68–24.19%, respectively. It is verified that R^2 gradually decreases, whereas RMSE and NRMSE gradually increase.

3.2. Estimation of AGB Using Spectral-Position Features

The twenty-eight spectral position characteristic parameters extracted at each growth period were correlated with the measured potato AGB, and the results are shown in Figure 9. A correlation analysis reveals that the responses of the various parameters differ significantly from that of the AGB, and the correlation between the de-envelope absorption valley parameters and AGB over the whole growth period is stronger than that of the "four sides, one valley, and one peak". Comparing the correlation of different parameters in each growth period shows that twelve parameters (DP1, DP2, A1, A2, W2, SL1, SR2, SDy, Dre, SDre, λre, and λrv) are significantly correlated with AGB over the four growth periods, and the correlation remains excellent. Compared with the correlations between the remaining parameters and AGB in each growth period, the number of positional parameters that correlate significantly with AGB differs greatly. Only the parameters Dy and SL2 correlate significantly with AGB at the budding period and tuber-formation period, respectively. The number of parameters (W1, Db, Dg, Dy, Dgp, Drv, and SDrv) that correlate significantly with AGB in the tuber-growth stage increases, as does the correlation, whereas the number of parameters significantly correlated with AGB in the starch-accumulation stage also increased (SL2, Db, Dg, SDg, Dy, Dgp, Drv, SDrv), but the correlation started to decrease. A comprehensive analysis of the correlation of location parameters over the whole growth period shows that the correlation from the budding period to the starch-accumulation period generally increases first and then decreases.

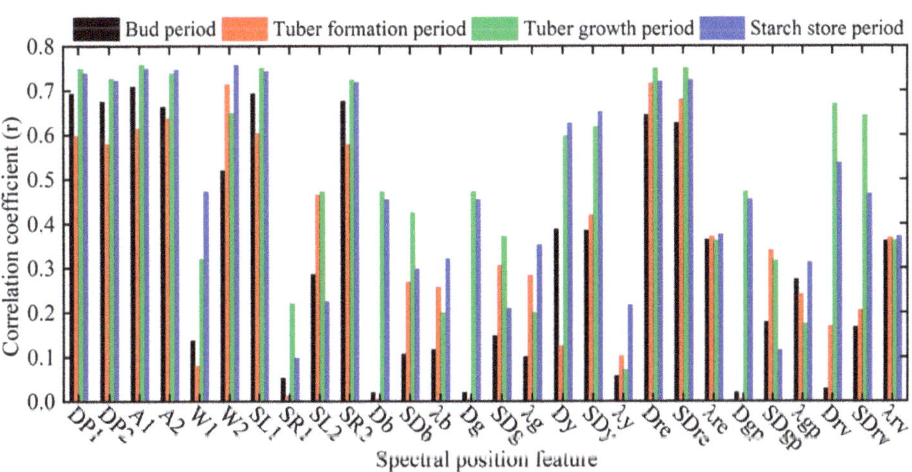

Figure 9. Correlation coefficients of spectral-position features for measured potato AGB at bud period, tuber-formation period, tuber-growth period, and starch-storage period.

The MWPLS, MC-UVE, and RF methods were also used to optimize and screen the location parameters that correlate significantly in each potato-growth period. According to the three selected indexes (RMSEcv, RI, and SP) in Figure 5 and based on the analysis of the test results, the thresholds for selecting model parameters for each growth period under each method are finally determined to be 120, 260, 275, and 360 kg/hm^2 (based on MWPLS), 1.0, 2.0, 1.0, and 1.5 (based on MC-UVE), 0.1, 0.2, 0.2, and 0.4 (based on RF). According to the selected threshold results, the model parameters used to estimate AGB in each potato-growth period were selected (Table 3). The model parameters listed in Table 3 show that the number and category of location parameters selected for each growth period vary greatly. Overall, the importance of the absorption valley parameters for AGB estimation is greater than that of the "four sides, one valley, and one peak".

Table 3. Location parameters of AGB estimation model selected by MWPLS, MC-UVE, and RF methods for each potato growth period.

Growth Stage	Variable Selection Method		
	MWPLS	**MC-UVE**	**RF**
Bud period	DP2, A1, A2, W2, SL1	A1, SR2, Dre, SDre, λrv	DP1, SR2, SDy, Dre, SDre
Tuber formation period	W2, SL1, SL2, SR2, SDy, Dre, SDre, λre	DP2, A2, W2, SL2, SR2, SDy, Dre, SDre	W1, DP2, SL1, SR2, SDy, Dre, SDre
Tuber growth period	DP2, A1, A2, SL1, SR2, Db	W1, W2, λre, λrv	DP1, DP2, A1, W1, W2, SL1, SR2, SDy
Starch store period	SR2, Db, Dg	DP2, A1, A2, SDg Drv, SDrv	SL1, SL2, SR2, SDy, Drv, SDrv

Using the PLSR and GPR methods based on the model parameters listed in Table 3 as independent variables, we obtain the relationship between the spectral location characteristic parameters and AGB in the four potato-growth periods. The indicators (R^2, RMSE, and NRMSE) used to evaluate the model accuracy in each growth stage are shown in Figure 10. Based on the evaluation index of the AGB estimation model, the estimation obtained by using the model parameters selected by different methods as variables is consistent with the spectral reflection characteristics, which shows that the variables selected by the RF method produce the best estimates, followed by MC-UVE, whereas the worst estimates are produced by the MWPLS method. Compared with the GPR regression technique, the model constructed by PLSR using the same variables in each growth period produces a larger R^2 and smaller RMSE and NRMSE, indicating that the PLSR method improves the accuracy of the AGB estimation model, which is consistent with the results given in Figures 7 and 8.

Comparing the results in Figures 7 and 8 with those in Figure 10 shows that the AGB estimate based on the spectral-location features is more accurate than the AGB estimate based on the corresponding spectral reflectance features. Similarly, the estimation quality first improves and then decreases from the budding period to the starch-accumulation period. In each growth period, the estimation model obtained by the RF-PLSR method is also more accurate and more reliable. From the budding period to the tuber-growth period, R^2 increases from 0.52 to 0.63, and RMSE and NRMSE decrease from 220.06 to 201.72 kg/hm^2 and from 25.92% to 18.87%, respectively. Validation R^2 also increases gradually and RMSE and NRMSE decrease gradually so that the estimate gets better and better. From the tuber-growth period to the starch-accumulation period, R^2 varies over the range 0.63–0.53, RMSE varies over the range 201.72–214.36 kg/hm^2, and NRMSE varies over the range 18.87–21.33%. The validation results are like the modeling results, and the estimation gradually deteriorates.

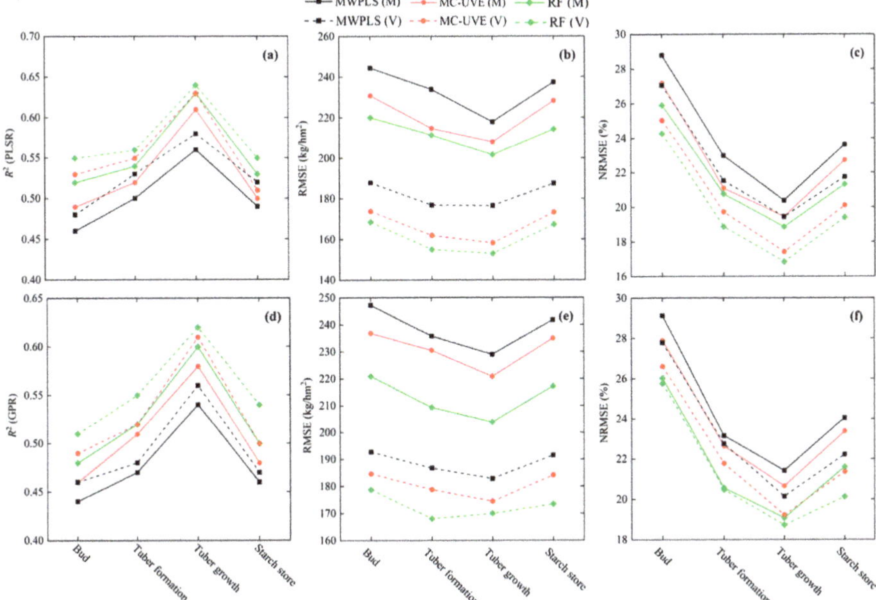

Figure 10. Accuracy of estimating AGB based on the feature variables screened from spectral-position parameters by MWPLS, MC-UVE and RF. (**a–c**) The R^2, RMSE (kg/hm^2) and NRMSE (%) of AGB were estimated using the PLSR method. (**d–f**) The R^2, RMSE (kg/hm^2) and NRMSE (%) of AGB were estimated using the GPR method.

3.3. Estimation of AGB Using Vegetation Indexes

The twenty selected vegetation indexes and the measured potato AGB in each growth period were used for a Pearson correlation analysis. The specific correlation size is shown in Figure 11, where the selected vegetation index and the measured AGB in each growth period correlate very strongly. Comparing the results in Figures 4, 9 and 11 shows that the correlations between the three types of model parameters and AGB differ significantly. In general, the vegetation index produces the best results, followed by the spectral-location feature parameter, and the spectral-reflectance feature parameter produces the worst results. Similarly, the variation of the correlation between vegetation indexes and AGB in the four growth periods is consistent with the characteristics of spectral position and spectral reflectance, which gradually improve from the budding period to the tuber-growth period and then decrease.

The vegetation indexes that do not correlate significantly in the four growth periods are MCARI and TCARI for the budding period, TCARI for the tuber-formation period, MSR and CRI for the tuber-growth period, and CRI for the starch-accumulation period. This indicates that the correlation between vegetation index and AGB depends on the growth period.

To reduce information redundancy, the MWPLS, MC-UVE, and RF methods were also used to optimize and screen the vegetation indexes that reached extremely significant correlation levels in each potato-growth period. Using the quantitative indicators RMSEcv, RI, and SP as criteria, the thresholds for selecting model parameters for each growth period under each method are determined. Based on MWPLS, they are 125, 260, 275, and 360 kg/hm^2, respectively. Based on MC-UVE, they are 1.0, 1.0, 1.0, and 1.5, respectively. Based on RF, they are 0.3, 0.2, 0.3, and 0.4, respectively. According to the threshold results, the vegetation index used to estimate AGB in each potato-growth period was selected (Table 4). From the results in Table 4, the vegetation indexes obtained in each growth period also differ, with NDVI, SIPI, NDRE, and WDRVI most likely to be selected by the three methods, indicating that these vegetation indexes are important for estimating potato AGB.

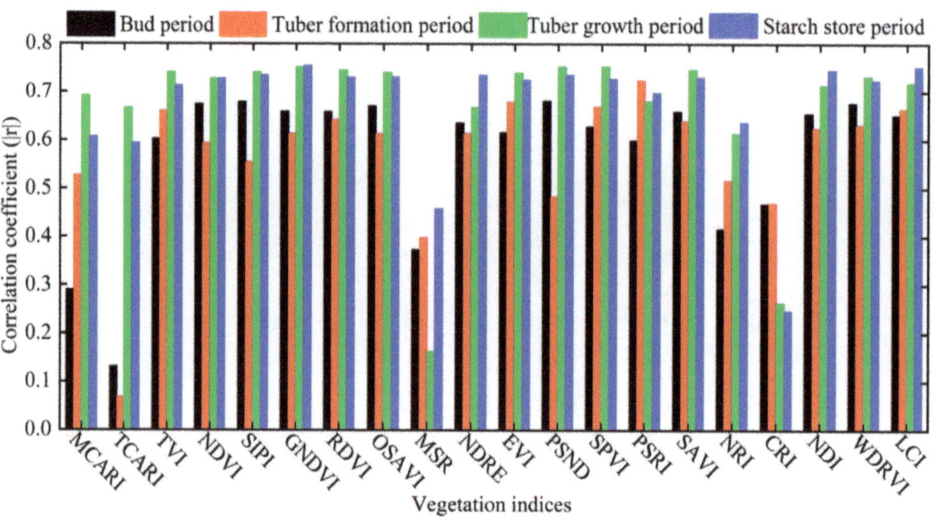

Figure 11. Correlation coefficients of vegetation indexes with measured potato AGB at bud period, tuber-formation period, tuber-growth period, and starch-storage period.

Table 4. Vegetation indexes of AGB estimation model were selected by the MWPLS, MC-UVE, and RF methods for each growth period.

Growth Stage	Variable Selection Method		
	MWPLS	**MC-UVE**	**RF**
Bud period	NDVI, SIPI, GNDVI, NDRE, EVI, PSND	SIPI, PSND, NDI, WDRVI, LCI	SIPI, RDVI, NDRE, EVI, SAVI, NDI
Tuber formation period	NDVI, SIPI, GNDVI, RDVI, OSAVI, MSR, NDRE, EVI, PSND, SPVI	TVI, SIPI, EVI, SPVI, PSRI, NRI, CRI, WDRVI	NDVI, SIPI, MSR, PSRI
Tuber growth period	TVI, NDVI, SIPI, GNDVI, RDVI, OSAVI	MCARI, TCARI, NDVI, NDRE, PSRI	MCARI, NDVI, SIPI, GNDVI, NDRE, PSRI
Starch store period	TVI, NDVI, SIPI, SAVI, NRI, NDI	NDVI, SIPI, SPVI, NRI, NDI, WDRVI	TCARI, EVI, SPVI, NDI, WDRVI, LCI

Based on the screening results in Table 4, the potato AGB estimate for each growth period is constructed by combining the PLSR and GPR methods, respectively. The modeling and validation indicators (R^2, RMSE, and NRSE) of each model are listed in Figure 12. From the perspective of the size of the model evaluation indicators, under the same method, the vegetation index selected by RF in each growth period has the best effect for estimating AGB, followed by MC-UVE, whereas the vegetation index selected by MWPLS produces the worst modeling and verification results. These results are consistent with those listed in Figures 7, 8 and 10. For a given variable, R^2 of the AGB estimation model constructed by PLSR for each growth period is greater than the R^2 of the AGB estimation model constructed by the GPR method, and the RMSE and NRMSE are smaller, indicating that the PLSR method is more conducive to AGB estimation, which is also consistent with the results in Figures 7, 8 and 10.

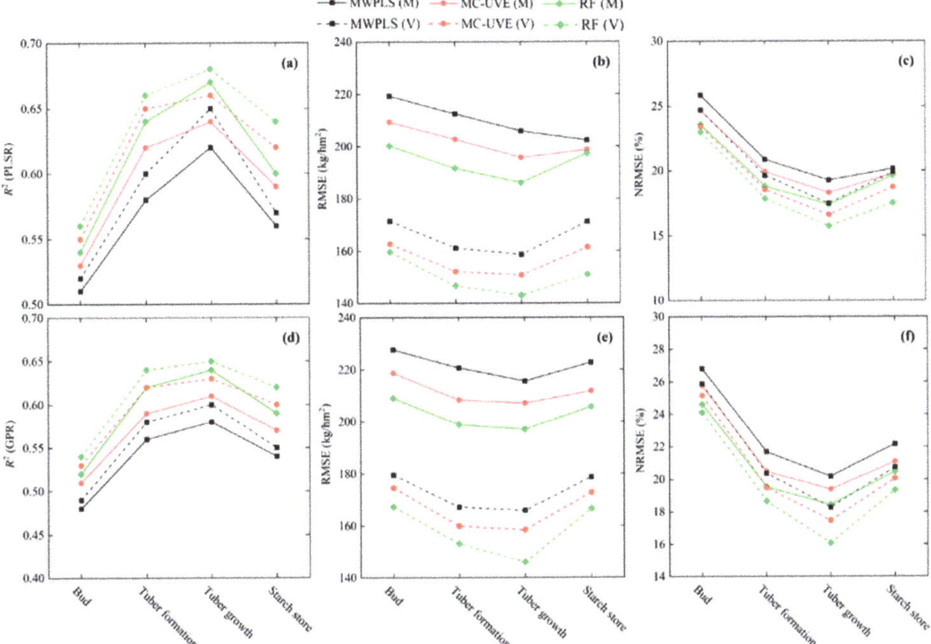

Figure 12. Accuracy of estimating AGB based on the feature variables screened from vegetation indexes by MWPLS, MC-UVE and RF. (**a–c**) The R^2, RMSE (kg/hm^2) and NRMSE (%) of AGB was estimated using the PLSR method. (**d–f**) The R^2, RMSE (kg/hm^2) and NRMSE (%) of AGB was estimated using the GPR method.

Comparing the results listed in Figures 7, 8, 10 and 12 shows that, under the same conditions, the estimates of AGB based on vegetation indexes are more accurate, followed by spectral position features, whereas estimating AGB based on spectral-reflection features is the least accurate. However, the accuracy of the AGB estimates based on three categories goes from high to low from the budding period to the starch-accumulation period. In each growth period, the AGB estimation model established by the RF-PLSR method is also more accurate and more stable. From the budding period to the tuber-formation period, the modeling R^2 increases from 0.54 to 0.67, the RMSE decreases from 200.27 to 186.21 kg/hm^2, and the NRMSE decreases from 23.59% to 17.42%. The verification results are consistent with the modeling results: R^2 continues to increase, RMSE and NRMSE gradually decrease, and the estimation improves. From the tuber-formation period to the starch-accumulation period, R^2 decreases from 0.67 to 0.60, and the RMSE and NRMSE increase from 186.21 to 197.28 kg/hm^2 and from 17.42% to 19.63%, respectively. The verification R^2 also gradually decreases, and the RMSE and NRMSE gradually increase, so the estimation deteriorates.

3.4. Estimation of AGB Using Composite Variables

The model variables (Tables 3 and 4 and Figure 6) extracted by the MWPLS, MC-UVE, and RF screening methods based on three types of spectral features were formed into a new data set, and the PLSR and GPR methods were used to estimate the potato AGB in each growth period. The R^2, RMSE, and NRMSE of the regression results appear in Figure 13, and the fitted scatterplots are shown in Figures 14 and 15. These results show that, under the same conditions, the comprehensive variables serve as model input parameters in each growth period, R^2 is maximal for modeling and verification, and the RMSE and NRMSE are minimal, so the estimation is the best (Figures 7, 8, 10, 12 and 13).

Comparing the influence of the three variable-screening methods on AGB estimation through a comparative analysis of the evaluation indicators shows that the variables selected by the RF method are the best in each growth period, followed by MC-UVE, and those selected by MWPLS are the worst, which is consistent with the results listed in Figures 7, 8, 10 and 12. For the same variables, the AGB estimation model constructed by the PLSR method in each growth period fits slightly more accurately (Figures 14 and 15) and is more stable than the GPR method (Figure 13). The estimations from the budding period to starch-accumulation period first improve and then decrease in accuracy, which is consistent with the estimation of AGB based on three types of spectral characteristics. In each growth period, the RF-PLSR method produces the best AGB estimates, and the model is the most stable. R^2 continues to increase (0.65–0.72) from the budding period to the tuber-growth period, and the corresponding RMSE (167.76–160.77 kg/hm^2) and NRMSE (19.76–15.04%) continue to decrease, so the estimates gradually improve. The verification results (Figures 13 and 14) are similar to the modeling results. R^2 (0.68–0.74) increases, RMSE (136.57–125.48 kg/hm^2) and NRMSE (19.68–13.82%) decrease, and the model gradually improves. From the tuber-growth period to the starch-accumulation period, R^2 decreases from 0.72 to 0.67, RMSE increases from 160.77 kg/hm^2 to 169.24 kg/hm^2, NRMSE increases from 15.04% to 16.84%, so the estimation deteriorates. We verified that the trends of R^2 (0.74–0.70), RMSE (125.48–135.16 kg/hm^2), and NRMSE (13.82–15.68%) are consistent with the modeling set, where R^2 decreases, RMSE and NRMSE increase, and the estimation deteriorates.

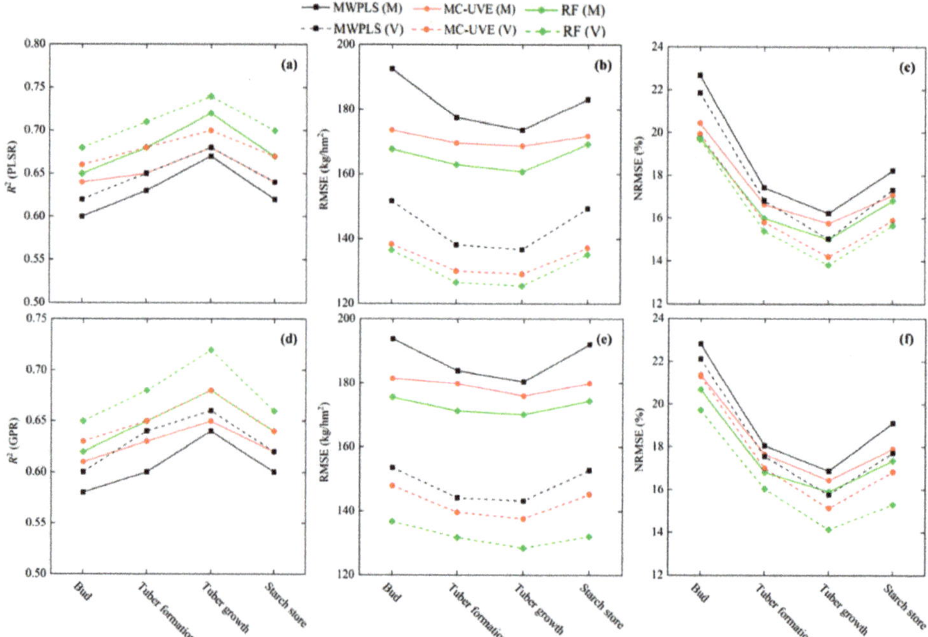

Figure 13. Accuracy of estimating AGB based on the feature variables screened from composite variables by MWPLS, MC-UVE and RF. (**a–c**) The R^2, RMSE (kg/hm^2) and NRMSE (%) of AGB was estimated using the PLSR method. (**d–f**) The R^2, RMSE (kg/hm^2) and NRMSE (%) of AGB was estimated using the GPR method.

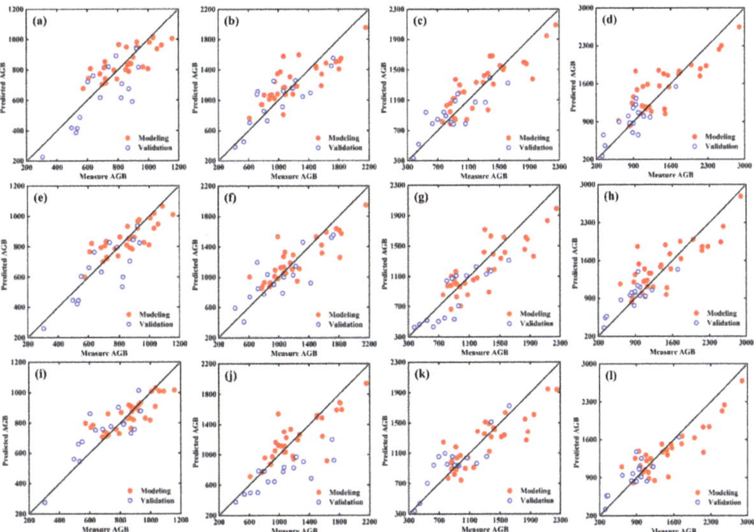

Figure 14. Scatter plots between measured and predicted values of potato AGB (kg/hm²) for modeling and verification data sets using PLSR-based composite variables screened by MWPLS, MC-UVE, and RF at different growth periods. (**a–d**) Relationship between measured and predicted values of potato AGB based on MWPLS-PLSR for bud period, tuber-formation period, tuber-growth period, and starch-storage period, respectively. (**e–h**) Same as panels (**a–d**) but based on MC-UVE-PLSR. (**i–l**) show the same as panels (**a–d**) but based on RF-PLSR.

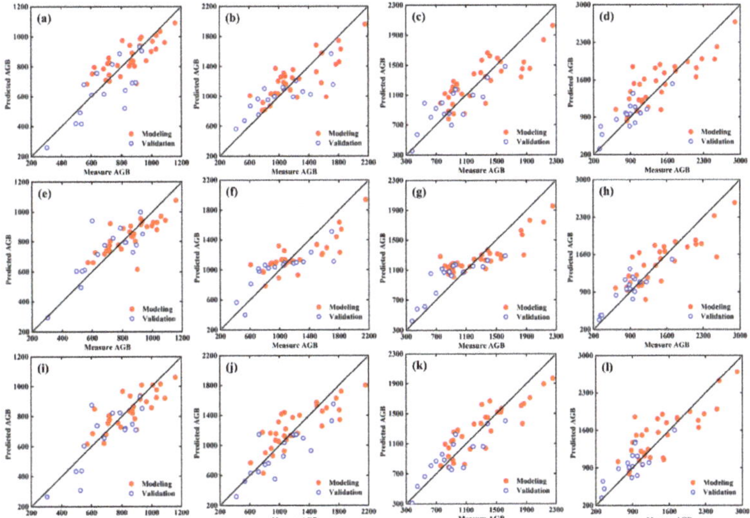

Figure 15. Scatter plots between measured and predicted values of potato AGB (kg/hm²) for modeling and verification data sets using GPR-based composite variables by MWPLS, MC-UVE, and RF at different growth periods. (**a–d**) Relationship between measured and predicted values of potato AGB based on MWPLS-GPR for bud period, tuber-formation period, tuber-growth period, and starch-storage period, respectively. (**e–h**) Same as panels (**a–d**) but based on MC-UVE-PLSR. (**i–l**) show the same as panels (**a–d**) but based on RF-GPR.

4. Discussion

4.1. Correlation between Three Types of Spectral Features and AGB

The analysis of the correlation between spectral variables and physicochemical parameters (such as AGB and LAI) is the premise of building the estimation model. Therefore, this study first applies a Pearson correlation analysis between (i) spectral-reflectance features, spectral-location features, and vegetation indexes and (ii) AGB in each potato-growth period. The results show that differences exist in the correlation between various types of parameters and AGB (Figures 4, 9 and 11), and the number of parameters reaching the extremely significant correlation level is also different, which shows that AGB can be estimated by using different types of spectral variables in each growth period, which is necessary for growth monitoring [34,37]. Overall, the correlation between the three types of spectral variables and AGB first increases and then decreases upon progressing from the budding period to the starch-accumulation period, which may be related to the potato-growth cycle. In the early stage, it is manifested by vegetative and reproductive growth, which is reflected in the elongation of stem nodes and the expansion of leaves. In the tuber-growth period, the fresh weight of stems and leaves reaches a maximum. At this time, the vegetation coverage reaches a maximum, and the acquisition of canopy spectral information is not contaminated by the ground soil. The extracted spectral characteristic parameters fully reflect the AGB. At the later stage of growth, due to the continuous expansion of underground tubers, nutrients accumulated in the early stage of aboveground stems and leaves need to be transferred underground. At the same time, due to the rainy weather, aboveground leaves wither and rapidly fall off, and the growth of potato crops becomes worse, so that the reflected energy of soil pixels contributes greatly to the spectrum of the potato canopy, making it difficult for spectral characteristic parameters to characterize the true AGB situation, thereby reducing the correlation between the two [39].

From the correlation of Figure 4, it is found that the correlation between FDS and AGB in each growth period is greater than that of CRS, which is consistent with the research results of Wang et al. [31], Fan et al. [41], and Chen et al. [52], and also shows that the correlation between FDS and AGB is higher, mainly because the interference of a part of the soil background can be eliminated through spectral differential transformation, the signal-to-noise ratio of canopy spectral information is improved, and the correlation with AGB is enhanced [32]. The correlation of Figure 9 shows that the correlation between the absorption valley parameter and AGB is greater than that of "four sides, one valley and one peak," which is consistent with the research results of Fu et al. [34] and Han et al. [45]. This result is mainly attributed to the spectrum reflected from the potato canopy being unified into the same background through de-envelope line processing, which highlights the absorption and reflection characteristics of the crop canopy from the visible to the near-infrared and increases the difference in canopy spectrum in each plot, making it closely related to AGB. An analysis of the correlation results in Figure 11 shows that most vegetation indexes correlate very significantly with AGB, with TVI, RDVI, SAVI, and NDI having a higher correlation with AGB in each growth period than the other vegetation indexes, mainly because these four vegetation indexes are related to the radiation absorbed by crops, and solar radiation provides input energy for crop photosynthesis. Therefore, the photosynthetic radiation intercepted by crops is closely related to the production of crop dry weight (AGB), which increases the correlation between the four vegetation indexes and AGB [52].

Comparing the correlation analysis results of Figures 4, 9 and 11 shows that the vegetation index and AGB are most strongly correlated, followed by the spectral-location feature, and the spectral-reflectance feature is the most poorly correlated, which is consistent with the research results of Tao et al. [33] and Gong et al. [53]. This is because most of the vegetation indexes selected in this study combine the red and near-infrared bands, and previous studies show that the red–near-infrared band vegetation indexes are most effective for estimating AGB, so these vegetation indexes are more closely related to AGB [54,55]. Due to the unique absorption valley and reflection peak of vegetation, the characteristic parameters of spectral positions are closely related to the growth of potato crops. Therefore, these positions contain significant spectral information, which can also reflect the AGB [44]. When the spectral information is collected, the jitter of the sensor and interference due to the external reflected signal may cause the potato-canopy spectrum to deviate from the actual spectrum, in which case the analysis of correlations with AGB is insufficient to reflect the real situation of the AGB, which may reduce the correlation between (i) FDS and CRS and (ii) AGB [52].

4.2. Estimation of AGB Effect Based on Three Spectral Features

Ultralow-altitude UAV remote-sensing platforms have become the most frequently used technical tool for quantitative monitoring of AGB in precision agriculture due to their high mobility, affordable price, and simple operation [12,28,56]. At present, the sensors on the UAV remote sensing platform are mainly digital cameras, multispectral, hyperspectral, and lidar. Although the price of digital cameras and multispectral sensors is relatively low, they accept only a small number of bands, which limits the spectral information of the crop canopy that can be obtained, making it difficult to interpret the subtle differences between optical spectra. Data obtained by lidar are of high precision but too expensive, which limits its promotion and application for monitoring AGB of crops [40]. Hyperspectral sensors have a significant application potential for making rapid, nondestructive, and accurate estimates of crop AGB because of their high spectral resolution and their ability to obtain images and spectra simultaneously [57]. The estimation of AGB based on UAV hyperspectral image data takes three main forms: spectral-reflectance features [30], spectral-location features [34], and vegetation indexes [36]. Numerous research results only use one type of spectral variable as model-input parameters but fail to fully consider the continuity and diversity of canopy spectral information obtained by hyperspectral sensors, which may prevent the full use of hyperspectral data, thereby restricting the accuracy of the estimation model [50]. Therefore, in this study, the spectral-reflectance feature, spectral-position feature, vegetation index, and their combinations are used as model-input parameters and combined with different modeling methods. This approach allows the potato AGB to be estimated at each growth period (Figures 7, 8, 10, 12 and 13).

From the evaluation indicators of the estimation model in Figures 7, 8, 10, 12 and 13, the estimations based on the four variables in each growth period differ significantly, but the common point is that the estimations based on the four variables gradually improve from the budding period to the tuber-growth period before deteriorating. This trend is the same as that of the correlation between the model variables and AGB. When estimating potato AGB based on a single type of spectral parameters, the R^2 of the model constructed with vegetation indexes as the variable are larger, and the RMSE and NRMSE are smaller (Figure 12), indicative of an improved estimate. The next-best estimates are based on spectral-location characteristic parameters (Figure 10), and the worst estimates are based on spectral-reflection characteristic parameters (Figures 7 and 8), which is related to the correlation between the three spectral parameters and AGB (Figures 4, 9 and 11). When estimating AGB based on spectral-reflection characteristics, the estimation based on FDS is significantly more accurate than that based on CRS, which is consistent with the results of Acquach et al. [57]. This result is mainly because processing hyperspectral data through first-order differentiation refines the spectral information and deeply mines the information hidden in the spectrum, in addition to expanding the selection space of the characteristic wavelength [19]. When estimating AGB based on spectral position features, R^2 of the estimation model in each growth period increases significantly, and the RMSE and NRMSE decrease, which is consistent with the research results of Sun et al. [58]. However, the accuracy of the estimation model constructed in this study for each growth period is slightly low, which is mainly because the estimated physical and chemical parameters differ. The goal of this study is estimating potato AGB, while the goal of Sun et al. was the total nitrogen content of cotton leaves. The second reason is that different types of sensors are used. In this study, the UAV was equipped with a UHD 185 sensor that was used to obtain the spectral data of the field canopy, whereas Sun et al. obtained their spectrum by using an ASD spectrometer in an artificial backpack. The range of field angle and the interference of the environment are easy to control when collecting the data, and the final measurement result is the weighted mean of the many collected spectra. When estimating AGB based on a vegetation index, the estimation model for each growth period is more accurate than the model based on spectral reflection characteristics and spectral position characteristics, which is consistent with the results of Yang et al. [46] and Kang et al. [50], mainly because the vegetation index combines two or more narrow bands in a certain mathematical way, eliminating or reducing the impact of the background and enhancing the vegetation information [11,12]. At the same time, it is the most effective factor for estimating AGB [52]. In this study, taking full account of the diversity of spectral data, three spectral features are combined into a new data set, and the same method is used to estimate the potato AGB for each growth period (Figure 13). Given the same conditions, the R^2 constructed for each growth period is maximal, the RMSE and NRMSE are minimal, and the estimation is the most accurate, which confirms that the accuracy of AGB estimates can be improved by using multiple types of variables from hyperspectral images, thereby compensating for a lack of single-variable estimation accuracy.

4.3. Estimation of AGB Based on Different Variable-Screening Methods

Significant progress has been made in estimating the AGB of winter wheat [33], cotton [58], corn [59], rice [60], soybean [61], and other crops based on UAV hyperspectral image data. However, few reports are available for estimating potato AGB over multiple growth periods. At the same time, most of these studies seek an appropriate vegetation index as the model-input parameter but do not explore how the introduction of new variables affects AGB estimates. Due to the wide spectral range and high spectral resolution of hyperspectral data, rich crop canopy spectral information is obtained. If only vegetation indexes are used to estimate AGB, more sensitive spectral variables may be missed, thereby limiting the accuracy of the estimation model [50]. At the same time, most studies choose the model parameters according to the correlation between the spectral variables and AGB, which leads to the subjective determination of the number of parameters input into the model, thereby preventing an optimal estimation model. Therefore, to better estimate the potato AGB for each growth period, we select three characteristic spectral parameters as the model-input parameters, and the MWPLS, MC-UVE, and RF methods are used to optimize and screen the model parameters of each growth period. Finally, taking the value of RMSECV, RI, and SP as the standard, we determine the optimized model variables for each growth period and estimate the potato AGB (Figures 7, 8, 10, 11 and 13–15).

Based on different spectral characteristics in each growth period, MWPLS, MC-UVE, and RF methods are used to screen variables, and the final models constructed differ significantly in accuracy. However, under the same conditions, the variables obtained by the RF method prove to be the best to estimate AGB, followed by the MC-UVE method, whereas the variables obtained by the MWPLS method are the worst for estimating AGB. This is because the variables screened by the RF method have a wide span, weak autocorrelation between variables, and rich information, Therefore, the accuracy of the model is high, which is consistent with the results of Sun et al. [58]. The performance of the MWPLS method is the worst, mainly because the variables selected by this method are related to the size of the moving window, and the selected variables have strong autocorrelation, which reduces the accuracy of the estimation model, making it consistent with the results of Yang et al. [62]. According to the results in Figure 6 and Tables 3 and 4, the number of variables screened by the RF method and based on the three spectral characteristics is less than the number produced by MWPLS and MC-UVE, but the model constructed in the former case remains the most accurate, indicating that this method may serve to eliminate the "while removing uncorrelated variables", and it can significantly improve the predictive ability and robustness of the AGB estimation model [49]. When using the three variable-screening methods to optimize the spectral-reflectance features, it was found that the final model parameters were mainly located in the position of "four sides, one valley, and one peak" (Figure 6), indicating that these spectral positions contain more information related to potato AGB, which allows for better AGB estimates [32]. When using the same method to screen sensitive model parameters for spectral-location features, the probability of selecting absorption valley parameters to estimate AGB in each growth period is greater than that of the location parameters in "four sides, one valley, and one peak" (Table 3), mainly because the difference in potato-canopy absorption characteristics is more prominent and more potential information may be mined after the continuum-removing transformation, which enhances the differential characteristics of spectral curves of different AGB contents, thereby improving the correlation with AGB [46]. When the MWPLS, MC-UVE, and RF methods are used to screen the vegetation index, NDVI, SIPI, NDRE, and WDRVI have a high probability of being selected by the three methods (Table 4), indicating that these vegetation indexes play an important role in the estimation of potato AGB. This is consistent with the conclusions of Tao et al. [33] and Liu et al. [63], who report that these vegetation indexes are important parameters for estimating AGB.

4.4. Estimation of AGB Based on Different Modeling Methods

In this study, MWPLS, MC-UVE, and RF methods are used to optimize and screen the spectral-reflectance features, spectral-location features, and vegetation indexes of potatoes for each growth stage and then combine with PLSR and GPR methods to construct potato AGB estimation models and finally obtain the R^2, RMSE, and NRMSE of each model (Figures 7, 8, 10 and 12). The results show that, given the same variable, although the estimation models constructed by PLSR and GPR methods for each growth period differ significantly, they trend from good to bad upon going from the budding period to the starch-accumulation period, which is the same trend seen in the correlation between the model parameters and AGB (Figures 4, 9 and 11). Comparing and analyzing how the PLSR and GPR methods affect AGB estimates based on three spectral variables show that the modeling and verification results obtained by the two methods are similar. If R^2 is large, the RMSE and NRMSE are smaller, indicating that the two methods used to build AGB estimation models are reliable and stable [50,64]. Based on three spectral variables, we find that the PLSR method is better than the GPR method for estimating the AGB, which indicates that the PLSR method can effectively improve the AGB estimation accuracy. This is because the PLSR method reduces the dimensions and decomposes the data according to the number of input samples. The estimation model is established after the optimal principal component, which effectively solves the problem of collinearity between variables [37]; the problem of collinearity between variables in the GPR model reduces the prediction accuracy and stability of the model [64].

Analyzing the accuracy indicators of each model (Figures 7, 8, 10 and 12) shows that the model constructed by RF-PLSR method in each growth period is the most accurate. The best R^2 based on spectral-reflection characteristics is 0.49, 0.53, 0.57, and 0.50, and the NRMSE is 29.69%, 23.88%, 21.68%, and 24.19%, which is lower than the accuracy obtained by Sun et al. [65] using this method to estimate the chlorophyll content in potato leaves. The main reasons for this discrepancy are that the physical and chemical parameters estimated by the two approaches differ, and the types of sensors used to collect spectral information differ. In this study, the data were obtained in the field by the UAV platform equipped with a UHD 185 imaging spectrometer, whereas Sun et al. studied data obtained by a Gaia hyperspectral imaging system on a closed laboratory platform. In addition, to monitor the AGB potato canopy plants, Sun et al. used only single leaves as targets to estimate crop parameters. The best R^2 values obtained based on the spectral-location features are 0.52, 0.54, 0.63, and 0.53, and the NRMSE values are 25.92%, 20.77%, 18.87%, and 21.33%, which is more accurate than the results of Tao et al. [33] for estimating the AGB of winter wheat over multiple growth periods with red-edge parameters, mainly because this study extracts all the location parameters related to crop growth in the region "four sides, one valley, and one peak" (Table 1). At the same time, the continuum division method serves to analyze the band depth of the two absorption valleys (Figure 3), which increases the diversity of spectral parameters. The best R^2 values for models based on vegetation indexes are 0.54, 0.64, 0.67, and 0.60 and the corresponding NRMSE values are 23.59%, 18.83%, 17.42%, and 19.63%, which is consistent with the accuracy of crop AGB estimated by Hansen et al. [20] and Liu et al. [63], mainly because the vegetation index selected in this study also includes important model parameters such as NDVI, SIPI, NDRE, and WDRVI.

In this study, by fully considering the diversity of hyperspectral data, three spectral characteristic variables are formed into a new data set to estimate the potato AGB for each growth period. The results show that the accuracy of the models established by PLSR and GPR methods increase significantly (Figure 13), and the model constructed by the PLSR method is more accurate and stable (Figures 13–15). R^2 values for the estimation models for each growth period are 0.65, 0.68, 0.72, and 0.67, and NRMSE values are 19.76%, 16.01%, 15.04%, and 16.84%, respectively, which is consistent with the results of Fan et al. [42], who used a variety of hyperspectral data variables to improve the nitrogen content of summer maize. Future studies should focus on considering the fusion of image features and spectral features for AGB estimation, and test the accuracy of models for each reproductive period at different locations and times.

5. Conclusions

The use of imaging hyperspectral sensors on the UAV platform has great potential for crop AGB monitoring because it allows convenient and fast data acquisition and high spectral and spatial resolution, all at a reasonable price. In view of the continuity and diversity of hyperspectral data, this study extracts spectral-reflectance feature parameters, spectral-location feature parameters, and vegetation indexes. Based on an analysis of the correlation between these parameters and AGB, and to reduce data redundancy and improve the stability of the model, the MWPLS, MC-UVE, and RF methods are used to optimize and screen the spectral parameters to ensure a significant correlation. Finally, these are combined with PLSR and GPR methods to estimate the potato AGB at the budding period, tuber-formation period, tuber-growth period, and starch-accumulation period.

The results show that the vegetation index has the strongest correlation during each growth period and AGB, followed by the spectral-position feature parameter, and then by the spectral-reflectance feature parameter. The change trend of correlation first increases and then deteriorates from the budding stage to starch-accumulation period. When estimating AGB based on different spectral characteristics in each growth period, the variables screened by the RF method are the best, followed by those screened by the MC-UVE method, whereas those screened by the MWPLS method are the worst. When estimating AGB based on a given variable for each growth period, the estimation model obtained by using the PLSR method is more accurate than that obtained by using the GPR method. However, the models constructed by the two methods and the correlations between the model parameters and AGB remain the same: both go from good to bad as the budding stage progresses to the starch-storage period. The model constructed from the comprehensive variables is the most accurate, followed by the model constructed from the vegetation indexes, the spectral-position feature parameters, and the spectral-reflection feature parameters. The RF-PLSR method is optimal for estimating AGB in each growth period based on comprehensive variables, with R^2 = 0.65, 0.68, 0.72, and 0.67, and NRMSE = 19.76%, 16.01%, 15.04%, and 16.84%, respectively.

Author Contributions: H.F., Y.L., G.Y. and J.Y. designed the experiments. H.F., Z.F. and Y.F. collected the AGB and UAV hyperspectral images. Y.L. and H.F. analyzed the data and wrote the manuscript. X.J. and Z.L. made comments and revised the manuscript. All authors have read and agreed to the published version of the manuscript.

Funding: This study was supported by the Key Scientific and Technological Projects of Heilongjiang Province (2021ZXJ05A05), the National Natural Science Foundation of China (41601346), the Platform

Construction Funded Program of Beijing Academy of Agriculture and Forestry Sciences (No. PT2022-24), and the Key Field Research and Development Program of Guangdong Province (2019B020216001).

Acknowledgments: We thank Bo Xu, Hong Chang, Weiguo Li, Yang Meng and Yu Zhao for field management and data collection. We thank the National Precision Agriculture Experiment Station for providing the test site and employees.

Conflicts of Interest: The authors declare no conflict of interest.

References

1. Banerjee, B.P.; Spangenberg, G.; Kant, S. Fusion of spectral and structural information from aerial images for improved biomass estimation. *Remote Sens.* **2020**, *12*, 3164. [CrossRef]
2. Banerjee, B.P.; Raval, S.; Cullen, P.J. UAV-hyperspectral imaging of spectrally complex environments. *Int. J. Remote Sens.* **2020**, *41*, 4136–4159. [CrossRef]
3. Yang, Y.; Liu, B.M.; Ni, X.Y.; Tao, L.Z.; Yu, L.X.; Yang, Y.; Feng, M.X.; Zhong, W.J.; Wu, Y.J. Rice productivity and profitability with slow-release urea containing organic-inorganic matrix materials. *Pedosphere* **2021**, *31*, 511–520. [CrossRef]
4. Morier, T.; Cambouris, A.N.; Chokmani, K. In-season nitrogen status assessment and yield estimation using hyperspectral vegetation indices in a potato crop. *Agron. J.* **2015**, *107*, 1295–1309. [CrossRef]
5. Mahlein, A.K.; Rumpf, T.; Welke, P.; Dehne, H.W.; Plumer, L.; Steiner, U.; Oerke, E.C. Development of spectral indices for detecting and identifying plant diseases. *Remote Sens. Environ.* **2013**, *128*, 21–30. [CrossRef]
6. Thenkabail, P.S.; Smith, R.B.; Pauw, E.D. Hyperspectral vegetation indices and their relationships with agricultural crop characteristics. *Remote Sens. Environ.* **2000**, *71*, 158–182. [CrossRef]
7. Kumar, A.; Tewari, S.; Sing, H.; Kumar, P.; Kumar, N.; Bisht, S.; Devi, S. Biomass accumulation and carbon stock in different agroforestry systems prevalent in the Himalayan foothills, India. *Curr. Sci.* **2021**, *120*, 1083–1088. [CrossRef]
8. Virlet, N.; Sabermanesh, K.; Sadeghi, P.; Hawkesford, M.J. Field Scanalyzer: An automated robotic field phenotyping platform for detailed crop monitoring. *Funct. Plant Biol.* **2016**, *44*, 143–153. [CrossRef]
9. Zhang, H.Y.; Du, H.Y.; Zhang, C.K.; Zhang, L.P. An automated early-season method to map winter wheat using time-series Sentinel-2 data: A case study of Shandong, China. *Comput. Electron. Agric.* **2021**, *182*, 105962–105977. [CrossRef]
10. Huang, J.X.; Sedano, F.; Huang, Y.B.; Ma, H.Y.; Li, X.L.; Liang, S.L.; Tian, L.Y. Assimilating a synthetic Kalman filter leaf area index series into the WOFOST model to improve regional winter wheat yield estimation. *Agric. For. Meteorol* **2016**, *216*, 188–202. [CrossRef]
11. Tao, H.L.; Feng, H.K.; Xu, L.J.; Miao, M.K.; Yang, G.J.; Yang, X.D. Estimation of the yield and plant height of winter wheat using UAV-based hyperspectral images. *Sensors* **2020**, *20*, 1231. [CrossRef] [PubMed]
12. Yue, J.B.; Feng, H.K.; Yang, G.J.; Li, Z.H. A comparison of regression techniques for estimation of above-ground winter wheat biomass using near-surface spectroscopy. *Remote Sens.* **2018**, *10*, 66. [CrossRef]
13. Zhang, C.; Kovacs, J.M. The application of small unmanned aerial systems for precision agriculture: A review. *Precis. Agric.* **2012**, *13*, 693–712. [CrossRef]
14. Yu, N.; Li, L.; Schmitz, N.; Tian, L.F.; Greenberg, J.A.; Diers, B.W. Development of methods to improve soybean yield estimation and predict plant maturity with an unmanned aerial vehicle based platform. *Remote Sens. Environ.* **2016**, *187*, 91–101. [CrossRef]
15. Lydia, S.; Iolanda, F.; Josep, P. Remote sensing of biomass and yield of winter wheat under different nitrogen supplies. *Crop Sci.* **2000**, *40*, 723–731. [CrossRef]
16. David, A.J.; Hernan, D.B.; Jocelyn, C. Graph-based data fusion applied to: Change detection and biomass estimation in rice crops. *Remote Sens.* **2020**, *12*, 2683. [CrossRef]
17. Kanemasu, E.T. Seasonal canopy reflectance patterns of wheat, sorghum, and soybean. *Remote Sens. Environ.* **1974**, *3*, 43–47. [CrossRef]
18. Kooistra, L.; Clevers, J.W. Estimating potato leaf chlorophyll content using ratio vegetation indices. *Remote Sens. Lett.* **2016**, *7*, 611–620. [CrossRef]
19. Daughtry, C.T.; Walthall, C.L.; Kim, M.S.; Colstoun, E.B.; Mcmurtrey, J.E. Estimating corn leaf chlorophyll concentration from leaf and canopy reflectance. *Remote Sens. Environ.* **2000**, *74*, 229–239. [CrossRef]
20. Poley, L.G.; McDermid, G.J. A systematic review of the factors influencing the estimation of vegetation aboveground biomass using unmanned aerial systems. *Remote Sens.* **2020**, *12*, 1052. [CrossRef]
21. Behmann, J.; Mahlein, A.K.; Rumpf, T.; Romer, C.; Plumer, L. A review of advanced machine learning methods for the detection of biotic stress in precision crop protection. *Precis. Agric.* **2015**, *16*, 239–260. [CrossRef]
22. Andres, V.; Gitelson, A.A.; Nguy-Robertson, A.L.; Peng, Y. Comparison of different vegetation indices for the remote assessment of green leaf area index of crops. *Remote Sens. Environ.* **2011**, *115*, 3468–3478. [CrossRef]
23. Melian, J.M.; Jimenez, A.; Diaz, M.; Morales, A.; Horstrand, P.; Guerra, R.; Lopez, S. Real-time hyperspectral data transmission for UAV-based acquisition platform. *Remote Sens.* **2021**, *13*, 850. [CrossRef]
24. Guerra, R.; Lopez, S.; Sarmiento, R. A computationally efficient algorithm for fusing multispectral and hyperspectral images. *IEEE Trans. Geosci. Remote Sens.* **2016**, *54*, 5712–5728. [CrossRef]

25. Guo, A.T.; Huang, W.J.; Dong, Y.Y.; Ye, H.C.; Ma, H.Q.; Liu, B. Wheat yellow rust detection using UAV-based hyperspectral technology. *Remote Sens.* **2021**, *13*, 123. [CrossRef]

26. Li, C.C.; Ma, C.Y.; Cui, Y.Q.; Lu, G.Z. UAV hyperspectral remote sensing estimation of soybean yield based on physiological and ecological parameter and meteorological factor in China. *J. Indian Soc. Remote Sens.* **2021**, *49*, 873–886. [CrossRef]

27. Liu, T.; Shi, T.Z.; Zhang, H.; Wu, C. Detection of rise damage by leaf folder (Cnaphalocrocis medinalis) using unmanned aerial vehicle based hyperspectral data. *Sustainability* **2020**, *12*, 9343. [CrossRef]

28. Pugh, N.A.; Horne, D.W.; Murray, S.C.; Carvalho, G.; Malambo, L.; Jung, J.H. A Temporal estimates of crop growth in sorghum and maize breeding enabled by unmanned aerial systems. *Plant Phenome J.* **2017**, *28*, 170006–170016. [CrossRef]

29. Astor, T.; Dayananda, S.; Nautiyal, S. Vegetable crop biomass estimation using hyperspectral and RGB 3D UAV data. *Agronomy* **2020**, *10*, 1600. [CrossRef]

30. Wang, Y.; Li, F.L.; Wang, W.D.; Chen, X.K.; Chang, Q.R. Hyperspectral remote sensing of shoot biomass of winter wheat based on SPA and transformation spectra. *J. Triticeae Crops* **2020**, *40*, 1389–1398. [CrossRef]

31. Jia, M.; Li, W.; Wang, K.; Zhou, C.; Cheng, T.; Tian, Y.; Zhu, Y.; Cao, W. A newly developed method to extract the optimal hyperspectral feature for monitoring leaf biomass in wheat. *Comput. Electron. Agric.* **2019**, *165*, 104942. [CrossRef]

32. Kong, Q.M.; Su, Z.B.; Shen, W.Z.; Zhang, B.F.; Wang, J.B.; Ji, N. Research of straw biomass based on NIR by wavelength selection of IPLS-SPA. *Spectrosc. Spectr. Anal.* **2015**, *35*, 1233–1238. [CrossRef]

33. Tao, H.L.; Feng, H.K.; Xu, X.J.; Miao, M.K.; Long, H.L.; Yue, J.B.; Li, Z.H. Estimation of crop growth parameters using UAV-based hyperspectral remote sensing data. *Sensors* **2020**, *20*, 1296. [CrossRef] [PubMed]

34. Fu, Y.Y.; Wang, J.H.; Yang, G.J.; Song, X.Y.; Xu, X.G.; Feng, H.K. Band depth analysis and partial least square regression based winter wheat biomass estimation using hyperspectral measurements. *Spectrosc. Spectr. Anal.* **2013**, *33*, 1315–1319. [CrossRef]

35. Gnyp, M.L.; Bareth, G.; Li, F.; Lenz-Wiedemann, V.S.; Koppe, W.G.; Miao, Y.X.; Hennig, S.D. Development and implementation of a multiscale biomass model using hyperspectral vegetation indices for winter wheat in the North China Plain. *Int. J. Appl. Earth Obs. Geoinf.* **2014**, *33*, 232–242. [CrossRef]

36. Jin, X.L.; Li, Z.H.; Feng, H.K.; Ren, Z.B.; Li, S.K. Estimation of maize yield by assimilating biomass and canopy cover derived from hyperspectral data into the Aqua Crop model. *Agric. Water Manag.* **2019**, *227*, 105846–105856. [CrossRef]

37. Hansen, P.M.; Schjoerring, J.K. Reflectance measurement of canopy biomass and nitrogen status in wheat crops using normalized difference vegetation indices and partial least squares regression. *Remote Sens. Environ.* **2003**, *86*, 542–553. [CrossRef]

38. Liu, C.; Liu, Y.; Lu, Y.H.; Liao, Y.L.; Nie, J.; Yuan, X.L. Use of a leaf chlorophyll content index to improve the prediction of above-ground biomass and productivity. *PeerJ* **2019**, *6*, 6240–6255. [CrossRef]

39. Liu, Y.; Feng, H.K.; Yue, J.B.; Li, Z.H.; Yang, G.J. Remote-sensing estimation of potato above-ground biomass based on spectral and spatial features extracted from high-definition digital camera images. *Comput. Electron. Agric.* **2022**, *198*, 107089–107099. [CrossRef]

40. Li, T.S.; Zhu, Z.; Cui, J.; Chen, J.H.; Shi, X.Y.; Zhao, X. Monitoring of leaf nitrogen content of winter wheat using multi-angle hyperspectral data. *Int. J. Remote Sens.* **2021**, *42*, 4676–4696. [CrossRef]

41. Arroyo-Mora, J.P.; Kalacska, M.; Loke, T.; Schlapfer, D.; Coops, N.C.; Lucanus, O. Assessing the impact of illumination on UAV pushbroom hyperspectral imagery collected under various cloud cover conditions. *Remote Sens. Environ.* **2021**, *258*, 112396–112410. [CrossRef]

42. Fan, L.L.; Zhao, J.L.; Xu, X.G.; Liang, D.; Yang, G.J.; Feng, H.K.; Yang, H. Hyperspectral-based estimation of leaf nitrogen content in corn using optimal selection of multiple spectral variables. *Sensors* **2019**, *19*, 2898. [CrossRef] [PubMed]

43. Marabel, M.; Alvarez-Taboada, F. Spectroscopic determination of aboveground biomass in grasslands using spectral transformations, support vector machine and partial least squares regression. *Sensors* **2014**, *13*, 10027. [CrossRef] [PubMed]

44. Kokaly, R.F.; Clark, R.N. Spectroscopic determination of leaf biochemistry using band-depth analysis of absorption features and stepwise multiple linear regression. *Remote Sens. Environ.* **1999**, *67*, 267–287. [CrossRef]

45. Han, X.; Wong, Y.S.; Song, M.H.; Tam, N.Y. Feasibility of using microalgal biomass cultured in domestic wastewater for the removal of chromium pollutants. *Water Environ. Res.* **2008**, *80*, 647–653. [CrossRef]

46. Yang, H.B.; Li, F.; Wang, W.; Yu, K. Estimating above-ground biomass of potato using random forest and optimized hyperspectral indices. *Remote Sens.* **2021**, *13*, 2339. [CrossRef]

47. Zhe, L.; Lee, Y.S.; Chen, J.H.; Qian, Y.W. Developing variable moving window PLS models: Using case of NOx emission prediction of coal-fired power plants. *Fuel* **2021**, *296*, 120441–120457. [CrossRef]

48. Zhang, J.; Cui, X.Y.; Cai, W.S.; Shao, X.G. A variable importance criterion for variable selection in near-infrared spectral analysis. *Sci. China Chem.* **2019**, *62*, 271–279. [CrossRef]

49. Fan, N.Y.; Liu, G.S.; Zhang, J.J.; Zhang, C.; Yuan, R.R.; Ban, J.J. Hyperspectral model optimization for protein of tan mutton based on Box-Behnken. *Spectrosc. Spectr. Anal.* **2021**, *41*, 918–923. [CrossRef]

50. Kang, X.Y.; Zhang, A.W.; Pang, H.Y. Estimation of grassland above ground biomass from UAV-mounted hyperspectral image by optimized spectral reconstruction. *Spectrosc. Spectr. Anal.* **2021**, *41*, 250–256. [CrossRef]

51. Verrelst, J.; Rivera, J.P.; Gitelson, A.; Delegido, J.; Moreno, J.; Camps-Valls, G. Spectral band selection for vegetation properties retrieval using Gaussian processes regression. *Int. J. Appl. Earth Obs. Geoinf.* **2016**, *52*, 554–567. [CrossRef]

52. Chen, J.; Gu, S.; Shen, M.G.; Tang, Y.H.; Matsushita, B. Estimating above ground biomass of grassland having a high canopy cover: An exploratory analysis of in situ hyperspectral data. *Int. J. Remote Sens.* **2009**, *30*, 6497–6517. [CrossRef]

53. Gong, Z.; Kawamura, K.; Ishikawa, N.; Inaba, M.; Alateng, D. Estimation of herbage biomass and nutritive status using band depth features with partial least squares regression in Inner Mongolia grassland, China. *Grassl. Sci.* **2016**, *62*, 45–54. [CrossRef]
54. Cho, M.A.; Skidmore, A.; Corsi, F.; Sobhan, I. Estimation of green grass/herb biomass from airborne hyperspectral imagery using spectral indices and partial least squares regression. *Int. J. Appl. Earth Obs. Geoinf.* **2007**, *9*, 414–424. [CrossRef]
55. Fu, Y.Y.; Yang, G.J.; Wang, J.; Feng, H.K.; Xu, B. Winter wheat biomass estimation based on spectral indices, band depth analysis and partial least squares regression using hyperspectral measurements. *Comput. Electron. Agric.* **2014**, *100*, 51–59. [CrossRef]
56. Niu, Y.X.; Zhang, L.Y.; Zhang, H.H.; Han, W.T.; Peng, X.S. Estimating above-ground biomass of maize using features derived from UAV-based RGB imagery. *Remote Sens.* **2019**, *11*, 1261. [CrossRef]
57. Acquach, G.E.; Via, B.K.; Fasina, O.; Eckhardt, L.G. Non-destructive prediction of the properties of forest biomass for chemical and bioenergy applications using near infrared spectroscopy. *J. Near Infrared Spectrosc.* **2015**, *23*, 93–102. [CrossRef]
58. Sun, L.; Chen, X.; Wu, J.J.; Feng, X.W.; Bao, A.M.; Ma, Y.Q. Study on the biomass change derived from the hyperspectral data of cotton leaves in canopy under moisture stress. *Chin. Sci. Bull.* **2006**, *51*, 173–178. [CrossRef]
59. Li, C.C.; Cui, Y.Q.; Ma, C.Y.; Niu, Q.L.; Li, J.B. Hyperspectral inversion of maize biomass coupled with plant height data. *Crop Sci.* **2021**, *61*, 2067–2079. [CrossRef]
60. Kanke, Y.; Tubana, B.; Dalen, M.; Harrell, D. Evaluation of red and red-edge reflectance-based vegetation indices for rice biomass and grain yield prediction models in paddy fields. *Precis. Agric.* **2016**, *17*, 507–530. [CrossRef]
61. Akhtar, K.; Wang, W.Y.; Khan, A.; Ren, G.X.; Afridi, M.Z.; Feng, Y.Z. Wheat straw mulching offset soil moisture deficient for improving physiological and growth performance of summer sown soybean. *Agric. Water Manag.* **2019**, *211*, 16–25. [CrossRef]
62. Yang, B.H.; Chen, J.L.; Chen, L.H.; Cao, W.X.; Yao, X.; Zhu, Y. Estimation model of wheat canopy nitrogen content based on sensitive bands. *Trans. Chin. Soc. Agric. Eng.* **2015**, *31*, 176–182. [CrossRef]
63. Liu, Y.; Feng, H.K.; Huang, J.; Yang, F.Q.; Wu, Z.C.; Sun, Q.; Yang, G.J. Estimation of potato above-ground biomass based on hyperspectral characteristic parameters of UAV and plant height. *Spectrosc. Spectr. Anal.* **2021**, *41*, 903–911. [CrossRef]
64. Fu, Y.Y.; Yang, G.J.; Li, Z.H.; Song, X.Y.; Li, Z.H.; Xu, X.G.; Wang, P.; Zhao, C.J. Winter wheat nitrogen status estimation using UAV-based RGB imagery and Gaussian processes regression. *Remote Sens.* **2020**, *12*, 3778. [CrossRef]
65. Sun, H.; Zheng, T.; Liu, N.; Li, M.Z.; Zhang, Q. Vertical distribution of chlorophyll in potato plants based on hyperspectral imaging. *Trans. Chin. Soc. Agric. Eng.* **2018**, *34*, 149–156. [CrossRef]

Article

Building Extraction and Floor Area Estimation at the Village Level in Rural China Via a Comprehensive Method Integrating UAV Photogrammetry and the Novel EDSANet

Jie Zhou [1,2], Yaohui Liu [3,4,5,*], Gaozhong Nie [1,2], Hao Cheng [6], Xinyue Yang [3], Xiaoxian Chen [3] and Lutz Gross [5]

[1] Institute of Geology, China Earthquake Administration, Beijing 100029, China
[2] Key Laboratory of Seismic and Volcanic Hazards, China Earthquake Administration, Beijing 100029, China
[3] School of Surveying and Geo-Informatics, Shandong Jianzhu University, Jinan 250101, China
[4] College of Geodesy and Geomatics, Shandong University of Science and Technology, Qingdao 266590, China
[5] School of Earth and Environmental Sciences, The University of Queensland, Brisbane, QLD 4072, Australia
[6] School of Remote Sensing and Information Engineering, Wuhan University, Wuhan 430079, China
* Correspondence: liuyaohui20@sdjzu.edu.cn

Abstract: Dynamic monitoring of building environments is essential for observing rural land changes and socio-economic development, especially in agricultural countries, such as China. Rapid and accurate building extraction and floor area estimation at the village level are vital for the overall planning of rural development and intensive land use and the "beautiful countryside" construction policy in China. Traditional in situ field surveys are an effective way to collect building information but are time-consuming and labor-intensive. Moreover, rural buildings are usually covered by vegetation and trees, leading to incomplete boundaries. This paper proposes a comprehensive method to perform village-level homestead area estimation by combining unmanned aerial vehicle (UAV) photogrammetry and deep learning technology. First, to tackle the problem of complex surface feature scenes in remote sensing images, we proposed a novel Efficient Deep-wise Spatial Attention Network (EDSANet), which uses dual attention extraction and attention feature refinement to aggregate multi-level semantics and enhance the accuracy of building extraction, especially for high-spatial-resolution imagery. Qualitative and quantitative experiments were conducted with the newly built dataset (named the rural Weinan building dataset) with different deep learning networks to examine the performance of the EDSANet model in the task of rural building extraction. Then, the number of floors of each building was estimated using the normalized digital surface model (nDSM) generated from UAV oblique photogrammetry. The floor area of the entire village was rapidly calculated by multiplying the area of each building in the village by the number of floors. The case study was conducted in Helan village, Shannxi province, China. The results show that the overall accuracy of the building extraction from UAV images with the EDSANet model was 0.939 and that the precision reached 0.949. The buildings in Helan village primarily have two stories, and their total floor area is 3.1×10^5 m^2. The field survey results verified that the accuracy of the nDSM model was 0.94; the RMSE was 0.243. The proposed workflow and experimental results highlight the potential of UAV oblique photogrammetry and deep learning for rapid and efficient village-level building extraction and floor area estimation in China, as well as worldwide.

Keywords: building extraction; floor area estimation; rural China; deep learning; UAV

Citation: Zhou, J.; Liu, Y.; Nie, G.; Cheng, H.; Yang, X.; Chen, X.; Gross, L. Building Extraction and Floor Area Estimation at the Village Level in Rural China Via a Comprehensive Method Integrating UAV Photogrammetry and the Novel EDSANet. *Remote Sens.* **2022**, *14*, 5175. https://doi.org/10.3390/rs14205175

Academic Editors: Giovanni Laneve, Chenghai Yang, Wenjiang Huang and Yingying Dong

Received: 4 September 2022
Accepted: 13 October 2022
Published: 16 October 2022

Publisher's Note: MDPI stays neutral with regard to jurisdictional claims in published maps and institutional affiliations.

1. Introduction

Homesteads are an important part of basic rural geographic information and multifunctional complex spaces for rural residents [1–3]. With the advancement of urban–rural economic integration in China, many farmers have migrated to cities. From 2000 to 2016, the rural resident population in China decreased from 808 to 589 million (a decline of 27.1%) [4]. The migration of rural residents to cities reduces the area of rural homestead land. However, due to the free acquisition and use of the homestead system, local governments launch new rural construction without proper scientific planning, which has increased the area of idle rural homesteads by 20.6% [5], from 0.99 to 1.21 million km^2 [4]. Compared to developed cities, rural areas are dominated by low-rise buildings, and the excessive occupation of land resources by farmers affects land-use efficiency [6]. To promote rural development, the Chinese government has proposed "beautiful countryside" construction. In-depth investigations should be conducted on the living conditions of farmers, and land-use areas in rural areas should be rationally planned. Field surveys can provide accurate information about the residents of farms but require time and labor. Moreover, land use for rural homesteads in developing countries is usually scattered, resulting in barriers to the acquisition of rural building information. Therefore, additional methods should be proposed to quickly and accurately extract building information and estimate floor area in rural environments.

To ameliorate adverse social problems, building density regulations (such as those for building heights or floor area ratios) are common practices in urban planning and management worldwide [6]. Various remote sensing products and classification methods have been used to extract building coverage areas [7,8] and building heights [9,10]; the nDSM [11] (the difference between a DSM and a digital terrain models (DTM)) is widely used in height estimation [12]. Ji and Tang [13] proposed three methods for gross floor area estimation from monocular optical imagery using the NoS R-CNN model. Given the densely populated villages and scattered land-use layout in China, UAVs have become the latest trend in rural homestead detection because of their flexibility, low cost, real-time results, and high resolution [14]. Nyaruhuma et al. [15] used oblique photogrammetry to reconstruct 3D buildings on an urban scale. High-resolution UAV images can obtain sufficiently detailed information and provide new challenges to existing methods of building extraction [16,17]. Previous studies have primarily focused on the extraction of architectural features based on machine learning, including maximum likelihood classification [18], support vector machines [19], and object-based classification methods [20]. However, machine learning algorithms based on feature extraction rely heavily on manual parameter setting and expert knowledge, which usually leads to poor generalization with different environmental backgrounds [21,22]. In rural areas with more complex surface compositions, the use of traditional algorithms for ground object classification can be improved further [23].

Owing to the complexity of image backgrounds and the semantic texture of buildings, automatic and high-precision building extraction from UAV images presents uncertainty [24,25]. Recently, scholars have employed deep learning technology to identify building contour information [26–29]. Long et al. proposed the FCN model for pixel-level semantic segmentation, which is the first end-to-end fully convolutional network that accepts any size input for image segmentation, and it has successfully led to a new wave of semantic segmentation tasks [30]. Subsequently, many variant FCN-based models have improved the feature expression capabilities to obtain better experimental results (such as SegNet [31], U-Net [32], and ERFNet [33]). Liu et al. [34] proposed a novel convolutional neural network combined encoder–decoder and spatial pyramid pooling module named USPP for building extraction from high-resolution remote sensing images. Konstantinidis et al. [35] proposed a modular CNN to improve the performance of building detectors by employing a histogram of oriented gradients and local binary patterns in a remote sensing dataset. Zhang [36] developed a method for estimating homestead areas based on UAV images and the U-Net algorithm. The results demonstrate that, in rural areas with complex surface compositions, the deep learning method can achieve fast, stable, and high-precision results. Liao et al. [37] proposed a boundary-preserved model that works by jointly learning the contours and structures of buildings. Experiments on the WHU, Aerial, and Massachusetts Building Datasets showed that the proposed model outperformed other state-of-the-art methods. Xiao et al. [38] proposed a shifted-window transformer-based encoding booster to capture the semantic information of large buildings in high-resolution remote sensing images. Li et al. [39] proposed a novel end-to-end network integrating lightweight spatial and channel attention modules to refine features adaptively for building extraction tasks. Wei et al. [40] proposed a multi-branch network for the extraction of rural homesteads based on aerial images. Jing et al. [41] proposed an efficient memory module to enhance the learning ability of deep learning models in building extraction. Li et al. [42] proposed a global style and local matching contrastive learning model for image-level and pixel-level representation. However, most existing deep learning models focus on stacking complex architectures and parameter settings to improve accuracy, which also has disadvantages, such as requiring extensive calculations and slow iteration speed [43]. Moreover, in the extraction of comprehensive building information, high-resolution remote sensing images cannot directly identify the numbers of floors in homesteads. The use of remote sensing data with high spatiotemporal resolution to estimate the area of village-level homesteads at the pixel level still remains challenging. Comprehensive methods and models should be combined with building extraction and floor area estimation at the village level.

Here, we propose a comprehensive method for building extraction and floor area estimation of village-level homesteads by combining UAV oblique photogrammetry and deep learning technology. First, the footprint of buildings is identified using the novel EDSANet model proposed, which employs dual attention extraction and attention feature refinement to enhance the accuracy of building extraction. Then, the number of floors of each building is estimated using the nDSM generated from UAV remote sensing. The total floor area of the homestead is rapidly calculated by multiplying the floor area of each building by the number of floors. A case study was conducted in Helan village, Shaanxi province, China. The experiments demonstrate that the proposed method can achieve rapid and low-cost results in building extraction and floor area estimation in rural villages. To summarize, the main contributions of this paper are as follows:

(1) We propose a comprehensive method combining UAV oblique photogrammetry and deep learning technology for building extraction and floor area estimation of village-level homesteads. A novel EDSANet model is proposed to tackle the problem of complex surface feature scenes in remote sensing images and improve performance in building extraction;

(2) We designed a semantic encoding module by applying three down-sample stages (with atrous convolution) to enlarge the receptive field and a spatial information encoding module with only six layers and three stages using one eighth of the original input to enrich spatial details and improve the accuracy in building extraction;

(3) A dual attention module is proposed to extract useful information from the kernel and channel, respectively. To adjust the excessive convergence of building feature information after attention extraction, we propose an attention feature refinement module to further improve the extraction effect of the model for useful features by redefining the attention features, thereby improving the accuracy.

The remainder of this paper is organized as follows: Section 2 describes the study area and data. Section 3 presents deep learning methods for building extraction and the UAV oblique photogrammetry method for floor area estimation. Section 4 introduces the results of the building extraction and floor area estimation. The discussion and conclusions are presented in Sections 5 and 6, respectively.

2. Study Area and Data

2.1. Study Area

Weinan City is located in Shaanxi province, China, from $34°13'E$ to $35°52'E$ and $108°50'N$ to $110°38'N$. According to the 2017 census, the total population of the city is approximately 5.38 million, and it has an area of 13,134 km^2. Since 2018, Shaanxi province has vigorously promoted rural innovation and reform and accelerated the implementation of the rural revitalization strategy. The pilot reform project in Weinan city achieved remarkable success. Based on a field survey, this research selected Helan village, Fuping county, Weinan city, Shaanxi province as the research area. The village is located between the Guanzhong Plain and the northern Shaanxi Plateau. The village has an area of

3.88×10^4 m^2 with 205 households (of which 151 are residents) and a registered population of 321. The buildings are densely distributed in the research area, the village roads are planted with regular arbor forests, and parts of the homesteads are shaded by tall trees or shrubs. An overview of the study area is presented in Figure 1.

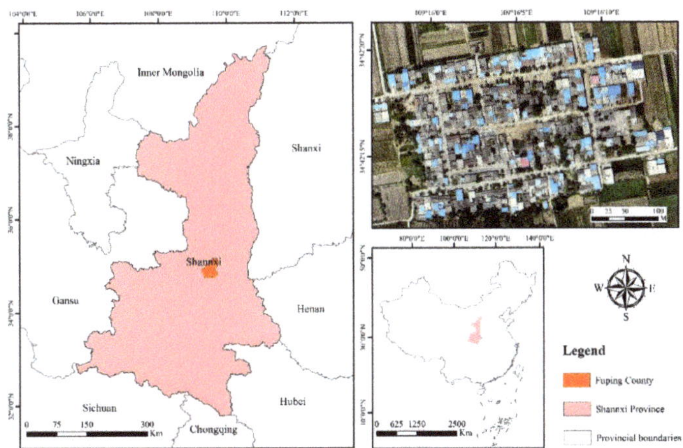

Figure 1. The geographical location of the study area.

2.2. UAV Data

The experimental data utilized in this research were tokens from a small four-rotor unmanned aerial vehicle (UAV). The drone model was an INSPIRE 2 (Shenzhen DJI Innovation Technology Co., Ltd., Shenzhen, China) equipped with a Zenmuse X5s HD camera, an effective pixel count of 16 million for the four thirds CMOS, and a built-in optical imaging lens camera composed of nine glass sheets in seven groups. The UAV was equipped with GPS and GLONASS dual satellite navigation systems, which can be used to autonomously plan the flight path in a study area. Table 1 presents detailed information on the UAV equipment.

Table 1. Detailed information on the UAV equipment.

Parameters	Value
Takeoff Weight	1280 g
Image Size	4608×3456
Flight Duration	27 min
Focal Length	15 mm
Ground Sample Distance	0.23 cm
Spectral Range	0.38–0.76 μm
Working Temperature	0–40°
Maximum Flight Altitude	6000 m
Maximum Horizontal Flight Speed	18 m/s
GPS Module	GPS/GLONASS dual mode
Image Coordinate System	WGS 84/UTM Zone 49N
UAV Flight Permission	Needed

A warm, clear, and windless day (2 August 2018) was chosen to ensure stability for the UAV photography. The flight track ranged from 108°50′E to 110°38′E and 34°13′N to 35°52′N (Figure 2). The flight route was from the southeast corner to the northwest corner of the study area, and pictures were taken along the S route. To construct photogrammetric stereo pairs, the two adjacent images were set with an 85% heading overlap and 75% inside overlap. The spatial resolution of the UAV data reached 2.3 cm.

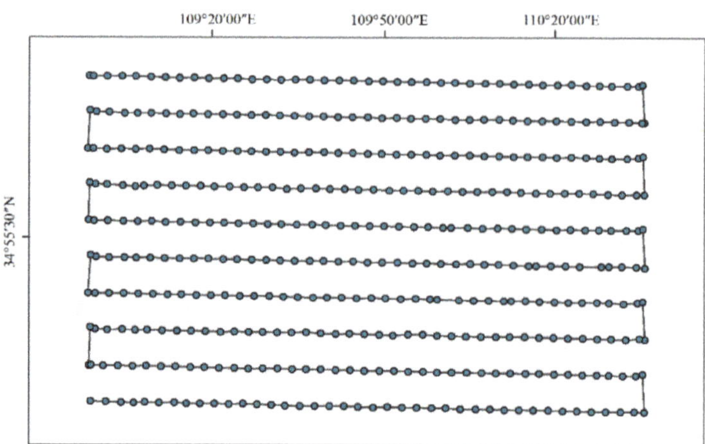

Figure 2. Flight route map in the research area.

3. Methodology

Figure 3 illustrates the detailed workflow, which includes seven principal steps. The first and second steps consisted of obtaining the orthophoto of the research area from the aerial UAV images. The orthophoto of the research area and the building sample dataset were produced through data preprocessing and augmentation. The proposed EDSANet model was then used to extract the building footprint of the study area, the accuracy was evaluated using five metrics, and the segmented images were merged into an entire image. Based on the UAV point cloud data, the tilt photogrammetry method was applied to generate the DSM, DTM, and nDSM to determine the building height. Lastly, the floor area of the homesteads in the study area was calculated based on the building footprints and the numbers of floors.

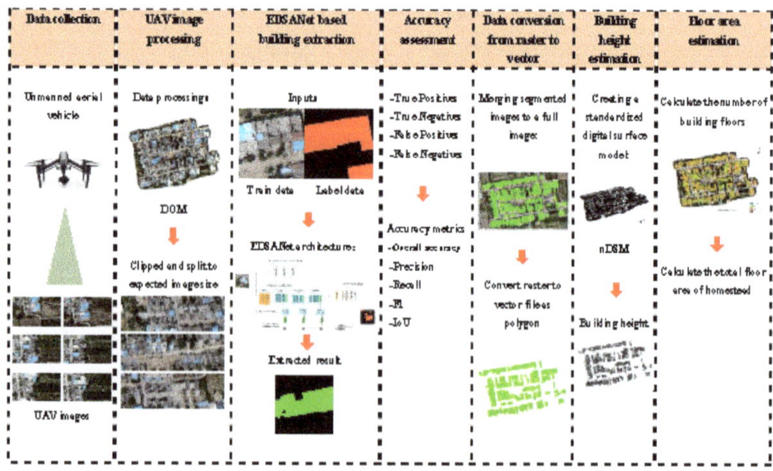

Figure 3. Flowchart of building extraction and floor area estimation in this research.

3.1. Methodology

3.1.1. EDSANet Architecture

We propose a novel fully connected network named the Efficient Deep-wise Spatial Attention Network (EDSANet) to tackle the problem of complex surface feature scenes in remote sensing images and improve the efficiency and accuracy of building extraction tasks. Figure 4 shows an overview of the EDSANet architecture, including two branch networks composed of four units. (1) We first designed a semantic encoding module (SEM, Figure 4b), which employs channel splitting and shuffling to reduce computation and maintain higher segmentation accuracy. (2) A dual attention module (DAM, Figure 4d), consisting of spatial attention and channel attention, and an attention feature refinement module (AFRM, Figure 4e) were designed to make full use of the multi-level feature maps simultaneously, which helps predict the pixel-wise labels in each stage. (3) A spatial information encoding module (SIEM, Figure 4a) was used to enhance spatial semantic information and preserve spatial details. (4) We developed a simple feature fusion module (FFM, Figure 4c) to better aggregate the context information and spatial information [44].

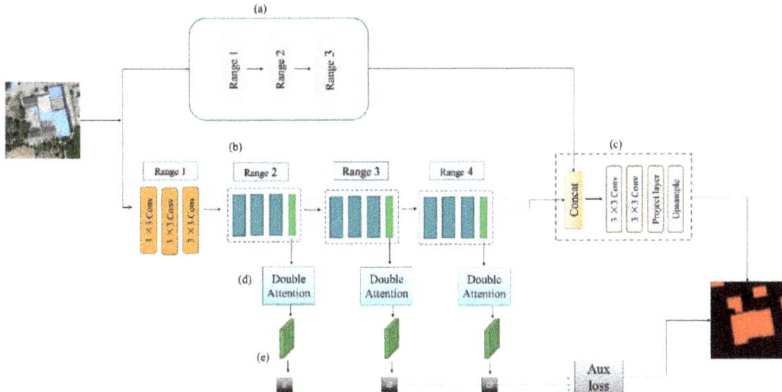

Figure 4. The architecture of the EDSANet model consists of two parts: the semantic encoding branch and the spatial information encoding branch. (**a**) Spatial information encoding module, (**b**) semantic encoding module, (**c**) feature fusion module, (**d**) dual attention module, and (**e**) attention feature refinement module.

First, input images are fed into the SEM to generate four feature maps ($F_{h,1}$, $F_{h,2}$, $F_{h,3}$, $F_{h,4}$) with decreasing spatial resolution. The feature maps $F_{h,3}$ and $F_{h,4}$ have the same numbers of channels, with different dilation rates, to enlarge the receptive field convolutional filters. Then, inspired by the efficiency of dilated convolution [45], we adopted a one-eighth down-sample strategy. As Equation (1) shows, the final segmentation $FFM_{h,s}$ is obtained by combining the high-resolution feature map F_h with the spatial feature map F_s from SIEM.

$$FFM_{h,s} = F_{up}(conv([F_h, F_s])),\qquad(1)$$

3.1.2. Semantic Encoding Module (SEM)

This building block was designed with inspiration from lightweight image classification model strategies, such as in Ma et al. [46], Zhang et al. [47], and Sandler et al. [48]. The models mentioned above set the ratio of the input image resolution by applying five down-samplings and the size of the final output is only 1/32 of the input image size, which can lead to a significant loss in the spatial details. As Table 2 shows, our proposed SEM is based on this building block and applies three down-samplings (the output resolution is only one eighth of the original image resolution with 32, 64, and 128 channels). In stages three and four, atrous convolution is introduced to increase the receptive field.

3.1.3. Spatial Information Encoding Module (SIEM)

To improve the performance of semantic segmentation, the model aimed to effectively combine high-level semantics and low-level details. As the SEM was not designed for spatial details or low-level information, in the shallow SIEM, which has only six layers and three stages, each layer consists of a convolution operation (Conv), batch normalization

(BN), and a parametric rectified linear unit (PReLU) [49]. The first and second layers of each stage have the same number of filters (stride of 2) and output feature map size. Therefore, one eighth of the original input is extracted by the SIEM, which enriches the spatial details due to the high channel capacity.

Table 2. SEM is used to extract high-level semantic information.

Stage	Type	Filters
Input		
Stage 1	3 × 3 Conv	32
Stage 2	Down-sample	64
Stage 3	Down-sample	128
Stage 4	Building block	128

3.1.4. Dual Attention Module (DAM)

For the spatial dimension, we designed an attention mechanism based on kernel attention named the kernel attention module (KAM). For the channel dimension, the number of input channels C is normally far less than the number of pixels contained in the feature maps (i.e., $C \ll N$). Therefore, the complexity of the Softmax function for channels is not high. Thus, we utilized a channel attention mechanism based on the dot-product [50] named the channel attention module (CAM). As Figure 5 shows, using the KAM, which models the long-range dependencies of positions, and CAM, which models the long-range dependencies of channels, we designed the dual attention module (DAM) to enhance the discriminative ability of the feature maps extracted by each layer.

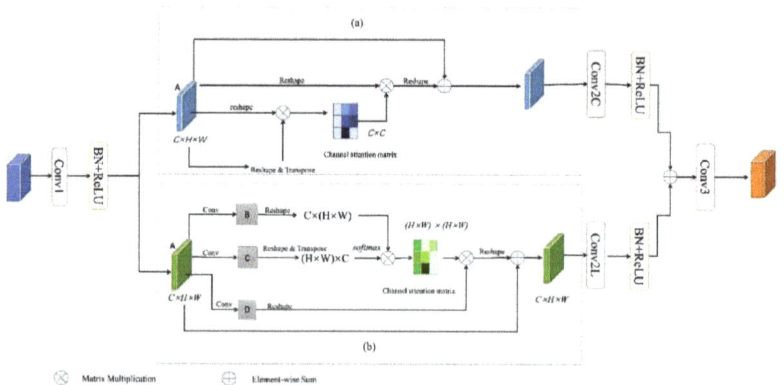

Figure 5. The architecture of the dual attention module consists of two branches: the kernel attention module and channel attention module. (**a**) Kernel attention module, and (**b**) channel attention module.

3.1.5. Deep Supervision

As providing supervision to the hidden layer reduces classification errors [21], researchers have adopted similar strategies [51] to ease the loss propagation in shallow layers. Therefore, we adopted auxiliary losses (Equation (2)) in stages two to four to supervise the predictions:

$$L_t = \alpha L_f \beta \sum_{i=1}^{n} L_i, \tag{2}$$

where α and β are the weights of the main loss function and auxiliary loss, with both weights set to 1; L_t is the total loss; L_f represents the loss for the output layer; and L_i represents the loss of the j-th stage after applying dual attention and feature refinement.

3.1.6. Loss Function

The loss function has an essential impact on the model accuracy and, usually, the most suitable loss function depends on the data properties and the class definitions [28]. Cross-entropy loss is a widely used loss function in two-dimensional semantic segmentation tasks. The aim of the learning-based remote sensing building extraction task is to train a binary classifier. The positive samples are pixels representing the buildings, whereas the negative samples are pixels containing the background. We here employed binary cross-entropy loss (Equation(3)) [52] in the training process:

$$H_p(q) = -\frac{1}{N} \sum_{i=1}^{N} y_i \cdot log(p(y_i)) + (1 - y_i) \cdot log(1 - p(y_i)), \tag{3}$$

where y is the label (1 for green points and 0 for red points) and $p(y_i)$ is the predicted probability of the point being green for all N points.

3.2. Data Preprocessing

The data preprocessing in the deep learning technology primarily consists of image clipping and image labeling. Building segmentation is a binary classification task involving buildings and non-building elements [21]. The building samples were intended to contain various types of buildings in the study area. The building labels were manually completed in ArcGIS 10.2. The pixel values of each image were scaled to the interval [0,1] by dividing by 255. To facilitate the deep learning calculation, the original image was uniformly cropped to generate 256 × 256 pixels with an overlap of 56 pixels between two adjacent images.

Data augmentation is an effective way to enlarge a dataset and avoid overfitting [53]. As presented in Figure 6, the images were rotated by 90°, 180°, and 270°. Random horizontal and vertical flipping were performed with a probability of 0.5. After data augmentation, 4980 images with 256 × 256 pixels were generated. The spatial resolution of these images was about 2.3 to 5.3 cm. A total of 30% of the images were randomly selected as the test set, while the rest of the images were the training set. The final results of the building extraction were obtained by further applying a threshold of 0.5. No additional post-processing was performed in this study.

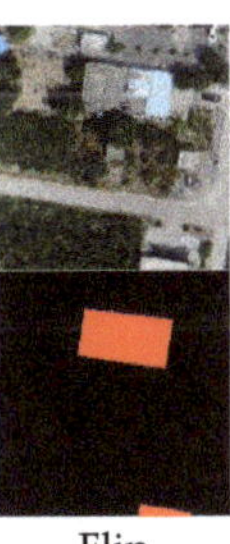

Figure 6. An example of data augmentation by rotating and flipping the rural Weinan building dataset.

3.3. Experimental Setting

The experiments were conducted using the PyTorch deep learning framework. All experiments were conducted on servers with 12th Gen Intel(R) Core™ i9-12900KF (3.20 GHz) and NVIDIA GeForce RTX 3090 (24 GB). All deep learning models were trained for 100 epochs, and 16 batches were randomly selected as the input data. The Adam optimizer was applied with an initial learning rate of 0.0001 and a weight decay of 0.0001. Figure 7 presents the dynamic changes in the accuracy and loss of the EDSANet model during the training process with the rural Weinan building dataset: the loss decreased and the accuracy increased as the training epochs increased; after the number of epochs reached 60, the model training tended to stabilize, and the accuracy remained high.

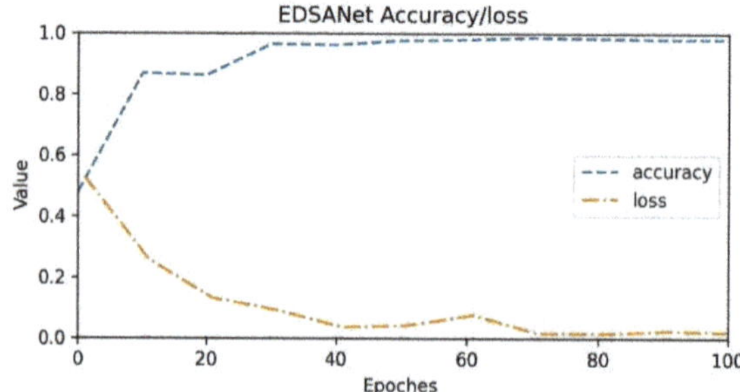

Figure 7. Changes in accuracy and loss for the EDSANet model in the training process.

3.4. Evaluation Metrics

Five common evaluation metrics were employed for quality evaluation in this research: overall accuracy (OA) (Equation (4)), precision (Equation (5)), recall (Equation (6)), F1-score (F1) (Equation (7)), and intersection-over-union (IoU) (Equation (8)). The five metrics are calculated as follows:

$$Overall\ Accuracy = \frac{TP + TN}{TP + TN + FP + FN}, \tag{4}$$

$$Precision = \frac{TP}{TP + FP}, \tag{5}$$

$$Recall = \frac{TP}{TP + FN}, \tag{6}$$

$$F1 = \frac{2 \times Precision \times Recall}{Precision + Recall}, \tag{7}$$

$$IoU = \frac{TP}{TP + FP + FN}, \tag{8}$$

where P is the number of positive samples, N is the number of negative samples, TP is the number of true positives, TN is the number of true negatives, FP is the number of false positives, and FN is the number of false negatives.

3.5. Building Height and Floor Area Estimation

Figure 8 shows the workflow for the building height and floor area estimation. The UAV images obtained from field surveys were first fed into the Pix4Dmapper software (version 4.5.6) [54]. This software contains three image processing steps: initial processing, point cloud and mesh, and DSM orthomosaic [55]. The DSM can be extracted from overlapping aerial images obtained with photogrammetry technology using location information stored in the header file of each aerial image from the UAV flight [56]. Based on the DSM, the point-cloud filtering algorithm with mathematical morphology was employed to identify whether the filter window was the ground point. Subsequently, the ground objects on the surface (including buildings, trees, and other non-ground points) were eliminated. The DTM data, which represent the terrain elevation information, were then formed [57–59]. The difference between the DSM and the DTM is referred to as the nDSM, which is widely used in height estimation. The height of the rural elements above the terrain was then generated [60]. Based on field surveys of the usual heights of local buildings, a threshold was set to estimate the number of floors in rural buildings. Using the footprint area of the buildings identified by the EDSANet model, the numbers of floors of each building in the nDSM were extracted and estimated in the ArcGIS environment (version 10.2). Lastly, the total floor area of the homesteads in the study area was obtained by counting the construction areas of each floor. The formula is as follows:

$$Area_{floors} = \sum_{i=1}^{floors_{max} \Sigma} Area_{gird} \times N_i,\tag{9}$$

where $Area_{floors}$ is the total floor area of the homesteads, $Area_{grid}$ is the area of the grid resolution of nDSM, N is the number of floors, and i ranges from 1 to $floors_{max}$ for each building.

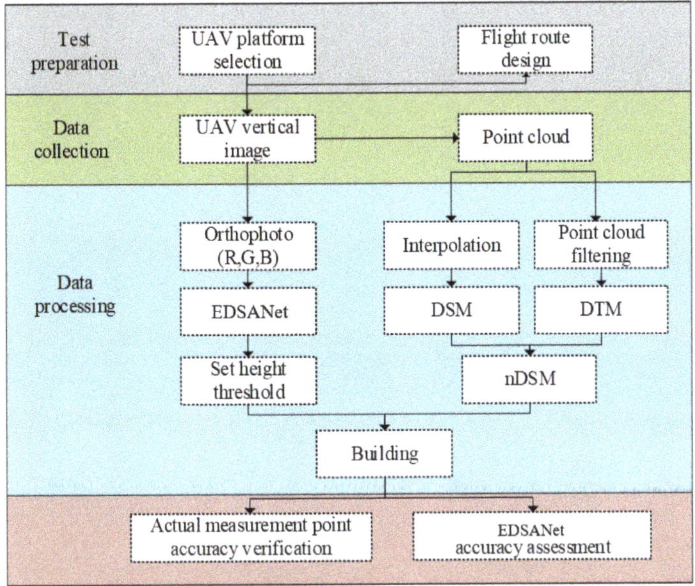

Figure 8. Flowchart of building height and floor area estimation.

4. Results

4.1. Building Extraction Using Deep Learning Models

Five classic and state-of-the-art deep learning models, including SegNet [31], UNet [30,32], Deeplabv3+ [61], MAP-Net [62], ARC-Net [23], and AGs-Unet [21], were compared to verify the performance and efficiency of the proposed EDSANet model with the rural Weinan building dataset. Figure 9 presents the qualitative results of building extraction using different deep learning models. SegNet returned too many false positives and false negatives exhibiting the worst performance with the dataset. Deeplabv3+, MAP-Net, and AGs-Unet presented quite similar performances in building extraction. For the proposed EDSANet model, the building segmentation results were satisfactory, and most buildings were generally well-segmented regardless of the type of roof (e.g., colored steel tile or sloped tile); additionally, the building footprints were very clear. However, the deep learning model could not clearly separate the boundaries between households in connected buildings (Figure 9a,b).

The specific analysis of the figure is as follows: Columns (a)–(d) represent four images randomly selected to show the test results. In (a) and (b), the proposed EDSANet model achieved effective completeness in extracted results for the whole single building. In the second column of buildings in (c), compared with Ags-Unet and ARC-Net, EDSANet clearly extracted the boundary of the buildings and showed the distinct gap between the buildings. Moreover, EDSANet was more advanced in the expression of the surrounding details of the building gap than Unet, MAP-Net, and Deeplabv3+, as shown in the lower right corner of the image building extraction results in (d), but it was not as good as the boundary smoothness that the ARC-Net model achieved.

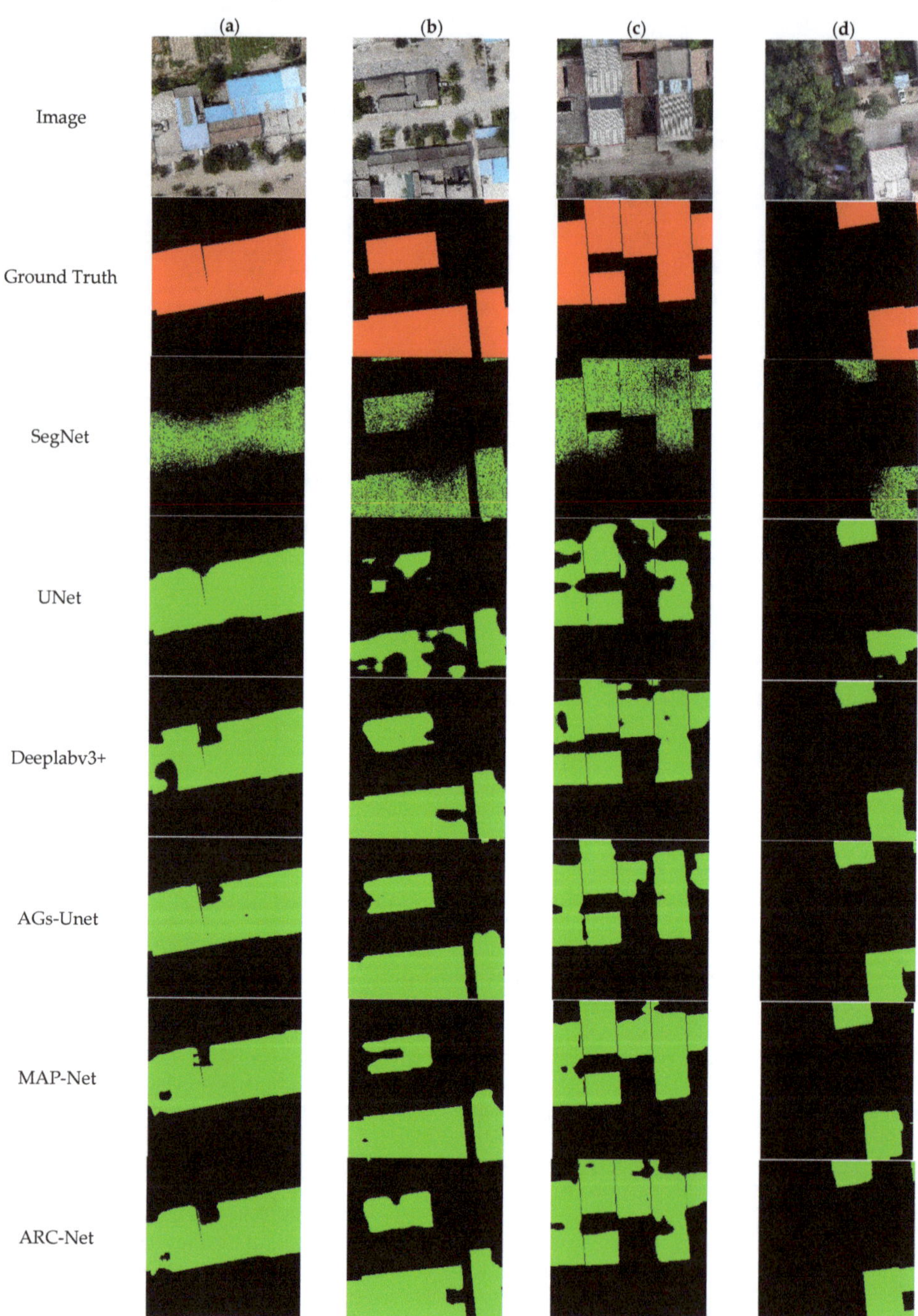

Figure 9. *Cont.*

EDSANet

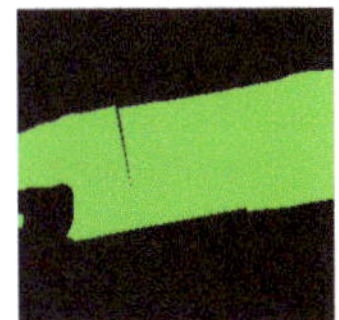

Figure 9. Building extraction results of different deep learning models with the rural Weinan building dataset. (**a–d**) Four images randomly selected to show the test results. SegNet, UNet, Deeplabv3+, Ags-Unet, MAP-Net, ARC-Net, and EDSANet, respectively, are represented by the building extraction results from the four groups of comparison experiments. Green represents the buildings and black represents the background. In the ground truth, red represents the buildings and black represents the background.

Table 3 presents the quantitative results of the building segmentation with the rural Weinan building dataset. SegNet obtained an overall accuracy of 0.740, while other models were all above 0.80. ARC-Net obtained an overall accuracy of 0.929 with a precision of 0.876, while EDSANet obtained an overall accuracy of 0.939 with an IoU of 0.848. In the experiments with the rural Weinan dataset, our proposed EDSANet model better balanced efficiency and accuracy compared to the MAP-Net and the ARC-Net models and achieved optimality for four evaluation metrics but not for recall, where Deeplabv3+ held the highest score of 0.946. Both the qualitative and quantitative experiment results demonstrate that EDSANet can effectively extract and fuse the features of rural buildings, improving the extraction accuracy for rural buildings. The results of the building extraction using the EDSANet model in Helan village are presented in Figure 10.

Table 3. Building extraction results with rural Weinan building dataset using different CNN models.

Models	OA	Precision	Recall	F1	IoU
SegNet	0.740	0.759	0.698	0.723	0.568
UNet	0.876	0.774	0.939	0.848	0.738
Deeplabv3+	0.899	0.813	**0.946**	0.872	0.777
AGs-Unet	0.907	0.864	0.911	0.887	0.798
MAP-Net	0.916	0.877	0.888	0.891	0.799
ARC-Net	0.929	0.876	0.921	0.902	0.822
EDSANet	**0.939**	**0.949**	0.887	**0.916**	**0.848** [1]

[1] Bold items in each column indicate the highest value.

(**a**)

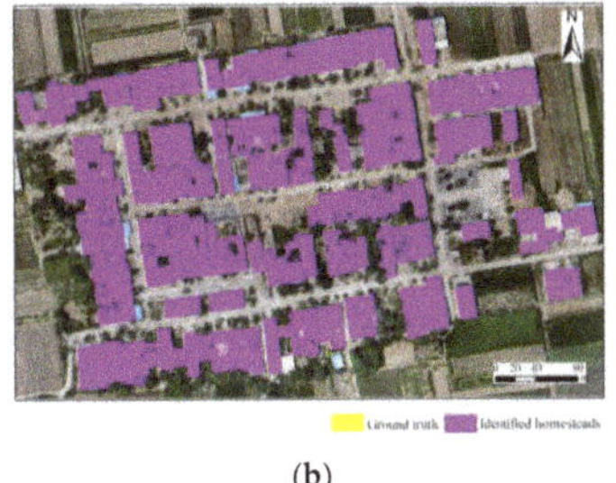

(**b**)

Figure 10. Spatial distribution of rural buildings in Helan village. (**a**) Ground truth of homesteads and (**b**) identification results based on the EDSANet model.

4.2. Building Height Estimation

Figure 11a shows the DSM extracted from the overlapping aerial images using photogrammetry technology. The DTM, based on morphological filtering, was utilized to obtain the ground area in the DSM (Figure 11b). The pixel values of the nDSM represent the height of the rural elements above the terrain (Figure 11c) and were calculated using the difference between the DSM and the DTM. The enclosed building area and the vegetation area on the ground cannot be correctly distinguished based only on the difference in height data. In the extraction of ground objects from high-resolution remote sensing data, the building segmentation precision obtained from the combination of spectral and height information is generally higher than that obtained using only spectral information or only height information.

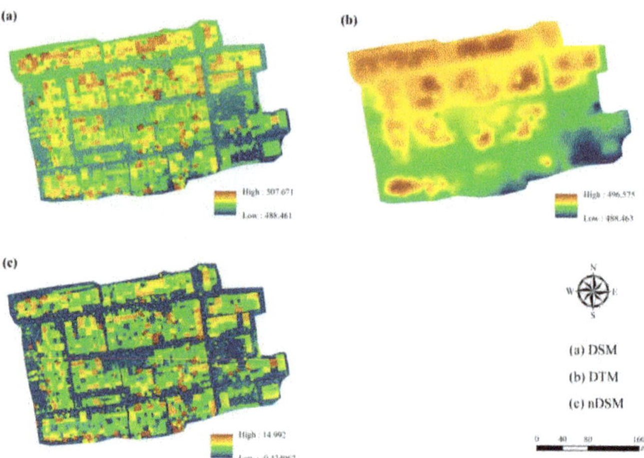

Figure 11. UAV-based estimation of the number of floors in rural buildings. (**a**) The DSM based on the photogrammetry workflow with the overlapping UAV images, (**b**) the DTM based on the point cloud filtering algorithm with the DSM images, and (**c**) the DTM subtracted from the DSM to create the nDSM.

The frequency distribution of the nDSM pixel values (Figure 12) was then calculated to obtain the building height information. Two pixel-value peaks were distributed near the 0.3 m and 4 m height differences. The pixel value of 0.3 m represents farmland crops and country roads, and the height difference of 4 m primarily represents the height of the roofs of one-story buildings or the walls of courtyards. All pixels in the nDSM grid with values less than 0.3 m were removed to avoid interference when extracting the building height. Moreover, because of the low reflectivity of the vegetation in the red band, the vegetation was well-extracted in the red band of the DOM image; the vegetation pixels in the nDSM were then detached with a raster operation in ArcGIS.

The building segmentation results of the deep learning method were a set of architectural and non-architectural images without a spatial reference. To facilitate the calculation and to display the results, the raster-based building footprint was converted into vector data after map projection in ArcGIS software to the coordinate system consistent with the reference image, which also made it possible to further calculate the floor area. Furthermore, the nDSM model of the interference pixels, including vegetation and roads, was removed with the mask of the homestead area. In accordance with the results of the field survey, the floors of the buildings in the study area were set at 4 m intervals. The height difference of 1–12 m was then set as indicating the first, second, and third floors (Table 4). In contrast, areas with floor heights of less than 1 m were set as indicating the courtyard height. Seventeen field survey buildings were randomly selected and utilized to examine the accuracy of the floor classification from the nDSM.

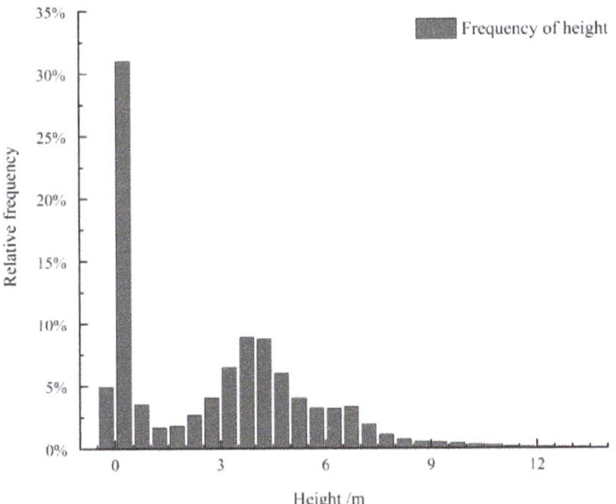

Figure 12. Frequency distribution diagram of the nDSM pixel values.

Table 4. Classification rules for the vegetation and the number of building floors.

Parameter	Threshold	Class
Brightness	≤ 60	Vegetation
Height	≤ 1 m	Courtyard
Height	$1\ \text{m} \leq \text{nDSM} \leq 4\ \text{m}$	One floor
Height	$4\ \text{m} \leq \text{nDSM} \leq 8\ \text{m}$	Two floors
Height	$8\ \text{m} \leq \text{nDSM} \leq 12\ \text{m}$	Three floors

Figure 13 displays the classification results for the building floors; 16 buildings were correctly classified and the height of 1 building was overestimated. The accuracy of the floor classification from the nDSM was 0.94, and the RMSE was 0.243 (Table 5). Field verification showed that the abnormal point was a canopy built by residents. The canopy was low and easily covered by the tall arbor canopy on one side. Therefore, the canopy height was calculated as the height of the vegetation canopy.

Figure 13. Classification results of building floors with nDSM.

Table 5. Confusion matrix for the number of floors divided by the nDSM model.

		Prediction			
		Courtyard	Courtyard	Courtyard	Courtyard
Actual	Courtyard	1	0	0	0
	One floor	0	3	0	0
	Two floors	0	1	11	0
	Three floors	0	0	0	1

4.3. Floor Area Estimation

The area of the homesteads was computed by multiplying the building area by the number of floors. We calculated the total construction area of the homesteads based on the reclassified results for the building heights from Section 4.2. The results showed that the total area of the homesteads was 3.1×10^5 m^2. Specifically, the homestead area for one-floor buildings was approximately 1.14×10^5 m^2 and accounted for 37.3% of the total homestead area; the homestead area for two-floor buildings with heights of 4–8 m was approximately 1.78×10^5 m^2 and accounted for 58.2% of the total construction area of the homesteads; three-floor buildings with a height difference of more than 8 m had a homestead area of approximately 3.33×10^3 m^2 and accounted for approximately 1.1% of the total construction area of the homesteads. The construction area for courtyards and low-rise shanty households accounted for only 3.4% of the total construction area. Figure 14 displays the frequency histogram for the building height; the average value of the pixels reached 4.45 m, and the standard deviation was 1.62. In conclusion, the average height and pixel frequency distribution indicated that residential buildings in the research area are primarily composed of two floors; this was consistent with the field survey results.

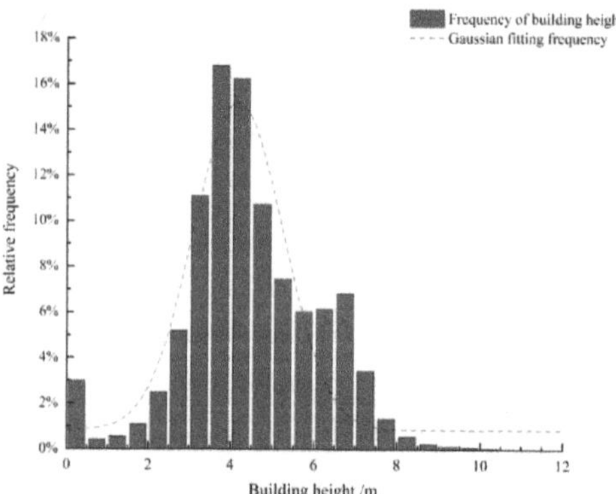

Figure 14. Frequency distribution diagram of building heights.

5. Discussion

5.1. Ablation Experiments

To further verify the feasibility of the DAM consisting of kernel attention and channel attention, the effectiveness of the atrous convolution in the SEM and the extraction precision of the different modules and fusion strategies were evaluated here in ablation experiments. The backbone model included the SIEM, SEM (without atrous convolution), and FFM. The benchmark models and strategies contained the backbone, the backbone + SEM (with atrous convolution), the backbone + DAM, the backbone + AFRM, and the backbone + SEM (with atrous convolution) + DAM + AFRM. Some of the ablation results for building extraction are presented in Figure 15. EDSANet (Figure 15b) showed the best performance in building extraction, with clear boundary identification compared to the model without DAM (Figure 15c) or AFRM (Figure 15d).

(a)   (b)   (c)   (d)

Figure 15. Example of extracted results from ablation experiment with the Weinan building dataset. (a) The input images, (**b**) results extracted with the proposed EDSANet model, (**c**) EDSANet without DAM, and (**d**) EDSANet without AFRM.

The quantitative comparison results for the different combinations are shown in Table 6. Based on the reference network as the backbone, both the dual attention module consisting of the kernel and the channel attention module and attention feature refinement module improved the representation ability for the features extracted from the network. Compared with the backbone, the accuracy was significantly improved. However, the recall performance of the backbone, at 0.907, was better than that of the backbone + SEM (atrous convolution), the backbone + DAM, and the backbone + AFRM, individually. It can be concluded from Table 4 that adding the DAM or AFRM modules reduced the accuracy of the model based on the backbone, and all the OA, precision, recall, F1, and IoU results for the backbone + SEM (atrous convolution) + DAM + AFRM network, with AFRM, were improved. This indicates that AFRM can adjust the excessive convergence of building feature information after attention extraction with DAM, thereby improving the accuracy of building extraction in remote sensing. The last row in Table 6 shows the results for the proposed EDSANet model, which achieved the best performance in all evaluation metrics except for recall.

Table 6. Building extraction accuracy for modules and variants of the model.

Models	OA	Precision	Recall	F1	IoU
Backbone	0.911	0.862	**0.907**	0.883	0.783
Backbone + SEM (atrous convolution)	0.905	0.855	0.889	0.870	0.771
Backbone + DAM	0.906	0.847	0.899	0.870	0.773
Backbone + AFRM	0.914	0.878	0.882	0.879	0.787
Backbone + SEM (atrous convolution) + DAM + AFRM	**0.939**	**0.949**	0.887	**0.916**	0.848 [1]

[1] Bold items in each column indicate the highest value.

5.2. Summaries and Limitations

Recent years have witnessed widespread application of deep learning in building extraction and other tasks owing to advancements in automatic learning features and strong adaptability. Previous studies have primarily focused on urban building extraction, which lacks application in rural China. In this study, we proposed the EDSANet model to extract buildings from UAV imagery in rural Weinan, China. The overall accuracy of the building extraction achieved by EDSANet was 0.929, and the precision was 0.876. Buildings were well-identified with clear boundaries regardless of the type of roof (e.g., colored steel tile or sloped tile). Buildings in rural areas mostly have one or two floors and are generally made of adobe, brickwood, and brick-concrete. The rural area selected in this research has fewer and more consistent building structure types than urban areas, which facilitates building extraction. However, for some irregularly arranged rural areas, the performance of building extraction with the EDSANet should be further analyzed.

Consumer-grade drones are flexible and have high spatial resolution, which can ensure the clear boundaries of buildings and the accuracy of the three-dimensional point cloud model. Series of 3D products based on UAV flight data, including DSM and DTM, were here generated using oblique photogrammetry technology. The nDSM was used to remove the vacant rural plots, and the heights of the buildings were extracted from the nDSM model. However, classifying different types of ground objects with complex spectral information from high-resolution UAV images is difficult. In addition, 17 buildings field-surveyed on the ground, accounting for 8.3% of buildings and covering all numbers of floors in this village, were randomly selected and employed to verify the classification results of building heights in this study. As mentioned in Section 4.2, the height of the building at one sample point was overestimated because the roof was covered by the vegetation canopy. The overall accuracy of the classification results was 0.94. As the property rights and structures of the rural buildings were investigated and confirmed in the field survey, the building height error was primarily due to the instability in the drone flight conditions and the overestimation of the roof height caused by trees. In future studies, we will adopt mathematical morphological methods to eliminate interference factors and to further optimize the accuracy of the building boundaries identified by deep learning methods and elevation extraction using UAV oblique photogrammetry.

6. Conclusions

Rapid and accurate building extraction and floor area estimation at the village level are of great significance for the overall planning of rural development and intensive land use. In this study, we proposed a comprehensive method to estimate village-level homestead areas by combining UAV remote sensing and deep learning technology. First, the building footprints were identified using the proposed EDSANet model, which merged dual attention extraction and attention feature refinement to aggregate multi-level semantics and enhance the performance of building extraction, especially for high-spatial-resolution images. Then, the number of floors of each building was estimated using the nDSM model generated from UAV oblique photogrammetry. The floor area of the entire village was estimated by multiplying the floor area of each building by the number of floors in the village. The case study was conducted in Helan village, Shaanxi province, China. The results show that the overall accuracy of the building extraction with the EDSANet model from UAV images was 0.929, with the precision reaching 0.876. The buildings in Helan village are primarily composed of two stories and have a total floor area of 3.1×10^5 m^2. The field survey verified that the accuracy of the nDSM model was 0.94; the RMSE was 0.243. The experimental results demonstrate that the proposed workflow, combining UAV remote sensing and deep learning technology, can aid in rapid and efficient building extraction and floor area estimation at the village level in China, as well as worldwide.

Author Contributions: Conceptualization, J.Z. and Y.L.; methodology, J.Z.; software, X.C.; validation, X.Y.; formal analysis, X.Y.; investigation, X.Y.; resources, H.C.; data curation, H.C.; writing—original draft preparation, J.Z.; writing—review and editing, Y.L.; visualization, X.C. and L.G.; supervision, L.G.; project administration, Y.L. and G.N.; funding acquisition, Y.L. and G.N. All authors have read and agreed to the published version of the manuscript.

Funding: This research was jointly supported by the National Natural Science Foundation of China, grant numbers 42201077 and 42177453; the Natural Science Foundation of Shandong Province, grant number ZR2021QD074; the Shandong Top Talent Special Foundation; and the National Nonprofit Fundamental Research Grant of China, Institute of Geology, China Earthquake Administration, grant number IGCEA2106.

Data Availability Statement: The codes are available at: https://github.com/Avery1991/2022 EDSANet (accessed on 4 September 2022).

Acknowledgments: We would like to thank the editors and the anonymous reviewers for their insightful comments and suggestions.

Conflicts of Interest: The authors declare no conflict of interest.

References

1. Li, X.; Li, Z.; Yang, J.; Li, H.; Liu, Y.; Fu, B.; Yang, F. Seismic vulnerability comparison between rural Weinan and other rural areas in Western China. *Int. J. Disaster Risk Reduct.* **2020**, *48*, 101576. [CrossRef]
2. Liu, Y.; So, E.; Li, Z.; Su, G.; Gross, L.; Li, X.; Qi, W.; Yang, F.; Fu, B.; Yalikun, A.; et al. Scenario-based seismic vulnerability and hazard analyses to help direct disaster risk reduction in rural Weinan, China. *Int. J. Disaster Risk Reduct.* **2020**, *48*, 101577. [CrossRef]
3. Zhu, Q.; Li, Z.; Zhang, Y.; Guan, Q. Building Extraction from High Spatial Resolution Remote Sensing Images via Multiscale-Aware and Segmentation-Prior Conditional Random Fields. *Remote Sens.* **2020**, *12*, 3983. [CrossRef]
4. Liu, S.Y.; Xiong, X.F. Property rights and regulation: Evolution and reform of China's homestead system. *China Econ. Stud.* **2019**, *6*, 17–27.
5. Liu, Y.; Fang, F.; Li, Y. Key issues of land use in China and implications for policy making. *Land Use Policy* **2014**, *40*, 6–12. [CrossRef]
6. Yu, B.; Liu, H.; Wu, J.; Hu, Y.; Zhang, L. Automated derivation of urban building density information using airborne LiDAR data and object-based method. *Landsc. Urban Plan.* **2010**, *98*, 210–219. [CrossRef]
7. Liu, Y.; Zheng, X.; Ai, G.; Zhang, Y.; Zuo, Y. Generating a High-Precision True Digital Orthophoto Map Based on UAV Images. *ISPRS Int. J. Geo Inf.* **2018**, *7*, 333. [CrossRef]
8. Allouche, M.K.; Moulin, B. Amalgamation in cartographic generalization using Kohonen's feature nets. *Int. J. Geogr. Inf. Sci.* **2005**, *19*, 899–914. [CrossRef]
9. Dandabathula, G.; Sitiraju, S.R.; Jha, C.S. Retrieval of building heights from ICESat-2 photon data and evaluation with field measurements. *Environ. Res. Infrastruct. Sustain.* **2021**, *1*, 011003. [CrossRef]
10. Kamath, H.G.; Singh, M.; Magruder, L.A.; Yang, Z.-L.; Niyogi, D.J. GLOBUS: GLObal Building heights for Urban Studies. *arXiv* **2022**, arXiv:2205.12224.
11. Weidner, U.; Förstner, W. Towards automatic building extraction from high-resolution digital elevation models. *ISPRS J. Photogramm. Remote Sens.* **1995**, *50*, 38–49. [CrossRef]
12. Sefercik, U.G.; Karakis, S.; Bayik, C.; Alkan, M.; Yastikli, N. Contribution of Normalized DSM to Automatic Building Extraction from HR Mono Optical Satellite Imagery. *Eur. J. Remote Sens.* **2014**, *47*, 575–591. [CrossRef]
13. Ji, C.; Tang, H. Gross Floor Area Estimation from Monocular Optical Image Using the NoS R-CNN. *Remote Sens.* **2022**, *14*, 1567. [CrossRef]
14. Toth, C.; Jozkow, G. Remote sensing platforms and sensors: A survey. *Isprs J. Photogramm. Remote Sens.* **2016**, *115*, 22–36. [CrossRef]
15. Colomina, I.; Molina, P. Unmanned aerial systems for photogrammetry and remote sensing: A review. *ISPRS J. Photogramm. Remote Sens.* **2014**, *92*, 79–97. [CrossRef]
16. Wang, J.Z.; Lin, Z.J.; Li, C.M.; Hong, Z.G. 3D Reconstruction of Buildings with Single UAV Image. *Remote Sens. Inf.* **2004**, *4*, 11–15.
17. Ma, Y.; Wu, H.; Wang, L.; Huang, B.; Ranjan, R.; Zomaya, A.; Jie, W. Remote sensing big data computing: Challenges and opportunities. *Futur. Gener. Comput. Syst.* **2015**, *51*, 47–60. [CrossRef]
18. Zhong, Y.; Ma, A.; Ong, Y.S.; Zhu, Z.; Zhang, L. Computational intelligence in optical remote sensing image processing. *Appl. Soft Comput.* **2018**, *64*, 75–93. [CrossRef]
19. Meng, Y.; Peng, S. Object-Oriented Building Extraction from High-Resolution Imagery Based on Fuzzy SVM. In Proceedings of the 2009 International Conference on Information Engineering and Computer Science, Wuhan, China, 19–20 December 2009.
20. Dahiya, S.; Garg, P.K.; Jat, M.K. Object Oriented Approach for Building Extraction from High Resolution Satellite Images. In Proceedings of the 2013 3rd IEEE International Advance Computing Conference (IACC), Ghaziabad, India, 22–23 February 2013.
21. Yu, M.; Chen, X.; Zhang, W.; Liu, Y. AGs-Unet: Building Extraction Model for High Resolution Remote Sensing Images Based on Attention Gates U Network. *Sensors* **2022**, *22*, 2932. [CrossRef]
22. Liu, Y.; Zhang, W.; Chen, X.; Yu, M.; Sun, Y.; Meng, F.; Fan, X. Landslide Detection of High-Resolution Satellite Images Using Asymmetric Dual-Channel Network. In Proceedings of the 2021 IEEE International Geoscience and Remote Sensing Symposium IGARSS, Brussels, Belgium, 11–16 July 2021; pp. 4091–4094. [CrossRef]
23. Liu, Y.; Zhou, J.; Qi, W.; Li, X.; Gross, L.; Shao, Q.; Zhao, Z.; Ni, L.; Fan, X.; Li, Z. ARC-Net: An Efficient Network for Building Extraction From High-Resolution Aerial Images. *IEEE Access* **2020**, *8*, 154997–155010. [CrossRef]
24. Boonpook, W.; Tan, Y.; Xu, B. Deep learning-based multi-feature semantic segmentation in building extraction from images of UAV photogrammetry. *Int. J. Remote Sens.* **2020**, *42*, 1–19. [CrossRef]
25. Trevisiol, F.; Lambertini, A.; Franci, F.; Mandanici, E. An Object-Oriented Approach to the Classification of Roofing Materials Using Very High-Resolution Satellite Stereo-Pairs. *Remote Sens.* **2022**, *14*, 849. [CrossRef]
26. Yuan, J. Learning Building Extraction in Aerial Scenes with Convolutional Networks. *IEEE Trans. Pattern Anal. Mach. Intell.* **2018**, *40*, 2793–2798. [CrossRef] [PubMed]

27. Vakalopoulou, M.; Karantzalos, K.; Komodakis, N.; Paragios, N. Building Detection in Very High Resolution Multispectral Data with Deep Learning Features. In Proceedings of the 2015 IEEE International Geoscience and Remote Sensing Symposium (IGARSS), Milan, Italy, 26–31 July 2015.

28. Touzani, S.; Granderson, J. Open Data and Deep Semantic Segmentation for Automated Extraction of Building Footprints. *Remote Sens.* **2021**, *13*, 2578. [CrossRef]

29. Chen, J.; Yuan, Z.; Peng, J.; Chen, L.; Huang, H.; Zhu, J.; Liu, Y.; Li, H. DASNet: Dual Attentive Fully Convolutional Siamese Networks for Change Detection in High-Resolution Satellite Images. *IEEE J. Sel. Top. Appl. Earth Obs. Remote Sens.* **2021**, *14*, 1194–1206. [CrossRef]

30. Long, J.; Shelhamer, E.; Darrell, T. Fully Convolutional Networks for Semantic Segmentation. In Proceedings of the IEEE Conference on Computer Vision and Pattern Recognition (CVPR), Boston, MA, USA, 7–12 June 2015; pp. 3431–3440.

31. Badrinarayanan, V.; Kendall, A.; Cipolla, R. Segnet: A deep convolutional encoder-decoder architecture for image segmentation. *IEEE Trans. Pattern Anal. Mach. Intell.* **2017**, *39*, 2481–2495. [CrossRef]

32. Ronneberger, O.; Fischer, P.; Brox, T. U-Net: Convolutional Networks for Biomedical Image Segmentation. In Proceedings of the International Conference on Medical Image Computing and Computer-Assisted Intervention, Munich, Germany, 5–9 October 2015; pp. 234–241.

33. Romera, E.; Alvarez, J.M.; Bergasa, L.M.; Arroyo, R. ERFNet: Efficient Residual Factorized ConvNet for Real-Time Semantic Segmentation. *IEEE Trans. Intell. Transp. Syst.* **2017**, *19*, 263–272. [CrossRef]

34. Liu, Y.; Gross, L.; Li, Z.; Li, X.; Fan, X.; Qi, W. Automatic Building Extraction on High-Resolution Remote Sensing Imagery Using Deep Convolutional Encoder-Decoder With Spatial Pyramid Pooling. *IEEE Access* **2019**, *7*, 128774–128786. [CrossRef]

35. Konstantinidis, D.; Argyriou, V.; Stathaki, T.; Grammalidis, N. A modular CNN-based building detector for remote sensing images. *Comput. Netw.* **2020**, *168*, 107034. [CrossRef]

36. Zhang, X. Village-Level Homestead and Building Floor Area Estimates Based on UAV Imagery and U-Net Algorithm. *ISPRS Int. J. Geo-Inf.* **2020**, *9*, 403. [CrossRef]

37. Liao, C.; Hu, H.; Li, H.; Ge, X.; Chen, M.; Li, C.; Zhu, Q. Joint Learning of Contour and Structure for Boundary-Preserved Building Extraction. *Remote Sens.* **2021**, *13*, 1049. [CrossRef]

38. Xiao, X.; Guo, W.; Chen, R.; Hui, Y.; Wang, J.; Zhao, H. A Swin Transformer-Based Encoding Booster Integrated in U-Shaped Network for Building Extraction. *Remote Sens.* **2022**, *14*, 2611. [CrossRef]

39. Li, H.; Qiu, K.; Chen, L.; Mei, X.; Hong, L.; Tao, C. SCAttNet: Semantic Segmentation Network With Spatial and Channel Attention Mechanism for High-Resolution Remote Sensing Images. *IEEE Geosci. Remote Sens. Lett.* **2021**, *18*, 905–909. [CrossRef]

40. Wei, R.; Fan, B.; Wang, Y.; Zhou, A.; Zhao, Z. MBNet: Multi-Branch Network for Extraction of Rural Homesteads Based on Aerial Images. *Remote Sens.* **2022**, *14*, 2443. [CrossRef]

41. Jing, W.; Lin, J.; Lu, H.; Chen, G.; Song, H. Learning holistic and discriminative features via an efficient external memory module for building extraction in remote sensing images. *Build. Environ.* **2022**, *222*, 109332. [CrossRef]

42. Li, H.; Li, Y.; Zhang, G.; Liu, R.; Huang, H.; Zhu, Q.; Tao, C. Global and Local Contrastive Self-Supervised Learning for Semantic Segmentation of HR Remote Sensing Images. *IEEE Trans. Geosci. Remote Sens.* **2022**, *60*, 5618014. [CrossRef]

43. Lin, J.; Jing, W.; Song, H.; Chen, G. ESFNet: Efficient Network for Building Extraction from High-Resolution Aerial Images. *IEEE Access* **2019**, *7*, 54285–54294. [CrossRef]

44. Elhassan, M.A.; Huang, C.; Yang, C.; Munea, T.L. DSANet: Dilated spatial attention for real-time semantic segmentation in urban street scenes. *Expert Syst. Appl.* **2021**, *183*, 115090. [CrossRef]

45. Li, G.; Yun, I.; Kim, J.; Kim, J. Dabnet: Depth-wise asymmetric bottleneck for real-time semantic segmentation. *arXiv* **2019**, arXiv:1907.11357.

46. Ma, N.; Zhang, X.; Zheng, H.-T.; Sun, J. Shufflenet v2: Practical Guidelines for Efficient Cnn Architecture Design. In Proceedings of the European Conference on Computer Vision (ECCV), Munich, Germany, 8–14 September 2018; pp. 116–131.

47. Zhang, X.; Zhou, X.; Lin, M.; Sun, J. Shufflenet: An Extremely Efficient Convolutional Neural Network for Mobile Devices. In Proceedings of the IEEE Conference on Computer Vision and Pattern Recognition (CVPR), Salt Lake City, UT, USA, 18–23 June 2018; pp. 6848–6856.

48. Sandler, M.; Howard, A.; Zhu, M.; Zhmoginov, A.; Chen, L.-C. Mobilenetv2: Inverted Residuals and Linear Bottlenecks. In Proceedings of the IEEE Conference on Computer Vision and Pattern Recognition (CVPR), Salt Lake City, UT, USA, 18–23 June 2018; pp. 4510–4520.

49. He, K.; Zhang, X.; Ren, S.; Sun, J. Delving Deep into Rectifiers: Surpassing Human-Level Performance on Imagenet Classification. In Proceedings of the International Conference on Computer Vision, Las Condes, Chile, 11–18 December 2015; pp. 1026–1034.

50. Fu, J.; Liu, J.; Tian, H.; Li, Y.; Bao, Y.; Fang, Z.; Lu, H. Dual Attention Network for Scene Segmentation. In Proceedings of the IEEE/CVF Conference on Computer Vision and Pattern Recognition (CVPR), Long Beach, CA, USA, 15–20 June 2019; pp. 3146–3154.

51. Yu, M.; Zhang, W.; Chen, X.; Liu, Y.; Niu, J. An End-to-End Atrous Spatial Pyramid Pooling and Skip-Connections Generative Adversarial Segmentation Network for Building Extraction from High-Resolution Aerial Images. *Appl. Sci.* **2022**, *12*, 5151. [CrossRef]

52. De Boer, P.-T.; Kroese, D.P.; Mannor, S.; Rubinstein, R.Y. A Tutorial on the Cross-Entropy Method. *Ann. Oper. Res.* **2005**, *134*, 19–67. [CrossRef]

53. Zhang, Z.; Wang, Y. JointNet: A Common Neural Network for Road and Building Extraction. *Remote Sens.* **2019**, *11*, 696. [CrossRef]
54. Krause, S.; Sanders, T.G.M.; Mund, J.-P.; Greve, K. UAV-Based Photogrammetric Tree Height Measurement for Intensive Forest Monitoring. *Remote Sens.* **2019**, *11*, 758. [CrossRef]
55. Kameyama, S.; Sugiura, K. Effects of Differences in Structure from Motion Software on Image Processing of Unmanned Aerial Vehicle Photography and Estimation of Crown Area and Tree Height in Forests. *Remote Sens.* **2021**, *13*, 626. [CrossRef]
56. Karantzalos, K.; Koutsourakis, P.; Kalisperakis, I.; Grammatikopoulos, L. Model-based building detection from low-cost optical sensors onboard unmanned aerial vehicles. *ISPRS Int. Arch. Photogramm. Remote Sens. Spat. Inf. Sci.* **2015**, *XL-1/W4*, 293–297. [CrossRef]
57. Gevaert, C.; Persello, C.; Nex, F.; Vosselman, G. A deep learning approach to DTM extraction from imagery using rule-based training labels. *ISPRS J. Photogramm. Remote Sens.* **2018**, *142*, 106–123. [CrossRef]
58. Özcan, A.H.; Ünsalan, C.; Reinartz, P. Ground filtering and DTM generation from DSM data using probabilistic voting and segmentation. *Int. J. Remote Sens.* **2018**, *39*, 2860–2883. [CrossRef]
59. Serifoglu Yilmaz, C.; Gungor, O. Comparison of the performances of ground filtering algorithms and DTM generation from a UAV-based point cloud. *Geocarto Int.* **2018**, *33*, 522–537. [CrossRef]
60. Shukla, A.; Jain, K. Automatic extraction of urban land information from unmanned aerial vehicle (UAV) data. *Earth Sci. Inform.* **2020**, *13*, 1225–1236. [CrossRef]
61. Chen, L.-C.; Papandreou, G.; Kokkinos, I.; Murphy, K.; Yuille, A. DeepLab: Semantic Image Segmentation with Deep Convolutional Nets, Atrous Convolution, and Fully Connected CRFs. *IEEE Trans. Pattern Anal. Mach. Intell.* **2018**, *40*, 834–848.
62. Zhu, Q.; Liao, C.; Hu, H.; Mei, X.; Li, H. MAP-Net: Multiple Attending Path Neural Network for Building Footprint Extraction from Remote Sensed Imagery. *IEEE Trans. Geosci. Remote. Sens.* **2020**, *59*, 6169–6181. [CrossRef]

remote sensing

Article

Weakly Supervised Learning for Transmission Line Detection Using Unpaired Image-to-Image Translation

Jiho Choi and Sang Jun Lee *

Division of Electronic Engineering, Jeonbuk National University, 567 Baekje-daero, Deokjin-gu, Jeonju 54896, Korea; jihochoi@jbnu.ac.kr
* Correspondence: sj.lee@jbnu.ac.kr; Tel.: +82-63-270-2463

Abstract: To achieve full autonomy of unmanned aerial vehicles (UAVs), obstacle detection and avoidance are indispensable parts of visual recognition systems. In particular, detecting transmission lines is an important topic due to the potential risk of accidents while operating at low altitude. Even though many studies have been conducted to detect transmission lines, there still remains many challenges due to their thin shapes in diverse backgrounds. Moreover, most previous methods require a significant level of human involvement to generate pixel-level ground truth data. In this paper, we propose a transmission line detection algorithm based on weakly supervised learning and unpaired image-to-image translation. The proposed algorithm only requires image-level labels, and a novel attention module, which is called parallel dilated attention (PDA), improves the detection accuracy by recalibrating channel importance based on the information from various receptive fields. Finally, we construct a refinement network based on unpaired image-to-image translation in order that the prediction map is guided to detect line-shaped objects. The proposed algorithm outperforms the state-of-the-art method by 2.74% in terms of F1-score, and experimental results demonstrate that the proposed method is effective for detecting transmission lines in both quantitative and qualitative aspects.

Keywords: weakly supervised learning; image-to-image translation; transmission line detection; unmanned aerial vehicle; channel attention mechanism

Citation: Choi, J.; Lee, S.J. Weakly Supervised Learning for Transmission Line Detection Using Unpaired Image-to-Image Translation. *Remote Sens.* **2022**, *14*, 3421. https://doi.org/10.3390/rs14143421

Academic Editors: Yingying Dong, Chenghai Yang, Giovanni Laneve and Wenjiang Huang

Received: 7 June 2022
Accepted: 13 July 2022
Published: 16 July 2022

Publisher's Note: MDPI stays neutral with regard to jurisdictional claims in published maps and institutional affiliations.

1. Introduction

Recently, unmanned aerial vehicles (UAVs) have been widely utilized in many industrial fields. In the applications of construction site monitoring and infrastructure inspection, UAVs reduce the cost and inspection time while ensuring the safety of inspectors [1]. Furthermore, UAVs contribute to increasing the efficiency of precision agriculture by spreading seeds and monitoring the conditions of crops more effectively than human workers [2]. Beyond this, UAVs are also applied to military surveillance, aerial photography, search and rescue, and product delivery [3]. As technology advances, UAVs with high-level autonomy have been utilized in applications such as forest fire monitoring and security system.

For the reliable operation of autonomous UAVs, obstacle detection and avoidance are important functions. Most autonomous UAVs and drones equip cameras for visual recognition, and path planning and control are successfully conducted based on the accurate recognition of surrounding environments. Previous studies proposed several obstacle detection methods. In particular, Huang et al. [4] proposed an obstacle avoidance algorithm using a monocular camera and millimeter-wave radar together. Similarly, a deep learning-based recognition algorithm was employed to detect multiple obstacles in [5]. On the other hand, many researchers have conducted research on obstacle avoidance and path planning based on deep learning approaches [6–11]. Ou et al. [9] suggested a framework based on deep reinforcement learning to plan feasible global paths with an obstacle map. Yuan et al. [10] presented a path planning method based on a convolutional neural network (CNN) model that can detect and localize obstacles such as buildings.

Data collection has become more straightforward than ever, and with the support of extensive public datasets, deep learning techniques have shown promising results in various industrial fields. However, supervised learning methods require ground truth data, which cause expensive costs for manual labor and time-consuming tasks, especially for large datasets. To address these limitations, weakly supervised learning has received attention to train deep neural networks with weak supervision. In recent years, weakly supervised learning has been applied to various tasks such as object detection [12–14], segmentation [15–19], and localization [20–22]. Wang et al. [16] suggested a segmentation algorithm based on a combination of U-Net and class activation map and trained using only image-level labels. They demonstrated that training a CNN model with weak supervision can also segment cropland accurately.

Among several types of obstacles, the transmission line is a critical obstacle that must be avoided. A collision with transmission lines can damage the stable supply of electricity, and furthermore, crashed UAVs can cause secondary accidents. Deep learning methods have been successfully applied to the detection and localization of transmission lines [21–25]. In the case of transmission line datasets, class imbalance occurs due to the background area that occupies most of the aerial image. In [23], they proposed a generalized focal loss function to handle class imbalance in the transmission line detection task. Lee et al. [21] introduced a weakly supervised learning method for detecting transmission lines, and they employed the *VisualBackProp* algorithm proposed by Bojarski et al. [26] to localize transmission lines. After this study, Choi et al. [22] proposed an extended study for transmission line detection. They utilized only patch-level labels to reduce the cost for collecting pixel-level ground truth data. In [22], they used the assumption that transmission lines are partially straight, and their proposed method connects broken lines by utilizing the orientations of the line segments. However, the class labels of the patch images require rough location information of transmission lines in images.

In this paper, we propose a transmission line detection method based on weakly supervised learning and unpaired image-to-image translation. The main contribution of this paper is three-fold as follows.

- We develop a weakly supervised algorithm for detecting transmission lines in UAV images. Unlike the previous methods, which require pixel-level labels, our proposed method requires minimal labeling work for preparing training data, and therefore, it is easily applicable to real-world problems.
- We integrate a novel attention module into the classification network to obtain a robust localization mask. To incorporate the information from various receptive fields, we introduce a parallel dilated attention (PDA) module.
- For the training of the refinement network, we generate pseudo-line data and employ the cycle consistency loss, which was proposed in [27]. The refinement network enhances the line-shaped property of transmission lines, and therefore, the localization result is significantly improved in both quantitative and qualitative aspects.

The remainder of this paper is structured as follows: the related work is summarized in Section 2. The proposed method is presented in Section 3. Results and conclusions are presented in Sections 4 and 5, respectively.

2. Related Work

2.1. Attention Mechanism

The attention mechanism has attracted many deep learning researchers, and they have proposed the following mechanisms. The bottleneck attention module (BAM) [28] is an attention module that computes attention maps from the two separated spatial and channel attention branches. Unlike the parallel structure of BAM, the spatial and channel attention of convolutional block attention modules (CBAM) [29] are sequentially configured. Yang et al. [30] proposed SimAM, which is a parameter-free attention module to calculate 3D attention weights in the channel and spatial dimensions. Hu et al. [31] tried to reduce the amount of computation while performing recalibration with a squeeze and excitation

(SE) module, although this operation destroys the relationship between channels and weights due to channel reduction. The SE module utilizes the global average pooling (GAP) and two fully connected layers, which can be integrated into CNN architectures such as VGGNet, ResNet, and GoogLeNet. Wang et al. [32] proposed an efficient channel attention (ECA) to compute local cross-channel interaction by applying 1D convolution. The ECA module has shown significant improvements in state-of-the-art object detection, image classification, and object segmentation along with lightweight parameters. We compare the ECA module and other channel attention methods [30,31] with the proposed attention module and demonstrate the effectiveness of our proposed attention module in Section 4.5.

It has been used in a variety of applications concurrently with the progress of many studies on the attention mechanism. From the perspective of time series data, attention mechanisms can be implemented with sequence-to-sequence models with encoder and decoder architectures, to make models that pay attention to specific sequence data [33–35]. Furthermore, the attention module was applied to the CNN model which uses time-series data to estimate the blood pressure [36] and classifies the sleep stage in [37]. Many studies attempted to solve diverse problems on remote sensing image data such as classifications [38,39], ship detection [40,41], and semantic segmentation [42]. Ma et al. [39] implemented the channel and spatial attention module and integrated it into CNN architecture for the classification of the remote sensing scene images. Detecting small-scale ships is a challenging task in optical remote sensing images. Hu et al. [40] proposed detection models with the attention mechanism to suppress background while focusing on small ships to improve detection accuracy. Moreover, refs. [43,44] integrate attention modules in their deep neural network to solve segmentation tasks such as esophagus and lungs in medical images. Motivated by channel attention, we improved the localization mask by adopting an attention mechanism to focus on the important channels.

2.2. Image-to-Image Translation

Generative adversarial networks (GAN) [45] have made great strides in deep learning, and subsequent algorithms such as deep convolutional GAN (DCGANs) [46], conditional GAN (CGAN) [47], and InfoGAN [48] have been proposed. Isola et al. [47] proposed a supervised learning method that performs image-to-image translation using a paired dataset. Paired image-to-image translation is restrictive in its applications to real-world problems because it requires data correspondences. On the other hand, unpaired image-to-image translation techniques can address the limitations of paired image-to-image translation methods. There are several style transfer networks including CycleGAN [27], DiscoGAN [49], and DualGAN [50], and these methods translate the style of input images based on the unpaired datasets. Zhu et al. [27] proposed CycleGAN, which translates images from the source domain to the target domain with the cycle consistency loss, and this method shows remarkable results in the style transfer task. Researchers have extended unpaired image-to-image translation for several applications [51–58] to address issues such as data imbalance, lack of diversity, and limitation in collecting real paired dataset.

Zi et al. [51] constructed a modified CycleGAN to effectively generate clear images from cloudy images by utilizing unpaired image datasets. Furthermore, CycleGAN has been applied for data augmentation by translating synthetic images into realistic images in [52,53]. In particular, Mao et al. [53] improved the performance of classifying the actual wildfire smoke by utilizing images that were artificially generated images by utilizing [27]. To safely drive autonomous vehicles, road surface detection is essential as knowledge of road surface conditions (e.g., dry, wet, snowy) affects autonomous driving control [54]. Dry conditions can be collected more frequently, resulting in unbalanced data problems. To address this lack, they generated images of wet and snowy roads through an unpaired image-to-image translation method. Although biometric systems (e.g., fingerprint-based, and face-based) are widely used for security purposes, these recognition systems can be vulnerable to presentation attacks. A presentation attack aims to interfere with the normal functioning of a biometric recognition by presenting artifacts or biometric characteristics.

To prepare for such a presentation attack detection (PAD), Nguyen et al. [55] generalized the model to fake presentation attack face images obtained via CycleGAN. Inspired by these studies, we employed the image-to-image translation to refine the localization mask using pseudo-line data, and the effectiveness of a refinement network is shown in an ablation study later on.

3. Proposed Method

This section presents the proposed transmission line detection method. The weakly supervised learning framework proposed by Bojarski et al. [26] is employed to generate a localization mask for the transmission lines. Different from the previous work, we introduce a novel attention mechanism called PDA to improve the quality of the localization mask, and it is called attention localization mask (ALM). Furthermore, we develop a refinement network by utilizing an unpaired image-to-image translation technique between ALM and pseudo-line data. An overview of the proposed framework is shown in Figure 1. The upper and lower left parts present the backbone network for classifying images with and without transmission lines and the process for generating ALM from the hierarchical feature maps, respectively. The lower right part is the generator of the refinement network that produces the refined image, and the upper right part is a discriminator for adversarial training of the refinement network.

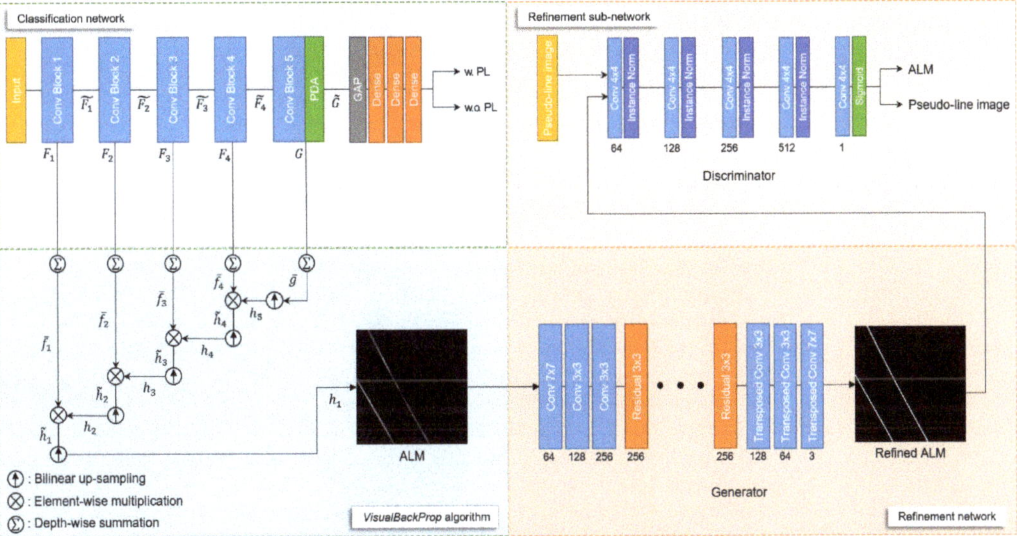

Figure 1. Overview of the proposed method. The (**left**) part shows the base network for extracting power line (PL) features and obtaining localization mask. The (**right**) part presents the refinement network and a sub-network for the adversarial learning of the refinement network.

3.1. Classification Network and VisualBackProp Algorithm

We constructed a classification network to implement the *VisualBackProp* algorithm [26], which can obtain localization masks using feature maps. The classification network was constructed based on the VGG16 architecture, which consists of five convolution blocks to classify images with and without transmission lines. A convolution block contains convolution layers, a rectified linear unit (ReLU), and a max pooling operation. Although similar models were utilized in [21,22] for localizing transmission lines in patch images, our proposed model is different from the previous methods in that transmission lines can be localized in the original images. We employed image-level labels of the same size as the original size of the 512×512 in the power line dataset.

After binary classification, the localization mask is generated by the *VisualBackProp* algorithm. In order to obtain the mask for localizing transmission lines, we employed the *VisualBackProp* algorithm for the last feature maps $\mathbf{F}_i \in \mathbb{R}^{H_i \times W_i \times C_i}$ of the convolution blocks. The i-th feature map $\mathbf{F}_i$ consists of C_i feature maps $\mathbf{f}_i{}^1, \cdots, \mathbf{f}_i{}^{C_i}$. The first process of the *VisualBackProp* algorithm is to compute a single feature map $\bar{\mathbf{f}}_i$ by accumulating the feature maps $\mathbf{f}_i{}^k \in \mathbb{R}^{H_i \times W_i}$ in the depth direction as (1).

$$\bar{\mathbf{f}}_i = \sum_{k=1}^{C_i} \mathbf{f}_i{}^k,$$ (1)

where i and c_i denote the number of convolution blocks and channels, respectively. The accumulated feature map is upsampled via bilinear interpolation to generate $\mathbf{h}_i$, and it is multiplied with the previous feature map $\bar{\mathbf{f}}_{i-1}$ to compute $\tilde{\mathbf{h}}_{i-1}$ as (2).

$$\tilde{\mathbf{h}}_{i-1} = \bar{\mathbf{f}}_{i-1} \otimes \mathbf{h}_i,$$ (2)

where $\otimes$ is the elementwise multiplication operation. Although the *VisualBackProp* algorithm provides reasonable localization masks in many cases, it fails when transmission lines have weak visual properties. To address this problem, we propose an attention mechanism for weakly supervised learning to enhance the responses of transmission lines in the localization map. In Figure 1, $\tilde{\mathbf{F}}_i \in \mathbb{R}_+^{\frac{H_i}{2} \times \frac{W_i}{2} \times C_i}$ is the output of i-th convolution block in VGG16, which can be obtained after the ReLU and max pooling operations for $\mathbf{F}_i$, and it is utilized as the input for the following convolution block.

3.2. Parallel Dilated Attention Module

Inspired by SE-Net [31] and ECA-Net [32], we introduce a novel channel attention module called PDA. SE-Net [31] conducts dimensionality reduction in fully connected layers to reduce computational load, and Wang et al. [32] utilizes 1D convolution instead of fully connected layers to reduce model complexity without dimensionality reduction. Although ECA-Net employs a kernel size, which can be adaptively determined through a mapping function, this module still has the limitation that its receptive field is fixed. Figure 2 presents the structure of the proposed PDA module. Compared to the previous methods, the PDA module consists of three lightweight 1D convolutions with different dilation ratios, and therefore, the proposed attention mechanism can merge the information from various receptive fields.

In PDA, the GAP is applied to the feature map to acquire a vector with the size of $1 \times 1 \times C$, where C denotes the number of channels. A feature vector of the identical length to the channel size of the previous feature map is diverged and provided as parallelized 1D convolutions. In Figure 2, D indicates the dilation ratio of the 1D convolution. The parallel structure of 1D convolutions is applied to obtain information from various receptive fields depending on dilation ratios. The various dilation ratios are advantageous for obtaining abundant features as they can collect information from narrow to broad receptive fields.

In this experiment, we set the dilation ratios to 1, 2, and 4, respectively, and padding was set equal to the dilation ratio to acquire an output vector of the same length as the input. The output vectors of the 1D convolution, which contain different locally correlated channel information, are concatenated together to aggregate the information of the parallel operation with a vector size of $1 \times 1 \times 3C$. A fully connected layer compresses meaningful information and learns interdependencies between channels. A sigmoid function is utilized to obtain an attention vector $\mathbf{a}$ that contains values between 0 and 1, and it is expressed as follows:

$$\mathbf{a} = \left[a_1, \cdots, a_{c_5}\right], \quad 0 \le a_k \le 1.$$ (3)

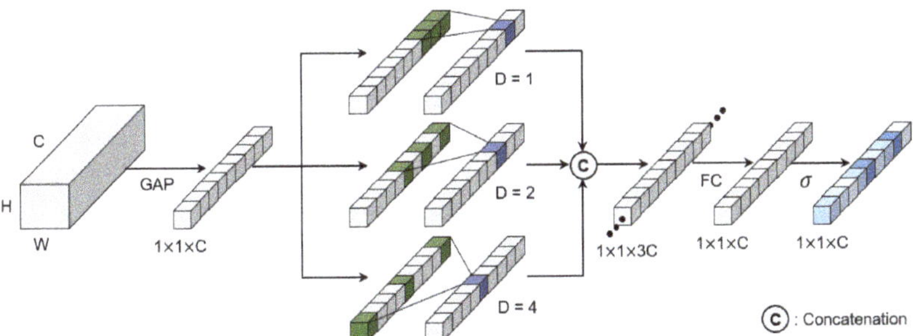

Figure 2. Structure of PDA module.

The attention vector **a** is computed from the last convolution block, and the length of the attention vector is equal to the number of channels. Each component of **a** indicates the significance of the corresponding channel, and therefore, PDA guides the network to focus on important features for classifying and localizing transmission lines. The localization mask based on PDA is called ALM.

Figure 3 shows details of the PDA module, which is placed between the last convolution feature map and ReLU operation. The last convolution feature map $\mathbf{F}_5$ consists of $\mathbf{f}_5^1, \cdots, \mathbf{f}_5^{c_5}$, and the k-th channel $\mathbf{g}^k$ of the weighted feature map $\mathbf{G} \in \mathbb{R}_+^{\frac{H_5}{2} \times \frac{W_5}{2} \times c_5}$ can be computed as

$$\mathbf{g}^k = a_k \otimes \mathbf{f}_5^k, \tag{4}$$

where $\otimes$ is the elementwise multiplication.

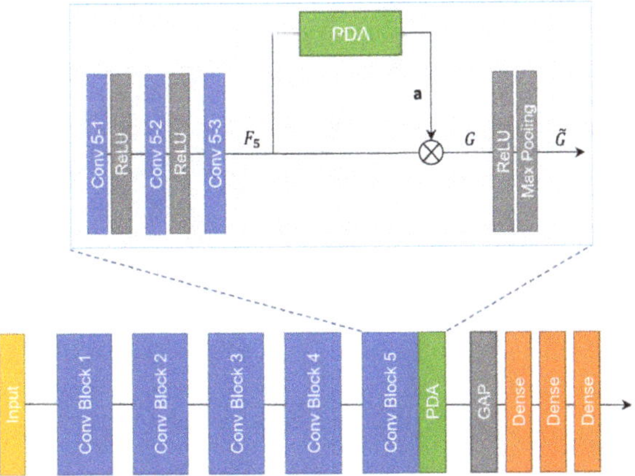

Figure 3. PDA module in the classification network.

The attention-weighted feature map is utilized for transmission line localization. Different from the summation operation in the *VisualBackProp* algorithm, we conducted the weighted summation with the components of the attention vector computed from the PDA module. The attention-weighted feature map is beneficial to focus on important features for localizing transmission lines. In the same way as (1), the weighted feature map **G** is accumulated in the depth direction to compute the accumulated feature map $\bar{\mathbf{g}}$ as

$$\bar{\mathbf{g}} = \sum_{k=1}^{c_5} \mathbf{g}^k. \tag{5}$$

Whereas feature maps are added in the channel direction in the original VBP algorithm, our PDA module computes a channel attention vector to more effectively aggregate the information in the feature maps. To incorporate $\bar{\mathbf{g}}$ with other feature maps, $\bar{\mathbf{g}}$ is upsampled to the identical size as the previous feature map $\bar{\mathbf{f}}_4$, and this procedure can be expressed as follows:

$$\mathbf{h}_5 = Upsampling(\bar{\mathbf{g}}), \tag{6}$$

where *Upsampling* indicates the bilinear interpolation, and the result is denoted by $\mathbf{h}_5$. Through the equation of (2), the elementwise multiplication is conducted between $\mathbf{h}_5$ and $\bar{\mathbf{f}}_4$, and the process is repeated until obtaining $\mathbf{h}_1$, which is called ALM.

3.3. Refinement Network via Image-to-Image Translation

The ALM in the previous step still has room for improvement due to blurred line responses and weak connections of transmission lines in the localization mask. Therefore, we constructed a refinement network to transfer visual characteristics of transmission lines. The refinement network employs a generator of an image-to-image transformation architecture. To transfer line-shaped properties, we generated a dataset based on a rule-based algorithm, and it is called the pseudo-line dataset. The pseudo-line dataset includes 250 binary pseudo-line images with the image size of 512×512, which is identical to the size of ALM. Figure 4 shows examples of pseudo-line data.

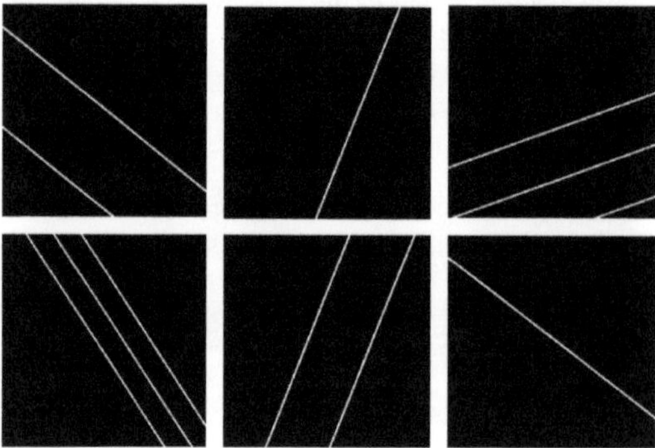

Figure 4. Examples of pseudo-line dataset.

Figure 5 presents the procedure for generating the pseudo-line dataset. We prepared an image filled with zeros and generated pseudo-lines with randomly selected pixel coordinates a and b, which are integers ranging from 0 to 512. Because multiple lines which are too close to each other are not desirable, we generated pseudo-lines that connect opposite sides of images. Since the power line dataset contains multiple transmission lines, we also generated the pseudo-line dataset with variable numbers of transmission lines. In Figure 5, i denotes an arbitrary line interval, and multiple lines are generated with the distance of i. By utilizing the properties of the pseudo-line dataset, we refine the ALM to improve the localization accuracy.

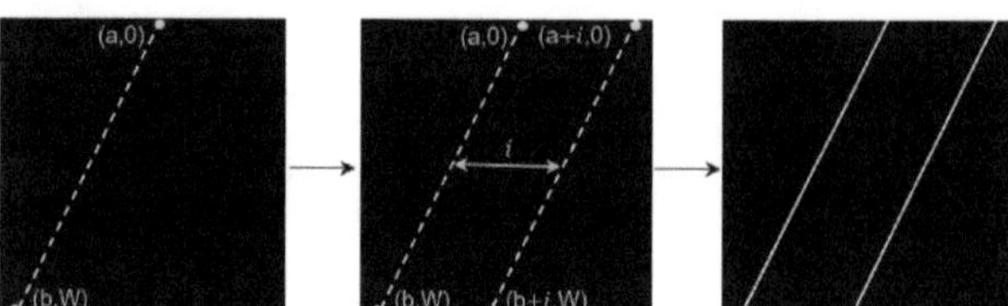

Figure 5. Procedure for generating pseudo-line dataset.

To construct a refinement network, we adopt the structure of CycleGAN [27]. We defined the ALM as a source domain S, and the pseudo-line dataset as a target domain T, respectively. The purpose of the refinement network is to obtain a mapping function from the source domain to the target domain. To train the refinement network, we constructed two generators and two discriminators for the adversarial training of the unpaired image-to-image translation framework. The first generator that maps from the source to the target domain is denoted by $G : S \rightarrow T$, and it is presented in the lower right part of Figure 1. In some images, ALMs show weak responses and missing parts for localizing transmission lines. To address these limitations, the generator G allows the ALMs to contain the properties of pseudo-line images, and it is utilized as the refinement network. The refinement network restores weak responses and missing parts of ALMs based on the line-shaped properties of pseudo-line images. The first discriminator D_T is trained to distinguish between $G(s)$ and pseudo-line dataset in the target domain T, where $G(s)$ is generated images from the source domain. As the adversarial training proceeds, the generator G becomes creating realistic line images, and these result images are called refined ALMs. The discriminator D_T is presented at the upper right part in Figure 1, and the adversarial loss for training G and D_T is defined as

$$L_{adv}(G, D_T, S, T) = E_{t \sim p_{data}(t)}[log(D_T(t)) + E_{s \sim p_{data}(s)}[log(1 - D_T(G(s)))], \qquad (7)$$

where $p_{data}(t)$ and $p_{data}(s)$ are the distributions of the source and target domains.

Similarly, the second generator $F : T \rightarrow S$ maps target domain data into the source domain, and the second discriminator D_S distinguishes ALMs and reconstructed ALMs in the source domain. Another adversarial loss for the training of F and D_S is defined as

$$L_{adv}(F, D_S, T, S) = E_{s \sim p_{data}(s)}[log(D_S(s)) + E_{t \sim p_{data}(t)}[log(1 - D_S(G(t)))]. \qquad (8)$$

Since using only adversarial losses can cause mode collapse, the cycle consistency loss is employed to train the generators in a constrained space. Based on the cycle consistency loss, which is defined as (9), the generated images can be translated back to the original images as shown in Figure 6.

$$L_{cycle}(G, F) = E_{s \sim p_{data}(s)}[\|F(G(s)) - s\|] + E_{t \sim p_{data}(t)}[\|G(F(t)) - t\|]. \qquad (9)$$

The total loss function for training the refinement network is formulated as the combination of the adversarial losses and the cycle consistent loss, and it is defined as

$$L(G, F, D_S, D_T) = L_{adv}(G, D_T, S, T) + L_{adv}(F, D_S, T, S) + \lambda L_{cycle}(G, F), \qquad (10)$$

where λ is a hyper-parameter for controlling the effect the cycle consistency loss, and we utilized $\lambda = 10$ in experiments.

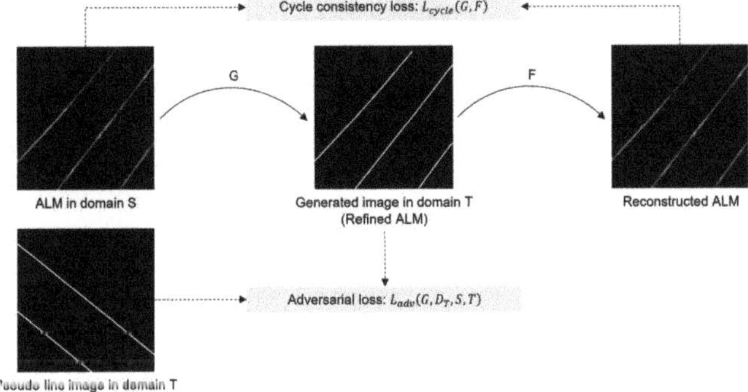

Figure 6. The process for generating refined ALMs. The generator G maps ALMs into the target domain, and therefore, responses of transmission lines are enhanced based on the line-shaped properties of the target domain data.

4. Experimental Results

We conducted experiments with the hardware environment including Intel Core i9-10900K CPU, 64 GB DDR4 RAM, and NVIDIA RTX 3090. The proposed algorithm was implemented based on PyTorch and OpenCV. To train the classification network, we utilized the Adam optimizer with the initial learning rate of 0.0001 and weight decay of 0.05. The refinement network was trained in the identical training setting, excluding the initial learning of 0.0002.

4.1. Dataset Description

We employed the public power line dataset consisting of 400 infrared (IR) and 400 visual light (VL) images. Figure 7 presents example images with and without transmission lines in the first and third columns, respectively. The ground truth corresponds to each image in the second and last columns. The dataset was collected for seasonal days in 21 different regions across Turkey in cooperation with the Turkish Electricity Transmission Company (TEIAS). The dataset is collected under diverse conditions, and therefore, it is challenging to recognize transmission lines in the image due to different backgrounds and illumination. In this study, 400 VL images with the size of 512 × 512 were used, of which 200 VL images contain transmission lines, and the others do not. For the training of the base classification network, we split the dataset into 300, 50, and 50 images for the training, validation, and test sets, respectively.

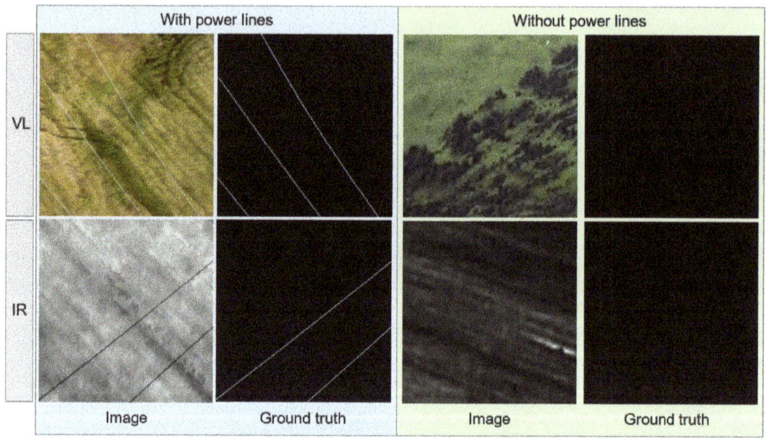

Figure 7. Example images and their corresponding ground truth data.

4.2. Evaluation Measure

The proposed method was evaluated based on the criterion proposed by Choi et al. [22]. Recall, precision, and F1-score were computed from the true positive (TP), false positive (FP), and false negative (FN). In [22], TP is defined as the number of cases where more than 50% of line pixels are correctly detected. To define false responses that occur near a transmission line, the FP is defined based on a tolerance range. By setting the tolerance range as 10 pixels on both sides of a transmission line, FP includes incorrect responses occurring in background regions and thick predictions on transmission lines. If a predicted line is composed of less than 10 pixels, then it is regarded as noise.

4.3. Quantitative Evaluation

Table 1 presents the quantitative comparison of the proposed method with several previous algorithms. To represent differences in approaches, we categorized the results based on the learning types and annotation levels, and more specifically, annotation levels are divided into pixel, patch, and image levels. The learning type of weakly supervised manner can be divided into patch and image-level annotations, which is a difference between the previous and our methods. In the previous methods, patch-level annotations were utilized by dividing the original images into 128 × 128 sub-images and assigning class labels for these patch images.

Table 1. Quantitative comparison with other methods. The best result is highlighted in bold, and the second-best result is underlined. S and WS indicate supervised learning and weakly supervised learning, respectively.

Methods	Learning Type	Annotation Level	Recall (%)	Precision (%)	F1-Score (%)
Long et al. (2015) [59]	S	Pixel	98.17	93.04	95.54
Li et al. (2019) [60]	S	Pixel	97.25	95.50	96.36
Bojarski et al. (2016) [26]	WS	Patch	41.28	28.13	33.46
Lee et al. (2017) [21]	WS	Patch	86.24	**100**	92.61
Choi et al. (2021) [22]	WS	Patch	**98.17**	90.68	<u>94.27</u>
Ours	WS	Image	<u>97.90</u>	<u>96.15</u>	**97.01**

Although annotating patch images require less burden compared to pixel-level labels, patch-level annotations have limitations with respect to it requiring the location information of transmission lines in the original images. On the other hand, we utilized the entire size of the image for the training of the classification network. In transmission line images, lines are composed of small numbers of pixels and most of the remaining area consists of the background, and therefore, the larger the image size, the more difficult it is to localize and detect transmission lines. Nevertheless, we achieved quantitatively significant improvements by utilizing the entire images.

Compared with Choi et al. [22], our proposed method improves the performance by 5.47% and 2.75% in terms of precision and F1-score, respectively, while the recall is slightly lower by 0.27%. We utilized the entire image size from the beginning to the end of the algorithm, while [21,22] divide the image into patches once in the middle and merge them back together. In addition, in the process of dividing into patch images, approximate location information of transmission lines is required to train the classification network. It is worth noting that our proposed method does not require any location information. Furthermore, our method is meaningful in that the localization masks obtained from the feature maps can be improved by the refinement network. In another weakly supervised learning method, Lee et al. [21] achieved substantial performance in precision and F1-score, but there was a trade-off between precision and recall. By contrast, our proposed method has satisfactory results in both recall and precision for detecting transmission lines.

Table 1 also presents the accuracy of the segmentation algorithms proposed in [59,60]. Although the supervised learning methods show satisfactory performance in both recall and precision, these algorithms require time-consuming work for generating pixel-level annotations. To reduce the cost for preparing ground truth data, our proposed method adopts a weakly supervised learning framework. It is noteworthy that even though we only utilized image-level annotations, the proposed algorithm shows higher accuracy in terms of precision and F1-score.

4.4. Ablation Study

We conducted an ablation study to demonstrate the effectiveness of each step of the proposed algorithm, and the results are presented in Table 2. We employed the algorithm proposed by Bojarski et al. [26], which is effective to represent clues in the feature map of a convolutional network, and it is considered as the baseline model and denoted as localization mask in Table 2. The attention vector obtained from the PDA contains the scores for each channel of the feature map, and the scores range from 0 to 1. This calculated score is multiplied by the feature map for channel-wise to give importance to each channel, and when performing binary classification of the input image, it is localized by focusing on the learned features. By adding the PDA to the baseline, we reached an F1-score of 92.35%, which improved performance by 1.22% compared to the existing localization mask. The localization mask performance after the refinement process showed that the refinement network, which is part of the proposed method, is meaningful. In Table 2, the performance improved by 3.59% and 2.96%, respectively, in terms of recall and precision. These results show that the adoption of CycleGAN is adequate as a way to improve the localization mask through the process of transferring the characteristics of an ideal line fully connected from one point to another. The best performance is the result of applying both steps, and the proposed model outperforms the baseline with recall and precision of 97.90% and 96.15%, respectively. In other words, the performance improved by 7.53% in recall and 4.25% in precision, and finally F1-score improved by 5.88% compared to the localization mask. We have shown that every step we suggest is beneficial in successfully detecting transmission lines.

Table 2. Ablation study for demonstrating the effectivness of PDA module and refinement network.

Localization Mask	PDA	Refinement	Recall Rate (%)	Precision (%)	F1-Score (%)
✓			90.37	91.90	91.13
✓	✓		92.03	92.67	92.35
✓		✓	93.96	94.86	94.41
✓	✓	✓	97.90	96.15	97.01

4.5. Comparison with Other Attention Modules

In this section, the proposed attention module is compared with previous attention mechanisms including SimAM [30], SE [31], and ECA [32]. Table 3 summarizes the results of comparative experiments. An identical base network of VGG16 was utilized to obtain localization masks based on the previous attention modules. The localization accuracy with the use of PDA is compared with the accuracies based on the previous methods. SE showed less than 90% accuracies in all evaluation metrics, and PDA was higher than SE by 3.15% in terms of precision. In addition, another channel attention module ECA showed better performance than SE, but it was still lower than the accuracy of ours. ECA utilizes only one selective kernel for 1D convolution, and it is limited to properly representing features of transmission lines due to small receptive field. As shown in Table 3, SimAM, which generates 3D weights, was unsuitable for our task, and this experiment showed that not all attention modules are effective in localizing transmission lines. The PDA module showed plausible performance in all evaluation metrics compared to the previous methods. The performance improvement is attributed to the advantage that PDA captures abundant feature representations and broadens the receptive field by utilizing three different 1D convolutions.

Table 3. Comparison with existing attention modules. These results indicate the performance before applying the refinement network.

Methods	Recall Rate (%)	Precision (%)	F1-Score (%)
SimAM [30]	89.88	88.97	89.42
SE [31]	89.89	89.52	89.71
ECA [32]	90.97	92.08	91.52
PDA (Ours)	92.03	92.67	92.35

4.6. Qualitative Evaluation

Figure 8 presents result images of the proposed algorithm for detecting transmission lines. Figure 8a,b show input images and the corresponding ground truth data. The ALM is the localization mask with the application of the attention vector obtained from the PDA, for focusing more on the important channels. As shown in Figure 8c, PDA is effective in localizing the lines in the input images, and the predicted structure and number of lines are similar to the ground truth even before applying the refinement network. However, most transmission lines in the ALMs are blurry, and several predictions contain disconnected or omitted parts. To address these limitations, the refinement network is applied to the ALMs, and the results are presented at Figure 8d. The refinement network generates refined ALMs, which complement smudged or indistinct lines to make the lines sharp and bold, based on the characteristics of the target domain data. The refinement network also connects broken lines to generate intersecting lines of images, providing qualitatively plausible results even if compared with ground truth data. In Figure 8e, original images are overlaid with the refined ALM. Figure 9 represents failure cases of the proposed algorithm. In the refined ALM of the second example, the red arrow indicates a merged prediction of two transmission lines close to each other. However, even though two close transmission lines are recognized as a single line, it does not critically affect UAVs to operate a collision avoidance function. The yellow arrows in the refined ALMs indicate missing parts of transmission lines. When the responses of transmission lines in ALM are too weak, the refinement network could not restore the corresponding parts of transmission lines. The green arrow in the last example in Figure 9d indicates a false positive case, and such a failure is usually isolated and occurs in a local area. Therefore, we expect that these types of failures can be recovered by applying a post-processing based on the properties of transmission lines.

Figure 8. Results of transmission line detection using our proposed method. (**a**) Original image. (**b**) Ground truth. (**c**) ALM. (**d**) Refined ALM. (**e**) Overlaid image with the refined ALM.

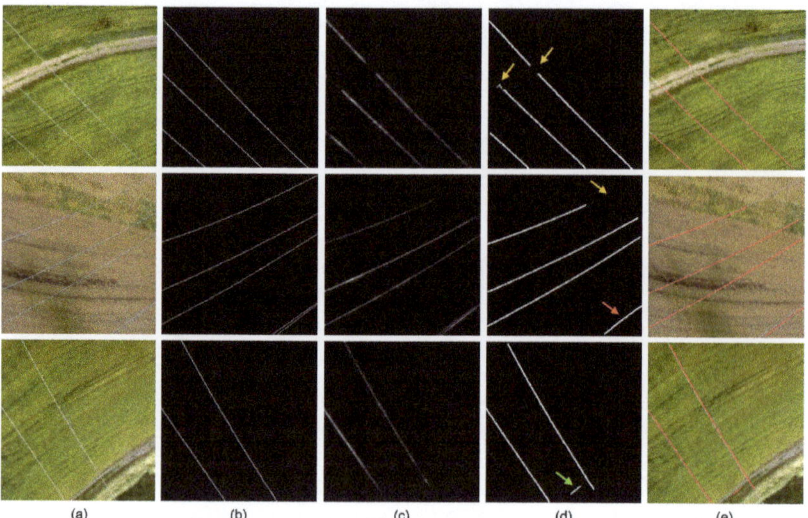

Figure 9. Examples of failure cases of the proposed method. The yellow and green arrows indicate missing parts of transmission lines and false positive responses, respectively. The red arrow indicates a false negative case for two transmission lines that are close to each other. (**a**) Original. (**b**) Ground truth. (**c**) ALM. (**d**) Refined ALM. (**e**) Overlaid image with the refined ALM.

5. Conclusions

In this paper, we propose a transmission line detection algorithm based on weakly supervised learning and image-to-image translation. By only utilizing image-level labels, the proposed algorithm can be trained with the minimal human involvement. The proposed method consists of two steps: (1) localization of transmission lines based on PDA and (2) refinement via image-to-image translation. The PDA module computes a score vector based on the information from various receptive fields. The attention vector provides the channel importance of object features, and it is utilized for generating ALM. Furthermore, we constructed a refinement network that transfers line-shaped properties of transmission lines to improve weak responses and disconnected components in the ALM. We demonstrated that the PDA module outperforms the previous attention methods for localizing transmission lines. Moreover, the refinement network significantly improved the accuracy of transmission line detection in both quantitative and qualitative aspects.

Author Contributions: Conceptualization, J.C.; methodology, J.C.; software, J.C.; validation, J.C.; formal analysis, J.C.; data curation, J.C.; writing—original draft preparation, J.C. and S.J.L.; writing—review and editing, J.C. and S.J.L.; visualization, J.C. All authors have read and agreed to the published version of the manuscript.

Funding: This work was supported by the National Research Foundation of Korea (NRF) grant funded by the Korea government (MSIT) (No. 2021R1G1A1009792).

Data Availability Statement: Publicly available datasets were analyzed in this study. These data can be found here: https://data.mendeley.com/datasets/twxp8xccsw/6, (accessed on 6 June 2022).

Conflicts of Interest: The authors declare no conflict of interest.

References

1. Mohamed, N.; Al-Jaroodi, J.; Jawhar, I.; Idries, A.; Mohammed, F. Unmanned aerial vehicles applications in future smart cities. *Technol. Forecast. Soc. Chang.* **2020**, *153*, 119293. [CrossRef]
2. Radoglou-Grammatikis, P.; Sarigiannidis, P.; Lagkas, T.; Moscholios, I. A compilation of UAV applications for precision agriculture. *Comput. Netw.* **2020**, *172*, 107148. [CrossRef]

3.   Shakhatreh, H.; Sawalmeh, A.; Al-Fuqaha, A.; Dou, Z.; Almaita, E.; Khalil, I.; Othman, N.S.; Khreishah, A.; Guizani, M. Unmanned aerial vehicles (UAVs): A survey on civil applications and key research challenges. *arXiv* **2018**, arXiv:1805.00881.

4.   Huang, X.; Dong, X.; Ma, J.; Liu, K.; Ahmed, S.; Lin, J.; Qiu, B. The Improved A* Obstacle Avoidance Algorithm for the Plant Protection UAV with Millimeter Wave Radar and Monocular Camera Data Fusion. *Remote Sens.* **2021**, *13*, 3364. [CrossRef]

5.   She, X.; Huang, D.; Song, C.; Qin, N.; Zhou, T. Multi-obstacle detection based on monocular vision for UAV. In Proceedings of the 2021 IEEE 16th Conference on Industrial Electronics and Applications (ICIEA), Chengdu, China, 1–4 August 2021; pp. 1067–1072. [CrossRef]

6.   Pedro, D.; Matos-Carvalho, J.P.; Fonseca, J.M.; Mora, A. Collision avoidance on unmanned aerial vehicles using neural network pipelines and flow clustering techniques. *Remote Sens.* **2021**, *13*, 2643. [CrossRef]

7.   González de Santos, L.M.; Frías Nores, E.; Martínez Sánchez, J.; González Jorge, H. Indoor path-planning algorithm for UAV-based contact inspection. *Sensors* **2021**, *21*, 642. [CrossRef]

8.   Dai, X.; Mao, Y.; Huang, T.; Qin, N.; Huang, D.; Li, Y. Automatic obstacle avoidance of quadrotor UAV via CNN-based learning. *Neurocomputing* **2020**, *402*, 346–358. [CrossRef]

9.   Ou, J.; Guo, X.; Lou, W.; Zhu, M. Quadrotor Autonomous Navigation in Semi-Known Environments Based on Deep Reinforcement Learning. *Remote Sens.* **2021**, *13*, 4330. [CrossRef]

10.   Yuan, S.; Ota, K.; Dong, M.; Zhao, J. A Path Planning Method with Perception Optimization Based on Sky Scanning for UAVs. *Sensors* **2022**, *22*, 891. [CrossRef]

11.   Wang, D.; Li, W.; Liu, X.; Li, N.; Zhang, C. UAV environmental perception and autonomous obstacle avoidance: A deep learning and depth camera combined solution. *Comput. Electron. Agric.* **2020**, *175*, 105523. [CrossRef]

12.   Ge, C.; Wang, J.; Wang, J.; Qi, Q.; Sun, H.; Liao, J. Towards automatic visual inspection: A weakly supervised learning method for industrial applicable object detection. *Comput. Ind.* **2020**, *121*, 103232. [CrossRef]

13.   Huang, B.; Chen, R.; Zhou, Q.; Xu, W. Eye landmarks detection via weakly supervised learning. *Pattern Recognit.* **2020**, *98*, 107076. [CrossRef]

14.   Zhang, F.; Du, B.; Zhang, L.; Xu, M. Weakly supervised learning based on coupled convolutional neural networks for aircraft detection. *IEEE Trans. Geosci. Remote Sens.* **2016**, *54*, 5553–5563. [CrossRef]

15.   Fu, K.; Lu, W.; Diao, W.; Yan, M.; Sun, H.; Zhang, Y.; Sun, X. WSF-NET: Weakly supervised feature-fusion network for binary segmentation in remote sensing image. *Remote Sens.* **2018**, *10*, 1970. [CrossRef]

16.   Wang, S.; Chen, W.; Xie, S.M.; Azzari, G.; Lobell, D.B. Weakly supervised deep learning for segmentation of remote sensing imagery. *Remote Sens.* **2020**, *12*, 207. [CrossRef]

17.   Kim, W.S.; Lee, D.H.; Kim, T.; Kim, H.; Sim, T.; Kim, Y.J. Weakly supervised crop area segmentation for an autonomous combine harvester. *Sensors* **2021**, *21*, 4801. [CrossRef]

18.   Wang, P.; Yao, W. Weakly Supervised Pseudo-Label assisted Learning for ALS Point Cloud Semantic Segmentation. *arXiv* **2021**, arXiv:2105.01919.

19.   Blaga, B.-C.-Z.; Nedevschi, S. Weakly Supervised Semantic Segmentation Learning on UAV Video Sequences. In Proceedings of the 2021 29th European Signal Processing Conference (EUSIPCO), Dublin, Ireland, 23–27 August 2021; pp. 731–735. [CrossRef]

20.   Yang, Z.; Zhao, L.; Wu, S.; Chen, C.Y.C. Lung lesion localization of COVID-19 from chest CT image: A novel weakly supervised learning method. *IEEE J. Biomed. Health Inform.* **2021**, *25*, 1864–1872. [CrossRef]

21.   Lee, S.J.; Yun, J.P.; Choi, H.; Kwon, W.; Koo, G.; Kim, S.W. Weakly supervised learning with convolutional neural networks for power line localization. In Proceedings of 2017 IEEE Symposium Series on Computational Intelligence (SSCI), Honolulu, HI, USA, 27 November–1 December 2017; pp.1–8. [CrossRef]

22.   Choi, H.; Koo, G.; Kim, B.J.; Kim, S.W. Weakly supervised power line detection algorithm using a recursive noisy label update with refined broken line segments. *Expert Syst. Appl.* **2021**, *165*, 113895. [CrossRef]

23.   Jaffari, R.; Hashmani, M.A.; Reyes-Aldasoro, C.C. A Novel Focal Phi Loss for Power Line Segmentation with Auxiliary Classifier U-Net. *Sensors* **2021**, *21*, 2803. [CrossRef]

24.   Hota, M.; Kumar, U. Power Lines Detection and Segmentation In Multi-Spectral Uav Images Using Convolutional Neural Network. In Proceedings of the 2020 IEEE India Geoscience and Remote Sensing Symposium (InGARSS), Ahmedabad, India, 1–4 December 2020; pp.154–157. [CrossRef]

25.   Vemula, S.; Frye, M. Mask R-CNN Powerline Detector: A Deep Learning approach with applications to a UAV. In Proceedings of the 2020 AIAA/IEEE 39th Digital Avionics Systems Conference (DASC), San Antonio, TX, USA, 11–15 October 2020; pp. 1–6. [CrossRef]

26.   Bojarski, M.; Choromanska, A.; Choromanski, K.; Firner, B.; Jackel, L.; Muller, U.; Zieba, K. VisualBackProp: visualizing CNNs for autonomous driving. *arXiv* **2016**, arXiv:1611.05418.

27.   Zhu, J.Y.; Park, T.; Isola, P.; Efros, A.A. Unpaired image-to-image translation using cycle-consistent adversarial networks. In Proceedings of the IEEE International Conference on Computer Vision, Venice, Italy, 22–29 October 2017; pp. 2223–2232. [CrossRef]

28.   Park, J.; Woo, S.; Lee, J.Y.; Kweon, I.S. Bam: Bottleneck attention module. *arXiv* **2018**, arXiv:1807.06514.

29.   Woo, S.; Park, J.; Lee, J.Y.; Kweon, I.S. Cbam: Convolutional block attention module. In Proceedings of the European Conference on Computer Vision (ECCV), Munich, Germany, 8–14 August 2018; pp. 3–19. [CrossRef]

30. Yang, L.; Zhang, R. Y.; Li, L.; Xie, X. Simam: A simple, parameter-free attention module for convolutional neural networks. In Proceedings of the 38th International Conference on Machine Learning (PMLR), Virtual, 18–24 July 2021; pp. 11863–11874.

31. Hu, J.; Shen, L.; Sun, G. Squeeze-and-excitation networks. In Proceedings of the IEEE Conference on Computer Vision and Pattern Recognition, Salt Lake City, UT, USA, 18–23 June 2018; pp. 7132–7141. [CrossRef]

32. Wang, Q.L.; Wu, B.G.; Zhu, P.F.; Li, P.H.; Zuo, W.M.; Hu, Q.H. ECA-Net: Efficient Channel Attention for Deep Convolutional Neural Networks. In Proceedings of the IEEE/CVF Conference on Computer Vision and Pattern Recognition (CVPR), Seattle, WA, USA, 13–19 June 2020; pp. 11531–11539. [CrossRef]

33. Aguirre, N.; Grall-Maës, E.; Cymberknop, L.J.; Armentano, R.L. Blood pressure morphology assessment from photoplethysmogram and demographic information using deep learning with attention mechanism. *Sensors* **2021**, *21*, 2167. [CrossRef] [PubMed]

34. Bahdanau, D.; Cho, K.; Bengio, Y. Neural machine translation by jointly learning to align and translate. *arXiv* **2014**, arXiv:1409.0473.

35. Vaswani, A.; Shazeer, N.; Parmar, N.; Uszkoreit, J.; Jones, L.; Gomez, A.N.; Kaiser, L.; Polosukhin, I. Attention is all you need. In *Advances in Neural Information Processing Systems*; MIT Press: Cambridge, MA, USA, 2017; pp. 5998–6008.

36. Eom, H.; Lee, D.; Han, S.; Hariyani, Y.S.; Lim, Y.; Sohn, I.; Park, K.; Park, C. End-to-end deep learning architecture for continuous blood pressure estimation using attention mechanism. *Sensors* **2020**, *20*, 2338. [CrossRef]

37. Eldele, E.; Chen, Z.; Liu, C.; Wu, M.; Kwoh, C.K.; Li, X.; Guan, C. An attention-based deep learning approach for sleep stage classification with single-channel eeg. *IEEE Trans. Neural Syst. Rehabil. Eng.* **2021**, *29*, 809–818. [CrossRef]

38. Shi, C.; Zhang, X.; Sun, J.; Wang, L. A Lightweight Convolutional Neural Network Based on Group-Wise Hybrid Attention for Remote Sensing Scene Classification. *Remote Sens.* **2022**, *14*, 161. [CrossRef]

39. Ma, W.; Zhao, J.; Zhu, H.; Shen, J.; Jiao, L.; Wu, Y.; Hou, B. A spatial-channel collaborative attention network for enhancement of multiresolution classification. *Remote Sens.* **2021**, *13*, 106. [CrossRef]

40. Hu, J.; Zhi, X.; Shi, T.; Zhang, W.; Cui, Y.; Zhao, S. PAG-YOLO: A portable attention-guided YOLO network for small ship detection. *Remote Sens.* **2021**, *13*, 3059. [CrossRef]

41. Chen, L.; Shi, W.; Deng, D. Improved YOLOv3 based on attention mechanism for fast and accurate ship detection in optical remote sensing images. *Remote Sens.* **2021**, *13*, 660. [CrossRef]

42. Seong, S.; Choi, J. Semantic segmentation of urban buildings using a high-resolution network (HRNet) with channel and spatial attention gates. *Remote Sens.* **2021**, *13*, 3087. [CrossRef]

43. Tran, M.T.; Kim, S.H.; Yang, H.J.; Lee, G.S.; Oh, I.J.; Kang, S.R. Esophagus segmentation in CT images via spatial attention network and STAPLE algorithm. *Sensors* **2021**, *21*, 4556. [CrossRef] [PubMed]

44. Kim, M.; Lee, B.D. Automatic lung segmentation on chest X-rays using self-attention deep neural network. *Sensors* **2021**, *21*, 369. [CrossRef] [PubMed]

45. Goodfellow, I.; Pouget-Abadie, J.; Mirza, M.; Xu, B.; Warde-Farley, D.; Ozair, S.; Courville, A.; Bengio, Y. Generative adversarial nets. In Proceedings of the Advances in Neural Information Processing Systems, Montreal, QC, Canada, 8–13 December 2014; pp. 2672–2680.

46. Radford, A.; Metz, L.; Chintala, S. Unsupervised representation learning with deep convolutional generative adversarial networks. *arXiv* **2015**, arXiv:1511.06434.

47. Isola, P.; Zhu, J.Y.; Zhou, T.; Efros, A.A. Image-to-image translation with conditional adversarial networks. In Proceedings of the IEEE Conference on Computer Vision and Pattern Recognition, Honolulu, HI, USA, 21–26 July 2017; pp. 1125–1134. [CrossRef]

48. Chen, X.; Duan, Y.; Houthooft, R.; Schulman, J.; Sutskever, I.; Abbeel, P. Infogan: Interpretable representation learning by information maximizing generative adversarial nets. In *Advances in Neural Information Processing Systems*; MIT Press: Barcelona, Spain, 2016; pp. 2172–2180.

49. Kim, T.; Cha, M.; Kim, H.; Lee, J.; Kim, J. Learning to discover cross-domain relations with generative adversarial networks. In Proceedings of the International Conference on Machine Learning (PMLR), Sydney, Australia, 6–11 August 2017; pp. 1857–1865.

50. Yi, Z.; Zhang, H.; Tan, P.; Gong, M. DualGAN: Unsupervised Dual Learning for Image-to-Image Translation. In Proceedings of the 2017 IEEE International Conference on Computer Vision (ICCV), Venice, Italy, 22–29 October 2017; pp. 2868–2876. [CrossRef]

51. Zi, Y.; Xie, F.; Song, X.; Jiang, Z.; Zhang, H. Thin Cloud Removal for Remote Sensing Images Using a Physical Model-Based CycleGAN with Unpaired Data. *IEEE Geosci. Remote Sens. Lett.* **2021**, *19*, 1–5. [CrossRef]

52. Liu, W.; Luo, B.; Liu, J. Synthetic Data Augmentation Using Multiscale Attention CycleGAN for Aircraft Detection in Remote Sensing Images. *IEEE Geosci. Remote Sens. Lett.* **2022**, *19*, 1–5. [CrossRef]

53. Mao, J.; Zheng, C.; Yin, J.; Tian, Y.; Cui, W. Wildfire Smoke Classification Based on Synthetic Images and Pixel-and Feature-Level Domain Adaptation. *Sensors* **2021**, *21*, 7785. [CrossRef]

54. Choi, W.; Heo, J.; Ahn, C. Development of Road Surface Detection Algorithm Using CycleGAN-Augmented Dataset. *Sensors* **2021**, *21*, 7769. [CrossRef]

55. Nguyen, D.T.; Pham, T.D.; Batchuluun, G.; Noh, K.J.; Park, K.R. Presentation attack face image generation based on a deep generative adversarial network. *Sensors* **2020**, *20*, 1810. [CrossRef]

56. Sandouka, S.B.; Bazi, Y.; Alajlan, N. Transformers and Generative Adversarial Networks for Liveness Detection in Multitarget Fingerprint Sensors. *Sensors* **2021**, *21*, 699. [CrossRef]

57. Gao, P.; Tian, T.; Li, L.; Ma, J.; Tian, J. DE-CycleGAN: An object enhancement network for weak vehicle detection in satellite images. *IEEE J. Sel. Top. Appl. Earth Obs. Remote Sens.* **2021**, *14*, 3403–3414. [CrossRef]

58. Noh, K.J.; Choi, J.; Hong, J.S.; Park, K.R. Finger-vein recognition using heterogeneous databases by domain adaption based on a cycle-consistent adversarial network. *Sensors* **2021**, *21*, 524. [CrossRef] [PubMed]
59. Long, J.; Shelhamer, E.; Darrell, T. Fully convolutional networks for semantic segmentation. In Proceedings of the IEEE conference on computer vision and pattern recognition (CVPR), Boston, MA, USA, 7–12 June 2015; pp. 3431–3440. [CrossRef]
60. Li, Y.; Xiao, Z.; Zhen, X.; Cao, X. Attentional information fusion networks for cross-scene power line detection. *IEEE Geosci. Remote Sens. Lett.* **2019**, *16*, 1635–1639. [CrossRef]

remote sensing

Article

Wide-Area and Real-Time Object Search System of UAV

Xianjiang Li [1], Boyong He [1], Kaiwen Ding [1], Weijie Guo [2], Bo Huang [1] and Liaoni Wu [1,*]

[1] School of Aerospace Engineering, Xiamen University, Xiamen 361102, China; lixianjiang@stu.xmu.edu.cn (X.L.); heboyong0220@stu.xmu.edu.cn (B.H.); dingkaiwen@stu.xmu.edu.cn (K.D.); huangbo@stu.xmu.edu.cn (B.H.)

[2] School of Informatics, Xiamen University, Xiamen 361005, China; ccgwj@stu.xmu.edu.cn

* Correspondence: wuliaoni@xmu.edu.cn

Abstract: The method of collecting aerial images or videos by unmanned aerial vehicles (UAVs) for object search has the advantages of high flexibility and low cost, and has been widely used in various fields, such as pipeline inspection, disaster rescue, and forest fire prevention. However, in the case of object search in a wide area, the scanning efficiency and real-time performance of UAV are often difficult to satisfy at the same time, which may lead to missing the best time to perform the task. In this paper, we design a wide-area and real-time object search system of UAV based on deep learning for this problem. The system first solves the problem of area scanning efficiency by controlling the high-resolution camera in order to collect aerial images with a large field of view. For real-time requirements, we adopted three strategies to accelerate the system, as follows: design a parallel system, simplify the object detection algorithm, and use TensorRT on the edge device to optimize the object detection model. We selected the NVIDIA Jetson AGX Xavier edge device as the central processor and verified the feasibility and practicability of the system through the actual application of suspicious vehicle search in the grazing area of the prairie. Experiments have proved that the parallel design of the system can effectively meet the real-time requirements. For the most time-consuming image object detection link, with a slight loss of precision, most algorithms can reach the 400% inference speed of the benchmark in total, after algorithm simplification, and corresponding model's deployment by TensorRT.

Keywords: unmanned aerial vehicles (UAVs); wide-area and real-time object search; aerial image; object detection

Citation: Li, X.; He, B.; Ding, K.; Guo, W.; Huang, B.; Wu, L. Wide-Area and Real-Time Object Search System of UAV. *Remote Sens.* **2022**, *14*, 1234. https://doi.org/10.3390/rs14051234

Academic Editors: Yingying Dong, Chenghai Yang, Giovanni Laneve and Wenjiang Huang

Received: 12 January 2022
Accepted: 28 February 2022
Published: 2 March 2022

Publisher's Note: MDPI stays neutral with regard to jurisdictional claims in published maps and institutional affiliations.

1. Introduction

The development of technology in recent years has promoted the popularization of UAV in object search tasks. For instance, many teams use the UAV to carry out ship detection [1,2], object tracking [3], and emergency search and rescue [4,5] in the ocean environment. In the field of agriculture, the UAV has a wide range of applications, such as weed recognition [6,7], pest detection [8], and precision pesticide spraying [9], which effectively promotes the intelligentization of agriculture. Forests are also one of the typical scenarios of object search, and common applications include fire monitoring and prevention [10,11], ecological protection [12,13], etc.

In most of these tasks, object search is mainly performed by collecting images or videos and processing them through a search algorithm. For the real-time object search of UAV aerial images or videos, the traditional feature extraction algorithms have high computational complexity and are time consuming, which cannot meet the real-time requirements. In 2012, with the outstanding performance of the AlexNet [14] deep learning model based on convolutional neural network (CNN) in the ImageNet [15] image classification competition, people began to realize the powerful capabilities of CNN in image feature extraction and analysis. In the object-detection branch, benefitting from the continuous exploration of deep learning technology in recent years, rich and diverse algorithms have been proposed, including RCNN series [16–18], SSD [19], YOLO series [20–23], and excellent practical application effects have been achieved. The commonly used object detection algorithms can be roughly divided into two categories. The first category is a regression-based single-stage algorithm, such as SSD [19], YOLOv4 [23], and RetinaNet [24]; another category is a two-stage algorithm based on region proposal and regression, such as Faster RCNN [18] and Cascade RCNN [25]. Generally, the single-stage detection algorithms are efficient, but the precision is relatively low, while the two-stage algorithms are inefficient but have higher detection precision. Recently, anchor-free type algorithms, such as FCOS [26] and ATSS [27], have also emerged in the field of object detection, which has reduced the amount of calculation that uses anchors. The classification of commonly used object detection algorithms can be seen in Figure 1.

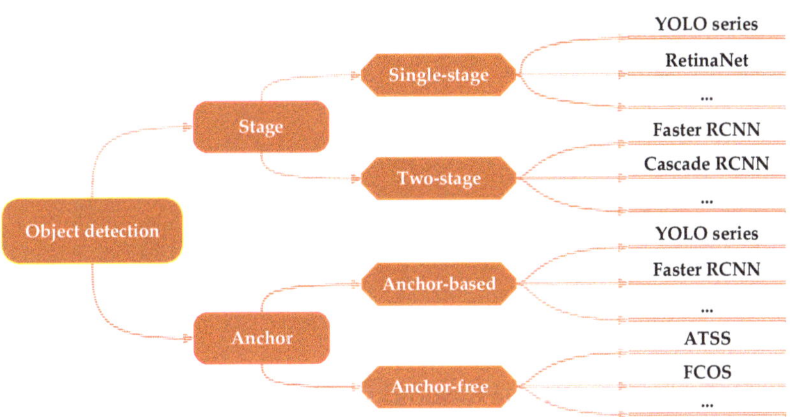

Figure 1. Classification of commonly used object detection algorithms.

When performing tasks that require UAV to search for objects in a wide area, two main issues of "area scanning efficiency" and "real-time detection" need to be considered. In terms of area scanning efficiency, an effective and easy-to-implement method is to use the UAV equipped with high-resolution cameras to take images of the ground surface at mid-to-high altitude. Figure 2 displays the characteristics of the images collected in this way. As shown in the figure, the aerial images collected at mid-to-high altitude have the characteristics of a large field of view and small objects. When the UAV's flight altitude relative to the ground is 900 m, and the camera is 42.4 megapixels (7952×5304 pixels), while the ground vehicle in the image only accounts for 0.0075% (about 80×40 pixels) of all of the pixels in the entire image. Since the cameras are just mounted on the UAV, the image data can only be taken out after the UAV has landed and then processed and filtered through the object detection algorithm. Undoubtedly this method takes much post-processing time and costs more to perform. In some time-sensitive task situations, it may also lead to missing the best time for the object search.

For the real-time detection issue, there have been many related kinds of research on the practical application of UAV object search. Some research mainly focus on simplifying a specific object detection algorithm, making it faster and more suitable for mobile devices with limited computing power. Zhan W. et al. [28] improved the YOLOv5 object detection algorithm from four aspects in order to achieve real-time detection of small objects, as follows: by redesigning the anchor size, adding attention module to the backbone, using CIOU loss function, and adding the P2 feature level [29] proposes ShuffleDet based on ShuffleNet [30], and a modified variant of SSD [19] to realize real-time vehicle detection by UAV. While improving the algorithm, there is also a part of research that describes the hardware composition and the workflow of the real-time object search system of UAV [31], which describes a real-time survivor detection system with a pruned object detection algorithm in a UAV, proposed in order to reduce the loss of lives caused by natural disasters. In [32], Chen, L. et al. collected video stream and executed real-time object detection by carrying a camera, then controlling the UAV to perform corresponding actions. Other related research includes [33–38].

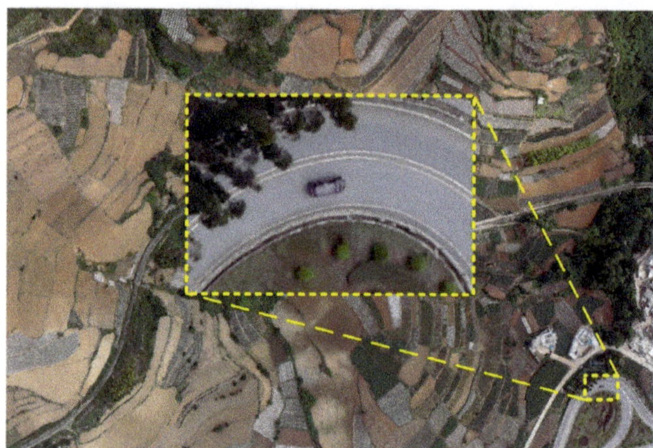

Figure 2. Small object in aerial image with a large field of view. The pixels of the vehicle in this image only account for 0.0075% of the whole image pixels. Although it is very efficient to search for objects by collecting large field view aerial images, it poses a challenge to real-time requirements.

Most of the UAV real-time object search systems that are involved in previous research [31,32,36–38] are more suitable for low-altitude and short-endurance flight tasks, and the equipped equipment is often a small-resolution camera for taking images or videos; there is little research on real-time systems when the UAV flies at mid-to-high altitude. While many UAVs can be used in parallel to achieve wide-area coverage, this requires the addition of multiple UAVs and ground crews, resulting in a significant increase in the operating costs.

The advantage of searching by collecting aerial images is that a large area is scanned at one time, which is highly efficient, but it is a challenge to the real-time performance of the system. In this paper, we have carried out the following points in order to design a system that satisfies both area scanning efficiency and real-time detection performance:

- We designed a wide-area and real-time object search system of UAV using a high-resolution camera and NVIDIA Jetson AGX Xavier embedded edge device, and verified the system's feasibility in an actual task. With the help of the parallel computing capabilities of the edge device, the system adopts the idea of parallel modular design to improve the system's performance. Based on this design, the system can also flexibly adapt different types of cameras and objection detection algorithms according to the task requirements, thereby increasing the versatility and configurability of the system;
- Considering the diversity of the object detection algorithm selection caused by the diversity of tasks, we adopted some general strategies to simplify the commonly used object detection algorithms in order to improve the inference speed instead of optimizing for a specific algorithm. Moreover, the TensorRT deployment method is used to further accelerate the object detection model on the NVIDIA Jetson AGX Xavier edge device.

Due to the broad coverage of a single aerial image, accurate positioning of the object in the image is essential. Otherwise, the object positioning will significantly deviate from the actual location. In many studies, such as [31,38], there are a lack of descriptions for object positioning, because the object occupies a large proportion of the image frame at a limited altitude, in which it can be approximately considered that the object location is equivalent to the location of the UAV. Ref. [37] mentioned the study of velocity estimation of the tracked object, but also no precise coordinate localization. This paper will also introduce the detailed calculation method to perfect the whole system.

For the selection of object detection algorithms, this paper chooses the mainstream algorithms that are commonly used in engineering as the benchmark, which have higher detection precision [32] and chose the lightweight YOLOv3-tiny algorithm, which is fast, but the detection precision is relatively low and will degrade the system's overall performance. After appropriate acceleration of the selected algorithms, their detection speed has significantly improved at the expense of a small amount of precision, allowing them to meet the real-time requirements in the system.

2. System Design and Optimization

The wide-area and real-time object search system comprises software and various hardware, as depicted in Figure 3. Simply put, if a wide-area object search task needs to be performed, the UAV will use the camera to scan the area to be searched at a mid-to-high altitude by taking images. The captured image will immediately be read into the onboard edge device (we chose the Jetson AGX Xavier, which will be introduced later), which applies the objection detection algorithm for object search. Then, the detected object image and corresponding positioning information will be transmitted to the ground control station (GCS), in real-time, for confirmation. After determining the suspicious object, the decision will be made as to whether to go to the scene for further inspection, according to the situation.

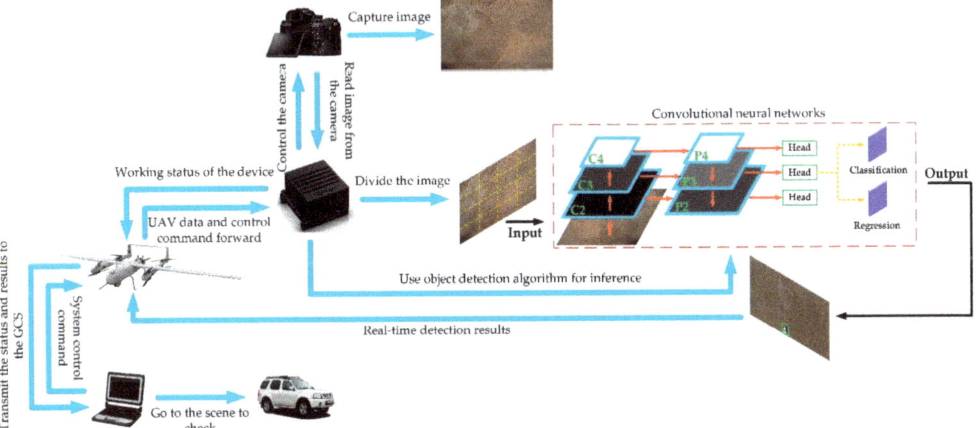

Figure 3. Wide-area and real-time object search system of UAV. It is a system composed of software and various hardware.

2.1. The Main Hardware of System

Most of the hardware that is used in this system is shelf products, which can be easily purchased without special customization, which saves the construction cost of the entire system, to a certain extent.

2.1.1. Jetson AGX Xavier

Jetson AGX Xavier is an embedded AI edge device produced by NVIDIA. The device is equipped with a Xavier processor with eight 64-bit ARM architecture CPUs, while the GPU uses NVIDIA Volta architecture with 512 NVIDIA CUDA cores and 64 Tensor cores. This device provides a peak computing power of up to 32 TOPS and a high-speed I/O transmission of 750 Gbps. Compared with the previous generation Jetson TX2, the performance has improved 20 times. In order to ensure the interaction with external devices in multiple ways simultaneously, the device integrates peripheral interfaces, such as PCIE, Gigabit Ethernet, UART serial port, type-c, and HDMI. Table 1 lists the performance comparison of some processors. We can see that Jetson AGX Xavier has the highest AI performance, and the memory also reaches 32 GB, which is very beneficial for model execution and large-volume image processing. Whether or not TensorRT can be used is also an important indicator because it can accelerate the model and achieve a high addition to the detection speed.

Table 1. Performance comparison of some processors. TFLOPS stands for tera floating point operations per second, and TOPS stand for tera operations per second.

Device Type	Memory Size	Max Power	AI Performance	TensorRT
Jetson AGX Xavier	32 GB	30 W	32 TOPS	Support
Jetson Xavier NX	8 GB	20 W	21 TOPS	Support
Jetson TX2	8 GB	15 W	1.33 TFLOPS	Support
Atlas 500	8 GB	40 W	22 TOPS	/

While having computing power comparable to a workstation, Jetson AGX Xavier also has a small size of $100 \times 100 \times 87$ mm^3 (see Figure 4 for the appearance), which is very suitable for deployment on mobile platforms, such as UAVs, with limited carrying capacity. In our system, the device is mainly responsible for controlling the camera, communicating with the UAV, and detecting and positioning the suspected objects in the image.

Figure 4. NVIDIA Jetson AGX Xavier edge device, which was selected as the system's central controller.

2.1.2. Camera

There are many types of cameras to choose from, but this paper mainly discusses high-resolution visible-light cameras. We have tested several common cameras on the market, and their appearances are shown in Figure 5. The comparison of these cameras is shown in Table 2.

(a)

(b)

Figure 5. (**a**) Sony A7R series cameras. They have a similar appearance; (**b**) The appearance of a Canon 5DS camera.

Table 2. Comparison of working performance of different cameras. The "Capture Interval" represents the average time from when the camera is triggered to when the image is transmitted to the edge device.

Camera Type	Camera Weight	Megapixels	Image Volume	Capture Interval
Sony A7R2	625 g	42.4	30.0 MB	3.6 s
Sony A7R3	657 g	42.4	30.0 MB	2.4 s
Sony A7R4	665 g	61.0	47.5 MB	2.8 s
Canon 5DS	930 g	50.6	35.0 MB	2.5 s

The most significant difference between aerial image capture and video capture is that there is a longer acquisition time interval between each aerial image. An acquisition delay of 2 to 4 s is acceptable in the actual application, as long as the images can cover the entire area that is to be searched. It can be seen from Figure 6 that the images taken along the flying direction of the UAV can cover the ground completely, while ensuring some overlap. Since the attitude of the UAV changes in real-time during the flight, there are some misalignments in the images, which can later be corrected by the positioning algorithm. We selected Sony A7R2 to perform the system verification in the actual task. From the data of Table 2, it seems that Sony A7R2 is not a good choice, but in fact, it already meets the working requirements, and our experiments will also prove that, within a certain range, the normal work of the system has nothing to do with the choice of camera, but the better the camera, the better the performance of the system.

Figure 6. Aerial images cover the ground with some overlap.

2.1.3. UAV

A strong endurance capability is essential for an UAV that conducts wide-area object searches. In addition, the UAV also needs to have an interface to communicate with the object search system. In order to meet these requirements, it is more suitable to use a small- or medium-sized industrial UAV of more than 25 kg, with a certain degree of carrying capacity. We used a 50 kg vertical take-off and landing (VTOL) hybrid UAV with a flying speed of 120 km/h as the carrying platform. Figure 7 shows the appearance of this UAV. The VTOL UAV uses batteries for vertical take-off and landing, and a gasoline engine provides the cruising power. This type of design significantly reduces the demand for take-off and landing sites while ensuring long-term cruises.

Figure 7. A VTOL industrial hybrid UAV weighs 50 kg, for carrying the entire real-time object search system.

2.2. Design of Software System

According to the functions, this section divides the software system into the following two subsystems for discussion and design: image acquisition and object detection.

2.2.1. Image Acquisition Subsystem

The image acquisition subsystem is mainly responsible for the camera connection check, camera control, and image transmission and storage. The workflow of this subsystem is shown in Figure 8. This subsystem focuses on the acquisition of original images and other useful information and requires efficient collaboration of various hardware. The work of each step will be explained in detail herein.

- **Camera Check**. Check the connection status between the camera and the embedded edge device. Until the device can successfully identify the camera, the subsystem will enter the next step. Otherwise, it will execute this step in a loop;
- **Image Capture**. The embedded edge device will control the camera to automatically capture images after the camera is successfully connected. The camera can capture images at its own fastest response or capture images according to a preset fixed time;
- **Read Image from Camera**. Read the currently captured image of the camera and transfer it to the embedded edge device. For the sake of efficiency, the image will be read directly from the camera's memory and then temporarily stored in the edge device's memory, without passing through low-speed links, such as the camera SD card storage;
- **Send Data to UAV**. The working status of the system is sent to the UAV and then transmitted to the GCS for real-time monitoring after being forwarded by the UAV;
- **Read Data from UAV**. Read data from the UAV's flight control computer (FCC). The FCC provides geographic coordinates (latitude and longitude), speed, altitude, and attitude (pitch, roll, and yaw) data for the edge device, and the edge device uses UAV data to match each aerial image in real-time. This step will also receive system control commands from GCS forwarded by the UAV, which can modify the camera trigger frequency, the threshold of the object detection algorithm, etc. online;
- **Write to Local Disk**. Save the original image in the memory of the embedded edge device to the local disk and store the corresponding UAV data simultaneously. This step retains all of the original data, which can be used for subsequent offline analysis;
- **Enqueue**. A queue is established in the memory of the edge device to be responsible for transmitting the data of the image acquisition subsystem to the object detection subsystem. In this step, the original image and corresponding UAV data in the memory are also put into the queue for processing by the following object detection subsystem.

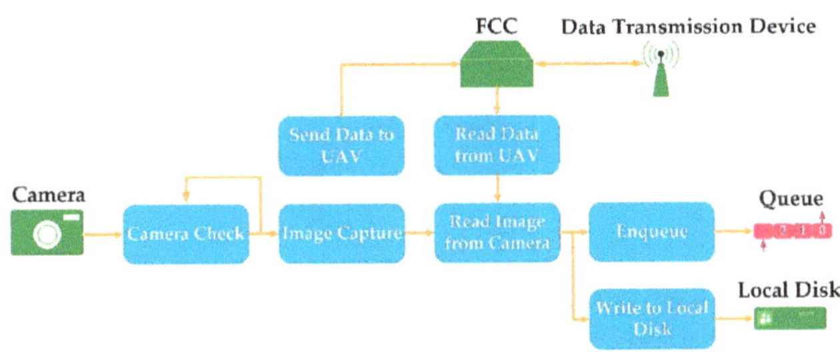

Figure 8. The detailed step flowchart of the image acquisition subsystem.

2.2.2. Object Detection Subsystem

The object detection subsystem is mainly responsible for image object detection, suspected object positioning, and real-time transmission to the GCS for display. The workflow of this subsystem is shown in Figure 9.

- **Dequeue.** After processing by the image acquisition subsystem, the original image and the corresponding UAV data have been temporarily stored in the memory queue. In this step, the temporarily stored data is dequeued in pairs according to the first-in-first-out principle for subsequent image object detection and suspect object positioning.
- **Object Detection.** Detect suspicious objects in the aerial image. In order to improve the object detection algorithm's inference speed and make it better applied on mobile edge devices, we have simplified the object detection algorithm and used the TensorRT to accelerate the object detection model. We will describe this part in detail later, here we are mainly concerned with how to detect aerial images with a large field of view. For the large field of view and small objects of aerial images, the object detection algorithm cannot process the entire image at once. In YOLT [39], a method for detecting small objects in satellite images is proposed, and our system refers to this method in the embedded edge device. In the inference stage, the image is first divided into $N \times N$ blocks of equal size and guarantees a certain degree of overlap in order to prevent the object from being split. The choice of N depends on the resolution of the image and the size of the object that is to be searched. If the object to be searched is relatively small, a large N should be selected for finer identification; otherwise, a small value can be selected for N to save the hardware resources. Then, object detection is performed on each block separately. After the detection is completed, merge all of the divided small images, as shown in Figure 10, and intercept the suspected object images and temporarily store them in the memory of the embedded edge device.
- **Calculate Object's Location.** The image acquisition subsystem has recorded a geographic coordinate for each image. However, this coordinate is only the location of the UAV at the moment the camera is triggered, not the suspected object (Figure 11a). Moreover, the attitude of the UAV when the camera is triggered will also cause the tilt and rotation of the image, which increases the complexity of the problem (Figure 11b). The height and width of each image usually represent the actual distance of hundreds of meters. If a large attitude accompanies the UAV at this time, the offset of the object may even reach more than 1 km, which significantly impacts the object's positioning.

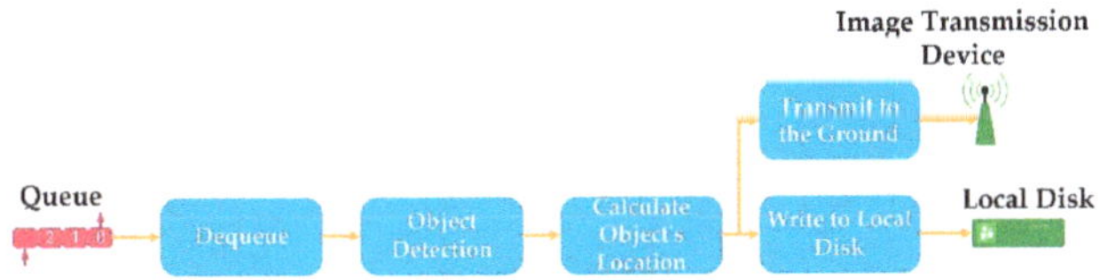

Figure 9. The detailed step flowchart of the object detection subsystem.

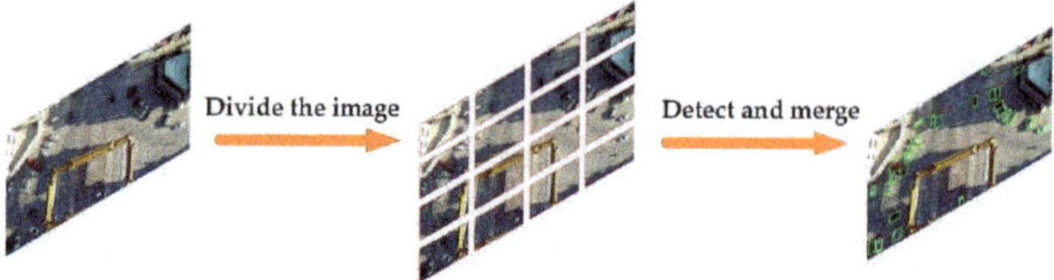

Figure 10. The aerial image is first divided into small images of equal size with overlap for detection and then merged after all of the detection is completed.

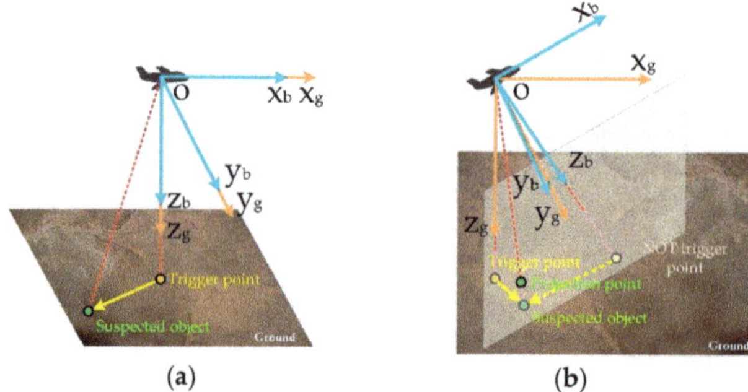

Figure 11. (**a**) The situation where the pitch, roll, and yaw are all zero during the flight of the UAV (basically impossible). The recorded coordinate is the exact center of the captured image; (**b**) When the pitch, roll, and yaw are not all zero, the airframe coordinate system does not coincide with the geodetic coordinate system, and the captured image is deformed.

To this end, we combine the UAV data to map the XY coordinate of the suspected object in the image to the actual geographic coordinate. Figure 11a depicts the ideal situation, when pitch, roll, and yaw are all zero. At this time, the UAV's airframe coordinate system (x_b, y_b, z_b axis) entirely coincides with the geodetic coordinate system (x_g, y_g, z_g axis), and the recorded geographic coordinate is the center of the image. It is easy to calculate the coordinate of the suspected object according to the flying altitude. When the pitch, roll, and yaw are not all zero, it becomes the situation shown in Figure 11b, and the airframe coordinate system is misaligned with the geodetic coordinate system. The transformation matrix can be used to transform the points on the projection plane (white translucent area) of the airframe coordinate system to the coordinates of the geodetic coordinate system. We define the pitch angle as θ, the roll angle as φ, and the yaw angle as ψ, the transformation matrix is defined as follows:

$$B = \begin{bmatrix} \cos\theta\cos\psi & \cos\theta\sin\psi & -\sin\theta \\ \sin\theta\sin\varphi\cos\psi - \cos\varphi\sin\psi & \sin\theta\sin\varphi\sin\psi + \cos\varphi\cos\psi & \cos\theta\sin\varphi \\ \sin\theta\cos\varphi\cos\psi + \sin\varphi\sin\psi & \sin\theta\cos\varphi\cos\psi - \sin\varphi\cos\psi & \cos\theta\cos\varphi \end{bmatrix} \quad (1)$$

Assuming the flying altitude relative to ground is H, for each point (x_b, y_b, H) on the projection plane of the airframe coordinate system, it can be transformed into the following geodetic coordinate (x_g, y_g, z_g):

$$\begin{bmatrix} x_g & y_g & z_g \end{bmatrix} = \begin{bmatrix} x_b & y_b & H \end{bmatrix} \cdot B \quad (2)$$

The coordinate can then be further projected to the true ground in geodetic coordinate system, and this is the work described in Figure 12. Suppose the final ground coordinate is (x, y, z), it can be solved according to the following formula:

$$\frac{z_g}{H} = \frac{x_g}{x}$$
$$\frac{z_g}{H} = \frac{y_g}{y}$$
$$z = H \tag{3}$$

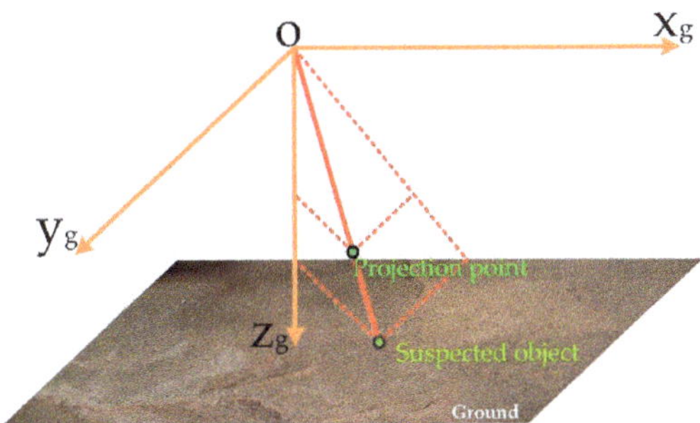

Figure 12. A schematic diagram of projecting the projection point to the actual position of the suspected object on the true ground.

After extracting the dependent variable, as follows:

$$x = \frac{x_g}{z_g} \cdot H$$
$$y = \frac{y_g}{z_g} \cdot H$$
$$z = H \tag{4}$$

- **Write to Local Disk**. Save the image and the positioning data of the suspected objects to the local disk for subsequent offline analysis.
- **Transmit to the Ground**. The suspected object images are further screened according to the image size and the score of the detection results to facilitate the real-time transmission to the GCS through the image transmission device. The image transmission device is a kind of equipment carried by the UAV for long-distance wireless transmission of images.

2.2.3. Optimization of Software System

From the above discussion of the two subsystems, it can be seen that there are many steps to go through from the acquisition of the image to the transmission of the detection results from the ground. If only the processing method of sequential execution is adopted, the system is not efficient enough. Benefitting from the parallel computing capabilities of the Jetson AGX Xavier, the two subsystems can further divide into several modules for multi-thread parallel processing in order to improve the running performance. Using this parallel pipeline work strategy, although the total processing time of a single image remains unchanged, the system's overall efficiency has been greatly improved.

The whole software system can be divided into six modules to run in parallel, as shown in Figure 13. For the image acquisition subsystem, the *Image Acquisition* module contains three steps of **Camera Check, Image Capture**, and **Read Image from Camera;** the *Communicate with UAV* module contains two steps of **Read Data from UAV** and **Send Data to UAV;** and the *Store Original Image* module contains two steps of **Write to Local Disk** and **Enqueue**. For the object detection subsystem, *the Object Detection* module contains two steps of **Dequeue** and **Object Detection;** the *Store Detect Results* module contains two steps of **Calculate Object's Location** and **Write to Local Disk;** and the *Results Real-time Transmit* module contains one step of **Transmit to the Ground**.

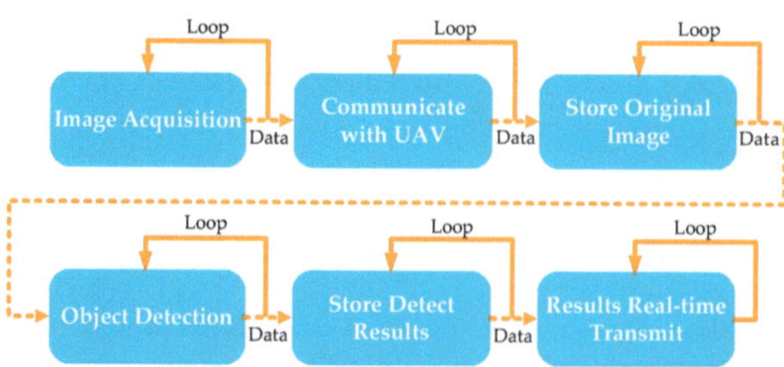

Figure 13. Divide the software system into modules that execute in parallel. After each module has finished running and the data is handed over to the next module, it will be executed in a loop immediately.

2.3. Object Detection Algorithm

2.3.1. Algorithm Simplification

Since the system needs to be applied to a mobile edge device with limited computing power, the speed of the object detection algorithm is the critical bottleneck of the system's performance. Given the characteristics of aerial images, we have carried out some general simplification strategies for different commonly used algorithms in order to weigh the precision and efficiency of object detection. We simplified the algorithms from the following three aspects:

- **Remove High-level Feature Maps**. High-level feature maps have a larger receptive field. Since the objects in the aerial image are mainly medium-sized (pixel size is between 32^2 and 96^2, COCO metrics), a situation in which the object occupies a large area of the image will hardly appear, so we try to remove some high-level feature maps. When the feature pyramid network (FPN) [40] is used as the neck of the model, only three low-level feature maps are retained, as shown in Figure 14. We also discussed the popular YOLOv5 algorithm with a different structure. The YOLOv5 algorithm has three detection heads with a stride of 8, 16, and 32 respectively, and the head of a stride 32 (P5) is removed for experimentation. The smaller S model (YOLOv5s) was selected as the benchmark, and the simplified version of YOLOv5s are shown in Figure 15.
- **Reduce the Channels of the Intermediate Layer**. The FPN's [40] output channels and the feature map channels of classification, regression, and region proposal network [18] (RPN, if it exists) are adjusted from 256 to 128. For YOLOv5s, we directly change the "width_multiple" parameter from 0.50 to 0.30 to reduce the channels. This simplified method can effectively reduce the number of parameters of the algorithm.
- **A Lightweight Backbone**. Using a lightweight backbone also helps to reduce the model's size and increase the inference speed. We tried to replace ResNet50 [41] with ResNet18, and YOLOv5s can directly modify "depth_multiple" to control the depth of the model. This paper has also performed an experiment with replacing CPSDarkNet53 in YOLOv5s with MobileNetV2 [42] directly.

2.3.2. TensorRT

It is often inefficient to use the deep learning framework for model deployment in actual applications directly. Instead, deploying object detection models trained on mainstream frameworks on the NVIDIA GPU through TensorRT can significantly improve inference speed, often at least one time faster than the original. TensorRT is a deep learning inference optimizer that transforms the trained model through a series of optimization techniques in order to enable it to run with higher performance on the NVIDIA GPU of a specific platform. Specifically, TensorRT has the following two main optimization strategies:

- **Lower Data Precision**. For most of the deep learning frameworks in training, the tensors in the network use 32-bit floating-point precision (FP32). Since the backpropagation is not required during the inference process, TensorRT supports FP16 and INT8 quantization to reduce data precision appropriately. Table 3 lists the value ranges of the different data types. Lower data precision represents lower memory usage and smaller model size.

- **Reconstruction and Optimization of Network.** For NVIDIA GPUs, TensorRT dramatically reduces the number of compute unified device architecture (CUDA) cores by merging layers, horizontally or vertically, as depicted in Figure 16. Horizontal merging can integrate the convolution, bias, and activation layers into one structure, while vertical merging can integrate the layers with the same structure but with different weights into a wider layer.

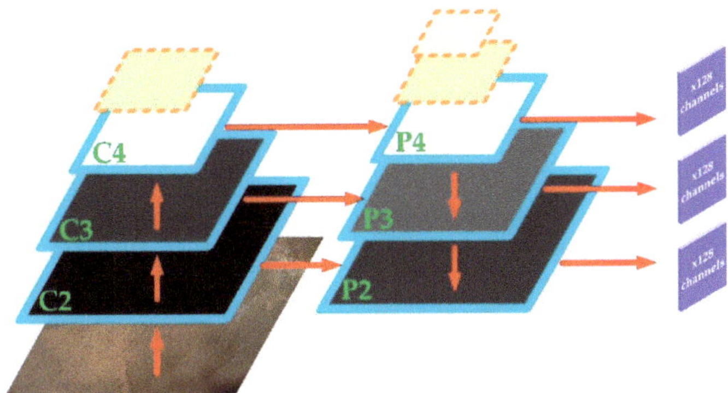

Figure 14. Retain three feature maps in FPN and change the output channels to 128. The RPN, classification, and regression feature map channels are also changed.

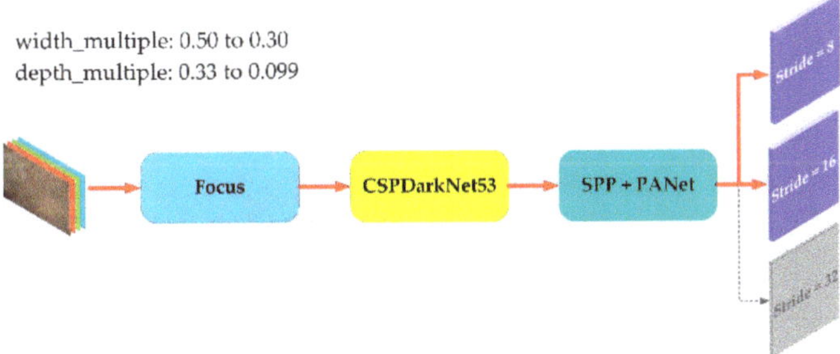

Figure 15. Schematic diagram of YOLOv5s. The simplification of YOLOv5s mainly includes removing the head whose stride is 32 (P5), reducing the parameters of "width_multiple" and "depth_multiple".

Table 3. Value ranges of different data types.

Data Type	Number of Bytes	Dynamic Range
FP32	4	-3.4×10^{38}~3.4×10^{38}
FP16	2	$-65{,}504$~$65{,}504$
INT8	1	-128~127

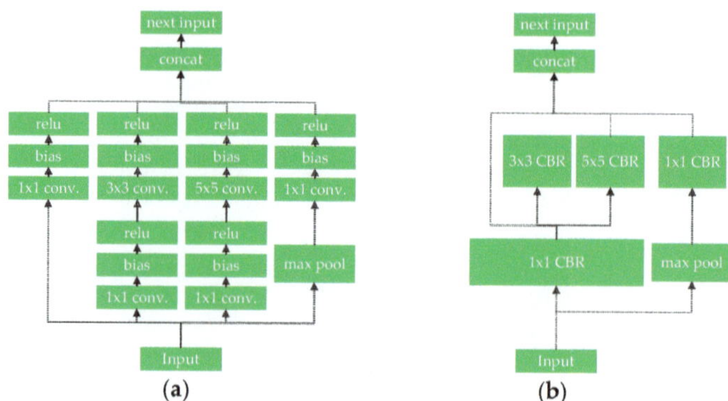

Figure 16. (**a**) The original network structure; (**b**) The network structure after reconstruction and optimization by TensorRT.

2.3.3. Data Augmentation

Some data augmentation methods that do not affect the inference speed are used to expand the existing training dataset in order to enhance the adaptability of the actual scene, according to the features of the aerial images that are collected by the UAV. As listed in Figure 17, the data augmentation methods adopted are as follows:

- **Blur**. Small objects in high-resolution aerial images collected by UAV will be blurred to some extent.
- **Ghosting**. Since the camera is mounted on a UAV that is flying at high speed, some of the captured images may show ghosting.
- **Lighting**. Depending on the date and weather when the task is being performed, the lighting conditions will vary.

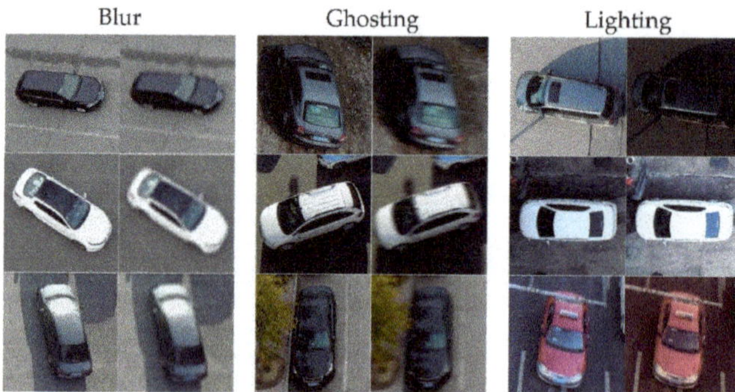

Figure 17. Data augmentation for aerial images.

3. Experiments

We trained and validated the simplified object detection algorithms for vehicle detection in aerial images and applied the entire wide-area and real-time object search system to a practical engineering task.

3.1. Simplification of Object Detection Algorithms

3.1.1. Dataset

We chose VisDrone2021-DET aerial images as the dataset. The VisDrone [43] dataset is collected and released by researchers in the Lab of Machine Learning and Data Mining, Tianjin University, China. All of the data are captured by drones at different locations and heights and consists of more than 400 video clips with 265,228 frames and 10,209 static images. VisDrone2021-DET is a dataset used by the VisDrone team for the image object detection challenge. The challenge used all 10,209 static images, of which 6471 were used for training, 548 were used for validating, and 3190 were used for testing. Figure 18 shows some of the images in the dataset. There are ten types of objects in the dataset, including the following: pedestrian, person, car, van, bus, truck, motor, bicycle, awning-tricycle, and tricycle, and our training was only conducted for the three categories of car, van, and truck. In addition to the original dataset, we also used the methods described in the previous chapter for data augmentation.

Figure 18. Some images of the VisDrone2021-DET dataset.

3.1.2. Benchmarks

We selected six common object detection algorithms as the benchmarks, including YOLOv5s, and Faster RCNN, Cascade RCNN, RetinaNet, FCOS, ATSS whose backbone is ResNet50 and neck is FPN.

3.1.3. Configuration

The training of all of our models was performed on a 2080 Ti graphics card with 11 GB of video memory. In order to facilitate the comparison of the results, we uniformly did not use the pre-trained model during training. The training epochs of the models were set to 20, and the batch size was eight. For YOLOv5s, the image size was 800 during training and testing, and the OneCycleLR learning rate policy was used, in which lr0 was 0.01 and lrf was 0.2, warmup was performed in the first three epochs. For other models, the image sizes were all set to 800 × 600, the initial learning rate was 0.005, the momentum was 0.9, and the weight decay was 0.0001. The learning rate warmup was performed in the first 500 iterations, and then the training was maintained at a fixed learning rate. When learning 14–18 epochs, the learning rate will be 1/10 of the original value. In the last two epochs, it was reduced to 1/100 of the original learning rate.

3.1.4. Metrics

We used the following six metrics to evaluate the performance of the model: mean average precision (mAP), mAP decline relative to the benchmark, trainable parameters, floating point operations (FLOPs), frames per second on the Jetson AGX Xavier embedded edge device (FPS), and FPS improving rate.

3.1.5. Results

The results of our experiments are shown in Table 4. For the algorithms using the ResNet-FPN structure, removing the high-level feature maps and reducing the number of channels in the intermediate layers had little effect on the model's mAP, and the mAP decline was within 1%. However, after replacing ResNet50 with ResNet18, the mAP loss was generally higher, reaching a maximum of 3.6%. The three simplification methods all improved the inference speed to varying degrees. For the YOLOv5s algorithm, removing the detection head of P5 basically did not affect the mAP, and the speed improvement was also very slight. Although reducing the "width_multiple" and "depth_multiple" parameters improved the detect FPS, they also sacrifice some mAP. MobileNetV2 replacement caused a large loss of mAP, which is not a good idea.

Table 4. Comparison results of various simplified methods of object detection algorithms. In this table, R50/R18 stands for ResNet50/ResNet18, and S1 represents the simplification of removing high-level feature maps on the model, only retaining the three low-level feature maps of the FPN. S2 represents the simplification of changing the number of channels of all intermediate layers, from 256 to 128. For the structure of YOLOv5s, Y1 means that the detection head with a stride of 32 (P5) is not used for detection, Y2 represents reducing the "width_multiple" from 0.50 of YOLOv5s to 0.30, and Y3 represents reducing the "depth_multiple" from 0.33 of YOLOv5s to 0.099. MobileNetV2 means that the backbone of YOLOv5s is replaced. The input size of YOLOv5s is 800, and the size of other models are 800 × 600. The calculation of FPS considers the time consumed by the preprocessing and postprocessing procedures.

Model	mAP	mAP Decline	Parameters	FLOPs	FPS
Faster-RCNN-R50-FPN	43.3%	/	41.12 M	104.59 G	4.06
Faster-RCNN-R50-FPN (S1)	42.8%	0.5%	40.01 M	103.70 G	4.17
Faster-RCNN-R50-FPN (S2)	43.0%	0.3%	31.99 M	60.89 G	5.53
Faster-RCNN-R18-FPN	41.2%	2.1%	28.12 M	79.66 G	5.18
Cascade-RCNN-R50-FPN	45.4%	/	68.93 M	132.39 G	3.01
Cascade-RCNN-R50-FPN (S1)	45.2%	0.2%	67.81 M	131.50 G	3.09
Cascade-RCNN-R50-FPN (S2)	45.2%	0.2%	46.95 M	75.85 G	4.16
Cascade-RCNN-R18-FPN	44.4%	1.0%	55.93 M	107.46 G	3.51
RetinaNet-R50-FPN	35.0%	/	36.10 M	96.34 G	4.66
RetinaNet-R50-FPN (S1)	34.9%	0.1%	32.04 M	95.50 G	4.93
RetinaNet-R50-FPN (S2)	34.2%	0.8%	27.92 M	54.71 G	5.89
RetinaNet-R18-FPN	32.1%	2.9%	19.61 M	72.42 G	6.12
FCOS-R50-FPN	41.1%	/	31.84 M	92.69 G	4.88
FCOS-R50-FPN (S1)	40.7%	0.4%	30.13 M	91.56 G	5.17
FCOS-R50-FPN (S2)	40.2%	0.9%	25.62 M	51.75 G	6.39
FCOS-R18-FPN	37.5%	3.6%	18.93 M	71.47 G	6.49
ATSS-R50-FPN	46.2%	/	31.89 M	94.92 G	4.73
ATSS-R50-FPN (S1)	46.0%	0.2%	30.18 M	93.79 G	4.97
ATSS-R50-FPN (S2)	45.7%	0.5%	25.67 M	53.98 G	5.96
ATSS-R18-FPN	43.2%	3.0%	18.94 M	71.47 G	6.16
YOLOv5s	44.0%	/	7.05 M	12.75 G	11.39
YOLOv5s (Y1)	43.9%	0.1%	5.27 M	11.64 G	11.66
YOLOv5s (Y2)	40.7%	3.3%	2.69 M	5.21 G	13.76
YOLOv5s (Y3)	42.8%	1.2%	6.64 M	11.11 G	13.18
YOLOv5s (MobileNetV2)	35.2%	8.8%	4.54 M	7.66 G	13.43

We integrated all of the simplification strategies in order to perform a comprehensive experiment and the results are shown in Table 5. After our simplification, most of the algorithms can achieve an efficiency improvement of more than 90%, with a slight loss of mAP. It can be seen from the table that the speed improvement of simplifying the YOLOv5s algorithm is relatively small, which may be due to YOLOv5 itself having already integrated so many optimization strategies.

3.2. TensorRT Acceleration Experiments

TensorRT will reconstruct the network structure by default. In addition, it also has a quantization strategy, which can convert the original model into two low-precision models of FP16 and INT8. FP16 and INT8 have only two and one bytes, respectively, and some information may be lost when they are used to represent a 4-byte FP32 value. The loss caused by the conversion to different precision models is shown in Table 6.

Table 5. The final result of object detection algorithm simplification. ALL means that the algorithm uses all of the simplification strategies mentioned in Table 3, and Y1 + Y2 + Y3 means that only the simplification strategy of a specific sign is used.

Model	mAP	mAP Decline	Parameters	FLOPs	FPS	Improving Rate
Faster-RCNN-R50-FPN	43.3%	/	41.12 M	104.59 G	4.06	
Faster-RCNN (ALL)	40.3%	3.0%	19.15 M	37.15 G	8.12	100.0%
Cascade-RCNN-R50-FPN	45.4%	/	68.93 M	132.39 G	3.01	
Cascade-RCNN (ALL)	43.5%	1.9%	34.11 M	52.11 G	5.75	91.0%
RetinaNet-R50-FPN	35.0%	/	36.10 M	96.34 G	4.66	
RetinaNet (ALL)	31.3%	3.7%	12.89 M	31.43 G	9.53	104.5%
FCOS-R50-FPN	41.1%	/	31.84 M	92.69 G	4.88	
FCOS (ALL)	36.4%	4.7%	12.69 M	30.92 G	9.80	100.8%
ATSS-R50-FPN	46.2%	/	31.89 M	94.92 G	4.73	
ATSS (ALL)	42.6%	3.6%	12.70 M	30.92 G	9.33	97.2%
YOLOv5s	44.0%	/	7.05 M	12.75 G	11.39	
YOLOv5s (Y1 + Y2 + Y3)	39.9%	4.1%	1.85 M	4.15 G	15.71	37.9%

Table 6. Performance comparison after converting the original model to FP16 or INT8 model using TensorRT. Cali. indicates that the model was calibrated before INT8 quantization, and engine represents a model file is suffixed with engine.

Model	Precision	mAP	mAP Decline	Model Volume
Faster-RCNN (ALL)	Original model	40.3%	/	120.4 MB
	FP16	38.9%	1.4%	37.7 MB
	INT8	37.4%	2.9%	24.3 MB
	INT8 (Cali.)	38.7%	1.6%	29.5 MB
Cascade-RCNN (ALL)	Original model	43.5%	/	240.0 MB
	FP16	40.7%	2.8%	82.8 MB
	INT8	39.4%	4.1%	47.4 MB
	INT8 (Cali.)	40.5%	3.0%	52.7 MB
RetinaNet (ALL)	Original model	31.3%	/	70.3 MB
	FP16	31.1%	0.2%	27.9 MB
	INT8	8.1%	23.2%	21.4 MB
	INT8 (Cali.)	31.0%	0.3%	21.7 MB
FCOS (ALL)	Original model	36.4%	/	68.8 MB
	FP16	36.3%	0.1%	28.2 MB
	INT8	4.5%	31.9%	22.2 MB
	INT8 (Cali.)	35.8%	0.6%	21.7 MB
ATSS (ALL)	Original model	42.6%	/	68.8 MB
	FP16	40.9%	1.7%	27.7 MB
	INT8	37.6%	5.0%	21.7 MB
	INT8 (Cali.)	40.7%	1.9%	21.3 MB
YOLOv5s (Y2 + Y3)	Original model	39.8%	/	5.6 MB
	FP16	38.5%	1.3%	6.4 MB (engine)
	INT8 (Cali.)	38.3%	1.5%	1.4 MB (engine)

It can be seen from the table that, after converting to the FP16 model, various algorithms have produced different degrees of mAP decline, but they were all within an acceptable range. Cascade RCNN, which had the least drop in mAP during model simplification, had the most drop during TensorRT deployment. Converting to the INT8 model requires the assistance of a calibration dataset. We performed a set of experiments without the calibration dataset and it turns out that the model loses too much mAP, which is predictable since INT8 had only 256 value representation ranges. More seriously, two of the models (RetinaNet and FCOS) diverged after INT8 quantization (marked in green font in Table 6). TensorRT takes this into account and provides a fully automatic calibration process for optimization in order to minimize the performance loss after INT8 quantization, which only requires some images in a similar style to the training dataset. We randomly selected 1000 images from the VisDrone2021-DET testing dataset as the calibration dataset. After calibration, the performance of the INT8 model was greatly improved. Most of the INT8 models had a mAP difference of only 0.2% from the FP16 model, but they were smaller and faster. When converting the YOLOv5s model, we retained the detection head of P5, which had little effect on precision and efficiency. The FP16 and INT8 model generated by YOLOv5s was a file with a engine suffix, but this did not affect the deployment effect.

We adopted the INT8 model for subsequent experiments. On the Jetson AGX Xavier edge device, whether TensorRT was used for deployment or not, and the comparison of the inference speeds of the different object detection algorithms are shown in Table 7. It can be seen that the inference speed of object detection models has been greatly improved after TensorRT deployment, reaching a maximum improvement rate of 140%, which is very critical for mobile edge devices with limited performance.

Table 7. Comparison of inference speed with and without TensorRT deployment (INT8).

Model	FPS (Original)	FPS (TensorRT)	Improvement Rate
Faster-RCNN (ALL)	8.12	18.72	130.5%
Cascade-RCNN (ALL)	5.75	13.85	140.8%
RetinaNet (ALL)	9.53	20.20	111.9%
FCOS (ALL)	9.80	20.46	108.7%
ATSS (ALL)	9.33	19.91	113.3%
YOLOv5s (Y2 + Y3)	15.48	27.20	75.7%

Table 8 summarizes all of the previous experiments. After model simplification and TensorRT deployment, most of the models sacrifice only about 5% of mAP, but in exchange for at least 400% of the operating efficiency of the benchmark model.

Table 8. Comparisons of the performance of the final model used on the edge device and the benchmark model.

Benchmark Model	mAP	mAP Final	mAP Decline	FPS	FPS Final	Improvement Rate
Faster-RCNN-R50-FPN	43.3%	38.7%	**4.6%**	4.06	18.72	**361.0%**
Cascade-RCNN-R50-FPN	45.4%	40.5%	**4.9%**	3.01	13.85	**360.1%**
RetinaNet-R50-FPN	35.0%	31.0%	**4.0%**	4.66	20.20	**333.4%**
FCOS-R50-FPN	41.1%	35.8%	**5.3%**	4.88	20.46	**319.2%**
ATSS-R50-FPN	46.2%	40.7%	**5.5%**	4.73	19.91	**320.9%**
YOLOv5s	44.0%	38.3%	**5.3%**	11.39	27.20	**138.8%**

3.3. Practical Application of the System

3.3.1. Task Analysis

Our target was to conduct a regular inspection task in a grazing area of the prairie with a range of about 7×20 km^2 in order to check for suspicious vehicles. We selected Sony A7R2 with a 55 mm prime lens as the image capture device, and the installation of this set of devices on the UAV is shown in Figure 19a. When flying at different altitudes relative to the ground, the coverage of a single acquisition is shown in Table 9.

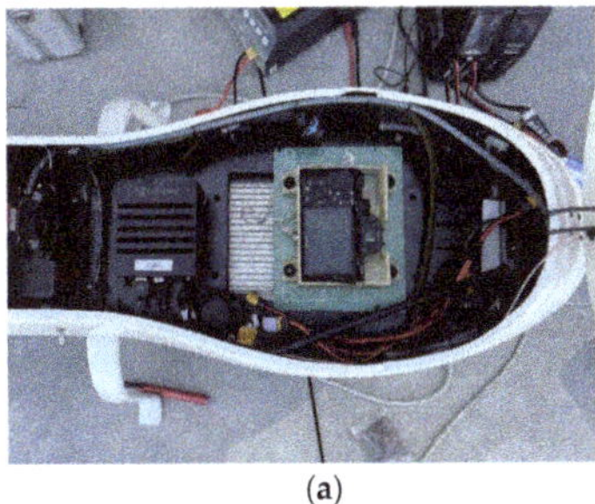
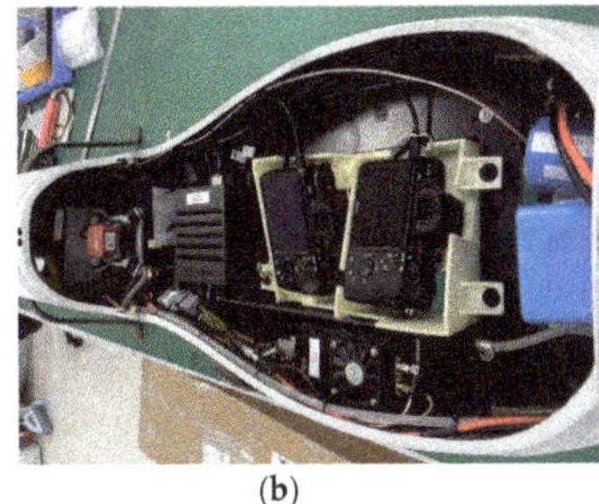

(a)   (b)

Figure 19. (**a**) Install a single camera with the angle of view facing directly below the UAV; (**b**) Install dual cameras, and the cameras deviate from the UAV by 15° to the left or right, respectively.

Table 9. The coverage of a single acquisition, using a Sony A7R2 camera with a 55 mm prime lens.

Fly Relative Altitude	Width Coverage	Height Coverage	Resolution Per Pixel
300 m	195.818 m	130.909 m	0.025 m
400 m	261.091 m	174.545 m	0.033 m
500 m	326.364 m	218.182 m	0.041 m
600 m	391.636 m	261.818 m	0.049 m
700 m	456.909 m	305.455 m	0.057 m
800 m	522.182 m	349.091 m	0.066 m
900 m	587.455 m	392.727 m	0.074 m
1000 m	652.727 m	436.364 m	0.082 m

We comprehensively considered the efficiency and effect and chose a flight altitude of 800 m relative to the ground for scanning. Figure 20 is a schematic diagram of a scanning route. With a 20% image side overlap rate, the distance between the two adjacent routes was calculated to be about 417 m. In the flying direction of the UAV, we considered 40% overlap, and the interval was 209 m. According to the UAV's flying speed (120 km/h, 33.3 m/s), it can be calculated that the capture interval between the two images should be less than about 6.27 s.

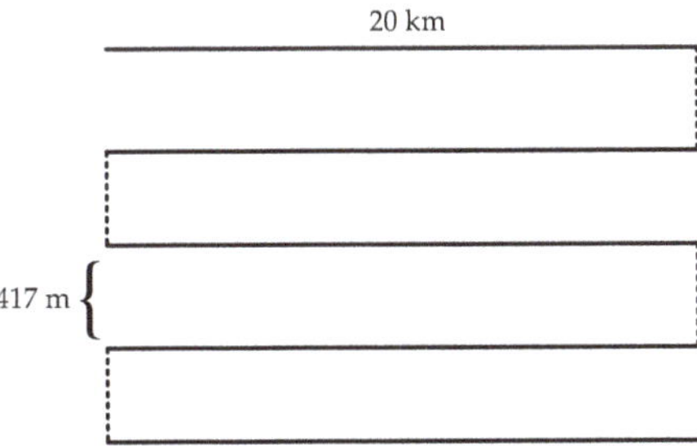

Figure 20. Schematic diagram of the scanning route.

3.3.2. System Performance

We already know that the entire system was executed in parallel by six modules. In the task of vehicle search in a prairie grazing area equipped with a single camera, the actual execution period of each module is shown in Table 10. We chose simplified Faster RCNN as the object detection algorithm of the system and selected $N = 5$ for each aerial image collected by the camera. That is, each aerial image needed to be detected 25 times in total. As we can see from the table, compared with sequential processing, modular parallel processing provided a speed increase of more than 100%. The average period of parallel execution was only 3.8 s, which is equal to the time consumption of the slowest Image Acquisition module in the system. Modular processing also brings good scalability and versatility to the system. In other practical applications, different cameras, different object detection algorithms, and different N values can be flexibly selected to adapt to different scenarios.

Table 10. Execution time of each module of the system.

Module	Average Execution Period
Image Acquisition (Camera = 1)	3.8 s
Communicate with UAV	Ignored
Store Original Image	0.9 s
Object Detection ($N = 5$)	2.4 s
Store Detect Results	1.1 s
Results Real-time Transmit	Ignored
Average sequential execution period	8.2 s
Average parallel execution period	3.8 s

3.3.3. Equipped with Dual Cameras

The system further improved the working efficiency by using two cameras. The installation method of the dual cameras is shown in Figure 19b, and the principle of image acquisition is shown in Figure 21. The system captures two images at a time while ensuring that the images from the two cameras partially overlap.

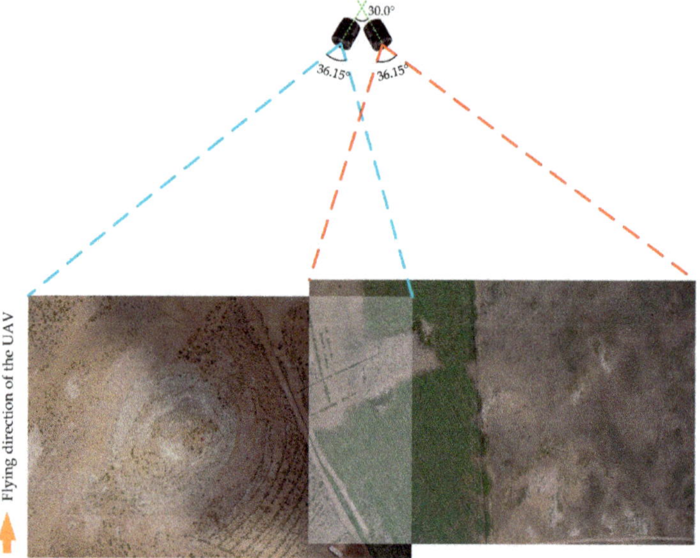

Figure 21. Images of two cameras with a certain degree of overlap.

When the UAV was equipped with dual cameras, the coverage of a single acquisition is shown in Table 11. As can be seen from the table, the width coverage was greatly improved. The tilt of the camera stretched the image, so the height coverage was calculated at the lowest value. After using two cameras, the resolution per pixel was reduced slightly at the same flying altitude, which lost some performance. Table 12 displays the actual execution period of each module. Since the two cameras can work in parallel, the bottleneck of the system at this time becomes the efficiency of the Object Detection module, which was about 0.8 s longer than that of a single camera. Although the average acquisition time increased, it significantly increased the coverage area of one acquisition, so the system's overall efficiency was still higher.

Table 11. The coverage of a single acquisition by using two cameras.

Fly Relative Altitude	Width Coverage	Height Coverage	Resolution Per Pixel
300 m	410.451 m	130.909 m	0.029 m
400 m	547.268 m	174.545 m	0.039 m
500 m	684.086 m	218.182 m	0.049 m
600 m	820.902 m	261.818 m	0.059 m
700 m	957.719 m	305.455 m	0.069 m
800 m	1094.537 m	349.091 m	0.078 m
900 m	1231.355 m	392.727 m	0.088 m
1000 m	1368.170 m	436.364 m	0.098 m

Table 12. Execution cycle of each module of the dual camera system.

Module	Average Execution Period
Image Acquisition (Camera = 2)	4.0 s
Communicate with UAV	Ignored
Store Original Image	1.4 s
Object Detection ($N = 5$)	4.6 s
Store Detect Results	1.7 s
Results Real-time Transmit	Ignored
Average sequential execution period	11.7 s
Average parallel execution period	4.6 s

Figure 22 displays some of the images containing vehicles that were successfully detected. Since a single image can reach a volume size of about 30 MB, the capacity of the image transmission equipment is difficult to meet the complete transmission requirements. Therefore, the system will only transmit the full image after compression, and together with the corresponding suspected object images and the positioning to the GCS. After post-processing on the ground, the results will be displayed on the computer screen in real-time.

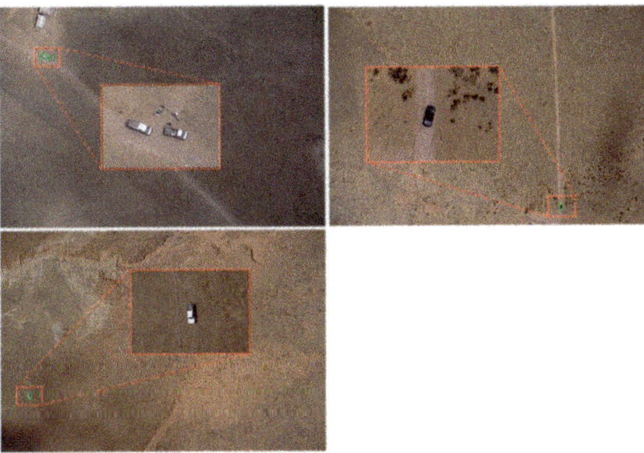

Figure 22. Images containing vehicles that were successfully detected.

4. Conclusions

This paper focuses on the real-time object search problem in a wide area. We selected the NVIDIA Jetson AGX Xavier edge device as the computing and control unit, and the high-resolution camera as the image acquisition device, to design a wide-area and real-time object search system of UAVs. The system considers both issues of real-time detection and area scanning efficiency in wide-area object search, which greatly reduces the cost of performing related tasks compared to other existing methods. Most of the hardware used in the system are shelf products, which do not need to be specially customized so that the entire system can be easily implemented.

The software part of the system is divided into the image acquisition subsystem and objection detection subsystem, which are designed respectively, and the realization and optimization scheme of each step in the subsystem are explained in detail. At the same time, the parallel multi-threading method is adopted in order to modularize the system so that the system's performance more than doubled. In this paper's discussion of the vehicle detection task, the system's execution period was reduced from 8.2 s to 3.8 s when a single camera was mounted on the UAV based on the parallel design of the software. The execution period was also compressed from 11.7 s to 4.6 s in the expansion scheme with dual cameras. The end result satisfies the time constraints calculated according to the UAV flight altitude and camera parameters in both cases. That is, the capture interval between the two images should be less than about 6.27 s. However, in the case of sequential execution, the system cannot operate normally.

For the most time-consuming object detection process, we adopted a variety of simplification strategies for the algorithms and used data augmentation on the training dataset in order to better adapt to the UAV aerial photography scene. We also adopted the TensorRT to optimize the object detection model, which significantly speeds up the detection speed and better applies to the embedded edge device with limited performance. After simplifying the algorithms and deploying the vehicle detection models on the edge device using the TensorRT tool, the detection speed of most of the models increased to 400% of the original, with only a loss of about 5% of the precision index, which solved the biggest limiting bottleneck of the system that the cost of detection time is too large.

The system design considers the scalability and versatility as much as possible, so different software and hardware modules can be flexibly selected according to different application scenarios, which provides a new idea for engineering applications in related fields.

Author Contributions: Conceptualization, X.L. and B.H. (Boyong He); methodology, X.L. and B.H. (Boyong He); software, X.L.; validation, K.D. and W.G.; investigation, K.D. and B.H. (Bo Huang); writing—original draft preparation, X.L.; writing—review and editing, X.L. and L.W.; supervision, L.W. All authors have read and agreed to the published version of the manuscript.

Funding: This research received no external funding.

Institutional Review Board Statement: Not applicable.

Informed Consent Statement: Not applicable.

Data Availability Statement: The data presented in this study are available upon request from the corresponding author.

Acknowledgments: The authors would like to thank the anonymous reviewers for their valuable comments and helpful suggestions.

Conflicts of Interest: The authors declare no conflict of interest.

References

1. Zhao, D.; Li, X. Ocean ship detection and recognition algorithm based on aerial image. In Proceedings of the 2020 Asia-Pacific Conference on Image Processing, Electronics and Computers (IPEC), Dalian, China, 14–16 April 2020; pp. 218–222.
2. Zhang, S.; Xin, M.; Wang, X.; Zhang, M. Anchor-free network with guided attention for ship detection in aerial imagery. *J. Appl. Remote Sens.* **2021**, *15*, 024511. [CrossRef]

3. Garcia-Aunon, P.; Peñas, M.; García, J. A new UAV ship-tracking algorithm. *IFAC-Pap.* **2017**, *50*, 13090–13095. [CrossRef]
4. Feraru, V.A.; Andersen, R.E.; Boukas, E. Towards an autonomous UAV-based system to assist search and rescue operations in man overboard incidents. In Proceedings of the 2020 IEEE International Symposium on Safety Security, and Rescue Robotics (SSRR), Abu Dhabi, United Arab Emirate, 4–6 November 2020; pp. 57–64.
5. Wang, S.; Han, Y.; Chen, J.; Zhang, Z.; Du, N. A deep-learning-based sea search and rescue algorithm by UAV remote sensing. In Proceedings of the 2018 IEEE CSAA Guidance, Navigation and Control Conference (GNCC), Xiamen, China, 10–12 August 2018; pp. 1–5.
6. Gašparović, M.; Zrinjski, M.; Barković, Đ.; Radočaj, D. An automatic method for weed mapping in oat fields based on UAV imagery. *Comput. Electron. Agric.* **2020**, *173*, 105385. [CrossRef]
7. Menshchikov, A.; Shadrin, D.; Prutyanov, V.; Lopatkin, D.; Somov, A. Real-time detection of hogweed: UAV platform empowered by deep learning. *IEEE Trans. Comput.* **2021**, *70*, 1175–1188. [CrossRef]
8. Lippi, M.; Bonucci, N.; Carpio, R.F.; Contarini, M.; Speranza, S.; Gasparri, A. A YOLO-based pest detection system for precision agriculture. In Proceedings of the 2021 29th Mediterranean Conference on Control and Automation (MED), Puglia, Italy, 22–25 June 2021; pp. 342–347.
9. Chen, C.J.; Huang, Y.Y.; Li, Y.S.; Chen, Y.C.; Chang, C.Y.; Huang, Y.M. Identification of fruit tree pests with deep learning on embedded drone to achieve accurate pesticide spraying. *IEEE Access* **2021**, *9*, 21986–21997. [CrossRef]
10. Jiao, Z.; Zhang, Y.; Xin, J.; Mu, L.; Yi, Y.; Liu, H.; Liu, D. A deep learning based forest fire detection approach using UAV and YOLOv3. In Proceedings of the 2019 1st International Conference on Industrial Artificial Intelligence (IAI), Shenyang, China, 22–26 July 2019; pp. 1–5.
11. Chen, Y.; Zhang, Y.; Xin, J.; Yi, Y.; Liu, D.; Liu, H. A UAV-based forest fire detection algorithm using convolutional neural network. In Proceedings of the 2018 37th Chinese Control Conference (CCC), Wuhan, China, 25–27 July 2018; pp. 10305–10310.
12. Wu, B.; Liang, A.; Zhang, H.; Zhu, T.; Su, J. Application of conventional UAV-based high-throughput object detection to the early diagnosis of pine wilt disease by deep learning. *For. Ecol. Manag.* **2021**, *486*, 118986. [CrossRef]
13. Moura, M.M.; de Oliveira, L.E.S.; Sanquetta, C.R.; Bastos, A.; Mohan, M.; Corte, A.P.D. Towards amazon forest restoration: Automatic detection of species from UAV imagery. *Remote Sens.* **2021**, *13*, 2627. [CrossRef]
14. Krizhevsky, A.; Sutskever, I.; Hinton, G.E. ImageNet classification with deep convolutional neural networks. *Commun. ACM* **2017**, *60*, 84–90. [CrossRef]
15. Russakovsky, O.; Deng, J.; Su, H.; Krause, J.; Satheesh, S.; Ma, S.; Huang, Z.; Karpathy, A.; Khosla, A.; Bernstein, M. ImageNet large scale visual recognition challenge. *Int. J. Comput. Vis.* **2015**, *115*, 211–252. [CrossRef]
16. Girshick, R.; Donahue, J.; Darrell, T.; Malik, J. Rich feature hierarchies for accurate object detection and semantic segmentation. In Proceedings of the IEEE Conference on Computer Vision and Pattern Recognition, Columbus, OH, USA, 24–27 June 2014; pp. 580–587.
17. Girshick, R. Fast R-CNN. In Proceedings of the IEEE International Conference on Computer Vision, Santiago, Chile, 13–16 December 2015; pp. 1440–1448.
18. Ren, S.; He, K.; Girshick, R.; Sun, J. Faster R-CNN: Towards real-time object detection with region proposal networks. *IEEE Trans. Pattern Anal. Mach. Intell.* **2017**, *39*, 1137–1149. [CrossRef] [PubMed]
19. Liu, W.; Anguelov, D.; Erhan, D.; Szegedy, C.; Reed, S.; Fu, C.Y.; Berg, A.C. SSD: Single shot multibox detector. In Proceedings of the European Conference on Computer Vision, Amsterdam, The Netherlands, 8–16 October 2016; pp. 21–37.
20. Redmon, J.; Divvala, S.; Girshick, R.; Farhadi, A. You only look once: Unified, real-time object detection. In Proceedings of the IEEE Conference on Computer Vision and Pattern Recognition, Las Vegas, NV, USA, 26 June–1 July 2016; pp. 779–788.
21. Redmon, J.; Farhadi, A. YOLO9000: Better, faster, stronger. In Proceedings of the IEEE Conference on Computer Vision and Pattern Recognition, Honolulu, HI, USA, 21–26 July 2017; pp. 7263–7271.
22. Redmon, J.; Farhadi, A. Yolov3: An incremental improvement. *arXiv* **2018**, arXiv:1804.02767,2018.
23. Bochkovskiy, A.; Wang, C.Y.; Liao, H. YOLOv4: Optimal speed and accuracy of object detection. *arXiv* **2020**, arXiv:2004.10934,2020.
24. Lin, T.Y.; Goyal, P.; Girshick, R.; He, K.; Dollár, P. Focal loss for dense object detection. In Proceedings of the IEEE International Conference on Computer Vision, Venice, Italy, 22–29 October 2017; pp. 2980–2988.
25. Cai, Z.; Vasconcelos, N. Cascade R-CNN: Delving into high quality object detection. In Proceedings of the IEEE Conference on Computer Vision and Pattern Recognition, Salt Lake City, UT, USA, 18–22 June 2018; pp. 6154–6162.
26. Tian, Z.; Shen, C.; Chen, H.; He, T. Fcos: Fully convolutional one-stage object detection. In Proceedings of the IEEE International Conference on Computer Vision, Seoul, Korea, 27 October–2 November 2019; pp. 9627–9636.
27. Zhang, S.; Chi, C.; Yao, Y.; Lei, Z.; Li, S.Z. Bridging the gap between anchor-based and anchor-free detection via adaptive training sample selection. In Proceedings of the IEEE/CVF Conference on Computer Vision and Pattern Recognition, Glasgow, UK, 16–18 June 2020; pp. 9759–9768.
28. Zhan, W.; Sun, C.; Wang, M.; She, J.; Zhang, Y.; Zhang, Z.; Sun, Y. An improved Yolov5 real-time detection method for small objects captured by UAV. *Soft Comput.* **2022**, *26*, 361–373. [CrossRef]
29. Azimi, S.M. ShuffleDet: Real-time vehicle detection network in on-board embedded UAV imagery. In Proceedings of the Computer Vision–ECCV 2018 Workshops, Munich, Germany, 8–14 September 2018; pp. 88–99.
30. Zhang, X.; Zhou, X.; Lin, M.; Sun, J. ShuffleNet: An extremely efficient convolutional neural network for mobile devices. In Proceedings of the 2018 IEEE/CVF Conference on Computer Vision and Pattern Recognition, Salt Lake City, UT, USA, 18–23 June 2018; pp. 6848–6856.

31. Dong, J.; Ota, K.; Dong, M. Real-time survivor detection in UAV thermal imagery based on deep learning. In Proceedings of the 2020 16th International Conference on Mobility, Sensing and Networking (MSN), Tokyo, Japan, 17–19 December 2020; pp. 352–359.
32. Chen, L.; Hu, J.; Li, X.; Quan, F.; Chen, H. Onboard real-time object detection for UAV with embedded NPU. In Proceedings of the 2021 IEEE 11th Annual International Conference on CYBER Technology in Automation, Control, and Intelligent Systems (CYBER), Jiaxing, China, 27–31 July 2021; pp. 192–197.
33. Zhang, Z.; Liu, Y.; Liu, T.; Lin, Z.; Wang, S. DAGN: A real-time UAV remote sensing image vehicle detection framework. *IEEE Geosci. Remote Sens. Lett.* **2020**, *17*, 1884–1888. [CrossRef]
34. Zhang, P.; Zhong, Y.; Li, X. SlimYOLOv3: Narrower, faster and better for real-time UAV applications. In Proceedings of the 2019 IEEE/CVF International Conference on Computer Vision Workshop (ICCVW), Seoul, Korea, 27–28 October 2019; pp. 37–45.
35. Sharma, R.; Pandey, R.; Nigam, A. Real time object detection on aerial imagery. In Proceedings of the CAIP 2019: Computer Analysis of Images and Patterns, Salerno, Italy, 3–5 September 2019; pp. 481–491.
36. Deng, J.; Shi, Z.; Zhuo, C. Energy-efficient real-time UAV object detection on embedded platforms. *IEEE Trans. Comput.-Aided Des. Integr. Circuits Syst.* **2020**, *39*, 3123–3127. [CrossRef]
37. Balamuralidhar, N.; Tilon, S.; Nex, F. MultEYE: Monitoring system for real-time vehicle detection, tracking and speed estimation from UAV imagery on edge-computing platforms. *Remote Sens.* **2021**, *13*, 573. [CrossRef]
38. Meng, L.; Peng, Z.; Zhou, J.; Zhang, J.; Lu, Z.; Baumann, A.; Du, Y. Real-time detection of ground objects based on unmanned aerial vehicle remote sensing with deep learning: Application in excavator detection for pipeline safety. *Remote Sens.* **2020**, *12*, 182. [CrossRef]
39. Etten, A.V. You only look twice: Rapid multi-scale object detection in satellite imagery. *arXiv* **2018**, arXiv:1805.09512,2018.
40. Lin, T.Y.; Dollar, P.; Girshick, R.; He, K.; Hariharan, B.; Belongie, S. Feature pyramid networks for object detection. In Proceedings of the IEEE Conference on Computer Vision and Pattern Recognition, Honolulu, HI, USA, 21–26 July 2017; pp. 2117–2125.
41. He, K.; Zhang, X.; Ren, S.; Sun, J. Deep residual learning for image recognition. In Proceedings of the 2016 IEEE Conference on Computer Vision and Pattern Recognition (CVPR), Las Vegas, NV, USA, 27–30 June 2016; pp. 770–778.
42. Sandler, M.; Howard, A.; Zhu, M.; Zhmoginov, A.; Chen, L.C. MobileNetV2: Inverted residuals and linear bottlenecks. In Proceedings of the 2018 IEEE/CVF Conference on Computer Vision and Pattern Recognition, Salt Lake City, UT, USA, 18–22 June 2018; pp. 4510–4520.
43. Zhu, P.; Wen, L.; Bian, X.; Ling, H.; Hu, Q. Vision meets drones: A challenge. *arXiv* **2018**, arXiv:1804.07437.

 *remote sensing*

Article

Occlusion and Deformation Handling Visual Tracking for UAV via Attention-Based Mask Generative Network

Yashuo Bai [1,2], Yong Song [1,2], Yufei Zhao [1,2,*], Ya Zhou [1,2], Xiyan Wu [1,2], Yuxin He [1,2], Zishuo Zhang [1,2], Xin Yang [1,2] and Qun Hao [1,2]

1   School of Optics and Photonics, Beijing Institute of Technology, Beijing 100081, China
2   Beijing Key Laboratory for Precision Optoelectronic Measurement Instrument and Technology, Beijing Institute of Technology, Beijing 100081, China
*   Correspondence: zhaoyufei@mail.tsinghua.edu.cn

Abstract: Although the performance of unmanned aerial vehicle (UAV) tracking has benefited from the successful application of discriminative correlation filters (DCF) and convolutional neural networks (CNNs), UAV tracking under occlusion and deformation remains a challenge. The main dilemma is that challenging scenes, such as occlusion or deformation, are very complex and changeable, making it difficult to obtain training data covering all situations, resulting in trained networks that may be confused by new contexts that differ from historical information. Data-driven strategies are the main direction of current solutions, but gathering large-scale datasets with object instances under various occlusion and deformation conditions is difficult and lacks diversity. This paper proposes an attention-based mask generation network (AMGN) for UAV-specific tracking, which combines the attention mechanism and adversarial learning to improve the tracker's ability to handle occlusion and deformation. After the base CNN extracts the deep features of the candidate region, a series of masks are determined by the spatial attention module and sent to the generator, and the generator discards some features according to these masks to simulate the occlusion and deformation of the object, producing more hard positive samples. The discriminator seeks to distinguish these hard positive samples while guiding mask generation. Such adversarial learning can effectively complement occluded and deformable positive samples in the feature space, allowing to capture more robust features to distinguish objects from backgrounds. Comparative experiments show that our AMGN-based tracker achieves the highest area under curve (AUC) of 0.490 and 0.349, and the highest precision scores of 0.742 and 0.662, on the UAV123 tracking benchmark with partial and full occlusion attributes, respectively. It also achieves the highest AUC of 0.555 and the highest precision score of 0.797 on the DTB70 tracking benchmark with the deformation attribute. On the UAVDT tracking benchmark with the large occlusion attribute, it achieves the highest AUC of 0.407 and the highest precision score of 0.582.

Keywords: visual object tracking; unmanned aerial vehicle; adversarial learning; convolutional neural network; attention mechanism

Citation: Bai, Y.; Song, Y.; Zhao, Y.; Zhou, Y.; Wu, X.; He, Y.; Zhang, Z.; Yang, X.; Hao, Q. Occlusion and Deformation Handling Visual Tracking for UAV via Attention-Based Mask Generative Network. *Remote Sens.* **2022**, *14*, 4756. https://doi.org/10.3390/rs14194756

Academic Editors: Wenjiang Huang, Giovanni Laneve, Yingying Dong and Chenghai Yang

Received: 4 August 2022
Accepted: 20 September 2022
Published: 23 September 2022

1. Introduction

The main purpose of visual object tracking (VOT) [1] is to estimate the position and scale of the target in each subsequent frame in the videos, given the ground truth of the first frame. Meanwhile, the motion trajectory could also be well described. Therefore, it has been widely used in various fields, especially in unmanned aerial vehicle (UAV) applications, such as air surveillance [2], target following [3], and visual navigation [4]. Nevertheless, UAV-based remote sensing images and videos have intrinsic properties, such as image degradation, uneven object intensity, and small object size, that make UAV-specific tracking more challenging.

Recently, discriminative correlation filter-based (DCF-based) and convolutional neural network-based (CNN-based) trackers have made up the two streams of VOT methods.

Since the application of the correlation filter in object tracking [5], many outstanding DCF-based algorithms have been proposed with balanced accuracy and low cost for UAV tracking [6–8]. Meanwhile, CNN-based trackers, which are typically based on a two-stage tracking-by-detection framework, have achieved state-of-the-art performance in terms of accuracy and robustness [9–12]. Although the current VOT method has grown considerably, robust and accurate tracking for UAVs has remained a demanding task due to occlusion, deformation, illumination variation, and other challenges. Among various factors, occlusion and deformation are two of the main causes of tracking failure.

Various strategies have been proposed to address these challenges. The most intuitive paradigm is to build a network for occlusion and deformation and collect a large-scale dataset of the objects in different conditions to train the network, expecting to learn the invariance of object features eventually. For example, Zhou et al. [13] proposed a deep alignment network for multiperson tracking with occlusion and motion reasoning. A deep alignment network-based appearance model and a Kalman filter-based motion model were adopted to handle the occlusion. Wu et al. [14] combined an adaptive Kalman filter with a Siamese region proposal network to make full use of the object's spatial–temporal information, thereby robustly dealing with complex tracking scenes, such as occlusion or deformation. Yuan et al. [15] adopted ResNet to extract more robust features, in which the response maps computed from ResNet were weighted and fused using to realize accurate localization during tracking under various conditions.

However, the occlusion and deformation always follow a long-tail distribution, some of which are rare or even nonexistent in large-scale datasets [16]. Therefore, learning invariance to such rare/uncommon occlusions and deformations needs to be addressed urgently. To alleviate this problem, one way is to dealing with different challenging situations without requiring more training samples by designing different coping strategies specifically for different situations. For example, ref. [17] designed an attribute-based CNN with multiple branches, each of which is used to classify objects with specific attributes, thereby reducing the diversity of object appearance under each challenge and reducing the demand for the amount of training data. Ref. [18] adaptively utilized level set image segmentation and bounding box regression techniques to deal with the deformation problem, while designing a CNN to classify objects as occluded or non-occluded during tracking, thereby avoiding collecting samples updated by the occlusion tracker. These methods achieved robust and accurate tracking in a variety of complex situations without requiring a larger sample size but may not be sufficient in the face of more complex and variable situations.

Another method is to enrich the expressive power of samples for different challenge scenarios without requiring more actual samples. Considering the advantages of generative adversarial networks (GANs) in sample generation, many works adopted GANs to increase the diversity of training samples, thereby improving the tracker's ability to cope with challenges, such as occlusion and deformation. Wang et al. [16] proposed to adopt the adversarial network to enrich data samples with occlusion and deformation. This approach essentially generates samples that are difficult to be classified by the target detector, driving the adversarial system to produce a better detection network. Chen et al. [19] further introduced GANs into the problem of face detection and proposed an adversarial occlusion-aware face detector (AOFD). The role of the generative model in the algorithm is also to cover the key features of the face by generating masks in the training set. Likewise, to increase positive samples, Song et al. [10] employed the generation network to generate masks randomly, which adaptively discarded the input features to capture various appearance changes. After the adversary learning, the network can identify the masks that maintain the most robust features of the target object for a long time. Similar thinking was utilized by Javanmardi [20] to reduce the influence of object deformation on tracking and detection. In image space, Souly et al. [21] developed a semi-supervised semantic segmentation approach, which employs GANs to generate plausible synthetic images, supporting the discriminator in the pixel-classification step. Differently, Wang et al. [22] skillfully combined the application in the image and feature space of GANs to further supplement the hard positive samples by using part of the image background to cover the target.

Like other methods partwise modeling object appearance [23], adversarial learning methods devote efforts to concentrate the classification network into some other features besides the visible parts of a target, which are more robust for giving reliable cues for tracking when the target is occluded and deformable. How to distinguish these features is the key. In the processes mentioned above, GAN predicts masks with 3×3 size to respectively cover the part of feature maps and dropout to adversarial training for the object tracking without these local features. Nevertheless, this mask is updated to cover only a portion of the features to select local features but is actually not enough to simulate occlusion and deformation. At the same time, inevitably, feature loss may make tracking drift in extreme situations, which here refers specifically to target occlusion and deformation, and the 3×3 feature maps from CNN contain less location and shape information of the object, which cannot give the object a thorough description. In this paper, we propose an attention-based mask generative network-based tracker, which we call the AMGN-based tracker, to address the above issues. The main contributions can be concluded as follows:

1. We propose an attention-based mask generative network-based (AMGN-based) tracker. First, we adopt a base deep CNN to extract the deep features of the candidate regions. Next, we use AMGN to generate a series of attention-based masks, which are applied to the deep feature to augment hard positive samples. Then, we design a feature fusion method to compensate for the possible over-subtraction of the features of hard positive samples by the masks and to compensate for target location information. Finally, these hard positive samples are used for subsequent generative adversarial learning, thereby improving the ability of the tracker to handle occlusion and deformation.
2. We develop an attention-based mask generative network (AMGN). After CNN extracts the deep features of the candidate region of which the salient positions are obtained through the attention module, masks for occluding the corresponding positions are generated. Multiply these masks with the deep features to simulate target occlusion and deformation in the feature space.
3. We design a feature fusion method. When multiplying the masks with the deep features, some features are discarded, and there is a chance that too many features are discarded in the process. To alleviate this problem, we incorporate shallower-layer features into deeper-layer features processed by masks, thus avoiding extreme cases of tracking drift due to excessive feature loss.

After the process of AMGN and feature fusion, many hard positive samples are generated. The enhancement of the occlusion and deformation training samples strengthens the object-tracking ability when the target is occluded and deformable by effectively covering the distinguishable features of the object and conducting confrontation training with the classification tracker as the discriminator. As a result, even if the target is occluded and deformable, the features of the unobstructed area assist target tracking. Figure 1 presents the principle of our method selecting local features and generating masks.

The rest of the paper is organized as follows. Section 2 covers related work. Section 3 describes the proposed method for tracking the occluded and deformed object, including the overall pipeline, base deep CNN, AMGN, and feature fusion method. Section 4 presents the comparison experiments and ablation studies. Finally, conclusive remarks and future research directions are given in Section 5.

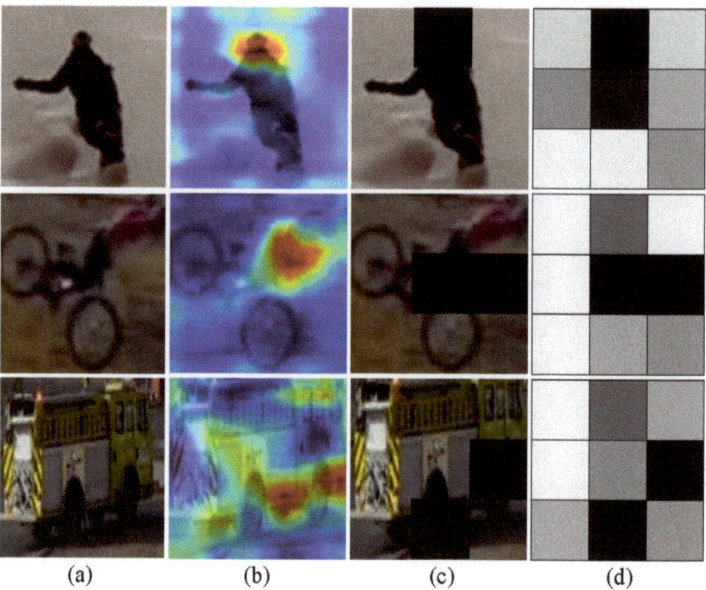

Figure 1. Principle of local feature selection and mask generation. In this paper, we propose to use an attention module and adversarial network to generate examples with occlusions and deformations that will be hard for the object tracker. The attention map is the visualization results of Grad-CAM [24] that learn the spatial attention of the target region. (**a**) Target region. (**b**) Attention map. (**c**) Mask template. (**d**) Generated mask.

2. Related Work

2.1. Occlusion and Deformation Handling in Visual Tracking

In general, visual tracking methods can be categorized as generative and discriminative. Generative methods extract target features before tracking to establish an appearance model that can represent the target. The model is then applied to pattern match the entire image and locate the most similar region. Typical generative model tracking algorithms include tracking algorithms based on Kalman filter [25], particle filter [26,27] and mean shift [28]. The generative methods only focus on tracking the target itself but ignore the background information, which is prone to tracking drift when the target is occluded or deforms drastically. The discriminative methods based on various approaches ranging from the traditional correlation filter (CF) [29,30], support vector machine (SVM) [31] to the currently widely used convolutional neural networks (CNNs) [32,33], GANs [10,34], recurrent neural networks (RNNs) [35,36], and especially Siamese neural networks [11,37] and other costume neural networks [38,39], always generate multiple suggestion boxes at first and then categorize each suggestion box into the target or background, employing offline pre-training and online learning. Nowadays, deep detection tracking methods, including multi-domain learning, ensemble learning, adversarial learning, reciprocating learning, and overlap maximization, have gradually become the mainstream of target tracking research due to various online update detector models, as they can better adapt to the complex changes of target objects in the tracking process.

Object tracking becomes a challenge when dealing with occluded and deformable objects, as they receive an incomplete description that does not resemble the patterns stored initially. Even if the selected candidate is indeed the target, the similarity between the features of the candidate image and the target image will not reach the threshold due to the effects of occlusion and deformation. Furthermore, the viability of deep learning relies on massive amounts of training data. When faced with the target occlusion and deformation problem, if the positive occlusion and deformation samples on each frame are highly overlapping, it is difficult for the deep learning model to capture the target features under large-scale occlusion and deformation.

To handle occlusion and deformation robustly, several strategies [40–44] have been used. In deep learning methods, data collection and annotation is the most straightforward way, while it seems impossible to collect data covering all potential occlusion and deformation, even for large-scale datasets. Considering the advantages of GAN in data generation, numerous works have attempted to use GAN to generate occlusion and deformation images that meet the requirements. In addition, modal segmentation is another approach to reducing the existence, degree, and contours of occlusion and deformation by exploiting its ability to infer the physical structure of objects. By the way, the modal training data are created by adding synthetic occlusion and deformation to the modal mask. As with conventional methods, it is also popular to divide the target image or region of interest into some cells or segments, and then analyze each segment individually to improve the accuracy of the tracking model. For example, Zhan et al. built a self-supervised framework for partially completed occluded objects for scene de-occlusion. Pathak et al. proposed a CNN that can generate missing paths of an image based on context. Nonetheless, human beings have a remarkable ability to detect and recognize objects when they are partially visible and deformable. Some human vision mechanisms are introduced to learn appropriate attention parameters in different channels and effectively handle different occlusion and deformation patterns [45,46]. Among various human vision mechanisms, the attention mechanism has shown to be effective in many computer vision tasks, for which we will make a brief review in the next subsection.

2.2. Attentional Mechanisms in Neural Networks

We aim to learn more robust target appearance models with the help of spatial and temporal attention. Informally, the neural attention mechanism enables a neural network to focus on a subset of its inputs (or features), i.e., it selects specific inputs. Let $x \in \mathcal{R}^d$ be the input vector, $z \in \mathcal{R}^k$ be the feature vector, $a \in [0,1]^k$ be the attention vector, $g \in \mathcal{R}^d$ be the attention glimpse, and $f_\phi(x)$ be the attention network with parameters ϕ. Typically, attention is implemented as

$$\mathbf{a} = f_\phi(\mathbf{x}), \tag{1}$$

$$\mathbf{g} = \mathbf{a} \odot \mathbf{z}, \tag{2}$$

where $\odot$ is the element-wise multiplication, and z is the output of another neural network $f_\theta(x)$ with parameters θ. In this case, the attention mechanism introduces multiplicative interactions into the neural network space, making it simple and compact. Taking matrix-valued images as an example, most of the research on the combination of deep learning and visual attention mechanism focuses on using masks to achieve an attention mechanism, identifying key features in images through another layer of weights, and learning what needs to be paid attention to, thereby forming attention. This idea has evolved into soft attention and hard attention. Relatively, soft attention is more applicable in the task of object tracking to obtain alignment weights [47].

Soft attention, attaching importance to the spatial scales and channel scales, could be explicitly determined through network learning. Moreover, its differentiable characteristic allows neural networks to calculate gradients and learn the weights of attention by forwarding propagation and backward feedback. Among them, SENet channel attention [48] is to allocate resources between each convolutional channel and selectively enhance the features with the largest amount of information so that subsequent processing can make full use of these features and suppress useless features. The residual attention network for image classification combines the attention of the spatial domain and the channel domain while combining the ideas of the residual network of ResNet. Subsequently, problems, such as rare information retained after mask processing and the difficulty of stacking deep network structures, would be well prevented. Based on SENet, CBAM [49] consists of two independent sub-modules, channel attention module (CAM) and spatial attention module (SAM), which realize channel attention and spatial attention, respectively. As a lightweight general-purpose module, it can be seamlessly integrated into any CNN architecture without the overhead and can be trained end-to-end with a basic CNN.

During VOT, the frequent disappearance, reappearance, and deformation of objects arouse tracking failures. Adopting an extra attention module can generate feature weights to select features and enhance the ability of feature expression. Combined with generative adversarial learning, the invariance of these important features can be effectively learned, thereby effectively improving the performance of target tracking algorithms.

2.3. Generative Adversarial Learning

GANs [50] have emerged as one of the hottest research fields in deep learning since they were proposed by Goodfellow et al. in 2014. Under the guidance of zero-sum game theory, the idea of a confrontation game runs through the whole training process of a GAN. It not only brings excellent generation quality to the model itself, but also is integrated into a series of traditional methods, forming a large number of new research directions. In terms of sample generation, the essence of GAN is a concept generation model, that is, to find out the statistical rules within a given observation data and generate new data similar to the observation data based on the probability distribution model obtained. On the other hand, GAN cleverly combines (self-)supervised learning and unsupervised learning, providing a new method for sample generation.

During the target tracking, the online training samples are not available before occlusion emerges. As a result, the tracking drift happens when the target is repeatedly blocked, and deforming for the tracking model is absent of the corresponding processing capacity. To tackle this problem, one solution is to furnish occlusion samples according to image synthesization. At present, there is a great deal of research work on image generation (pixel level) in various image generation algorithms [51]. Image generation technology based on the generative adversarial network has been able to generate real-like sample images with guaranteed quantity and diversity according to various requirements. Compared with other image generation networks, the generative adversarial network has lower complexity and higher flexibility. However, even if the sample images with occlusion can be supplemented in this way, it is still an arduous operation to provide the sample image with target tracking under complex background. A larger image sample database also meets the same problem for the long-tail problem and still has non-scalability.

In order to reduce the difficulty of sample replenishment, another solution is to add positive samples in the feature space to capture the appearance changes of the target in the time domain so as to improve the ability of the model to resist occlusion. Due to the flexibility of GAN, the training framework based on a generative adversarial mechanism can be combined with various types of loss functions according to specific tasks, and any differentiable function can be used as a generator and a discriminator. This way, there is no need to collect occlusion samples as the training base or consider the realistic rationality of generating samples, but there is a greater increase in the number of samples also containing as many diversities and features of occlusion.Consequently, the classification network, as a criterion, has stronger robustness in the process of confronting the generator network. It chiefly saves manpower, material, and financial resources.

3. Method
3.1. Overview

The proposed anti-occlusion and anti-deformation AMGN-based tracker consists of three modules. Firstly, the feature extraction of candidate regions is carried out through a base deep CNN. After that, AMGN utilizes feature maps and spatial attention weights of candidate regions to generate hard positive samples. Finally, the discriminant network distinguishes whether the features belong to the target or the background according to the fusion of the second and third convolution layer's features. Consequently, the tracking model captures the anti-occlusion and anti-deformation ability. Figure 2 shows the pipeline of our method, and the details are discussed below.

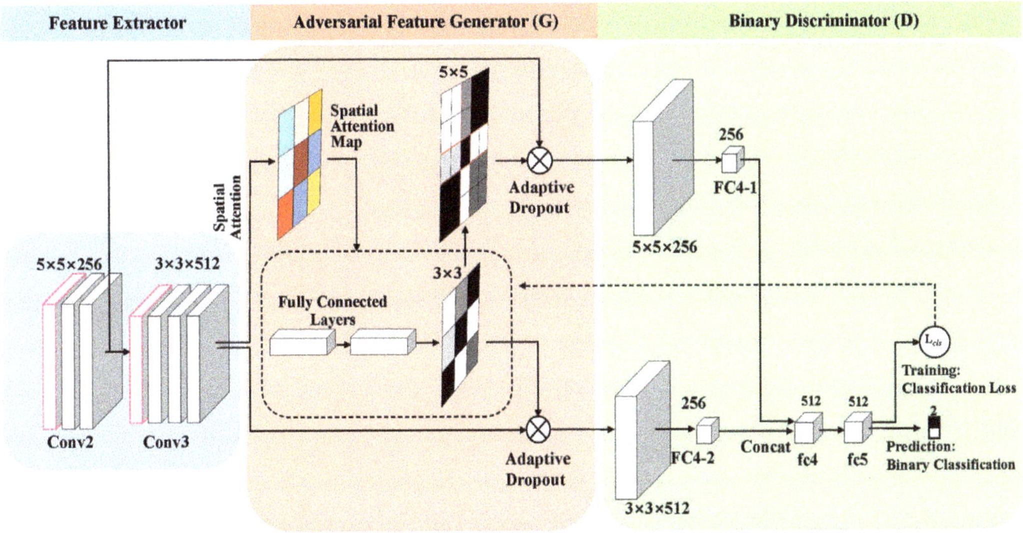

Figure 2. The architecture of AMGN-based object-tracking method.

3.2. Base Deep CNN and Tracking Network

Figure 3 shows the architecture of base deep CNN and tracking network. The first three convolution layers, Conv1-Conv3 from VGG-M, are used as the base deep CNN to extract the base deep features. The discriminant network takes over the fused features from Conv2 and Conv3 according to the form of fully connected layers and discriminates whether the feature belongs to the target. As we all know, lower-level CNN features have a higher spatial resolution to describe target locations, but show less semantic information, while higher-level CNN features are robust to target variations but with the absence of location information. For the purpose that the discriminator makes a better decision and the use of mask in AMGN makes up for much of the loss of features,we fuse Conv2 and Conv3 in the tracking network. In addition, in order to train the CBAM network's ability to recognize target robustness features independently, CBAM is placed after Conv3 for offline training and the parameters are retained.

3.3. Attention-Based Mask Generative Network

The attention mechanism can effectively focus eyes on areas of images that are discriminative to objects and backgrounds. Therefore, the human brain can devote more attention to these areas, obtain more details about the target, and suppress other useless information [52]. Attention weights acted on CNN feature maps also mark the most distinguishing feature that can assist the discriminator to make decisions, while the occluding of these features will always mislead the discriminator. However, combining the adversarial learning, the more these features are occluded, the more robust the discriminator can be.

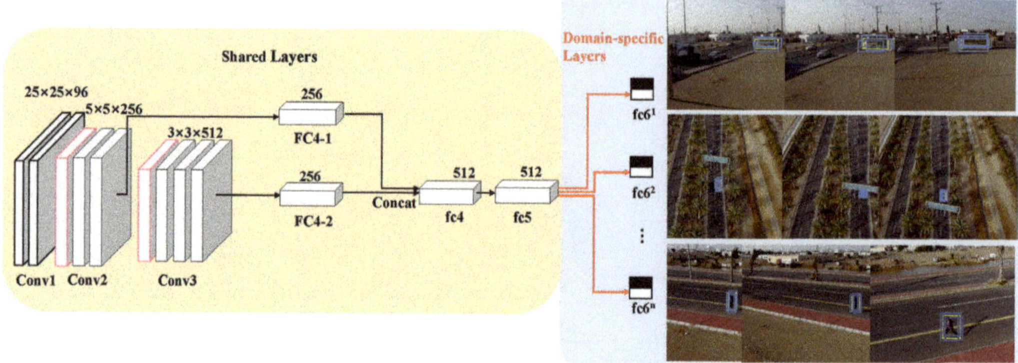

Figure 3. The architecture of base deep CNN and tracking network, consisting of shared layers and n branches of domain-specific layers.

In the proposed method, for all extracted positive sample feature maps, the result of the spatial attention module has the same size as M^3 of a single channel. C^3 is the output feature map of Conv3. Positions at which there is the maximum in the spatial attention weight matrices are set to zero, which compose the candidates' label of the mask. By selecting the template with the lowest classification score when it instructs C^3 to dropout features, it will be the final generation label M. In Figure 4, we give some examples of candidate labels generated based on spatial attention matrices.

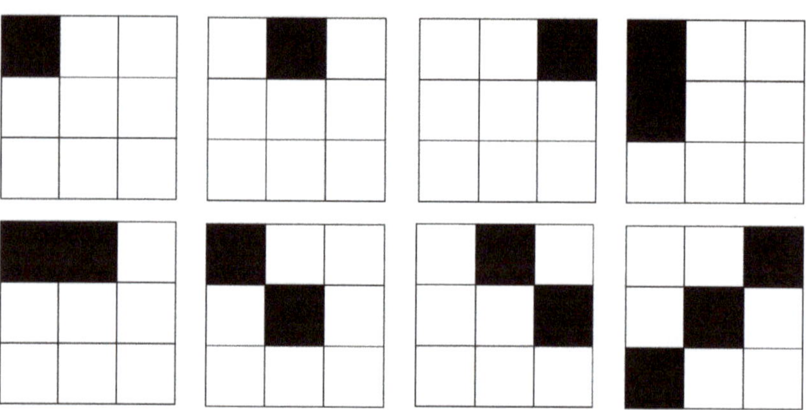

Figure 4. Examples of candidate labels generated based on spatial attention matrices.

Different input features will conduct different and continuous heatmaps under the guidance of the assigned label M with AMGN, which is composed of two fully connected layers. Here, mean squared error (MSE) is utilized to measure the difference between estimated generated masks and the assigned label. Given a feature map C with the size of $W \times H$, the MSE loss can be expressed as

$$L_{MSE} = \frac{1}{N}\sum_{j=1}^{N}\left(M_j - \hat{M}_j\right)^2, \tag{3}$$

where $\hat{M}$ and M denote the generated masks and assigned labels, respectively.

After the thresholding operation where we select the top $\frac{1}{3}$ pixels as 0 and others as 1, the generated masks become our attention-based masks. The dot product of the maskwith C^3 of multiple channels obtains the feature sample under occlusion, which is described by

$$C_M^3 = C^3 \cdot \hat{M}, \tag{4}$$

$$\hat{M} = G(C^3), \tag{5}$$

where C_M^3 is the output that the attention-based mask acts on C^3. G represents the generation operation of the AMGN. $\hat{M}$ is the attention-based mask.

3.4. Feature Fusion

Since the size of M is inconsistent with the size of C^2 in the feature graph after the second convolution, the mask needs to be processed. The weight values of rows and columns in M are dot multiplied by the corresponding rows and columns in C^2 to obtain C_M^2, that is, M^2 with the same size of C^2 is the mask processed by M. Finally, the two-layer feature images processed by the fully connected layer are cascaded and sent to the final target classification branch. The values at (r_A, c_A) in adjusted $H_A \times W_A$ mask M^2 are equal to the values at (r_B, c_B) in base $H_B \times W_B$ mask M, where $\{(r_A, c_A) | \lfloor r_1 \rfloor + 1 \leq r \leq \lfloor r_2 \rfloor, \lfloor c_1 \rfloor + 1 \leq c \leq \lfloor c_2 \rfloor; r, s \in \mathbb{Z}\}$. The r_1, r_2, c_1 and c_2 are calculated by

$$r_1 = ((r_B - 1)H_A/H_B) + (1/2), \tag{6}$$

$$r_2 = (r_B H_A/H_B) + (1/2), \tag{7}$$

$$c_1 = ((c_B - 1)W_A/W_B) + (1/2), \tag{8}$$

$$c_2 = (c_B W_A/W_B) + (1/2). \tag{9}$$

Mask M^2 is directly obtained by the transformation of mask M. Finally, the object function of AMGN is defined as

$$\mathcal{L}_{AMGN} = \min_G \max_D \mathcal{L}_1 + \mathcal{L}_2 + \lambda \mathbb{E}_{(C^3, M) \sim P(C^3, M)} \left\| G(C^3) - M \right\|^2, \tag{10}$$

$$\mathcal{L}_1 = \mathbb{E}_{(C^3, C^2, M) \sim P(C^3, C^2, M)} [\log D(M \cdot C^3, f^{C^2})], \tag{11}$$

$$\mathcal{L}_2 = \mathbb{E}_{(C^3, C^2) \sim P(C^3, C^2)} [\log(1 - D(G(C^3) \cdot C^3, f^{C^2}))], \tag{12}$$

where G represents the generative network, D represents the discriminative network, and M is the theoretically optimal mask matrix under the premise of a given feature map, which refers to the mask matrix that is most likely to make the D error, while the goal of G is to make it generate matrix $G(C^3)$, which is closest to the optimal matrix M, as the input is C^3. f denotes the operation described in Equations (6)–(11) that adjust the size of the mask and perform it on the feature map C^2.

In the process of online training, G is fixed at first. $\max_D \mathbb{E}_{(C^3, C^2, M) \sim P(C^3, C^2, M)} [\log D(M \cdot C^3, f^{C^2})] + \mathbb{E}_{(C^3, C^2) \sim P(C^3, C^2)} [\log(1 - D(G(C^3) \cdot C^3), f^{C^2})]$ has the requirements that increase $D(M \cdot C^3, f^{C^2})$ and decrease $1 - D(G(C^3) \cdot C^3, f^{C^2})$ at the same time, which means the demand toward D to distinguish the difference between $M \cdot C^3$ and $G(C^3) \cdot C^3$. Then, D is fixed and G is optimized.

To achieve $\min_G \mathbb{E}_{C^3 \sim P(C^3)} [\log(1 - D(G(C^3) \cdot C^3))] + \lambda \mathbb{E}_{(C^3, M) \sim P(C^3, M)} \left\| G(C^3) - M \right\|^2$, $D(G(C^3) \cdot C^3)$ should be increased, while $\mathbb{E}_{(C^3, M) \sim P(C^3, M)} \left\| G(C^3) - M \right\|^2$ should be decreased, and also G should be trained to make $G(C^3)$ close to the theoretical optimal mask matrix M. In this process, generative network G and discriminant network D play games with each other and evolve alternately to form a generative adversarial network structure. It should be noted that the optimal mask matrix M is the most error-prone matrix of discriminant network D, and discriminant network D treats $M \cdot C^3$ as a positive sample.

3.5. Tracking Process

The tracking process of our proposed AMGN tracker includes three parts: model initialization, online detection, and online model update.

(1) Model Initialization: The base CNN is initialized by VGG-M [53] trained in the classification task from ImageNet. The parameters in Conv1-Conv3 of the base CNN are fixed and the others are initiated according to offline pre-training by multidomain learning, which is fine-tuned online.

(2) Online Detection: Generated multiple candidate boxes on the first frame of the tracking sequence or previous frame and its predicted target position are sampled by base CNN in each and fed into the tracking network to obtain probability scores.

(3) Online Model Update: According to the target position given in the first frame and the predicted target position in other frames, we generate multiple candidate boxes around them and assign two-category labels divided by intersection-over-union (IoU) scores. The labeled samples are used to jointly train AMGN (as the generator G of GAN) and tracker (as the discriminator D of GAN) to complete the adversarial processes. AMGN produces the attention-based mask firstly as the C^3 input, the mask adjustment process then meets the size of the mask to C^2 and obtains the fused occluded feature maps. With the label unchanged, D is studied through supervised learning. After training, D will suffice for identifying the target features occluded. In return, D guides G to generate more difficult masks for D.

4. Experiments

To evaluate the performance of the proposed AMGN-based tracker, we perform extensive experiments in terms of accuracy and robustness.

At present, numerous VOT benchmark datasets have been introduced to provide fair and standardized evaluations of object-tracking algorithms. For UAV-specific tracking, commonly used datasets include UAV123 [54], DTB70 [55], UAVDT [56], UAV20L [54], VisDrone2018 [57], etc. For generic tracking, OTB2015 [58], VOT2016 [59], GOT10k [60], FaceTracking [61], etc., have been widely used for evaluation. In order to fully verify the performance of the proposed method in UAV-specific tracking, we select three datasets, including UAV123, DTB70 and UAVDT, for experimental verification. In addition, considering the application potential of our method in generic object tracking, we also conduct comparative experiments on VOT2016.

4.1. Implementation

In this work, the first three convolution layers are from VGG-M [53] trained on ImageNet and always fixed in the process of the online tracking process. After Conv3, a CBAM that is also trained offline is used to obtain spatial attention maps. Spatial attention maps with the same size with C^3 (3×3) are adjusted to the binary as candidate masks. We train D first by applying candidate masks independently to each fused feature and choosing the one with the lowest classification score. Here, we alert 3×3 masks to 5×5 and meet the size of C^2. The fused feature is from C^2 and C^3. Then, the trained D guides G to generate masks like the label but composed of numbers between 0 and 1. During the adversarial learning process, the SGD solver is iteratively applied to G and D. The learning rate is set to 10^{-3} and 10^{-4}, respectively. We update both networks every 10 frames using 10 iterations. The whole experiment is performed on a PC with an i7-8700 CPU and NVIDIA GeForce GTX 1660 Ti GPU.

4.2. Evaluation on UAV123

UAV123 [54] contains 123 UAV videos with 12 challenging attributes, including illumination variation, scale variation, full occlusion, partial occlusion, camera motion, etc. In this paper, we follow with interest the overcoming of tracking drift under full occlusion and partial occlusion. We use the one-pass evaluation (OPE) metrics to measure the tracking performance. The precision plot computes the percentages of frames whose estimated locations lie in a given threshold distance to ground truth centers. The typical threshold distance is 20 pixels. The success plot is set to measure the overlap score (OS) between the ground truth and the bounding box resulting from the tracker. Afterward, a frame whose OS is larger than a certain threshold is termed a successful frame, and the success rates under different thresholds constitute a success plot. The general threshold is set to 0.5.

We compare an AMGN-based tracker with 21 state-of-the-art trackers composed by CF-based and CNN-based trackers, including ECO [62], ECO-HC [62], SiamRPN [11], C-COT [63], STRCF [64], DeepSTRCF [64], TADT [65], SRDCF [66], SRDCFdecon [67], SAMF [68], Staple [69], Staple_CA [70], KCF [71], KCC [72], DSST [73], UDT [74], UDT-plus [74], BACF [75], CSRDCF [76], MEEM [77], and MOSSE [5]. We evaluate all the trackers on 123 video sequences through OPE with distance precision and overlap success metrics.

Figure 5 shows the success and precision plots for the top 10 of the 21 comparison trackers. The values listed in the legends are the AUCs of the success rates and the 20-pixel distance precision scores, respectively. It is evident that our AMGN-based tracker performs well compared to other state-of-the-art trackers, with a leading precision score of 0.779.

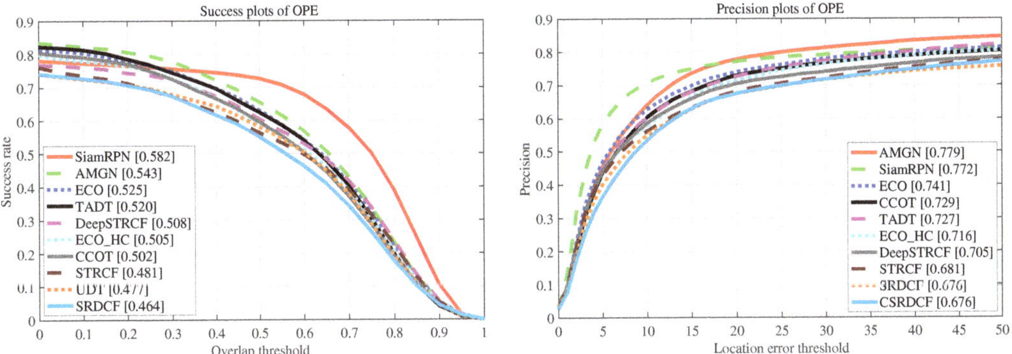

Figure 5. Success and precision plots on the UAV123 dataset using one-pass evaluation.

In Figure 6, we further show the success and precision plots under the two attributes: partial occlusion and full occlusion. The results show that the AMGN-based tracker achieves the best performance in handling occlusion challenges compared to the comparison trackers. The AUCs of the success plots under the two attributes lead the runner-up by 1.4% and 2.3%, respectively, and the precision scores lead the runner-up by 6.9% and 8.7%, respectively.

4.3. Evaluation on DTB70

We also evaluate 70 sequences from the DTB70 dataset [55] with 12 attributes, including scale variation, occlusion, deformation, fast camera motion, similar objects around, etc. Based on the same metrics as UAV123, we compare our proposed algorithm with 24 other state-of-the-art trackers: ECO [62], ECO-HC [62], C-COT [63], BACF [75], CoKCF [78], STRCF [64], DeepSTRCF [64], TADT [63], SRDCF [66], SRDCFdecon [67], SAMF [68], SAMF_CA [70], Staple [69], Staple_CA[70], KCF [71], KCC [72], MOSSE [5], MCCT [79], MCCT_H [79], MCPF [80], DSST [73], fDSST [81], UDT [74], and IBCCF [82].

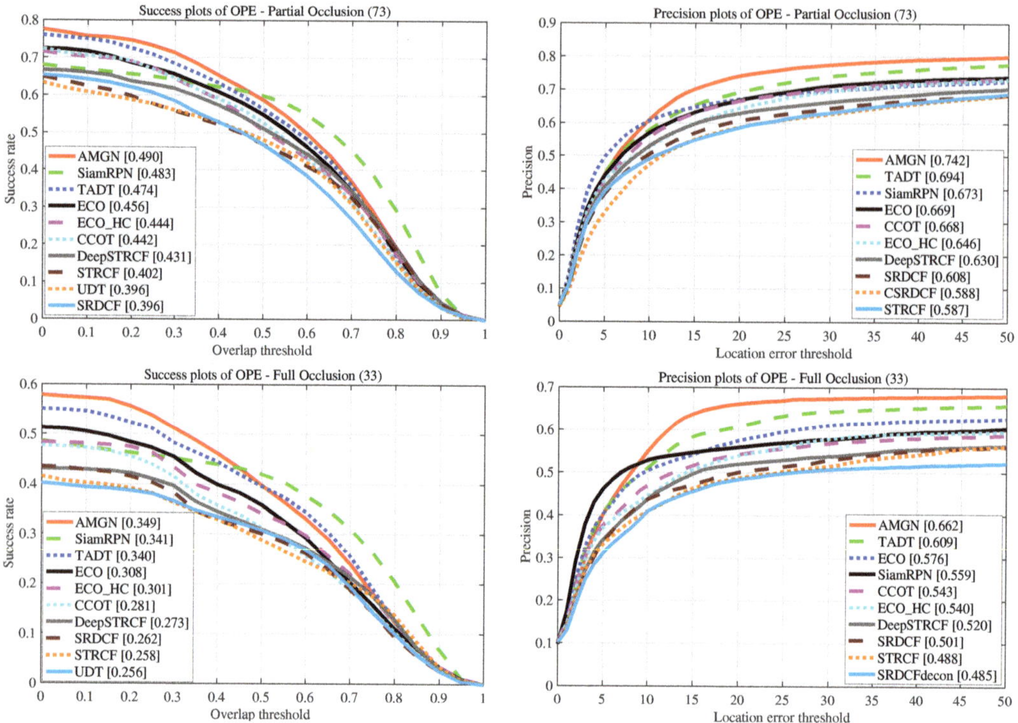

Figure 6. Success and precision plots on the UAV123 dataset using one-pass evaluation over tracking challenge occlusion.

Figure 7 shows the success and precision plots for the top 10 of the 21 comparison trackers. Similarly, the values listed in the legends are the AUCs of the success rates and the 20-pixel distance precision scores, respectively. Our AMGN-based tracker achieves the best AUC of 0.539 and the best precision score of 0.788 for overall videos.

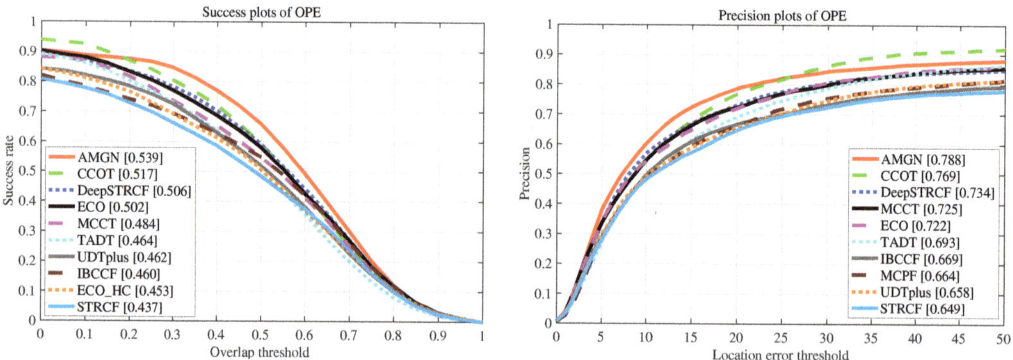

Figure 7. Success and precision plots on the DTB70 dataset using one-pass evaluation.

In Figure 8, we present the success and precision plots under the attribute, deformation. The results show that the AMGN-based tracker also achieves the best performance in handling deformation challenges. The AUC of the success plot leads the runner-up by 2.3%, and the precision score leads the runner-up by 9.1% .

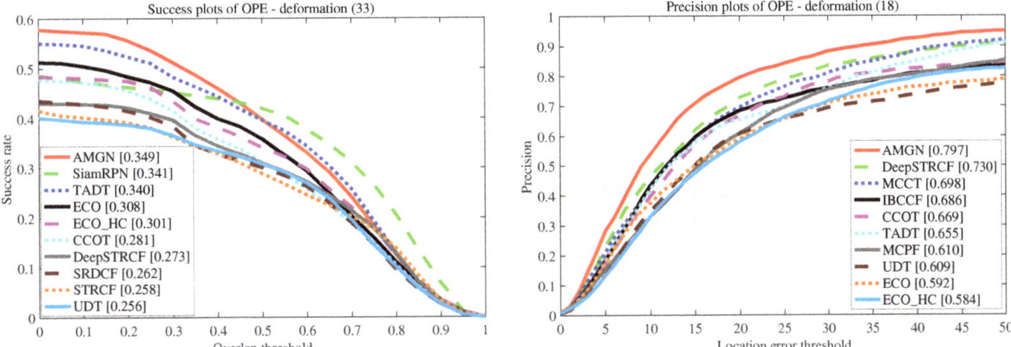

Figure 8. Success and precision plots on the DTB70 dataset using one-pass evaluation over tracking challenge Deformation.

4.4. Evaluation on UAVDT

For a more comprehensive evaluation, the proposed AMGN-based tracker is additionally compared with another 13 state-of-the-art trackers on UAVDT benchmark, including SiamFC [83], ECO [62], MDNet [34], CREST [84], C-COT [63], Staple_CA [70], SRDCFdecon [67], KCF [71], CFNet [85], MCPF [80], SRDCF [66], UDT [74], and SINT [22]. In the single object tracking task, UAVDT is composed of 50 aerial tracking videos with 8 attributes, including background clutter (BC), camera rotation (CR), object rotation (OR), small object (SO), illumination variation (IV), object blur (OB), scale variation (SV) and large occlusion (LO). Among these attributes, large occlusion is what we are concerned with. Similar to UVA123 and DTB70, we also conduct the evaluation through OPE with overlap success metrics and distance precision.

The success and precision plots for the top 10 of the 13 comparison trackers on the UAVDT dataset are presented in Figure 9. The proposed tracker exhibits the leading AUC of the success plot of 0.528 and the leading precision score of 0.771 for overall sequences.

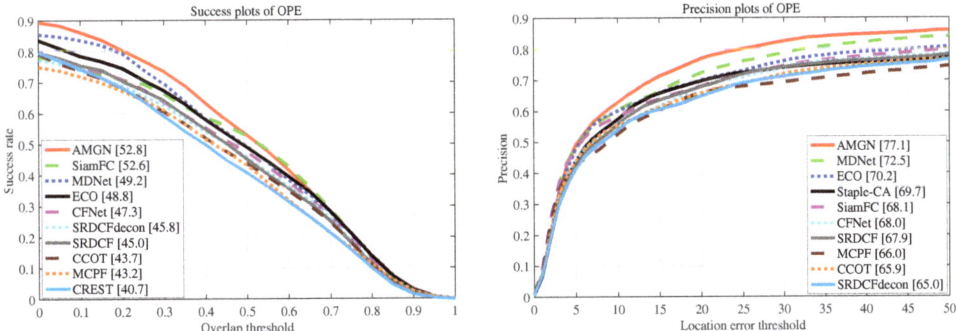

Figure 9. Success and precision plots on the UAVDT dataset using one-pass evaluation.

Meanwhile, Figure 10 presents the success and precision plots under the attribute, large occlusion (LO). The results show that our proposed AMGN-based tracker outperforms the second-best tracker 4% and 6.4% in AUC regarding the success plot and precision score, respectively.

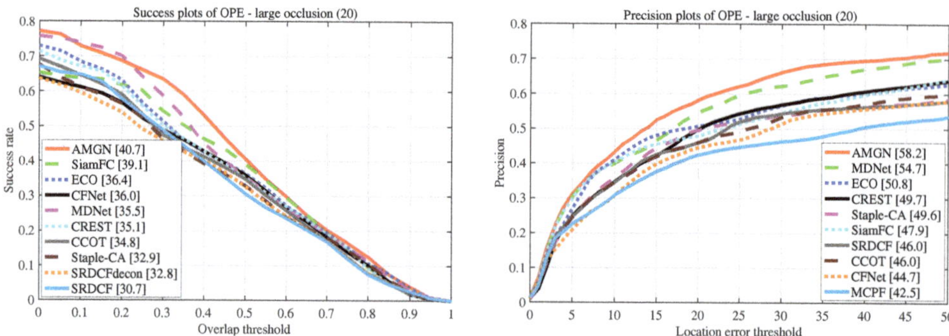

Figure 10. Success and precision plots on the UAVDT dataset using one-pass evaluation over tracking challenge large occlusion.

4.5. Evaluation on VOT2016

We also conduct a supervised evaluation on 60 sequences from the VOT2016 dataset [59]. Based on the VOT challenge protocol, trackers are re-initialized once a tracking drift is detected [86]. We use three metrics to evaluate the performance: expected average overlap (EAO) on short-term sequences, average overlap during the periods of successful tracking (accuracy), and the average number of failures during tracking (robustness). In Table 1, we compare our proposed algorithm with six other state-of-the-art trackers: ECO [62], CCOT [63], VITAL [10], MDNet [34], CREST [84], and Staple [69]. Our AMGN-based tracker achieves the best accuracy and robustness and the second EAO, which is sufficient to show its availability.

Table 1. Quantitative comparison results on the VOT2016 dataset. Values in red and green indicate the best and the second-best performance, respectively.

	ECO	CCOT	VITAL	MDNet	CREST	Staple	AMGN
EAO ↑	0.374	0.331	0.323	0.257	0.283	0.295	0.340
Accuracy ↑	0.54	0.52	0.55	0.57	0.51	0.54	0.576
Robustness ↓	0.72	0.85	0.98	1.20	1.08	0.35	0.191

4.6. Ablation Studies

4.6.1. Effectiveness of Attention Module and Adversarial Learning

In the AMGN-based tracker, we align the attention mechanism to the diversified hard positive samples as an AMGN module and train the classifier to overcome tracking drift adversarially. To validate the effectiveness of our AMGN module, we make experiments on the baseline tracker, baseline tracker with CBAM after the Conv3, and our proposed tracker. In addition, we add CBAM after C^3 to prove that an attention module in the high-level convolution layer can effectively avoid tracking drift, for it can benefit valid features and restrain the others. On the other hand, when we reversely drop out the target features of the focused parts of attention in our proposed method, the network concentrates attention on the whole target with more robustness characteristics, causing further improvement of the tracking effect. Figure 11 shows the ablation studies results on the DTB70 dataset. We observe that joint CBAM and adversarial learning produces significant improvements overall in both occlusion-tagged and deformation-tagged sequences. For example, compared with the baseline, the proposed AMGN-based tracker improves the precision scores by 18.4%, 11.4%, and 19.8% overall, in occlusion-tagged and deformation-tagged sequences, respectively.

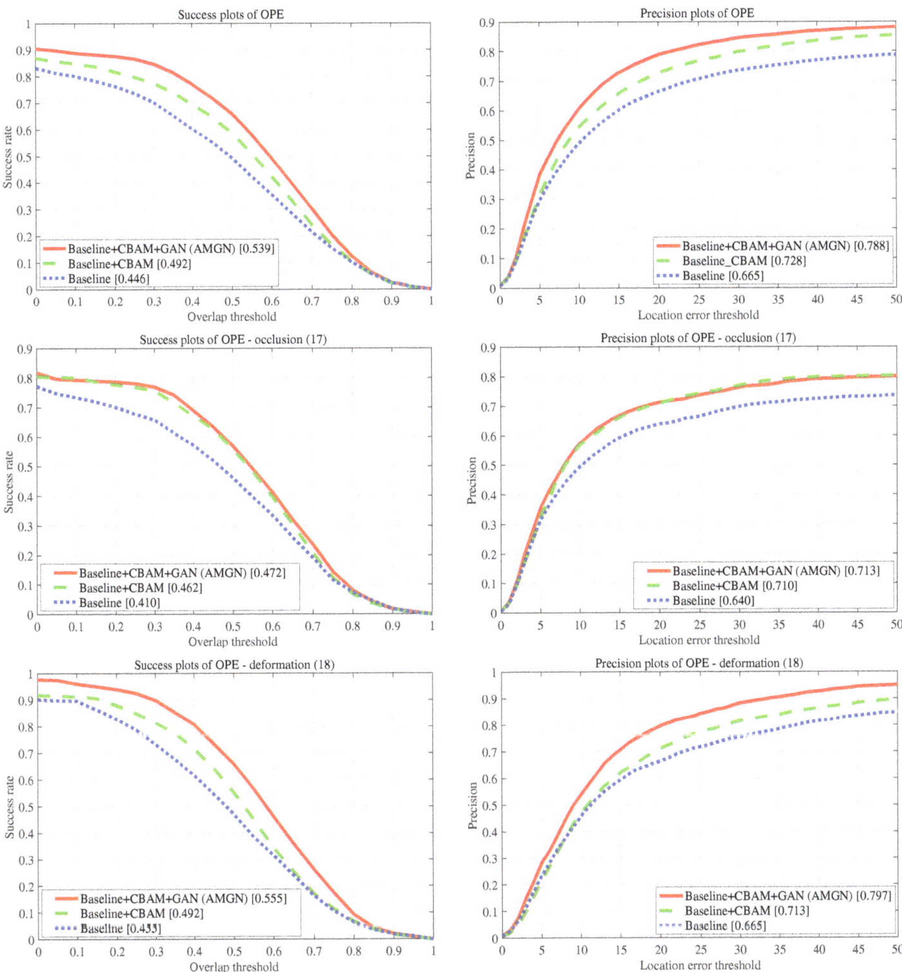

Figure 11. Ablation studies results on attention module and adversarial learning on the DTB70 dataset.

4.6.2. Effectiveness of Feature Fusion

To verify the effectiveness of our designed feature fusion method, we conduct further ablation studies on it. Since the original intention of the feature fusion method is to compensate for the excessive dropout of features that may be caused by the masks, in this experiment, we use the final formed AMGN-based tracker as the baseline and compare it with the configuration after only removing the feature fusion module. The results are shown in Figure 12. It can be seen that after removing the feature fusion method, the overall performance of the tracker and in both cases of occlusion and deformation has a significant decline, indicating the effectiveness of the feature fusion method. Specifically, after introducing the feature fusion method, the precision scores in the overall, occlusion and deformation sequences are improved by 10.8%, 5.6%, and 7.4%, respectively.

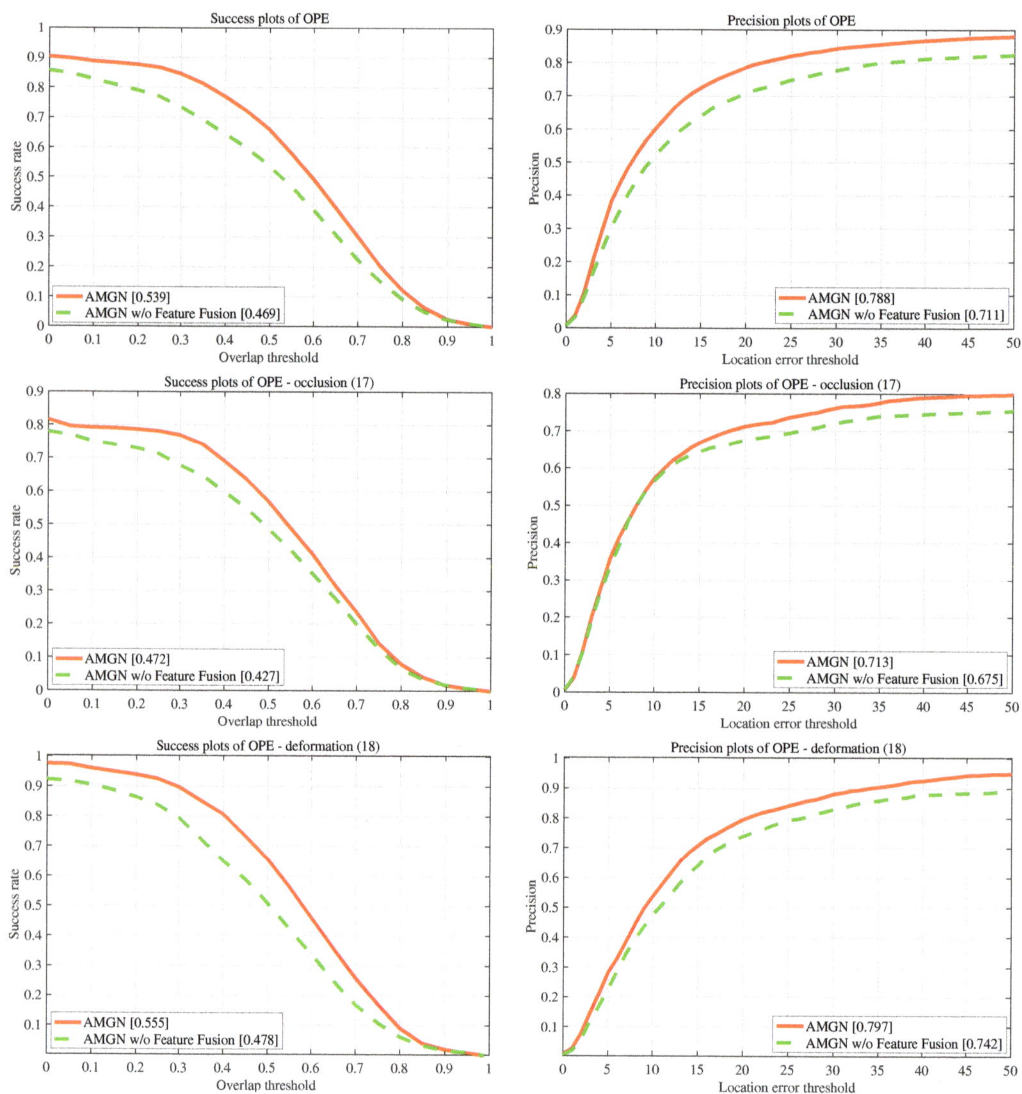

Figure 12. Ablation studies results on feature fusion on the DTB70 dataset.

4.7. Qualitative Evaluation

Figure 13 qualitatively compares the results of the top-performing trackers: ECO, CCOT, SiamRPN, SRDCFdecon, STRCF, BACF, and the proposed AMGN-based tracker on 13 challenging sequences. We choose six sequences from UAV123 with the attributes of full occlusion and partial occlusion, and another sequence occurs, deformation. In most sequences, SiamRPN, STRCF, and SRDCFdecon fail to locate the target with weak performance once there is occlusion. ECO and CCOT, despite unit CNN with correlation filtering and receiving richer feature representation, also lead to track failure when the target is fully blocked or there is interference of similar objects, as they do not take full advantage of the end-to-end deep architecture. The AMGN-based tracker keeps the best success rate under extreme conditions, especially in almost complete occlusion, and occlusion reappears because of our attention mechanism. It is noticed that the SimaRPN- and AMGN-based trackers show higher tracking precision under deformation. To verify the anti-deformation ability of our proposed tracker, we choose seven sequences from DTB70 that have bigger objects and obvious deformation from the qualitative results, even under the influence of strong deformation, clutter background, etc.

Figure 13. Qualitative results of the comparison tracker on challenging sequences from UAV123 and DTB70.

5. Conclusions

In this paper, we propose an AMGN-based tracker that leverages adversarial learning to enhance the tracker's resilience to occlusion and deformation. After extracting the deep features of the candidate target region through a base CNN, we first develop an attention-based mask generative network (AMGN), which adopts the attention mechanism to calculate the attention map of the deep features of the candidate target regions, and generates a series of masks according to the attention map. These masks are multiplied with deep features to simulate target occlusion and deformation in feature space. Then, to avoid over-dropping of target features by masks, we design a feature fusion method that incorporates shallower-layer features into deeper-layer features processed by masks, thus avoiding extreme cases of tracking drift due to excessive feature loss. After the above processing, the hard positive samples focusing on target occlusion and deformation are supplemented. Finally, we treat the tracker as the discriminator of GAN and use these hard positive samples for adversarial learning, thereby improving the tracker's ability to deal with occlusion and deformation. Comparative experiments show that our AMGN-based tracker achieves the highest AUC of 0.490 and 0.349, and the highest accuracy scores of 0.742 and 0.662 on the UAV123 benchmark with partial and full occlusion attributes, respectively. On the DTB70 benchmark with the deformation attribute, it achieves the highest AUC of 0.555 and the highest precision score of 0.797. On the UAVDT benchmark with the large occlusion attribute, it achieves the highest AUC of 0.407 and the highest precision score of 0.582.

Although the effectiveness of the proposed method is validated on several datasets, there are still some limitations. On the one hand, in order to improve the efficiency of the mask, we use the attention mechanism to generate the masks instead of doing so randomly, which will increase the computational complexity of the tracker; on the other hand, the method proposed in this paper focuses on considering the accurate target tracking under occlusion and deformation, so the overall performance improvement is not obvious. Further research will mainly focus on reducing the high computational complexity in terms of time and space, improving the flexibility of applying various basic CNNs, and considering more different challenging cases, thereby further improving the overall performance of the tracker.

Author Contributions: Conceptualization, Y.B.; methodology, Y.B. and Y.Z. (Yufei Zhao); software, Y.B., X.W., Z.Z. and Y.H.; validation, Z.Z. and Y.H.; formal analysis, Y.B.; investigation, Y.Z. (Ya Zhou) and X.Y.; resources, Y.S.; data curation, Y.B.; writing—original draft preparation, Y.B.; writing—review and editing, X.W., Y.H. and Y.B.; visualization, Y.H.; supervision, Y.S., Y.Z. (Yufei Zhao) and Q.H.; project administration, Y.S.; funding acquisition, Y.S. All authors have read and agreed to the published version of the manuscript.

Funding: This research was funded by the National Natural Science Foundation of China under Grant number 81671787.

Data Availability Statement: The current research was conducted on publicly available datasets.

Acknowledgments: The authors would like to thank all the colleagues who generously provided their image dataset with the ground truth. The authors would also like to thank the anonymous reviewers for their very competent comments and helpful suggestions.

Conflicts of Interest: The author declares no conflict of interest.

References

1. Lee, K.-H.; Hwang, J.-N. On-Road Pedestrian Tracking Across Multiple Driving Recorders. *IEEE Trans. Multimed.* **2015**, *17*, 1429–1438. [CrossRef]
2. Wu, Q.; Wang, H.; Liu, Y.; Zhang, L.; Gao, X. SAT: Single-shot adversarial tracker. *IEEE Trans. Ind. Electron.* **2019**, *67*, 9882–9892. [CrossRef]
3. Zhang, M.; Liu, X.; Xu, D.; Cao, Z.; Yu, J. Vision-based target-following guider for mobile robot. *IEEE Trans. Ind. Electron.* **2019**, *66*, 9360–9371. [CrossRef]
4. Guan, M.; Wen, C.; Shan, M.; Ng, C.L.; Zou, Y. Real-time event-triggered object tracking in the presence of model drift and occlusion. *IEEE Trans. Ind. Electron.* **2018**, *66*, 2054–2065. [CrossRef]
5. Bolme, D.S.; Beveridge, J.R.; Draper, B.A.; Lui, Y.M. Visual object tracking using adaptive correlation filters. In Proceedings of the 2010 IEEE Computer SOCIETY Conference on Computer Vision and Pattern Recognition, San Francisco, CA, USA, 13–18 June 2010; pp. 2544–2550.
6. Li, B.; Fu, C.; Ding, F.; Ye, J.; Lin, F. ADTrack: Target-aware dual filter learning for real-time anti-dark UAV tracking. In Proceedings of the 2021 IEEE International Conference on Robotics and Automation (ICRA), Xi'an, Shaanxi, China, 30 May–5 June 2021; pp. 496–502.
7. Çintaş, E.; Özyer, B.; Şimşek, E. Vision-based moving UAV tracking by another UAV on low-cost hardware and a new ground control station. *IEEE Access* **2020**, *8*, 194601–194611. [CrossRef]
8. Lin, F.; Fu, C.; He, Y.; Xiong, W.; Li, F. ReCF: Exploiting Response Reasoning for Correlation Filters in Real-Time UAV Tracking. *IEEE Trans. Intell. Transp. Syst.* **2021**, *23*, 10469–10480. [CrossRef]
9. Huang, B.; Chen, J.; Xu, T.; Wang, Y.; Jiang, S.; Wang, Y.; Wang, L.; Li, J. SiamSTA: Spatio-Temporal Attention based Siamese Tracker for Tracking UAVs. In Proceedings of the IEEE/CVF International Conference on Computer Vision, Montreal, BC, Canada, 11–17 October 2021; pp. 1204–1212.
10. Song, Y.; Ma, C.; Wu, X.; Gong, L.; Bao, L.; Zuo, W.; Shen, C.; Lau, R.W.; Yang, M.H. Vital: Visual tracking via adversarial learning. In Proceedings of the IEEE Conference on Computer Vision and Pattern Recognition, Salt Lake City, UT, USA, 18–23 June 2018; pp. 8990–8999.
11. Bo, L.; Yan, J.; Wei, W.; Zheng, Z.; Hu, X. High Performance Visual Tracking with Siamese Region Proposal Network. In Proceedings of the 2018 IEEE/CVF Conference on Computer Vision and Pattern Recognition (CVPR), Salt Lake City, UT, USA, 18–23 June 2018.

12. Li, Y.; Fu, C.; Ding, F.; Huang, Z.; Lu, G. AutoTrack: Towards high-performance visual tracking for UAV with automatic spatio-temporal regularization. In Proceedings of the IEEE/CVF Conference on Computer Vision and Pattern Recognition, Seattle, WA, USA, 13–19 June 2020; pp. 11923–11932.
13. Zhou, Q.; Zhong, B.; Zhang, Y.; Li, J.; Fu, Y. Deep Alignment Network Based Multi-Person Tracking With Occlusion and Motion Reasoning. *IEEE Trans. Multimed.* **2019**, *21*, 1183–1194. [CrossRef]
14. Wu, F.; Zhang, J.; Xu, Z. Stably Adaptive Anti-Occlusion Siamese Region Proposal Network for Real-Time Object Tracking. *IEEE Access* **2020**, *8*, 161349–161360. [CrossRef]
15. Yuan, Y.; Chu, J.; Leng, L.; Miao, J.; Kim, B.G. A scale-adaptive object-tracking algorithm with occlusion detection. *Eurasip J. Image Video Process.* **2020**, *2020*, 7. [CrossRef]
16. Wang, X.; Shrivastava, A.; Gupta, A. A-Fast-RCNN: Hard Positive Generation via Adversary for Object Detection. In Proceedings of the 2017 IEEE Conference on Computer Vision and Pattern Recognition (CVPR), Honolulu, HI, USA, 21–26 July 2017.
17. Qi, Y.; Zhang, S.; Zhang, W.; Su, L.; Huang, Q.; Yang, M.H. Learning attribute-specific representations for visual tracking. In Proceedings of the AAAI Conference on Artificial Intelligence, Honolulu, HI, USA, 27 January–1 February 2019; Volume 33, pp. 8835–8842.
18. Qi, Y.; Qin, L.; Zhang, S.; Huang, Q.; Yao, H. Robust visual tracking via scale-and-state-awareness. *Neurocomputing* **2019**, *329*, 75–85. [CrossRef]
19. Chen, Y.; Song, L.; Hu, Y.; He, R. Adversarial occlusion-aware face detection. In Proceedings of the 2018 IEEE 9th International Conference on Biometrics: Theory, Applications and Systems (BTAS), Redondo Beach, CA, USA, 22–25 October 2018; pp. 1–9.
20. Javanmardi, M.; Qi, X. Appearance variation adaptation tracker using adversarial network—ScienceDirect. *Neural Netw.* **2020**, *129*, 334–343. [CrossRef] [PubMed]
21. Souly, N.; Spampinato, C.; Shah, M. Semi Supervised Semantic Segmentation Using Generative Adversarial Network. In Proceedings of the 2017 IEEE International Conference on Computer Vision (ICCV), Venice, Italy, 22–29 October 2017.
22. Xiao, W.; Li, C.; Luo, B.; Jin, T. SINT++: Robust Visual Tracking via Adversarial Positive Instance Generation. In Proceedings of the 2018 IEEE/CVF Conference on Computer Vision and Pattern Recognition (CVPR), Salt Lake City, UT, USA, 18–23 June 2018.
23. Zhang, T.; Jia, K.; Xu, C.; Ma, Y.; Ahuja, N. Partial occlusion handling for visual tracking via robust part matching. In Proceedings of the IEEE Conference on Computer Vision and Pattern Recognition, Columbus, OH, USA, 23–28 June 2014; pp. 1258–1265.
24. Selvaraju, R.R.; Cogswell, M.; Das, A.; Vedantam, R.; Parikh, D.; Batra, D. Grad-cam: Visual explanations from deep networks via gradient-based localization. In Proceedings of the IEEE International Conference on Computer Vision, Venice, Italy, 22–29 October 2017; pp. 618–626.
25. Aa, A.; Js, B.; Knp, A. Sample-based adaptive Kalman filtering for accurate camera pose tracking. *Neurocomputing* **2019**, *333*, 307–318.
26. Cui, Y.; Zhang, J.; He, Z.; Hu, J. Multiple pedestrian tracking by combining particle filter and network flow model. *Neurocomputing* **2019**, *351*, 217–227. [CrossRef]
27. Xiao, Y.; Pan, D. Research on scale adaptive particle filter tracker with feature integration. *Appl. Intell.* **2019**, *49*, 3864–3880. [CrossRef]
28. Vojir, T.; Noskova, J.; Matas, J. Robust scale-adaptive mean-shift for tracking. *Pattern Recognit. Lett.* **2014**, *49*, 250–258. [CrossRef]
29. Zhang, J.; Liu, Y.; Liu, H.; Wang, J.; Zhang, Y. Distractor-aware visual tracking using hierarchical correlation filters adaptive selection. *Appl. Intell.* **2022**, *52*, 6129–6147. [CrossRef]
30. Zhang, Y.; Chen, D.; Zheng, Y. Satellite Video Tracking by Multi-Feature Correlation Filters with Motion Estimation. *Remote Sens.* **2022**, *14*, 2691. [CrossRef]
31. Qi, Y.; Yao, H.; Sun, X.; Sun, X.; Zhang, Y.; Huang, Q. Structure-aware multi-object discovery for weakly supervised tracking. In Proceedings of the 2014 IEEE International Conference on Image Processing (ICIP), Paris, France, 27–30 October 2014; pp. 466–470.
32. Yang, Y.; Li, G.; Qi, Y.; Huang, Q. Release the power of online-training for robust visual tracking. In Proceedings of the AAAI Conference on Artificial Intelligence, New York, NY, USA, 7–12 February 2020; Voume 34, pp. 12645–12652.
33. Danelljan, M.; Gool, L.V.; Timofte, R. Probabilistic regression for visual tracking. In Proceedings of the IEEE/CVF Conference on Computer Vision and Pattern Recognition, Seattle, WA, USA, 13–19 June 2020; pp. 7183–7192.
34. Nam, H.; Han, B. Learning multi-domain convolutional neural networks for visual tracking. In Proceedings of the IEEE Conference on Computer Vision and Pattern Recognition, Las Vegas, NV, USA, 27–30 June 2016; pp. 4293–4302.
35. Yang, T.; Chan, A.B. Learning dynamic memory networks for object tracking. In Proceedings of the European Conference on Computer Vision (ECCV), Munich, Germany, 8–14 September 2018; pp. 152–167.
36. Yang, T.; Chan, A.B. Recurrent filter learning for visual tracking. In Proceedings of the IEEE International Conference on Computer Vision Workshops, Venice, Italy, 22–29 October 2017; pp. 2010–2019.
37. Chen, Z.; Zhong, B.; Li, G.; Zhang, S.; Ji, R. Siamese box adaptive network for visual tracking. In Proceedings of the IEEE/CVF Conference on Computer Vision and Pattern Recognition, Seattle, WA, USA, 13–19 June 2020; pp. 6668–6677.
38. Danelljan, M.; Bhat, G.; Khan, F.S.; Felsberg, M. Atom: Accurate tracking by overlap maximization. In Proceedings of the IEEE/CVF Conference on Computer Vision and Pattern Recognition, Long Beach, CA, USA, 15–20 June 2019; pp. 4660–4669.
39. Dai, K.; Zhang, Y.; Wang, D.; Li, J.; Lu, H.; Yang, X. High-performance long-term tracking with meta-updater. In Proceedings of the IEEE/CVF Conference on Computer Vision and Pattern Recognition, Seattle, WA, USA, 13–19 June 2020; pp. 6298–6307.

40. Gilroy, S.; Jones, E.; Glavin, M. Overcoming occlusion in the automotive environment—A review. *IEEE Trans. Intell. Transp. Syst.* **2019**, *22*, 23–35. [CrossRef]

41. Mehmood, K.; Jalil, A.; Ali, A.; Khan, B.; Murad, M.; Cheema, K.M.; Milyani, A.H. Spatio-Temporal Context, Correlation Filter and Measurement Estimation Collaboration Based Visual Object Tracking. *Sensors* **2021**, *21*, 2841. [CrossRef]

42. Mehmood, K.; Ali, A.; Jalil, A.; Khan, B.; Cheema, K.M.; Murad, M.; Milyani, A.H. Efficient Online Object Tracking Scheme for Challenging Scenarios. *Sensors* **2021**, *21*, 8481. [CrossRef]

43. Kortylewski, A.; He, J.; Liu, Q.; Yuille, A.L. Compositional convolutional neural networks: A deep architecture with innate robustness to partial occlusion. In Proceedings of the IEEE/CVF Conference on Computer Vision and Pattern Recognition, Seattle, WA, USA, 13–19 June 2020; pp. 8940–8949.

44. Ren, Y.; Zhu, C.; Xiao, S. Deformable Faster R-CNN with Aggregating Multi-Layer Features for Partially Occluded Object Detection in Optical Remote Sensing Images. *Remote Sens.* **2018**, *10*, 1470. [CrossRef]

45. Li, C.; Chen, G.; Gou, R.; Tang, Z. Detector–tracker integration framework and attention mechanism for multi–object tracking. *Neurocomputing* **2021**, *464*, 450–461. [CrossRef]

46. Zeng, Y.; Wang, H.; Lu, T. Learning spatial-channel attention for visual tracking. In Proceedings of the 2019 IEEE/CIC International Conference on Communications in China (ICCC), Changchun, China, 11–13 August 2019; pp. 277–282.

47. Bahdanau, D.; Cho, K.; Bengio, Y. Neural machine translation by jointly learning to align and translate. *arXiv* **2014**, arXiv:1409.0473.

48. Hu, J.; Shen, L.; Sun, G. Squeeze-and-excitation networks. In Proceedings of the IEEE Conference on Computer Vision and Pattern Recognition, Salt Lake City, UT, USA, 18–23 June 2018; pp. 7132–7141.

49. Woo, S.; Park, J.; Lee, J.Y.; Kweon, I.S. *CBAM: Convolutional Block Attention Module*; Springer: Cham, Switzerland, 2018.

50. Goodfellow, I.; Pouget-Abadie, J.; Mirza, M.; Xu, B.; Warde-Farley, D.; Ozair, S.; Courville, A.; Bengio, Y. Generative adversarial nets. *Adv. Neural Inf. Process. Syst.* **2014**, *27*, 2672–2680.

51. Gui, J.; Sun, Z.; Wen, Y.; Tao, D.; Ye, J. A review on generative adversarial networks: Algorithms, theory, and applications. *IEEE Trans. Knowl. Data Eng.* **2021**. [CrossRef]

52. Olshausen, B.A.; Anderson, C.H.; Van Essen, D.C. A neurobiological model of visual attention and invariant pattern recognition based on dynamic routing of information. *J. Neurosci.* **1993**, *13*, 4700–4719. [CrossRef]

53. Chatfield, K.; Simonyan, K.; Vedaldi, A.; Zisserman, A. Return of the devil in the details: Delving deep into convolutional nets. *arXiv* **2014**, arXiv:1405.3531.

54. Mueller, M.; Smith, N.; Ghanem, B. A benchmark and simulator for uav tracking. In Proceedings of the European Conference on Computer Vision, Amsterdam, The Netherlands, 11–14 October 2016; pp. 445–461.

55. Li, S.; Yeung, D.Y. Visual object tracking for unmanned aerial vehicles: A benchmark and new motion models. In Proceedings of the Thirty-First AAAI Conference on Artificial Intelligence, San Francisco, CA, USA, 4–9 February 2017.

56. Du, D.; Qi, Y.; Yu, H.; Yang, Y.; Duan, K.; Li, G.; Zhang, W.; Huang, Q.; Tian, Q. The unmanned aerial vehicle benchmark: Object detection and tracking. In Proceedings of the European Conference on Computer Vision (ECCV), Munich, Germany, 8–14 September 2018; pp. 370–386.

57. Wen, L.; Zhu, P.; Du, D.; Bian, X.; Ling, H.; Hu, Q.; Liu, C.; Cheng, H.; Liu, X.; Ma, W.; et al. Visdrone-sot2018: The vision meets drone single-object tracking challenge results. In Proceedings of the European Conference on Computer Vision (ECCV) Workshops, Munich, Germany, 8–14 September 2018.

58. Wu, Y.; Lim, J.; Yang, M.H. Object Tracking Benchmark. *IEEE Trans. Pattern Anal. Mach. Intell.* **2015**, *37*, 1834–1848. [CrossRef]

59. Kristan, M.; Matas, J.; Leonardis, A.; Felsberg, M.; Cehovin, L.; Fernández, G.; Vojir, T.; Häger, G.; Lukežič, A.; Fernández, G.; et al. The visual object tracking vot2016 challenge results. In Proceedings of the European Conference on Computer Vision (ECCV) Workshops, Amsterdam, The Netherlands, 11–14 October 2016; Volume 2, p. 8.

60. Huang, L.; Zhao, X.; Huang, K. Got-10k: A large high-diversity benchmark for generic object tracking in the wild. *IEEE Trans. Pattern Anal. Mach. Intell.* **2019**, *43*, 1562–1577. [CrossRef]

61. Qi, Y.; Zhang, S.; Jiang, F.; Zhou, H.; Tao, D.; Li, X. Siamese local and global networks for robust face tracking. *IEEE Trans. Image Process.* **2020**, *29*, 9152–9164. [CrossRef]

62. Danelljan, M.; Bhat, G.; Shahbaz Khan, F.; Felsberg, M. ECO: Efficient convolution operators for tracking. In Proceedings of the IEEE Conference on Computer Vision and Pattern Recognition, Honolulu, HI, USA, 21–26 July 2017; pp. 6638–6646.

63. Danelljan, M.; Robinson, A.; Shahbaz Khan, F.; Felsberg, M. Beyond correlation filters: Learning continuous convolution operators for visual tracking. In Proceedings of the European Conference on Computer Vision, Amsterdam, The Netherlands, 11–14 October 2016; pp. 472–488.

64. Li, F.; Tian, C.; Zuo, W.; Zhang, L.; Yang, M.H. Learning spatial-temporal regularized correlation filters for visual tracking. In Proceedings of the IEEE Conference on Computer Vision and Pattern Recognition, Salt Lake City, UT, USA, 18–23 June 2018; pp. 4904–4913.

65. Li, X.; Ma, C.; Wu, B.; He, Z.; Yang, M.H. Target-aware deep tracking. In Proceedings of the IEEE/CVF Conference on Computer Vision and Pattern Recognition, Long Beach, CA, USA, 15–20 June 2019; pp. 1369–1378.

66. Danelljan, M.; Hager, G.; Shahbaz Khan, F.; Felsberg, M. Learning spatially regularized correlation filters for visual tracking. In Proceedings of the IEEE International Conference on Computer Vision, Santiago, Chile, 7–13 December 2015; pp. 4310–4318.

67. Danelljan, M.; Hager, G.; Shahbaz Khan, F.; Felsberg, M. Adaptive decontamination of the training set: A unified formulation for discriminative visual tracking. In Proceedings of the IEEE conference on Computer Vision and Pattern Recognition, Las Vegas, NV, USA, 27–30 June 2016; pp. 1430–1438.

68. Li, Y.; Zhu, J. A scale adaptive kernel correlation filter tracker with feature integration. In Proceedings of the European Conference on Computer Vision, Zurich, Switzerland, 6–12 September 2014; pp. 254–265.

69. Bertinetto, L.; Valmadre, J.; Golodetz, S.; Miksik, O.; Torr, P.H. Staple: Complementary learners for real-time tracking. In Proceedings of the IEEE Conference on Computer Vision and Pattern Recognition, Las Vegas, NV, USA, 27–30 June 2016; pp. 1401–1409.

70. Mueller, M.; Smith, N.; Ghanem, B. Context-aware correlation filter tracking. In Proceedings of the IEEE Conference on Computer Vision and Pattern Recognition, Honolulu, HI, USA, 21–26 July 2017; pp. 1396–1404.

71. Henriques, J.F.; Caseiro, R.; Martins, P.; Batista, J. High-speed tracking with kernelized correlation filters. *IEEE Trans. Pattern Anal. Mach. Intell.* **2014**, *37*, 583–596. [CrossRef] [PubMed]

72. Wang, C.; Zhang, L.; Xie, L.; Yuan, J. Kernel Cross-Correlator. In Proceedings of the Thirty-Second AAAI Conference on Artificial Intelligence (AAAI-18), New Orleans, LA, USA, 2–7 February 2018.

73. Danelljan, M.; Häger, G.; Khan, F.; Felsberg, M. Accurate scale estimation for robust visual tracking. In *Proceedings of the British Machine Vision Conference, Nottingham, UK, 1–5 September 2014*; BMVA Press: Nottingham, UK, 2014.

74. Wang, N.; Song, Y.; Ma, C.; Zhou, W.; Liu, W.; Li, H. Unsupervised deep tracking. In Proceedings of the IEEE/CVF Conference on Computer Vision and Pattern Recognition, Long Beach, CA, USA, 15–20 June 2019; pp. 1308–1317.

75. Kiani Galoogahi, H.; Fagg, A.; Lucey, S. Learning background-aware correlation filters for visual tracking. In Proceedings of the IEEE International Conference on Computer Vision, Venice, Italy, 22–29 October 2017; pp. 1135–1143.

76. Lukezic, A.; Vojir, T.; Čehovin Zajc, L.; Matas, J.; Kristan, M. Discriminative correlation filter with channel and spatial reliability. In Proceedings of the IEEE Conference on Computer Vision and Pattern Recognition, Honolulu, HI, USA, 21–26 July 2017; pp. 6309–6318.

77. Zhang, J.; Ma, S.; Sclaroff, S. MEEM: Robust tracking via multiple experts using entropy minimization. In Proceedings of the European Conference on Computer Vision, Zurich, Switzerland, 6–12 September 2014; pp. 188–203.

78. Zhang, L.; Suganthan, P.N. Robust Visual Tracking via Co-trained Kernelized Correlation Filters. *Pattern Recognit.* **2017**, *69*, 82–93. [CrossRef]

79. Wang, N.; Zhou, W.; Tian, Q.; Hong, R.; Wang, M.; Li, H. Multi-cue correlation filters for robust visual tracking. In Proceedings of the IEEE Conference on Computer Vision and Pattern Recognition, Salt Lake City, UT, USA, 18–23 June 2018; pp. 4844–4853.

80. Zhang, T.; Xu, C.; Yang, M.H. Multi-task correlation particle filter for robust object tracking. In Proceedings of the IEEE Conference on Computer Vision and Pattern Recognition, Honolulu, HI, USA, 21–26 July 2017; pp. 4335–4343.

81. Danelljan, M.; Häger, G.; Khan, F.S.; Felsberg, M. Discriminative scale space tracking. *IEEE Trans. Pattern Anal. Mach. Intell.* **2016**, *39*, 1561–1575. [CrossRef]

82. Li, F.; Yao, Y.; Li, P.; Zhang, D.; Zuo, W.; Yang, M.H. Integrating boundary and center correlation filters for visual tracking with aspect ratio variation. In Proceedings of the IEEE International Conference on Computer Vision Workshops, Venice, Italy, 22–29 October 2017; pp. 2001–2009.

83. Bertinetto, L.; Valmadre, J.; Henriques, J.F.; Vedaldi, A.; Torr, P. *Fully-Convolutional Siamese Networks for Object Tracking*; Springer: Cham, Switzerland, 2016.

84. Song, Y.; Ma, C.; Gong, L.; Zhang, J.; Lau, R.W.; Yang, M.H. Crest: Convolutional residual learning for visual tracking. In Proceedings of the IEEE International Conference on Computer Vision, Venice, Italy, 22–29 October 2017; pp. 2555–2564.

85. Valmadre, J.; Bertinetto, L.; Henriques, J.; Vedaldi, A.; Torr, P.H. End-to-end representation learning for correlation filter based tracking. In Proceedings of the IEEE Conference on Computer Vision and Pattern Recognition, Honolulu, HI, USA, 21–26 July 2017; pp. 2805–2813.

86. Kristan, M.; Matas, J.; Leonardis, A.; Vojíř, T.; Pflugfelder, R.; Fernandez, G.; Nebehay, G.; Porikli, F.; Čehovin, L. A novel performance evaluation methodology for single-target trackers. *IEEE Trans. Pattern Anal. Mach. Intell.* **2016**, *38*, 2137–2155. [CrossRef]

Article

AERO: AI-Enabled Remote Sensing Observation with Onboard Edge Computing in UAVs

Anis Koubaa *,†, Adel Ammar †, Mohamed Abdelkader, Yasser Alhabashi and Lahouari Ghouti

College of Computer & Information Sciences, Prince Sultan University, Riyadh 11586, Saudi Arabia; aammar@psu.edu.sa (A.A.)
* Correspondence: akoubaa@psu.edu.sa
† These authors contributed equally to this work.

Abstract: Unmanned aerial vehicles (UAVs) equipped with computer vision capabilities have been widely utilized in several remote sensing applications, such as precision agriculture, environmental monitoring, and surveillance. However, the commercial usage of these UAVs in such applications is mostly performed manually, with humans being responsible for data observation or offline processing after data collection due to the lack of on board AI on edge. Other technical methods rely on the cloud computation offloading of AI applications, where inference is conducted on video streams, which can be unscalable and infeasible due to remote cloud servers' limited connectivity and high latency. To overcome these issues, this paper presents a new approach to using edge computing in drones to enable the processing of extensive AI tasks onboard UAVs for remote sensing. We propose a cloud–edge hybrid system architecture where the edge is responsible for processing AI tasks and the cloud is responsible for data storage, manipulation, and visualization. We designed AERO, a UAV brain system with onboard AI capability using GPU-enabled edge devices. AERO is a novel multi-stage deep learning module that combines object detection (YOLOv4 and YOLOv7) and tracking (DeepSort) with TensorRT accelerators to capture objects of interest with high accuracy and transmit data to the cloud in real time without redundancy. AERO processes the detected objects over multiple consecutive frames to maximize detection accuracy. The experiments show a reduced false positive rate (0.7%), a low percentage of tracking identity switches (1.6%), and an average inference speed of 15.5 FPS on a Jetson Xavier AGX edge device.

Keywords: unmanned aerial vehicles; object detection; object tracking; remote sensing; object localization; edge computing; inspection; YOLOv4; YOLOv7; DeepSORT

Citation: Koubaa, A.; Ammar, A.; Abdelkader, M.; Alhabashi, Y.; Ghouti, L. AERO: AI-Enabled Remote Sensing Observation with Onboard Edge Computing in UAVs. *Remote Sens.* **2023**, *15*, 1873. https://doi.org/10.3390/rs15071873

Academic Editors: Wenjiang Huang, Giovanni Laneve, Yingying Dong and Chenghai Yang

Received: 3 March 2023
Revised: 27 March 2023
Accepted: 29 March 2023
Published: 31 March 2023

1. Introduction

The use of unmanned aerial vehicles (UAVs), also known as drones, in remote sensing has been increasingly beneficial as they help to speed up the data collection of assets of interest using aerial images. Drones make the data collection process cost-effective and flexible as drones can fly at low or high altitudes. It also helps missions to be more efficient as large regions can be precisely covered in short times thanks to the use of high-resolution cameras. Data collection also becomes safe as drones replace humans entering dangerous or difficult-to-access environments. These benefits are driving the remote sensing business to increasingly rely on drones. As a matter of fact, the market of commercial use of UAVs, including remote sensing applications, was valued at USD 5.85 billion in 2020 and is expected to have a compound annual growth rate of 14.2% [1].

1.1. Motivating Scenarios

AI-powered drones with onboard intelligence are becoming a game changer for many applications, including search and rescue, rapid infrastructure inspection, and remote sensing, to name a few. Integrating AI solutions directly on the drone's edge (compute)

devices can dramatically reduce the decision-making time and operational costs. In this paper, we will discuss several scenarios in different use cases' contexts, showing the limitation of existing solutions of UAVs for vision-based applications and discussing the advantages of real-time onboard AI.

1.1.1. Remote Sensing

Tree counting is one of the applications in remote sensing, where a drone surveys farm regions to count the number of trees. In [2], the authors proposed an offline counting and geo-localization of palm trees based on aerial images using deep learning. However, the processing was performed offline after collecting palm tree images from a UAV. The process of data collection and its offline processing takes a long time and needs to be performed in real time. Leveraging GPU-based edge devices on board the UAV enables the full automation of palm tree counting in real time. Furthermore, it helps to send each palm tree information (e.g., image and coordinates) to the cloud and store it in databases in real time. Naturally, the same concept can be applied to other remote sensing applications, such as gas leakage localization and mapping [3], flash flood real-time monitoring [4,5], and urban environment segmentation [6,7].

1.1.2. Search and Rescue

Consider a search-and-rescue mission where a drone is required to explore an extended region to search for a missing person in the desert or a forest, for example. It has been reported that more than 100 people get lost and die in the desert annually in Saudi Arabia alone [8]. The current practice for search-and-rescue using UAVs is to manually explore a region with human observers to find the target missing people. Using AI on board will help to automate the process as the UAV can execute specialized person detection models on board and automatically report their location in real time. It is also possible to use a swarm of drones to perform search-and-rescue missions in parallel, speeding up the search process and increasing the probability of finding and saving people [9].

1.1.3. Inspection and Surveillance

Surveillance and inspection using UAVs is one of the fastest businesses in the drone industry [10]. Drones are typically used to detect objects of interest in surveillance missions, such as vehicles [11], pedestrians [12], and buildings [13]. Traditional approaches either inspect real-time video streams by human observers or record scenes' videos and process them offline either manually or using AI techniques to extract target objects. The use of onboard AI processing in the UAVs will help to automate the inspection process and identify target objects in real time with a high accuracy, as will be demonstrated in this paper.

The automation of these applications on board UAVs is possible thanks to the evolution of edge devices and their support of advanced graphics processing units (GPUs), making it possible to process complex deep learning models in real time.

Before the evolution of edge computing, computation offloading has evolved as the prominent approach to processing heavy computation in the cloud instead of processing them on robots or drones. This concept has been known as cloud robotics. While computation offloading offers several advantages by leveraging the capabilities of the cloud resources to speed up the processing of deep learning models and computation-intensive applications, it suffers from high communication overhead. It also needs a large bandwidth and high-quality communication, which cannot always be afforded. In [14], the authors proposed a system architecture for computation offloading in Internet-connected drones and compared the performance of cloud computation offloading versus edge computing for deep learning applications. The study investigated the tradeoff between the communication cost and computation and found that computation offloading provides higher throughput despite larger communication delays.

1.2. Main Contributions

In this paper, we aimed to tackle the persisting challenge of deploying onboard artificial intelligence on the edge in commercial unmanned aerial vehicles (UAVs) that are primarily utilized for remote sensing applications. This predicament often necessitates laborious manual data observation or time-consuming offline processing, as cloud-based approaches are often impractical. There are a few recent works that tested onboard AI on edge in UAVs for detection and tracking, such as [15–17]. Nevertheless, they did not investigate the hybrid system architecture that we implemented in this work, and did not discuss the role of the cloud in their solution. To bridge this gap, we propose using edge computation on board drones to enable advanced observation and surveillance applications, involving object detection, multi-object-tracking, and real-time reporting of detected target objects to the cloud. In brief, the contributions of the paper can be summarized as follows:

- We propose a new approach to using edge computing in drones to enable the processing of extensive AI tasks on board UAVs for remote sensing. To overcome the limited connectivity and high latency of remote cloud servers, we propose a cloud–edge hybrid system architecture. In this architecture, the edge is responsible for processing AI tasks, and the cloud is responsible for data storage, manipulation, and visualization. Our proposed architecture can provide a more scalable and efficient solution for remote sensing applications.
- To implement our proposed architecture, we designed and developed AERO, a UAV brain system with onboard AI capability using GPU-enabled edge devices. AERO allows us to capture objects of interest with high accuracy and transmit data to the cloud in real time without redundancy. AERO processes the detected objects over multiple consecutive frames to maximize detection accuracy. AERO can be a significant advancement in the field of remote sensing as it enables UAVs to perform onboard AI tasks with high accuracy and real-time data transmission, providing a more efficient and cost-effective solution for remote sensing applications.

The remaining sections of the paper are organized as follows. Section 1.3 provides a review of the relevant literature and situates the contribution of the paper in comparison to previous work. Section 2 presents the architecture of the AERO system and describes the AERO AI Module. In Section 3, we detail the experimental study conducted to evaluate the AERO system's performance, and we discuss their results in Section 4. Finally, Section 5 concludes the paper and suggests potential future research directions for further improvements.

1.3. Related Works

The introduction of UAVs in remote sensing has paved the way for several promising applications that span a wide range of domains [18]. Impressive progress has been achieved in academic and industrial arenas. Diversity in available solutions is mainly attributed to the underlying technologies and modalities used in the data sense/acquisition processes [19]. The latter processes are domain-specific in nature [20,21]. Other techniques, including data preprocessing, feature extraction, and classification, are specifically designed for the application, whether civilian or military. UAV applications in remote sensing have been reviewed in [21,22].

1.3.1. Edge Computing and UAVs

Several recent works addressed the edge computing paradigm, which involves moving computational processing and storage closer to the end-users, devices, or sensors rather than relying solely on cloud-based solutions. Specifically, these works focused on leveraging UAVs to offload computation tasks to edge computing servers, which enables low-latency computations of specific tasks without noticeable delay.

In [23], Messous et al. proposed an evaluation mechanism of the integration of the computation offloading to edge computing servers for the efficient deployment of UAVs.

Based on the proposed evaluation, UAV-based models are able to decide whether to perform local processing, offload to an edge server, or delegate the computational tasks to the ground station. Informed decisions are based on low-latency computations of specific tasks without noticeable delay. Qian et al. [24] investigated the performance of a UAV-mounted mobile edge computing network where the UAV unit offloads and executes specific tasks that originate from some mobile terminal users. The trajectory planning problem was formulated as a Markov decision process (MDP) where optimal trajectories were obtained using a policy based on the double deep Q-network (DDQN) algorithm [25]. Thanks to the DDQN efficiency, higher throughput scores were attained.

A machine-learning-based solution for the planning of UAV trajectories is attributed to Afifi, and Gadallah [26]. Unlike many existing solutions, Afifi and Gadallah targeted missions with real-time navigation requirements in dense urban environments, where existing 5G infrastructures are astutely employed to ensure UAV navigation in complex environments through continuous interactions between the UAV units and the selected 5G network. Like [24], the proposed trajectory planning solution relies on deep reinforcement learning strategies, where the planning accuracy attains 99%.

In [27], Xia et al. proposed a flexible design of a wireless edge network using two UAV units. In this design, both units are restricted to operate at fixed altitudes with accelerated motions. Over a defined area, while the first UAV unit is in charge of forwarding downlink signals to the user terminals (UTs), the second unit is assigned to the collection of the uplink data. Using statistical information collected from the UT elements and UAVs, lower bounds on conditional average achievable rates are derived. The proposed scheme is demonstrated to attain an energy efficiency higher than existing ones.

Bin et al. [28] tackled the problem of the variability of user mobility and MEC environments, where they suggested a novel scheme for intelligent task offloading in UAV-enabled MEC systems using a digital twin (DT). At the core of the proposed scheme lies the DDQN model, which is specifically designed to effectively constrain multi-objective problems. The model was jointly optimized using closed-form and iterative procedures. The simulation results clearly indicate the convergence of the DDQN-based model while drastically minimizing the total energy consumption of the MEC system compared to existing optimization techniques.

A new aerial edge Internet of Things (EdgeIoT) system was contributed by Li et al. [29]. In this new EdgeIoT system, a UAV unit is operated as a mobile edge server for processing computational processes related to mission-critical tasks emanating from ground IoT devices. To capture the underlying feature correlations, a graph-based neural network architecture (GNN) was used for the supervised training of the A2C structure. The reported performance analysis highlights the superiority of the mixed GNN-A2C framework in terms of the convergence speed and missing task rates.

In [30], Qian et al. proposed a Monte Carlo tree search (MCTS)-based path planning technique assuming that a single UAV is deployed as a mobile service to provide computation tasks offloading services for a set of mobile users on the ground. The reported results show that the MCTS-based scheme outperforms state-of-the-art DQN-based planning algorithms in terms of the average throughput and convergence speed. In some instances, UAVs assist edge clouds (ECs) for the large-scale sparely distributed user equipment, which allows for wide coverage and reliable wireless communication. However, UAVs have limited computation and energy resources, which opens the floor for potential optimal resource allocation.

In [31], Wang et al. introduced a vehicular fog computing (VFC) system where unmanned ground vehicles (UGVs) perform the computation tasks offloaded from UAVs that are deployed in natural disaster areas. In these areas, UAVs are effectively used to survey disaster areas and even perform emergency missions, given their swift deployment and flexibility. However, this efficiency is hindered by the limited energy and computational capabilities of UAVs. These limitations are properly addressed by the VFC-based UAV system proposed by Wang et al., where UGVs may be assigned to perform the computation tasks offloaded from UAVs to save energy and computational power. To ensure a smooth and steady UAV–UGV collaboration and interaction, the computation task offloading problem was cast into a two-sided matching problem, where an iterative stable matching algorithm was used. This matching algorithm aims at assigning to each UAV the most suitable UGV among the available ones for offloading while maximizing the usage of both UAVs and UGVs and reducing the average delay.

Yang et al. [32] considered a UAV-enabled MEC platform where multiple mobile ground users move randomly and tasks arrive in a random fashion. To minimize the average weighted energy consumption of all users under constraints expressed in terms of data queue stability and average UAV energy consumption, Yang et al. suggested a multi-stage stochastic optimization scheme where Lyapunov optimization is converted into simpler per-slot deterministic problems vis-a-vis the number of optimizing variables. Based on their formulation, Yang et al. solved the resource allocation and the UAV movement problems using two reduced-complexity methods, either jointly or separately. The two methods not only satisfy the average UAV energy and queue stability constraints, but they also reconcile the length of the queue backlog and the user energy consumption bounds. The reported results show that the proposed joint and two-stage stochastic optimization schemes outperform existing learning-based solutions. Finally, it should be noted that the joint optimization scheme attains a better performance than its two-stage counterpart at the expense of an increased computational complexity. Most of the solutions discussed so far attempt to optimize the UAVs' total (or average) energy consumption and computational power allocation among mobile users using some type of learning-based strategy.

In their proposal, Lyi et al. [33] adopted a different approach to maximize the computation bits of the whole MEC system: the joint optimization of task offloading time allocation, bandwidth allocation, and the UAV trajectory under specific energy constraints of ground devices and maximal UAV battery energy. The proposed solution splits the overall optimization procedure into three stages, where successive convex optimization schemes are used. Once individual solutions are identified, a block coordinate descent (BCD) algorithm integrates the solution of the initial optimization problem. Such a formulation aims at obtaining alternating optimal solutions for the optimization variables considered (bandwidth allocation of ground devices, task offloading time, local computing time allocation, and UAV trajectory) at each time slot. Extensive simulation experiments were conducted to demonstrate the performance improvement attained by the proposed BCD-based solution.

Overall, the proposed solutions discussed in this section suggest that UAV-based edge computing systems have certain advantages over cloud-based techniques in terms of optimization, convergence speed, throughput, and energy efficiency. These advantages make UAV-based edge computing systems a promising solution for various applications, including precision agriculture, smart cities, and disaster management, where real-time data processing and optimization are critical.

1.3.2. Summary of Related Works

A summary of the current literature is provided in Table 1. Onboard AI edge computing is becoming increasingly important for UAV systems, especially those utilizing EMC-based solutions. While EMC-based UAV systems offer benefits such as flexibility, resilience, and swift deployment, they also present new challenges that can only be addressed by advanced AI-based solutions, such as reinforcement and deep learning frameworks.

One reason for why onboard AI edge computing is necessary for EMC-based UAV systems is the need for real-time decision making. In certain applications, such as emergency response, decisions need to be made quickly and accurately. Onboard AI edge computing can process data in real time, allowing the UAV to make decisions based on the information that it collects, without the need for remote servers. This reduces latency and ensures that decisions are made in a timely manner.

Another reason is the need for autonomy. UAVs equipped with onboard AI edge computing can perform tasks autonomously, without human intervention. This is important in applications where it may be dangerous or impractical for humans to be present, such as in disaster response or surveillance missions. The AI algorithms on board the UAV can analyze the data collected and make decisions based on pre-defined rules, allowing the UAV to carry out its tasks independently.

Table 1. Comparative analysis of related work.

Reference	Scope	Advantages	Limitations
[18]	Overview of current applications of UAVs in remote sensing (up to 2015)	Comprehensive review	Limited to remote sensing domain
[19]	Discussion of UAV usage in 3D mapping (up to 2014)	In-depth coverage of UAV applications	Limited to 3D mapping applications
[20]	Covers UAV-based platforms for glaciology investigations (up to 2016)	Detailed discussion of UAV systems in glaciology	Covers only one domain application (glaciology)
[21]	Review of UAV deployments in forestry (up to 2017)	In-depth analysis of UAV systems in forestry	Restricted to European systems
[22]	Review of UAV applications in remote sensing (up to 2019)	Discusses multi-sensor fusion	Imbalanced coverage of UAV sub-systems
[23]	Edge computing usage in UAV visual communication	Extensive simulation of proposed framework	Lack of detailed comparison with existing frameworks
[24]	Path planning algorithm using RL paradigms	Optimal planning and routing decisions	Does not analyze the stability of the RL-based policies
[26]	Autonomous trajectory planning for UAV missions	Use of 5G wireless infrastructure	Lack of comparison with existing SOTA solutions
[27]	Multi-agent Q-learning algorithm for energy optimization	Higher energy consumption efficiency in UAV systems	Does not discuss the tradeoffs required to achieve energy efficiency
[28]	A new solution for MTU association and UAV trajectory.	Efficient RL-based solution using DDQN algorithm	Missing analysis of the DDQN limitations
[29]	Joint optimization of UAV cruise control and task offloading allocation	Efficient advantage actor–critic (A2C) solution	No comparison with SOTA policy gradient algorithms
[30]	Path planning using Monte Carlo tree search (MCTS) algorithm	Constraint-based maximization of average throughput	Benchmarking with relatively basic RL-based solutions
[31]	Battery and computation resource optimization using fog computing solutions	Improved UAV usage and reduced average delay	Missing comparison with SOTA solutions
[32]	UAV energy efficiency using multi-stage stochastic optimization	Optimal resource allocation during UAV movement	Higher computational complexity
[33]	Constraint-based joint optimization of bandwidth allocation, task offloading time allocation, and UAV trajectory	Joint optimization using block coordinate descent (BCD) algorithm	Lack of in-depth comparison with SOTA solutions

Furthermore, onboard AI edge computing allows for a more efficient use of resources. With the computing power on board, data can be processed locally without the need for constant data transmission to remote servers. This saves time and energy, and allows for a more efficient use of the UAV's limited resources, such as its battery life.

Based on the previous review of the existing literature, there is a growing trend in adopting EMC-based UAV systems, given their flexibility, resilience, and swift deployment. However, new challenges emerge with the deployment of such systems that can be handled only by advanced AI-based solutions, including reinforcement and deep learning frameworks. In fact, the solutions reviewed in the previous section are founded on well-established algorithms that have shown promising results in other engineering and science fields, including the optimal policy for emergency situations, data fusion, and information retrieval [34–36].

UAVs are becoming increasingly prevalent across multiple industries due to their flexibility and resilience. MEC-enabled UAVs are capable of providing computing and communication services at the network edges, even for ground-based units in areas with limited network coverage. This is particularly important in the field of remote sensing, where data collected from sensors on board the UAV need to be processed and analyzed in real time to support timely decision making. The ability of MEC-equipped UAVs to handle computing tasks and communication services at the network edges can significantly improve the speed and accuracy of remote sensing data collection processes.

Adopting edge computing for the onboard processing on UAVs is a challenging problem, yet beneficial from several perspectives. Embedding computation-intensive applications on the UAV edge device requires sufficient energy, storage, and computation resources to manage the demanding requirements of AI tasks. However, with the evolution of edge devices' capabilities, most of these challenges are overcome to a large extent, which makes edge computing in UAVs possible.

2. Materials and Methods

2.1. The AERO System

2.1.1. Why AI-Enabled Edge Computing for UAVs?

AI-enabled edge computing for UAVs can provide several benefits, including a low latency, increased efficiency, improved reliability, and enhanced privacy, as described below.

1. Low Latency: with advances in graphics processing units (GPUs) for edge devices (e.g., NVIDIA's Jetson boards), edge computing enabled the real-time processing of AI tasks, such as object detection, recognition, and tracking. This was not possible a couple of years ago. Consequently, edge computing promotes the real-time processing of data on board by allowing the drone to make quick local decisions about detected objects (e.g., the detection of a person to rescue) before sending the information to the cloud, thus saving useless communication with the server.

2. Increased efficiency: this approach also improves efficiency by decreasing communication overhead, saving bandwidth usage, and reducing the latency and load of the cloud servers. In fact, in the case of the cloud computing approach, the drone has to stream images at a high frequency and offload AI computation to the cloud. This is greedy in terms of the bandwidth and communication overhead, induces more communication latencies, and lacks scalability and computation cost, as the cloud cannot tolerate massive video traffic with real-time data processing. Edge computing helps to reduce the amount of data to be transmitted over a network and sent to the server.

3. Improved Reliability: computation on edge also improves the reliability of AI-based UAV applications. First, the drone data collection process will be less affected by the possible loss of communication due to the increased autonomy of the drone by locally processing collected data. In case of total communication loss, the data of detected objects are still saved locally and transferred to the cloud when the communication is back or offline in the worst scenario. In addition, edge computing makes the processing of AI tasks distributed among the UAVs and not centralized in the cloud, which can be vulnerable to outages or other disruptions. There are two resulting benefits: (1) it avoids the single point of failure, and (2) it increases the system's scalability as computing is fully distributed.

4. Better privacy: the local processing of collected images and detected objects helps to enhance privacy preserving by reducing the amount of data that are transmitted and stored in centralized remote servers. Adopting strong encryption on individual detected object frames is more efficient than encrypting the whole video stream. In

addition, collected object images transmitted to the cloud will remain private and secure against unauthorized access, as they no longer require being processed as plain data.

2.1.2. AERO System Architecture

In this section, we present the system architecture of AERO, shown in Figure 1.

The objective of the AERO system is to provide an ecosystem for using an edge-device on UAVs to execute complex deep learning algorithms to help automate computer vision applications, including object detection and tracking, on board the UAVs.

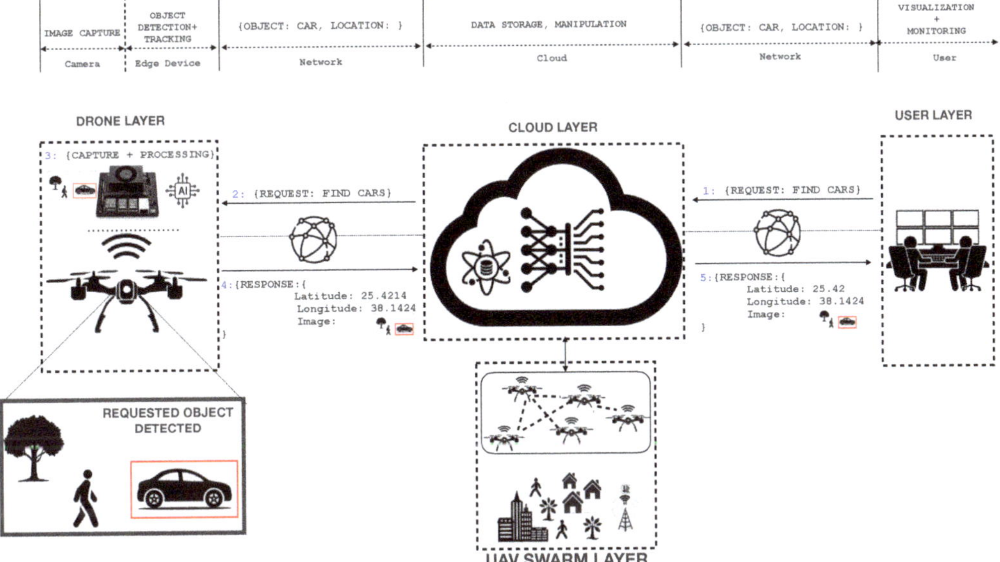

Figure 1. AERO system architecture.

The AERO system is composed of four layers:

- The Drone Layer: this represents the one UAV subsystem that is equipped with onboard processing and storage capabilities to perform AI tasks such as image and video analysis in real time. Edge computing is used to locally process collected raw data rather than sending them to a remote server as a video stream. In the UAV AERO, the edge device is a GPU-based embedded system (e.g., NVIDIA Jetson Xavier board) directly attached to the drone's camera through a proper channel (USB port, Ethernet (RTSP), or serial). The drone uses its network interfaces (e.g., 4G/5G cellular networks or WiFi) to communicate with and transmit detected objects' images to the cloud.

- The Swarm Layer: this layer consists of a cluster of UAVs equipped with camera sensors and AI-edge devices that coordinate together to perform a cooperative mission; for instance, distribute a search for lost people in a large area. In Figure 1, the UAVs swarm communicates with the cloud, which orchestrates their mission, rather than adopting ad hoc communication among the drones. The reasons are as follows:

    - Increased Reliability. the communication of UAVs with the cloud through cellular networks provides a more robust and stable connectivity compared to ad hoc swarms, which may be subject to interference and non-guaranteed message exchange, particularly in large-scale deployment. In critical applications such as search and rescue, it is essential to maintain reliable communication to ensure better coordination between drones through the cloud server.

- Interference: in ad hoc swarm communication, the drones have to contend for channel access (e.g., CSMA/CA). This will lead to interference and collision, which requires message retransmissions. This results in poor communication efficiency and increased delays. Other approaches involve the use of time synchronization (e.g., time division multiple access (TDMA))), but these techniques are challenging as they need to maintain synchronization among the UAVs. Clock drift, latency, interference, and the dynamic nature of the UAVs can all impact the accuracy of the transmissions, leading to disruptions in the synchrony of the TDMA system.
  - Global Knowledge: with all swarm UAVs communicating with the cloud, the latter maintains up-to-date information about all UAVs, including their positions, their states, and the list of detected objects. The cloud can make informed decisions in real time and an adjustment of the mission plan or resource allocations. For example, if a UAV experiences low battery levels, the cloud will be better positioned to reassign its tasks to other drones based on optimized criteria. The cloud can also optimize the task allocation among all drones and give its global knowledge to ensure that mission execution is completed effectively.

  Overall, these planes work together to support the operation and management of a fleet of drones. The data plane handles the collection and processing of data, the user plane enables human users to interact with the system, and the drone plane manages the operation of the drones themselves.

- The Cloud Layer: as the UAV edge device performs AI computation-intensive tasks, the cloud system does not require having sextensive/advanced computing resources (GPU-based cloud systems are not required), which reduces the deployment cost considerably, as GPU-based cloud systems tend to be more expensive than CPU-based cloud systems. The cloud is responsible for data storage, manipulation, and visualization. The cloud is organized into three planes.

  - UAV Plane: the UAV plane is primarily responsible for managing the operation of a fleet of drones, including overseeing and coordinating the drones' activities, managing the data collected by the drones, and performing mission planning to ensure compliance and safety. The fleet management system (FMS) plays a critical role in controlling and monitoring drones, scheduling their tasks and missions, and ensuring their compliance with airspace regulations. These benefits include improved efficiency, data management, and safety.
  - Data Plane: the data plane is responsible for handling the large amounts of data generated by the drones' sensors and onboard equipment. During operation, the drones collect a large amount of data and send them to the cloud for storage and processing using advanced data analytics frameworks, and visualize dashboards to end-users for quick analysis and decision making based on the data collected by the drones. The data plane also ensures the persistence and availability of the data when needed by the end users through replication, caching, and load balancing.
  - User Plane: the user plane in the AERO system is responsible for interacting with users, including mission planning, monitoring, and control. It allows users to access the system through various interfaces and applications, such as a web-based dashboard, mobile app, or API. Through the user plane, users can create and manage drone missions, view real-time drone data, and receive alerts and notifications. Users can monitor the status and performance of the operating drones in real time, providing important information such as flight paths, battery levels, and sensor data. This feature is essential in situations such as emergency response scenarios and surveillance operations. The user plane is a critical component of the AERO system, enabling efficient and effective drone operations by providing a user-friendly interface for mission management and real-time monitoring.

- The End-User Layer: the end-user layer in the AERO system enables end-users to access the system through the Internet using web service APIs. The end-users use interactive dashboards to monitor the status of their drones in real time, send commands, and receive real-time video streams that have been processed by deep learning applications located either at the edge or on the cloud. The end-user layer interacts with the cloud layer through its user plane, which provides access to authorized cloud resources and allows them to interact, monitor, and control drones for operation. The end-users can be of different types depending on their role.

- Authority: responsible for authorizing drone operations, managing the drone fleet, and ensuring compliance with regulations.
- Operator: responsible for managing and operating drone fleets, executing drone missions, and ensuring safety.
- User: requests drone operations for various purposes, such as aerial photography, surveying, or inspection.

2.1.3. AI-Enabled UAV

This section describes the UAV platform that we used to test the AERO system in practice. Figure 2 depicts our custom-built battery-powered hexacopter platform and highlights its main components. The hexacopter specifications are detailed in Table 2.

Figure 2. Top view of the custom hexacopter.

Table 2. Hexacopter specifications.

Number of motors	6
Motor type	T-Motor MN3508 KV380
Propeller size	15″
Wheelbase	850 mm
Battery	6200 mAh
Maximum takeoff weight	7 kg
Maximum flight time	15 min
Camera	ZR10 (30× zoom, 2K resolution)
Edge device	NVIDIA Xavier NX
Limited communication range	15 Km (2.4 Ghz)
Extended communication range	using 4/5G networks

The selected hexacopter platform was equipped with custom onboard electronics to enable edge computing as well as continuous cloud connectivity. The hardware architecture of the custom onboard electronics and communication systems are shown in Figure 3, and are described as follows.

- Gimbal–camera System: this is a camera–gimbal system which consists of the main vision sensor that is stabilized by a 3-axis gimbal. This system is called a SIYI ZR10 gimbal–camera system and has a 30× hybrid zoom (10× optical and 3× digital) and a 2K camera. The gimbal–camera system has its own microprocessor, which has an RTSP (real time streaming protocol) server that sends real-time image streams to clients (edge and communication devices) using Ethernet connections. In addition, the camera orientation is stabilized and controlled by a 3-axis gimbal to control the visual region of interest during flight.

- NVIDIA Jetson Xavier NX: this is the main computation board (edge device) and has adequate GPU power to perform real-time object detection and advanced autonomous surveillance mission planning. It is connected to the camera–gimbal system, via an Ethernet switch, to receive the real-time image stream and send camera–gimbal commands to control the camera orientation and zoom level. The Xavier NX runs our custom software, which performs real-time object detection and localization, which is described in Section 2.2. It also has a connected 4G module to enable extended communication with the cloud server to send information about the detected objects and receive surveillance mission requests.
- 4/5G communication: a 4/5G communication module is connected to the Xavier NX module to enable communication with the cloud server for an extended range. The communicated information includes the image frames with metadata (e.g., detected objects and their coordinates) sent from the edge device to the cloud server, and mission requests from the cloud server to the edge device.
- Ethernet switch: this hardware module is used to allowfor transmitting the camera image stream to the onboard computer (Jetson Xavier NX) for image processing, as well as the air unit transceiver, which communicates with a ground remote controller for visualization.
- Pixhawk Orange Cube flight controller: this is the autopilot hardware, which runs the well-known open-source PX4 autopilot firmware [37]. The autopilot stabilizes the drone's position and executes planned missions that are sent by the onboard computer.
- Air unit transceiver: this module exchanges image streams and UAV telemetry with a ground remote controller using a 2.4 GHz link.
- Remote controller: the ground remote controller is used by the UAV backup pilot to control the drone maneuvers, if needed, and have real-time visual feedback of the onboard camera stream.

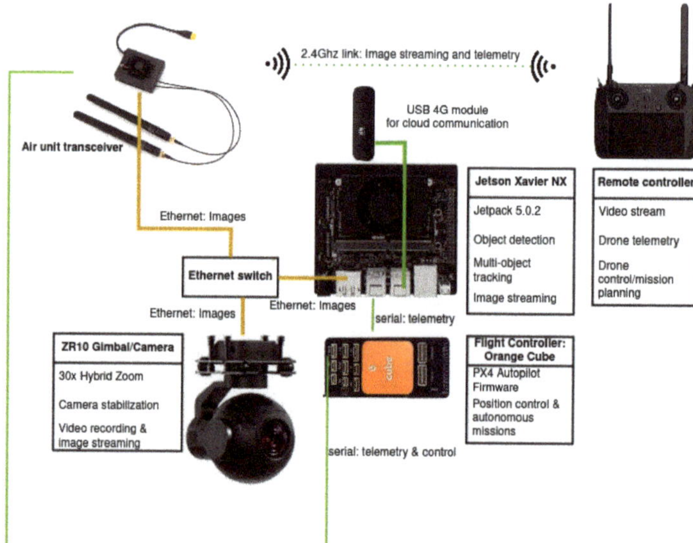

Figure 3. UAV onboard electronics.

2.2. The AERO AI Module

In this section, we present the AERO brain system that leverages YOLOv7 [38] for object detection, DeepSort [39] for object tracking, and TensorRT (TRT) [40] acceleration to ensure the real-time execution of the model on edge devices. The novelty of our approach is the design of a multi-stage deep learning model that allows for making object inferences over several consecutive frames to optimize the detection performance in two main aspects:

- Accuracy: typical object detection and tracking models perform inference on one static image from the video frame, which usually leads to high misclassification ratios. We dramatically improved the accuracy by considering several consecutive frames and using a voting approach to maximize the object recognition accuracy.
- Real Time: a multi-stage model uses several deep learning models in sequence. The deployment of a multi-stage model makes real-time inference more challenging, particularly on embedded edge devices, considering their lower capabilities. We overcame this issue by using TensorRT acceleration on NVIDIA's Jetson AXG to maintain a high frame rate for the AERO multi-stage inference model.

2.2.1. AERO Model Architecture

Figure 4 shows the main steps of the processing performed by the AERO AI module on edge. The AERO model is composed of three modules, namely the Detection Module, the Model Acceleration Module, and the Tracking Module, described as follows.

Detection Module

The detection module is based on YOLOv7, which is the latest version of the widely used YOLO family of single-stage object detectors. It established the state of the art both in terms of accuracy and speed, outperforming competitor models by a large margin. For comparison, we also tested YOLOv4 [41], which is still one of the most popular object detection models.

The DeepSORT tracker is an extension of the simple online and real-time tracking (SORT) algorithm [42], which is an efficient algorithm used for real-time object tracking. The key innovation of DeepSORT is the incorporation of a pre-trained deep association metric that utilizes object appearance information to improve the tracking performance. The deep association metric in DeepSORT uses a pre-trained deep neural network to encode the appearance information of objects. By comparing the features extracted from the neural network, DeepSORT is able to estimate the likelihood of two objects being the same. This allows DeepSORT to handle challenging scenarios such as occlusion, appearance changes, and the temporary disappearance of objects. Overall, DeepSORT provides a robust and accurate solution for tracking multiple objects in real time. Its ability to incorporate appearance information allows it to handle various challenging scenarios, making it an ideal solution for applications such as surveillance, robotics, and autonomous vehicles. For these reasons, and for its popularity in the literature, we opted for this particular tracker, although any other multi-object tracker could be used in our system. To integrate DeepSORT with the YOLO object detector and the other components of our system, we modified the implementation of the track class in a similar way to the one described in [43].

The object detection and tracking system processes each new frame by first applying YOLOv7 on the entire frame to obtain bounding boxes and confidence scores for all detected objects. These bounding boxes are then input to DeepSort, the multi-object tracker, which produces pairs of matched tracks and detections as well as lists of unmatched tracks and detections. For each track, the system checks whether it should be discarded, further processed, or sent to the server.

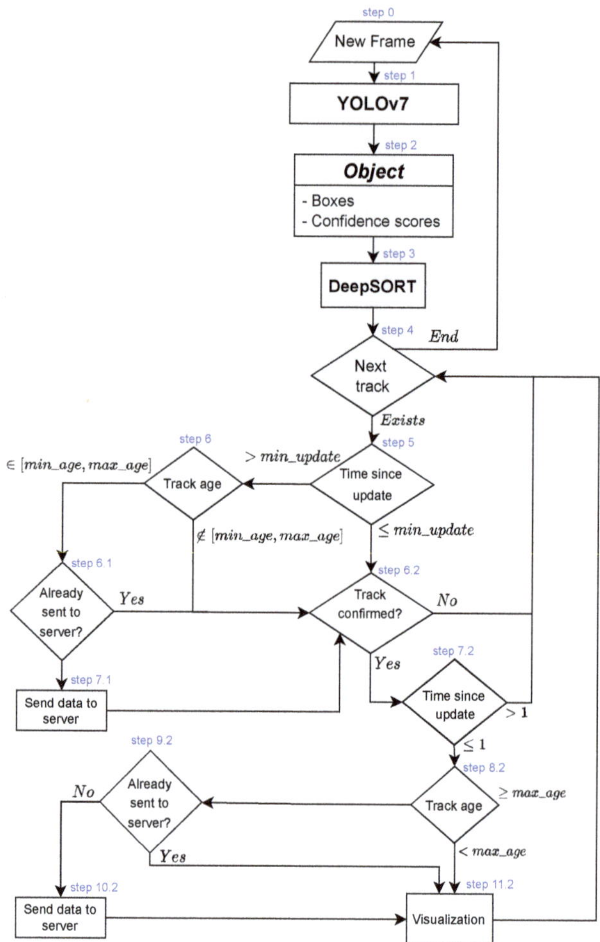

Figure 4. The AERO model architecture of the AI module.

First, the system checks if the track has not been matched with a detected bounding box for more than a predefined number of consecutive frames (default value of 10). If so, the system assumes that the object is no longer in the camera's field of view. Next, the system checks if the track's age (number of frames in which the same object has been detected) is within a predefined interval (default value of [2, 40]). A low value indicates that the track is unreliable, whereas a high value means that the object information has already been sent to the server. The default values of the minimum number of consecutive frames and the track's age interval were fixed empirically after a series of preliminary tests.

In all cases, the system checks if the current track has been confirmed by being observed in the required minimum number of consecutive frames (default value of 3) and has not been deleted due to missed detections. If the track is confirmed or has been matched with detected bounding boxes in the current or previous frames, the system checks its tracking age. If the age is equal to or greater than the maximum allowed age, the system sends its information to the server if it has not yet been sent. Finally, the system can optionally visualize the object's bounding box and information using the current attributes of the track instance.

If the track is not confirmed or has not been matched with bounding boxes for at least two consecutive frames, the system skips it and moves on to the next track. By following this process, the object detection and tracking system can accurately detect and track objects in real time while minimizing false detections and conserving computational resources.

Model Acceleration Module

While deep learning models can provide highly accurate results, they require significant computational and storage resources to train and run, even for YOLOv7, which is the fastest object detector to date. This makes deploying deep learning models on edge devices such as Jetson boards a challenging task as these devices often have limited resources in terms of memory and processing power.

To address this challenge, we leveraged the use of the TensorRT acceleration framework. TensorRT is a high-performance inference engine developed by NVIDIA that allows developers to optimize deep learning models for deployment on a range of NVIDIA platforms, including Jetson edge devices. It can optimize models by reducing the precision of model parameters and minimizing the memory required to store them, allowing the model to run more efficiently on edge devices with limited resources. TensorRT can also optimize models by using dynamic tensor memory allocation, which allocates memory dynamically during inference, reducing the overall memory usage.

The TensorRT optimization framework also optimizes models by fusing layers, which combines multiple layers in a neural network into a single layer to speed up model inference. This is particularly important for applications that require real-time processing on edge devices, where latency is critical, such as real-time surveillance applications. In a previous study [44], we have shown that TensorRT optimization provides the fastest execution on a wide variety of cloud and edge devices. This demonstrates the effectiveness of TensorRT in optimizing deep learning models for edge devices, achieving faster inference times and a lower latency.

Target Localization Module

In [2], we proposed a methodology for object detection and location estimation based on established photogrammetry concepts and metadata extracted from drone images, including EXIF and XMP data. This approach allows for accurately estimating the GPS location of detected objects within each frame. The use of metadata, such as the drone's altitude and GPS location, image size, and calibrated focal length, provides a demonstrably sound basis for determining the location of objects in the images.

To account for potential errors or uncertainty in the distance estimation, the algorithm also incorporates a correction factor based on the ratio between the drone's altitude and the estimated average height of the objects using the formula:

$$\begin{cases} D_x^c = \left(1 - \frac{h}{H}\right)D_x \\ D_y^c = \left(1 - \frac{h}{H}\right)D_y \end{cases} \tag{1}$$

where:

- D_x and D_y are the coordinates of the object's bounding box center before correction.
- D_x^c and D_y^c are the object's coordinates after correction.
- h is the estimated average object height.
- H is the drone altitude.

Additionally, the algorithm considers the yaw degree of the image to refine the location estimation of each object further. This approach allows for an accurate counting of objects even when there are overlaps between images, further demonstrating the scientific rigor of the methodology. This same methodology can be applied to the detected objects in the AERO system, although we did not include this target localization module in the experimental part of the current study.

3. Results

3.1. Experimental Setup

For the experimental evaluation, we tested two different object detection models (YOLOv4 and YOLOv7), two different implementations (PyTorch and TensorRT), three different video resolutions (1920 × 1080 for 2 videos, 2688 × 1512, and 3840 × 2160, see Figures 5 and 6), and three different devices (RTX8000, Jetson Xavier AGX, and Jetson Xavier NX). The videos' length ranges from 0.5 mn to 5.9 mn. Videos 1 and 3 were used for the detection of six classes of objects (car, person, bicycle, bus, monocycle, and truck), whereas videos 2 and 4 were used for the detection of a single class (car). On top of each bounding box, information is displayed about the detection class, the tracking ID, the number of frames in which the same object has been observed, and the object color. For videos 1 and 3, the number of objects of each class is also displayed on the top left corner. The outputs of videos 2 and 4 are available on this link: shorturl.at/nrzOY (accessed on 30 March 2023). As for videos 1 and 3, the original footage was provided by a third party that did not agree to disclose them.

Table 3 presents the conducted experiments that are analyzed below. Due to software environment limitations and compatibility issues, some frameworks did not work on some devices. We were able to run all configurations on Jetson Xavier NX (Jetson pack 5, TensortRT 8), whereas YOLOv7 did not work on Jetson Xavier AGX (Jetson pack 4.5, TensorRT 7) and the TRT versions of YOLOv4 and YOLOv7 did not work on the RTX8000 GPU (CUDA version 10.0).

Video **1** (3840×2160)

Video **2** (1920×1080)

Video **3** (2688×1512)

Video **4** (1920×1080)

Figure 5. Sample frames from the output of the four videos used for the evaluation of the AI-enabled system, with different resolutions.

We chose the YOLOv7 object detector because it was the state-of-the-art object detector in terms of accuracy and speed at the time of this study. As for YOLOv4, we tested it for comparison, seeing that it is still one of the most popular object detectors (YOLOv5 and YOLOv6 are not as popular in the literature). For our case study, we could not use the pre-trained models of YOLOv4 and YOLOv7 because they were mainly trained on ground-level images (COCO dataset or OpenImages dataset), and we are dealing with aerial images. Consequently, for training YOLOv7, we used the VisDrone dataset [45], which we filtered to keep only one class of vehicles (cars), and, for YOLOv4, we trained a model on a private dataset containing 940 UAV images showing six classes (car, person, bicycle, bus, monocycle, and truck) with a total of 33,088 instances. These images were captured in the Jeddah region in Saudi Arabia, in daylight and sunny conditions, and were manually labeled. Table 4 summarizes the main hyperparameters and results of the training of the YOLOv4 and YOLOv7 object detectors. Since we built our custom dataset gradually, we show the results of the training for several sizes of the dataset. We observe that there is a stagnation in terms of the mAP (mean average precision) when moving from 545 to 821 training images. YOLOv7 shows notably better results in terms of mAP but they are not directly comparable to YOLOv4's results since the number of classes is different.

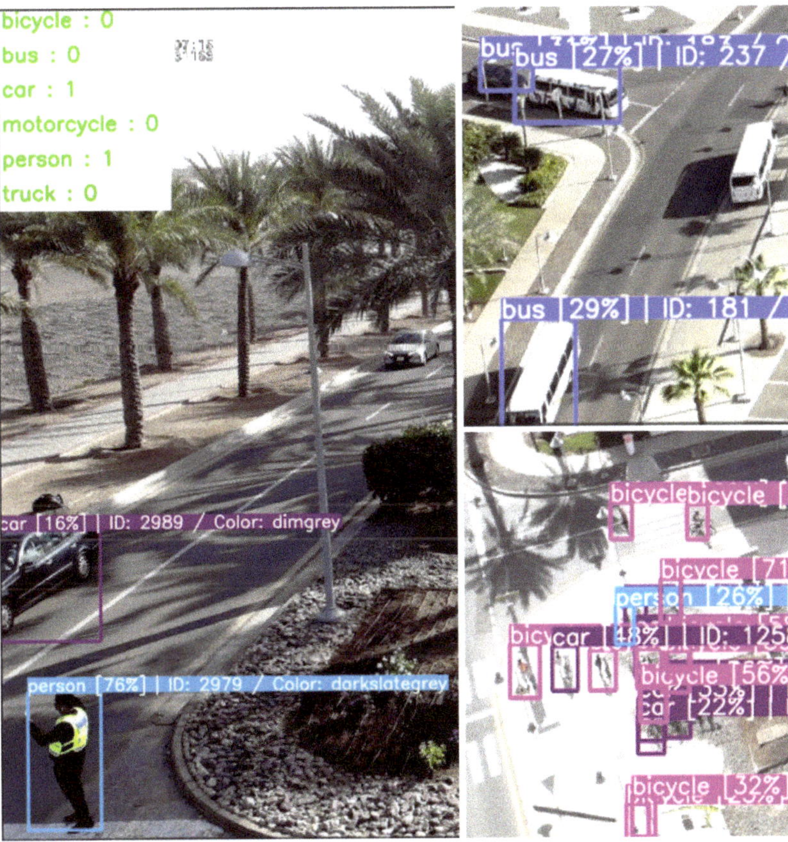

Figure 6. Close view of a sample of the output videos showing various classes, and some false negative and false positive detections.

Table 3. Conducted experiments on different devices, object detection models, frameworks, and input video resolutions.

Model	Implementation	Resolution	RTX8000	Jetson Xavier AGX	Jetson Xavier NX
YOLOv4	TRT	2688×1512		✓	✓
		3840×2160		✓	✓
YOLOv7	TRT	1920×1080			✓
	PyTorch	1920×1080	✓		✓

Table 4. Hyperparameters and results of the training of the YOLOv4 object detector for several configurations.

YOLO Version	Dataset	Nb Classes	Training Images	Validation Images	Input Size	Best mAP
v4	Custom	6	311	35	608×608	41.9%
v4	Custom	6	545	60	768×768	57.0%
v4	Custom	6	821	91	768×768	57.0%
v7	VisDrone	1	4935	617	640×640	91.3%

3.2. Performance Evaluation

We first analyzed the inference speed for each device and detection model using a series of box plots (Figures 7–9). Box plots are a useful way to visualize the distribution of data and compare data across multiple variables, and can provide insights into the central tendency, variability, and skewness of the data. The grey line inside each box represents the median value of the data. Half of the data points fall above this line and half fall below. The box itself represents the interquartile range (IQR), which contains the middle 50% of the data. The bottom of the box represents the first quartile (Q1), or the value at which 25% of the data fall below. The top of the box represents the third quartile (Q3), or the value at which 75% of the data fall below. The whiskers extend from the box to show the range of the data, excluding any outliers, while the individual blue points represent a 1D scatter plot of the data.

Figure 7 depicts the box plot of the inference speed in frames per second (FPS) for each device, detection model, and input video resolution. We observe that the TensorRT optimization of the YOLOv4 model provides the fastest inference speed, even on higher-resolution input videos, whereas, for YOLOv7, the TRT optimization provides no gain in speed. In contrast, the average inference speed deteriorates from 7.2 FPS (for the PyTorch implementation) to 2.8 FPS (for TRT). This is likely due to the fact that the new features introduced in YOLOv7 are not yet adequately optimized in the latest versions of TensorRT.

Figure 8 shows the box plot of the inference speed for each device and detection model in the case where the detected objects are sent to the cloud and in the case where the connection to the cloud is disabled. In all cases, the connection to the cloud significantly slows down the inference speed of the whole system. The average speeds drops from 12.3 FPS when no data are sent to the cloud to 5.0 FPS when sending data to the cloud. This highlights the importance of choosing a high-quality network and optimizing the edge–cloud communication.

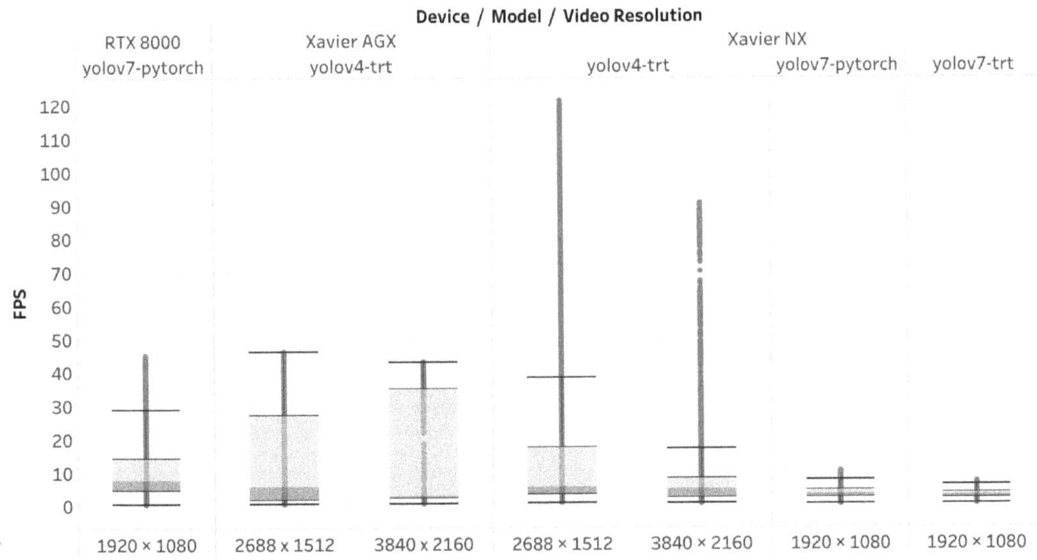

Figure 7. Inference speed per device, detection model, and video resolution.

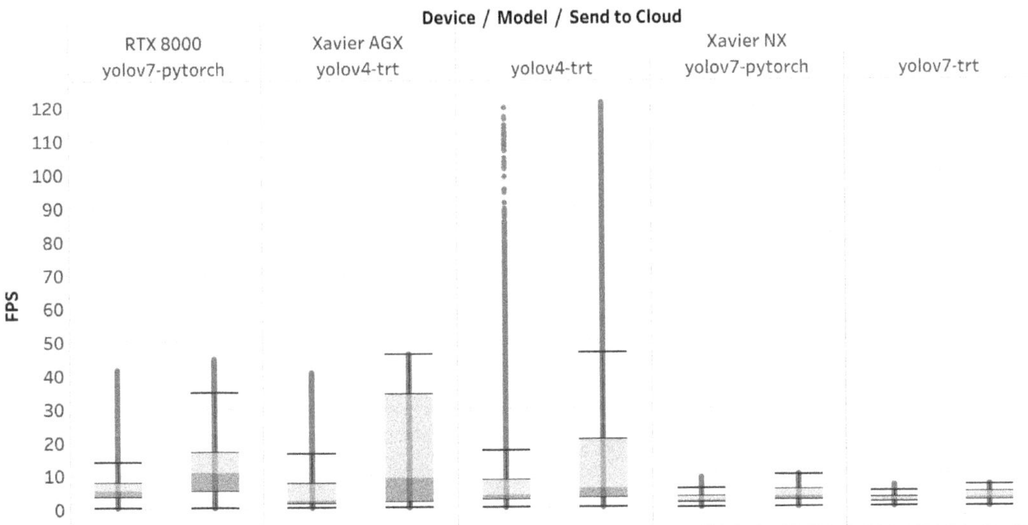

Figure 8. Inference speed per device, detection model, and connection to the cloud.

Figure 9 shows the box plot of the inference speed for each device and detection model in the case where the DeepSORT tracker is included or excluded. On all devices, and for all object detection models, the use of the tracker markedly decelerates the system. The average inference speed declines from 19.6 FPS (without tracker) to 5.0 FPS (with tracker). Nevertheless, the use of the tracker is necessary to correctly count the number of objects and send each object's information to the server only once. We should, however, investigate faster multi-object trackers to enhance the overall system speed.

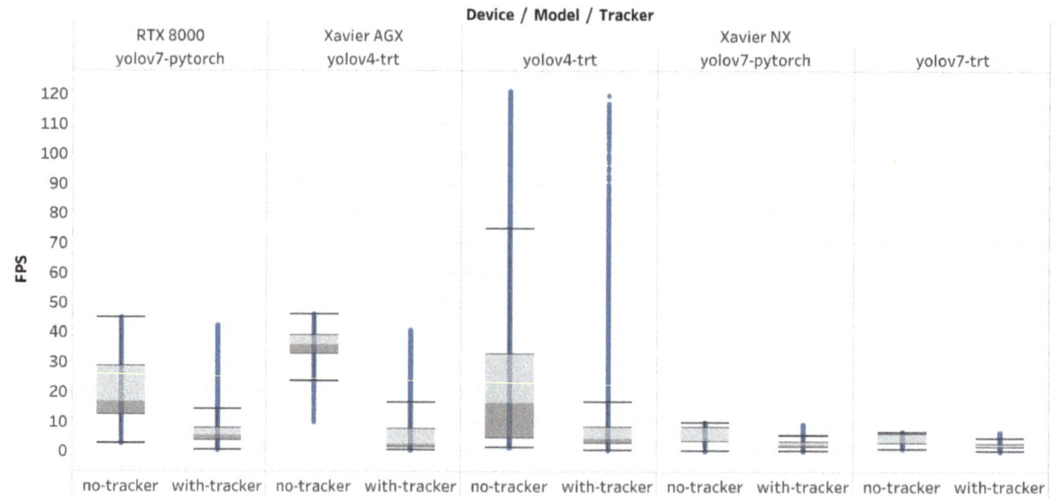

Figure 9. Inference speed per device, detection model, and use of tracker.

To analyze the influence of each component of the AI system and control for variability due to different devices and video resolutions, we generated a set of scatter plots to measure the inference speed on the Jetson Xavier NX device with an input video resolution of 1920×1080. Figure 10 illustrates the scatter plot of the inference speed per number of detected objects in each frame using both PyTorch and TRT versions of the YOLOv7 object detection model. As previously noted (about Figure 7), the PyTorch implementation achieved higher inference speeds compared to the TRT implementation. Figure 10 appears as a superimposition of three plots, which we will distinguish in subsequent figures.

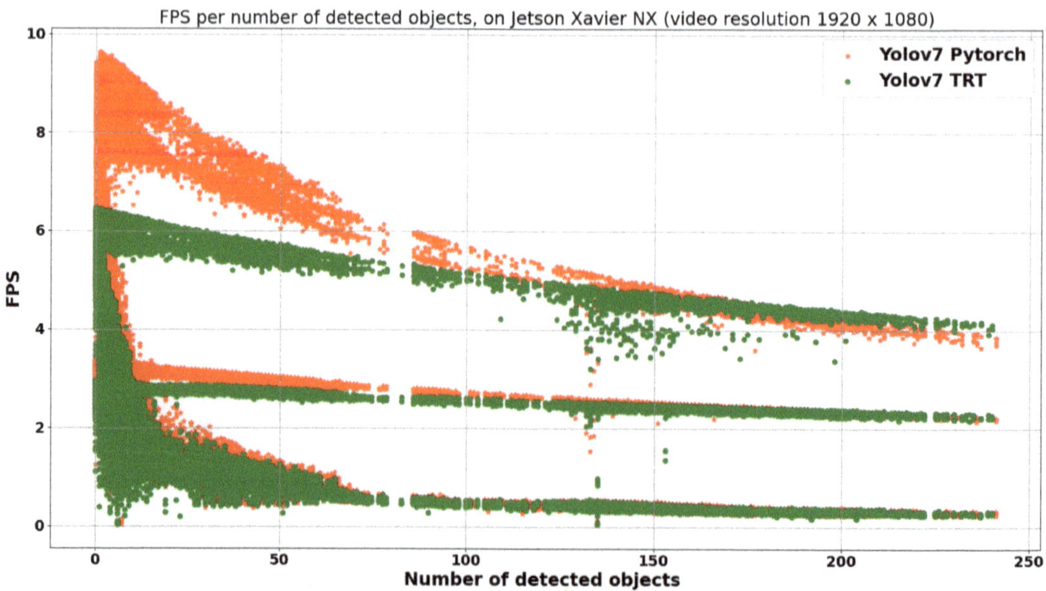

Figure 10. Inference speed (in FPS) per number of detected objects, on Jetson Xavier NX, with an input video resolution of 1920×1080, using the PyTorch and TRT version of the YOLOv7 object detection model.

Figure 11 presents the scatter plot of the inference speed per number of detected objects, on Jetson Xavier NX, with an input video resolution of 1920 × 1080, using the TRT version of the YOLOv7 object detection model, when including or excluding the tracker. As already noted in Figure 9, the use of the tracker significantly slows down the system performance. The blue dots in Figure 11 represent the measures that included the tracker, and correspond to the lower part of the plot in Figure 10. The magenta dots, corresponding to the inclusion of the tracker in the AI system, still appear as the superimposition of two plots. They will be distinguished in the next figure.

Figure 12 shows a similar scatter plot but with no tracker when including or excluding the local saving of the output video. It demonstrates that storing the resulting output video on the edge's disk consumes a significant amount of time and markedly slows down the overall inference speed. The system speed decreases from 12.9 FPS to 6.8 FPS on average over all devices and configurations. For the configuration shown in Figure 12 (Jetson Xavier NX, YOLOv7 TRT, no tracker, 1920 × 1080 video resolution), the average inference speed drops from 5.8 FPS to 2.7 FPS when saving the output video. Consequently, this local storage should not be used unless it is absolutely required for the application.

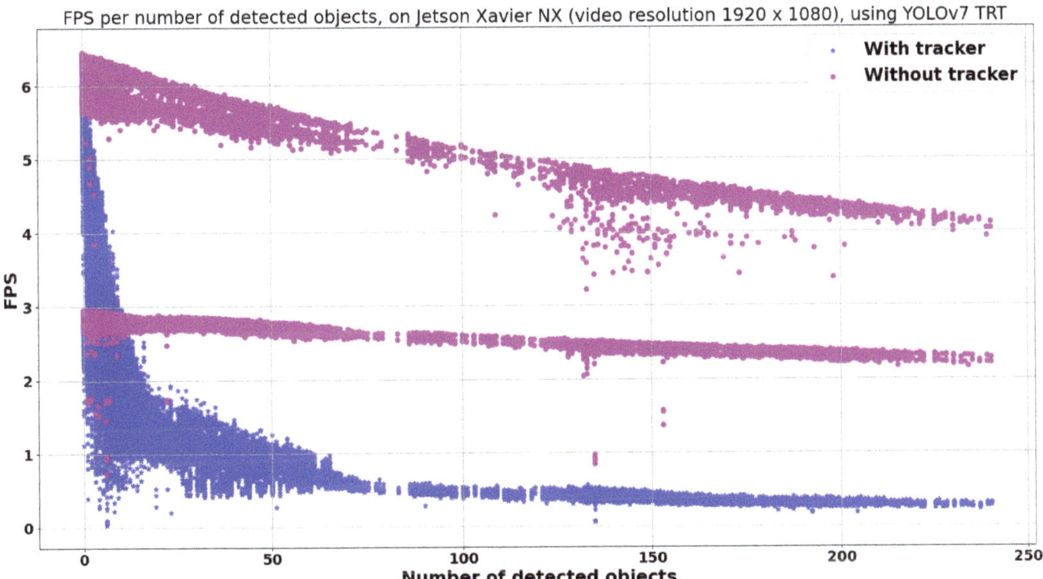

Figure 11. Inference speed (in FPS) per number of detected objects, on Jetson Xavier NX, with an input video resolution of 1920 × 1080, using the TRT version of the YOLOv7 object detection model, when including or excluding the tracker.

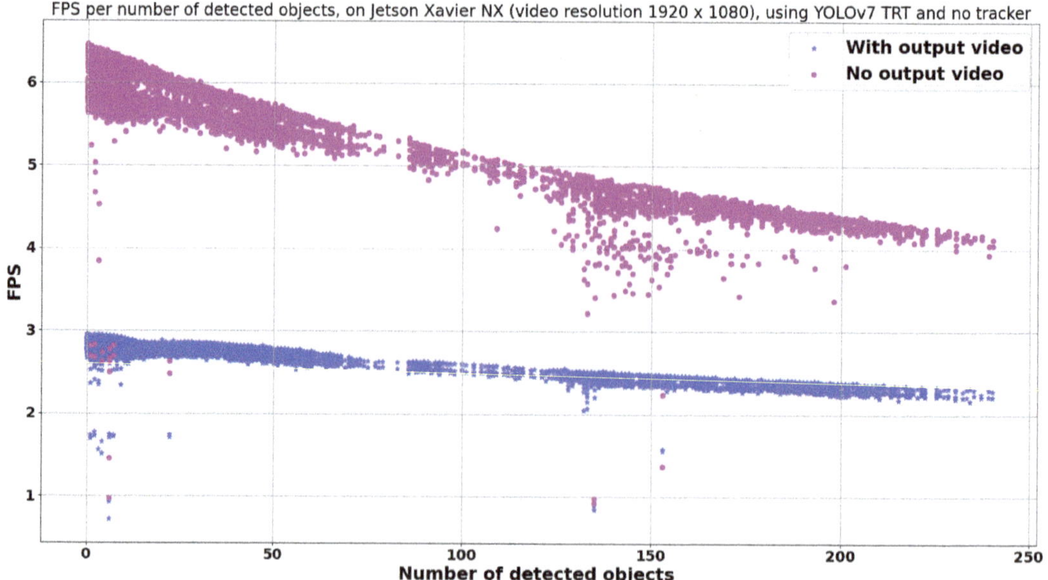

Figure 12. Inference speed (in FPS) per number of detected objects, on Jetson Xavier NX, with an input video resolution of 1920 × 1080, using the TRT version of the YOLOv7 object detection model, with no tracker, when including or excluding the saving of the video output.

4. Discussion

From Figures 7–12, we conclude that the inference speed of an AI system for object detection can be affected by various factors, including the device used, the detection model, the input video resolution, the use of cloud connectivity, and the inclusion of a tracker or local saving of output videos. The TensorRT optimization of the YOLOv4 model provides the fastest inference speed even on higher-resolution input videos. However, for YOLOv7, the TRT optimization did not provide any gain in speed due to an inadequate optimization of new features in the TensorRT version used. Sending data to the cloud significantly slows down the inference speed, highlighting the importance of choosing a high-quality network and optimizing edge–cloud communication. The use of a multi-object tracker is necessary to correctly count the number of objects and send each object's information to the server only once, but it markedly decelerates the system. Finally, avoiding the local saving of the output video can also help to improve the system's inference speed. Therefore, the best configuration for an AI system for object detection depends on the specific application requirements and hardware constraints.

To assess the accuracy of the object detector, the influence of the TRT optimization, and the multi-object tracker, we selected two test videos (see Figure 5):

- video 1: showing six classes (car, person, bicycle, bus, monocycle, and truck), with an average of six objects per frame, an input resolution of 3840 × 2160, a length of 50 s, and an FPS of 30.
- video 4: showing a single class of cars (with an average of six cars per frame), with an input resolution of 1920 × 1080, a length of 4 mn and 25 s, and an FPS of 24.

We manually counted the following metrics on still images extracted from the video every 20 frames (75 frames for video 1 and 319 frames for video 4):

- FP: number of false positives (objects incorrectly detected) generated by the object detection model.
- FN: number of false negatives (non-detected objects) generated by the object detection model.

- Precision: $Precision = \frac{TP}{TP+FP}$, where TP is the number of true positives (correctly detected objects).
- Recall: $Recall = \frac{TP}{TP+FN}$
- F1 score: $F1score = \frac{2 \times Precision \times Recall}{Precision + Recall}$
- Identity switches: number of switches between the IDs assigned by the tracker. This happens when the tracker conflates two objects that are too close.
- Identity changes: number of changes in the IDs assigned by the tracker to the same object. This happens when the tracker misinterprets a single moving object for two objects.

Table 5 summarizes the obtained results for these metrics when using the TRT implementations of the YOLOv4 object detector on video 1. The number of FNs is relatively low compared to the number of FPs due to the fact that most vehicles have a relatively large size (compared to video 4). The number of identity switches and changes is also reduced compared to video 4 because the distance between objects is markedly larger, which makes the tracker's task easier. Figure 13 shows two close frames from the output of video 1 where several detection and tracking errors appear. We notice one false positive in frame 240 ('person'), and two other false positives in frame 260 ('person' and 'truck'), as well as a misclassification (truck classified as 'person'). Between the two frames, there are three identity changes (4→34, 5→4, and 30→19). Identity switches often happen when two objects move close to each other, while identity changes may happen when the object's speed is relatively high.

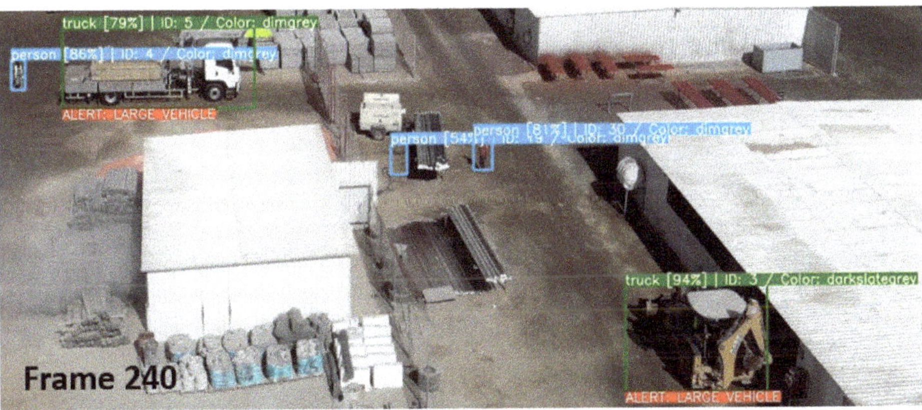

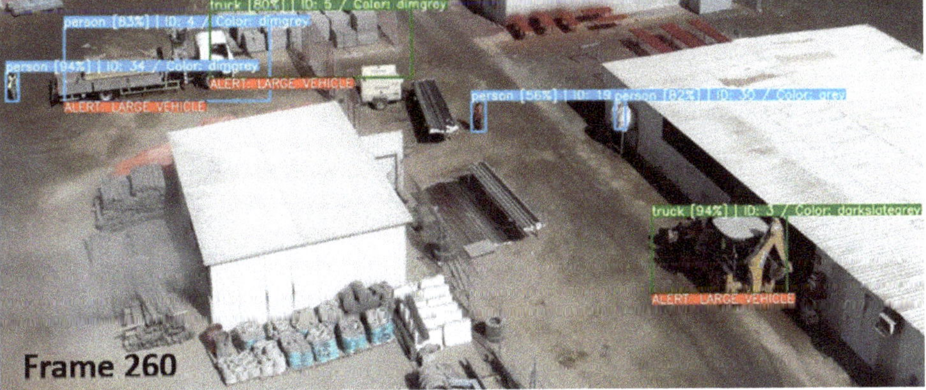

Figure 13. Sample frames from the output of video 1, showing frames number 240 and 260.

On the other hand, Table 6 summarizes the obtained results when using the PyTorch or the TRT implementations of the YOLOv7 object detector on video 4. The difference between the two implementations is relatively minor, except for identity switches, which double from 5 to 10 when converting the PyTorch model to TRT. This indicates a loss in precision in the converted detection model that impacts the tracker accuracy. Nevertheless, this figure remains relatively low (1.6% to 3.1% relative to the number of frames) considering the number of cars and the duration of the video. By contrast, the number of identity changes is much higher, both for the PyTorch and the TRT implementations. The tradeoff between the number of identity switches and identity changes can be modified by changing the tracker hyperparameters, but we consider the identity switches to be more critical because they entail the conflation of the information of different objects, whereas the identity changes only result in duplicate information sent to the server. On the other hand, we observe that the number of false negatives is much higher than the number of false positives. In fact, small or occluded objects are often missed by the object detector, as can be seen in Figure 5. Consequently, the precision is high (99.3% for both PyTorch and TRT implementations), whereas the recall is much lower (72.5% and 73.1% for PyTorch and TRT, respectively). This tradeoff can also be modified by changing the score threshold for the object detector.

Table 5. Number of false positive detections (FPs), false negative detections (FNs), precision, recall, F1 score, identity switches, and identity changes for the TRT implementation of the YOLOv4 object detection model on test video 1 (resolution of 3840×2160, length of 50 s, FPS of 30) captured by a drone, showing 6 classes (car, person, bicycle, bus, monocycle, and truck).

	FP	FN	Precision	Recall	F1 Score	Identity Switches	Identity Changes
YOLOv4 TRT	80	33	82.7%	92.1%	87.1%	16	26

Table 6. Number of false positive detections (FPs), false negative detections (FNs), precision, recall, F1 score, identity switches, and identity changes for the PyTorch and TRT implementation of the YOLOv7 object detection model on test video 4 (resolution of 1920×1080, length of 4 mn and 25 s, FPS of 24) captured by a drone, showing a single class of 'cars'.

	FP	FN	Precision	Recall	F1 Score	Identity Switches	Identity Changes
YOLOv7 PyTorch	20	1136	99.3%	72.5%	83.8%	5	184
YOLOv7 TRT	22	1099	99.3%	73.1%	84.2%	10	176

5. Conclusions

The commercial usage of UAVs is still largely limited by the lack of onboard AI on the edge, leading to manual data observation and offline processing after data collection. Alternatively, some approaches rely on the cloud computation offloading of AI applications, which can be unscalable and infeasible due to a limited connectivity and high latency of remote cloud servers. To address these issues, in this paper, we proposed a new approach that uses edge computing in drones to enable extensive AI task processing on board UAVs for remote sensing applications. The proposed system architecture involves a cloud–edge hybrid approach where the edge is responsible for processing AI tasks and the cloud is responsible for data storage, manipulation, and visualization.

To implement this architecture, coined AERO, we designed a UAV brain system with onboard AI capabilities that uses GPU-enabled edge devices. AERO is a novel multi-stage deep learning module that combines object detection (YOLOv4 and YOLOv7) and tracking (DeepSort) with TensorRT accelerators to capture objects of interest with a high accuracy and transmit data to the cloud in real time without redundancy. AERO processes the detected objects over multiple consecutive frames to maximize detection accuracy. The experiments show that the proposed approach is effective for utilizing UAVs equipped with onboard AI capabilities for remote sensing applications.

While the proposed system architecture and AERO module were designed to process visual data from UAVs, future work could explore the integration of other sensors, such as LiDAR or thermal cameras, to enhance the accuracy and efficiency of remote sensing applications. In addition, we plan to explore the integration of autonomous navigation capabilities to enable UAVs to navigate and collect data independently, without the need for manual control or intervention.

Another crucial aspect that needs to be considered in future works when designing drone systems with onboard AI capabilities is security, as highlighted in [46–48] . Drone communications are susceptible to cyber-attacks, making it crucial to protect the data being transmitted between the UAV and the cloud. Implementing security measures such as encryption and authentication protocols can protect the system from unauthorized access and data breaches. Additionally, implementing physical security measures such as tamper-proofing the onboard AI hardware can prevent malicious actors from tampering with the system. These security measures must be implemented at every stage of the system development and deployment to ensure the safety and privacy of data collected by UAVs. Nevertheless, these measures can affect the system's inference speed in a way that still has to be investigated.

Author Contributions: Conceptualization, A.K. and A.A.; methodology, A.K., A.A. and M.A.; software, Y.A. and A.A.; validation, A.K., A.A. and M.A.; formal analysis, A.K., A.A. and M.A.; investigation, A.K., A.A. and M.A.; resources, A.K.; data curation, A.A. and Y.A.; writing—original draft preparation, A.K., A.A., M.A. and L.G.; writing—review and editing, A.K., A.A., M.A. and L.G.; visualization, A.A. and M.A.; supervision, A.K., A.A. and M.A.; project administration, A.K.; funding acquisition, A.K. All authors have read and agreed to the published version of the manuscript.

Funding: The APC for this article was funded by Prince Sultan University.

Acknowledgments: We thank Prince Sultan University for facilitating the experiments on the university campus and financially supporting publication expenses.

Conflicts of Interest: The authors declare no conflict of interest. The funders had no role in the design of the study; in the collection, analyses, or interpretation of data; in the writing of the manuscript, or in the decision to publish the results

References

1. Zanelli, E.; Bödecke, H. *Global Drone Market Report 2022–2030;* Technical report; Drone Industry Insights: Hamburg, Germany, 2022.
2. Ammar, A.; Koubaa, A.; Benjdira, B. Deep-Learning-Based Automated Palm Tree Counting and Geolocation in Large Farms from Aerial Geotagged Images. *Agronomy* **2021**, *11*, 1458. [CrossRef]
3. Gallego, V.; Rossi, M.; Brunelli, D. Unmanned aerial gas leakage localization and mapping using microdrones. In Proceedings of the 2015 IEEE Sensors Applications Symposium (SAS), Zadar, Croatia, 13–15 April 2015 ; pp. 1–6. [CrossRef]
4. Abdelkader, M.; Shaqura, M.; Claudel, C.G.; Gueaieb, W. A UAV based system for real time flash flood monitoring in desert environments using Lagrangian microsensors. In Proceedings of the 2013 International Conference on Unmanned Aircraft Systems (ICUAS), Atlanta, GA, USA, 28–31 May 2013; pp. 25–34. [CrossRef]
5. Abdelkader, M.; Shaqura, M.; Ghommem, M.; Collier, N.; Calo, V.; Claudel, C. Optimal multi-agent path planning for fast inverse modeling in UAV-based flood sensing applications. In Proceedings of the 2014 International Conference on Unmanned Aircraft Systems (ICUAS), Orlando, FL, USA, 27–30 May 2014; pp. 64–71. [CrossRef]
6. Benjdira, B.; Bazi, Y.; Koubaa, A.; Ouni, K. Unsupervised Domain Adaptation Using Generative Adversarial Networks for Semantic Segmentation of Aerial Images. *Remote Sens.* **2019**, *11*, 1369. [CrossRef]
7. Benjdira, B.; Ammar, A.; Koubaa, A.; Ouni, K. Data-Efficient Domain Adaptation for Semantic Segmentation of Aerial Imagery Using Generative Adversarial Networks *Appl. Sci.* **2020**, *10*, 1092. [CrossRef]
8. Gulf News, Saudi Arabia: 131 People Went Missing In Desert Last Year. 2021. Available online: https://gulfnews.com/world/gulf/saudi/saudi-arabia-131-people-went-missing-in-desert-last-year-1.78403752 (accessed on 1 March 2023).
9. Kobaa, A. System and Method for Service Oriented Cloud Based Management of Internet-of-Drones. U.S. Patent US11473913B2, 15 October 2022.
10. Fortune Buisness Insights, Drone Surveillance Market. 2022. Available online: https://www.fortunebusinessinsights.com/industry-reports/drone-surveillance-market-100511 (accessed on 1 March 2023).

11. Ammar, A.; Koubaa, A.; Ahmed, M.; Saad, A.; Benjdira, B. Vehicle detection from aerial images using deep learning: A comparative study. *Electronics* **2021**, *10*, 820. [CrossRef]

12. Yeom, S.; Cho, I.J. Detection and tracking of moving pedestrians with a small unmanned aerial vehicle. *Appl. Sci.* **2019**, *9*, 3359. [CrossRef]

13. Ding, J.; Zhang, J.; Zhan, Z.; Tang, X.; Wang, X. A Precision Efficient Method for Collapsed Building Detection in Post-Earthquake UAV Images Based on the Improved NMS Algorithm and Faster R-CNN. *Remote Sens.* **2022**, *14*, 663. [CrossRef]

14. Koubaa, A.; Ammar, A.; Alahdab, M.; Kanhouch, A.; Azar, A.T. DeepBrain: Experimental Evaluation of Cloud-Based Computation Offloading and Edge Computing in the Internet-of-Drones for Deep Learning Applications. *Sensors* **2020**, *20*, 5240. [CrossRef] [PubMed]

15. Hossain, S.; Lee, D.j. Deep learning-based real-time multiple-object detection and tracking from aerial imagery via a flying robot with GPU-based embedded devices. *Sensors* **2019**, *19*, 3371. [CrossRef]

16. Queralta, J.P.; Raitoharju, J.; Gia, T.N.; Passalis, N.; Westerlund, T. Autosos: Towards multi-uav systems supporting maritime search and rescue with lightweight ai and edge computing. *arXiv* **2020**, arXiv:2005.03409.

17. Vasilopoulos, E.; Vosinakis, G.; Krommyda, M.; Karagiannidis, L.; Ouzounoglou, E.; Amditis, A. A Comparative Study of Autonomous Object Detection Algorithms in the Maritime Environment Using a UAV Platform. *Computation* **2022**, *10*, 42. [CrossRef]

18. Pajares, G. Overview and Current Status of Remote Sensing Applications Based on Unmanned Aerial Vehicles (UAVs). *Photogramm. Eng. Remote Sens.* **2015**, *81*, 281–330. [CrossRef]

19. Nex, F.; Remondino, F. UAV for 3D mapping applications: A review. *Appl. Geomat.* **2014**, *6*, 1–15. [CrossRef]

20. Bhardwaj, A.; Sam, L.; Martín-Torres, F.J.; Kumar, R. UAVs as remote sensing platform in glaciology: Present applications and future prospects. *Remote Sens. Environ.* **2016**, *175*, 196–204. [CrossRef]

21. Torresan, C.; Berton, A.; Carotenuto, F.; Di Gennaro, S.F.; Gioli, B.; Matese, A.; Miglietta, F.; Vagnoli, C.; Zaldei, A.; Wallace, L. Forestry applications of UAVs in Europe: A review. *Int. J. Remote Sens.* **2017**, *38*, 2427–2447. [CrossRef]

22. Yao, H.; Qin, R.; Chen, X. Unmanned Aerial Vehicle for Remote Sensing Applications—A Review. *Remote Sens.* **2019**, *11*, 1443–1464. [CrossRef]

23. Messous, M.A.; Hellwagner, H.; Senouci, S.M.; Emini, D.; Schnieders, D. Edge computing for visual navigation and mapping in a UAV network. In Proceedings of the ICC 2020–2020 IEEE International Conference on Communications (ICC), Dublin, Ireland, 7–11 June 2020; pp. 1–6.

24. Liu, Q.; Shi, L.; Sun, L.; Li, J.; Ding, M.; Shu, F. Path Planning for UAV-Mounted Mobile Edge Computing with Deep Reinforcement Learning. *IEEE Trans. Veh. Technol.* **2020**, *69*, 5723–5728. [CrossRef]

25. Mnih, V.; Kavukcuoglu, K.; Silver, D.; Graves, A.; Antonoglou, I.; Wierstra, D.; Riedmiller, M. Playing Atari with Deep Reinforcement Learning. *arXiv* **2013**, arXiv:1312.5602.

26. Afifi, G.; Gadallah, Y. Cellular Network-Supported Machine Learning Techniques for Autonomous UAV Trajectory Planning. *IEEE Access* **2022**, *10*, 131996–132011. [CrossRef]

27. Xia, W.; Zhu, Y.; De Simone, L.; Dagiuklas, T.; Wong, K.K.; Zheng, G. Multiagent Collaborative Learning for UAV Enabled Wireless Networks. *IEEE J. Sel. Areas Commun.* **2022**, *40*, 2630–2642. [CrossRef]

28. Li, B.; Liu, Y.; Tan, L.; Pan, H.; Zhang, Y. Digital twin assisted task offloading for aerial edge computing and networks. *IEEE Trans. Veh. Technol.* **2022**, *71*, 10863–10877. [CrossRef]

29. Li, K.; Ni, W.; Yuan, X.; Noor, A.; Jamalipour, A. Deep Graph-based Reinforcement Learning for Joint Cruise Control and Task Offloading for Aerial Edge Internet-of-Things (EdgeIoT). *IEEE Internet Things J.* **2022**, *9*, 21676–21686. [CrossRef]

30. Qian, Y.; Sheng, K.; Ma, C.; Li, J.; Ding, M.; Hassan, M. Path Planning for the Dynamic UAV-Aided Wireless Systems Using Monte Carlo Tree Search. *IEEE Trans. Veh. Technol.* **2022**, *71*, 6716–6721. [CrossRef]

31. Wang, Y.; Chen, W.; Luan, T.H.; Su, Z.; Xu, Q.; Li, R.; Chen, N. Task Offloading for Post-Disaster Rescue in Unmanned Aerial Vehicles Networks. *IEEE/ACM Trans. Netw.* **2022**, *30*, 1525–1539. [CrossRef]

32. Yang, Z.; Bi, S.; Zhang, Y.J.A. Online Trajectory and Resource Optimization for Stochastic UAV-Enabled MEC Systems. *IEEE Trans. Wirel. Commun.* **2022**, *21*, 5629–5643. [CrossRef]

33. Lyu, L.; Zeng, F.; Xiao, Z.; Zhang, C.; Jiang, H.; Havyarimana, V. Computation Bits Maximization in UAV-Enabled Mobile-Edge Computing System. *IEEE Internet Things J.* **2022**, *9*, 10640–10651. [CrossRef]

34. Hamasha, M.; Rumbe, G. Determining optimal policy for emergency department using Markov decision process. *World J. Eng.* **2017**, *14*, 467–472. [CrossRef]

35. El-Shafai, W.; El-Hag, N.A.; Sedik, A.; Elbanby, G.; Abd El-Samie, F.E.; Soliman, N.F.; AlEisa, H.N.; Abdel Samea, M.E. An Efficient Medical Image Deep Fusion Model Based on Convolutional Neural Networks. *Comput. Mater. Contin.* **2023**, *74*, 2905–2925. [CrossRef]

36. Sabry, E.S.; Elagooz, S.; El-Samie, F.E.A.; El-Shafai, W.; El-Bahnasawy, N.A.; El-Banby, G.; Soliman, N.F.; Sengan, S.; Ramadan, R.A. Sketch-Based Retrieval Approach Using Artificial Intelligence Algorithms for Deep Vision Feature Extraction. *Axioms* **2022**, *11*, 663–698. [CrossRef]

37. Meier, L.; Honegger, D.; Pollefeys, M. PX4: A node-based multithreaded open source robotics framework for deeply embedded platforms. In Proceedings of the 2015 IEEE International Conference on Robotics and Automation (ICRA), Seattle, WA, USA, 26–30 May 2015; pp. 6235–6240. [CrossRef]

38. Wang, C.Y.; Bochkovskiy, A.; Liao, H.Y.M. YOLOv7: Trainable bag-of-freebies sets new state-of-the-art for real-time object detectors. *arXiv* **2022**, arXiv:2207.02696. https://doi.org/10.48550/ARXIV.2207.02696.

39. Wojke, N.; Bewley, A.; Paulus, D. Simple online and realtime tracking with a deep association metric. In Proceedings of the 2017 IEEE International Conference on Image Processing (ICIP), Beijing, China, 17–20 September 2017; pp. 3645–3649.

40. Shafi, O.; Rai, C.; Sen, R.; Ananthanarayanan, G. Demystifying TensorRT: Characterizing Neural Network Inference Engine on Nvidia Edge Devices. In Proceedings of the 2021 IEEE International Symposium on Workload Characterization (IISWC), Storrs, CT, USA, 7–9 November 2021; pp. 226–237. [CrossRef]

41. Bochkovskiy, A.; Wang, C.Y.; Liao, H.Y.M. Yolov4: Optimal speed and accuracy of object detection. *arXiv* **2020**, arXiv:2004.10934.

42. Bewley, A.; Ge, Z.; Ott, L.; Ramos, F.; Upcroft, B. Simple online and realtime tracking. In Proceedings of the 2016 IEEE International Conference on Image Processing (ICIP), Phoenix, AN, USA, 25–28 September 2016; pp. 3464–3468.

43. Ammar, A.; Koubaa, A.; Boulila, W.; Benjdira, B.; Alhabashi, Y. A Multi-Stage Deep-Learning-Based Vehicle and License Plate Recognition System with Real-Time Edge Inference. *Sensors* **2023**, *23*, 2120. [CrossRef]

44. Koubaa, A.; Ammar, A.; Kanhouch, A.; AlHabashi, Y. Cloud Versus Edge Deployment Strategies of Real-Time Face Recognition Inference. *IEEE Trans. Netw. Sci. Eng.* **2022**, *9*, 143–160. [CrossRef]

45. Zhu, P.; Wen, L.; Du, D.; Bian, X.; Fan, H.; Hu, Q.; Ling, H. Detection and Tracking Meet Drones Challenge. *IEEE Trans. Pattern Anal. Mach. Intell.* **2021**, *44*, 7380–7399. [CrossRef]

46. Krichen, M.; Adoni, W.Y.H.; Mihoub, A.; Alzahrani, M.Y.; Nahhal, T. Security Challenges for Drone Communications: Possible Threats, Attacks and Countermeasures. In Proceedings of the 2022 2nd International Conference of Smart Systems and Emerging Technologies (SMARTTECH), Riyadh, Saudi Arabia, 22–24 May 2022; pp. 184–189.

47. Ko, Y.; Kim, J.; Duguma, D.G.; Astillo, P.V.; You, I.; Pau, G. Drone secure communication protocol for future sensitive applications in military zone. *Sensors* **2021**, *21*, 2057. [CrossRef]

48. Khan, N.A.; Jhanjhi, N.Z.; Brohi, S.N.; Nayyar, A. Emerging use of UAV's: Secure communication protocol issues and challenges. In *Drones in Smart-Cities*; Elsevier: Amsterdam, The Netherlands, 2020; pp. 37–55.

Article

Multiple Object Tracking of Drone Videos by a Temporal-Association Network with Separated-Tasks Structure

Yeneng Lin [1], Mengmeng Wang [1], Wenzhou Chen [1], Wang Gao [2], Lei Li [2] and Yong Liu [1,*]

[1] Institute of Cyber-Systems and Control, Zhejiang University, Hangzhou 310027, China
[2] Science and Technology on Complex System Control and Intelligent Agent Cooperation Laboratory, Beijing 100191, China
* Correspondence: yongliu@iipc.zju.edu.cn

Abstract: The task of multi-object tracking via deep learning methods for UAV videos has become an important research direction. However, with some current multiple object tracking methods, the relationship between object detection and tracking is not well handled, and decisions on how to make good use of temporal information can affect tracking performance as well. To improve the performance of multi-object tracking, this paper proposes an improved multiple object tracking model based on FairMOT. The proposed model contains a structure to separate the detection and ReID heads to decrease the influence between every function head. Additionally, we develop a temporal embedding structure to strengthen the representational ability of the model. By combing the temporal-association structure and separating different function heads, the model's performance in object detection and tracking tasks is improved, which has been verified on the VisDrone2019 dataset. Compared with the original method, the proposed model improves MOTA by 4.9% and MOTP by 1.2% and has better tracking performance than the models such as SORT and HDHNet on the UAV video dataset.

Keywords: deep learning; multi-object tracking; data association; object detection; temporal information; remote sensing data; video sequence

Citation: Lin, Y.; Wang, M.; Chen, W.; Gao, W.; Li, L.; Liu, Y. Multiple Object Tracking of Drone Videos by a Temporal-Association Network with Separated-Tasks Structure. *Remote Sens.* **2022**, *14*, 3862. https://doi.org/10.3390/rs14163862

Academic Editors: Yingying Dong, Chenghai Yang, Giovanni Laneve and Wenjiang Huang

Received: 13 June 2022
Accepted: 6 August 2022
Published: 9 August 2022

Publisher's Note: MDPI stays neutral with regard to jurisdictional claims in published maps and institutional affiliations.

1. Introduction

In recent years, with the rapid development of artificial intelligence technology, computer vision [1–3] has also penetrated into various fields in society, and the application of remote sensing data has become more and more popular [4–6]. Making full use of remote sensing videos and computer vision technology can greatly improve the efficiency of environmental monitoring and security monitoring. The use of computer vision technology to accomplish multi-object detection and tracking tasks in UAV images has gradually become one of the research hotspots. Multiple object tracking (MOT) refers to the object detection of all targets in each frame in the continuous frame sequence of the videos and obtaining the positions of targets in the images, the sizes of bounding boxes, and the speed of each object, as well as assigning individual ID identifications for every object in each frame. In the current mainstream research, the MOT model system can often be divided into two different paradigms, tracking by detection (TBD) [7–10] and joint detection and tracking (JDT) [11–14].

In the task of tracking video sequences, the accuracy of target detection [15–28] and data association can affect the performance of the model. Our research mainly pays attention to these two aspects. In the current research, target detection based on deep learning has been an important part of MOT. The anchor-free target detection algorithm has also become one of the more popular frameworks recently. The main idea of the algorithm is to perform target detection based on the center point of every object, such as CenterNet [29], FCOS [30], and Centerpoint [31]. The task of multi-object tracking can also be achieved based on the anchor-free framework. In the framework of JDT, such as RetinaTrack [32] and CenterTrack [33], these two methods combine detection and tracking, simplifying the model structure and improving the real-time performance of calculation. Figure 1 is the process of two paradigms of MOT. Compared with our research, it is mainly the use of a single frame without using temporal information. Although the combination of detection and tracking tasks can improve training efficiency, the two tasks have different concerns, affecting the accuracy of detection or tracking performance. The model in our paper separates the detection and tracking tasks to a certain extent in the phases of training and inference for this problem and achieves end-to-end multi-object tracking. In addition, the use of temporal information will also affect the performance of the model on the MOT. Wu [34] and Liang [35] proposed models to improve the detection ability by using temporal information with multiple frames. Except for the detection performance, our research mainly focuses on the use of temporal information to improve the feature representation of the objects so as to improve the accuracy of the data association.

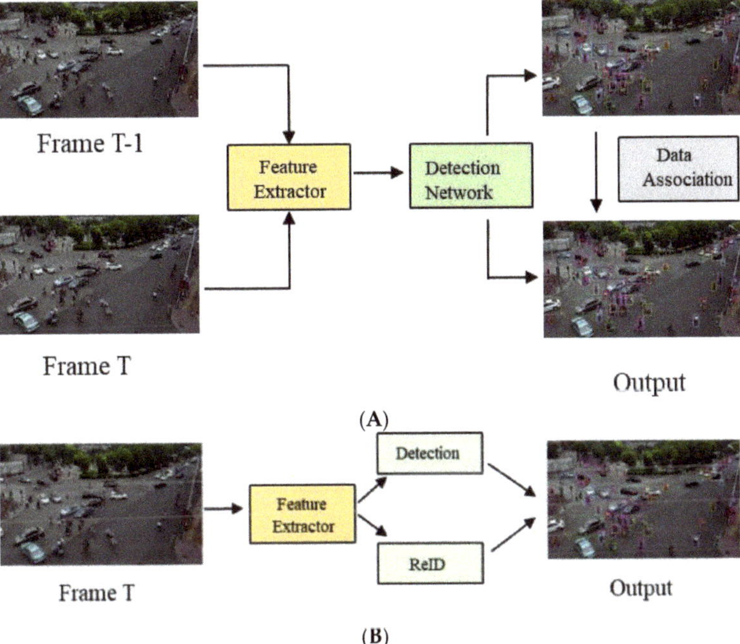

Figure 1. The process of two paradigms of MOT. (**A**) is the process of tracking by detection, and (**B**) is the process of joined detection and tracking.

MOT based on deep learning has also been used in the field of remote sensing, and the video interpretation of drones has been used as the research direction. This research direction is the main topic of this article, which is also a current research hotspot. Some research [36–42] combines trackers and detection models to realize the MOT tasks. Although separating the detection and tracking tasks can train models independently, it is more difficult to adjust one of the structures of a part to adapt the another. Our research aims to set up an improved end-to-end MOT model for remote sensing data. Furthermore, to improve the problems of small size and diverse backgrounds in remote sensing data, Jin [43] and Kraus [44] made use of features in different aspects of the tracked objects to improve the problems. Compared with the above research, temporal information can also be used in the MOT for remote sensing videos except for the feature of every object in our research.

MOT is a cross-frame video interpretation task, and most models in the current research do not make good use of the information in time series. There are certain limitations in relying only on the information of a single frame. The object lacks the links between the frames. For example, if an object is occluded in a certain frame, and if the model only relies on a single frame of information for data association, there will often be situations where the same object has different characterization information, which may lead to the ID switch (IDS) problem, thereby reducing the accuracy of the model. Therefore, using temporal information can significantly improve the model's performance for this task. In addition, although the JDT paradigm model combines object detection and data association for joint training to achieve end-to-end MOT, object detection and tracking are often two different vision tasks. Object detection needs to distinguish multiple categories, which needs to maximize the distance between different types and minimize the distance between the same type, to improve object detection accuracy. However, object tracking needs to maximize the distance among all objects in the same category. Therefore, if the two subtasks share many parameters during training, the training efficiency of the model may be reduced, and the performance of the trained model, in some cases, may get worse.

Given using temporal information and the conflict of two subtasks during the training phase, we propose an improved MOT model, using FairMOT [45] as the baseline. For the overall model structure, the detection part and the ReID part are disassembled. Compared with the detection part, the generation part of embedding is on an additional branch. A feature enhancement structure based on temporal information is added to the branch to improve the model's ability to discriminate ReID information. In the calculation process of model loss, compared with single-frame loss calculations, we perform double-frame output in the output part of the model detection and perform loss calculations on the output of two adjacent frames simultaneously.

Our contributions can be summarized as follows:

1. We change the single frame output of the original model to an output of two adjacent frames to improve the training efficiency.
2. We improve the conflict problem of two subtasks, including object detection and tracking during the training phase.
3. We constructed a feature enhancement structure based on temporal information to improve the representation of ReID information, enhancing the training efficiency of the ReID head of the model and ensuring data association performance.

2. Materials and Methods

This chapter mainly explains the detailed structure of the model proposed in this paper and the overall process of using the model to complete the MOT task. The model in this study uses FairMOT [45] as the baseline, and we improve it to achieve the functions mentioned in this article. Compared with the original structure, we have improved the feature extraction part of the model for the UAV videos, separated its detection and embedding parts, and added a feature enhancement structure in the ReID head. The detection head is changed to generate outputs of two adjacent frames for loss calculation. The following parts will describe each block in detail as a subsection.

2.1. The Structure of FairMOT

In the research field of MOT, many tracking algorithms can achieve MOT tasks based on detection results. The authors believe that object detection and ReID should be parallel visual tasks, so an anchor-free multi-target tracking algorithm called FairMOT is constructed. FairMOT is an end-to-end anchor-free MOT framework built on CenterNet. The model adopts a simple network framework, which is mainly composed of detection and ReID modules. FairMOT contains an anchor-free target detection framework, which can output heatmaps, sizes of bounding boxes information and offset information. In the Reid branch, the appearance feature of each pixel can be obtained, and it is used as the feature of the object with the pixel as the center point. The two functions can be performed during stages of training and inference at the same time, which achieves a balance between detection and ReID functions and has a better MOT performance. The feature extraction part of the model is shown in Figure 2. DLA-34 [46] is used as the backbone network to perform feature extraction on two-dimensional video images. The structure of the encoder-decoder network is shown in Figure 2B. Then, multiple heads will be used according to different vision tasks, namely the heatmap head, offset head, the size head for bounding boxes, and ReID head. These branches share the same feature map after the feature extraction structure.

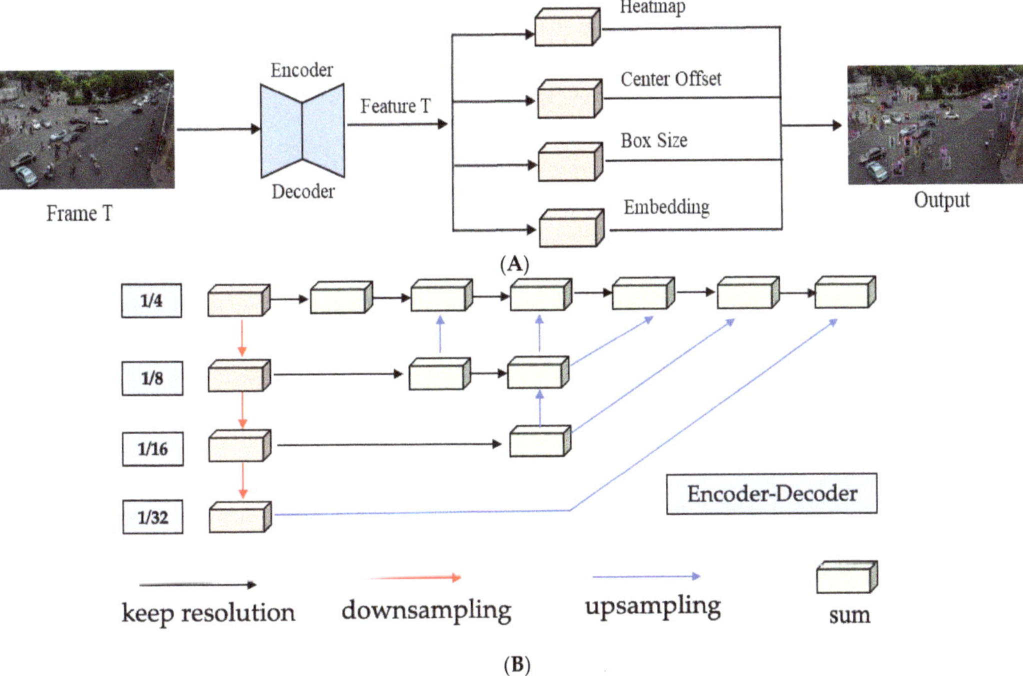

Figure 2. (**A**) is the process and structure of FairMOT. The blue part is the backbone of FairMOT, and the orange part includes different heads to form detection and ReID functions. (**B**) is the structure of the encoder-decoder network, which is used as the encoder and decoder of the FairMOT (the figure is revised from [45]).

The heatmap branch is used to generate a response map of the center point of every object, which represents the probability of objects. The size of the heatmap is HxWxC, where C represents the number of object categories. The response value of each pixel in the heatmap can reflect the probability of the object appearing, and the value of the pixel response value decreases as the distance from the center point increases. The offset branch represents the possible offset of each point compared to the original point after being encoded from the original image to the feature maps. This branch is responsible for accurately locating objects. Since the size of the feature map in the calculation graph varies greatly, this will produce some quantization errors, which will affect the locations of predicted objects and the extraction accuracy of ReID features of different objects. The size branch represents the width and height of the bounding box corresponding to each object. The three branches consist of the part responsible for object detection in the MOT framework.

The ReID branch generates an embedding of a specific length vector for each point, which is used to characterize the unique information of each point. Each pixel on the feature map contains a vector with a depth of 128 to represent the appearance features of the object for that point, and finally, an embedding map with the size of $128 \times W \times H$ will be obtained. In the training phase of the model, each head performs loss calculations separately. The three branches of the detection part, including the heatmap, offset, and size head, are mainly calculated by focal loss and L1 regression loss. In contrast, the loss of the ReID branch is calculated through classifiers.

2.2. The Structure of the Proposed Model

Compared with the feature extraction part of the FairMOT framework, we consider that there is a certain degree of conflict between target detection and ReID tasks during training; that is, the target detection task is to maximize the distance of different categories and minimize the distance of the same category, and the ReID task is to maximize the distance of objects with the same category. Therefore, it is necessary to adjust the model structure for this problem.

Given the conflict between ReID and target detection, the model in this paper is mainly to separate the two branches. The functions of the two structures on the FairMOT are realized through four branches, and these branches share the same encoder and decoder. The adjustment of the model in this article is mainly to improve the decoder part of the backbone network. Because the original model with this decoder can realize detection and ReID functions well, we change it to two decoders with the same structure as the original model, which is used for target detection and ReID, respectively. Furthermore, the same structure can meet conditions of finetuning. Still, there is no parameter sharing in these two decoders, which reduces the mutual influence between parameters of two tasks during model training. The encoder and decoder structures have been shown in Figure 2B, which is used to extract image features. The overall structure of the model is shown in Figure 3. In general, the model in this paper mainly uses two adjacent frames as input and can achieve multiple associations and utilization of temporal information in object detection and tracking tasks. From the overall structure of the model, the separation of the object detection and the ReID structure can make the two parts of the structure better realize their respective functions. In the ReID part, the temporal feature association structure is added to the ReID structure. This structure mainly integrates the historical frame information with the current frame to improve the robustness of the model while processing temporal series information. Compared with the single-frame input, the proposed model changes the input to two adjacent frames, which can make good use of temporal information. Furthermore, the output of two adjacent frames can improve the training efficiency of the model compared with the single-frame input. In the test phase, the model has two adjacent frames as inputs, and if it is the first frame of the video sequence, the model's input will be two images of the first frame. In the input part, the processing method of two frames is parameter sharing.

After the feature extraction of the encoder, the obtained features are simultaneously input into two decoders, and the processing of target detection and ReID information are performed, respectively. In the target detection part, the functions of the heatmap branch, offset branch, and size branch are similar to the original model, which is indicated in Section 2.1. Compared with the original structure, the structure of the heatmap branch is adjusted. Firstly, the feature of the previous frame obtained by the decoder is followed by a multi-layer convolution, which generates a centered map, and we combine this map with the features of the current frame obtained by decoder B. Then, the output of the branch can be obtained through the heatmap branch. The only difference is the output results of these two branches. Compared with the single-frame output, the model's output in this paper includes two predicted results of adjacent frames in two branches, which perform loss calculations simultaneously during training.

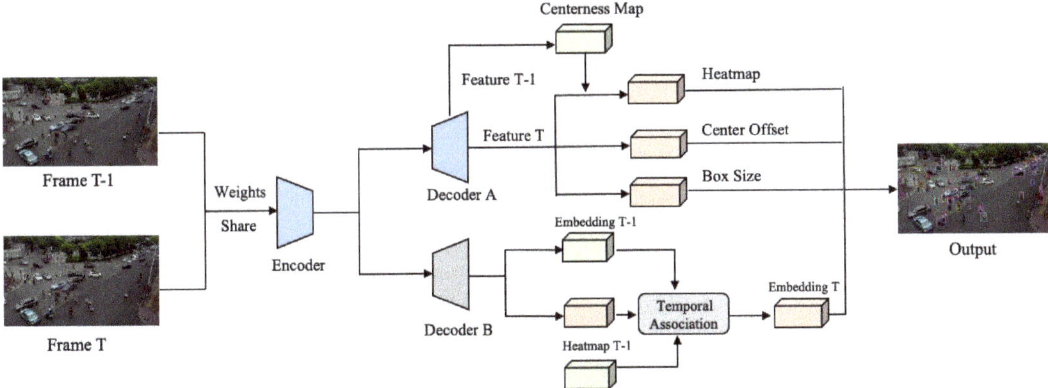

Figure 3. The structure of the proposed model. The blue part is the backbone of the model to extract features of the inputs, and there are two decoders with blue and grey colors. The orange part includes different heads to form detection and ReID function. To distinguish the information from different frames, the green part mainly makes use of the information from the T-1 frame. The detailed structure and process of the temporal association are explained in Figure 4, which can integrate the temporal information of two adjacent frames.

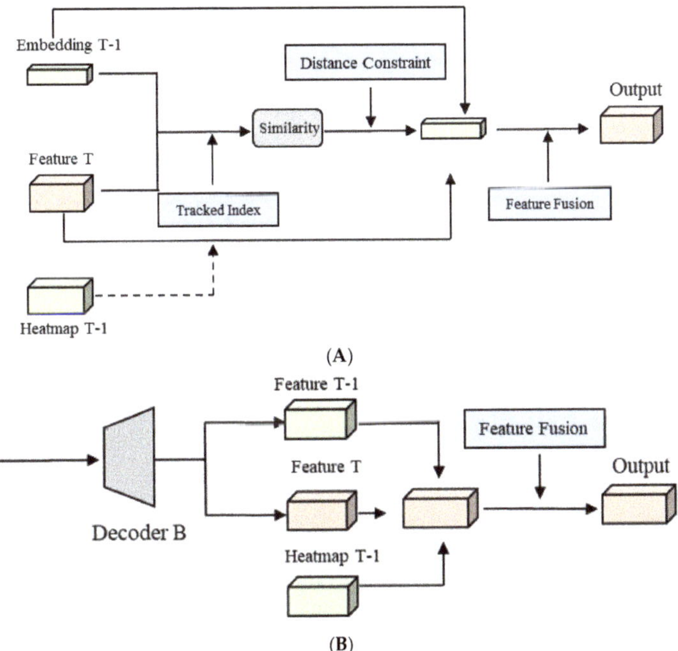

Figure 4. The two detailed structures of the module for the temporal association are in Figure 3. (**A**) is the indication of structure A, and (**B**) is the indication of structure B. Structure A can add to the model directly, and feature T and embedding T-1 are obtained by decoder B. The dotted arrow means that heatmap T-1 is not regarded as one of the inputs which can provide tracked indexes during the testing phase. Structure B combines the information of feature T-1, feature T obtained by decoder B, and heatmap T-1.

On the ReID branch, the branch adds a feature enhancement structure for the temporal association, which uses the features of two adjacent frames obtained by decoder B and the heatmap of the previous frame as the input information of the feature module. The structure and process of the temporal association are in the following sections. The temporal association can match and integrate the information of two adjacent frames to enhance the robustness of the ReID branch to get the final output of the ReID branch. Making use of temporal information on the ReID branch can improve the robustness of the appearance features generated by the model. Objects in a single frame may be deformed or occluded by other objects and background information, and the appearance features will change greatly, which will affect subsequent data associations. The use of temporal information can make some objects fused with historical feature information, which makes it possible to reduce the IDS problems caused by the influence of appearance features when problems such as occlusion and deformation of predicted objects occur in the current frame.

2.2.1. The ReID Branch with Structure A

As shown in Figure 4A, firstly, we associate the obtained feature T-1 with feature T. In the training phase, the calculation process is performed by inputting label information. The input includes the number of objects that exist simultaneously and the corresponding location index in two adjacent frames. Tracked indexes in Figure 4A mean that the heatmap T-1 can provide the detailed location and number of detected objects in the last frame, which can be used to extract the embedding information. We use this information to obtain the feature at the corresponding position of feature T-1 and calculate the similarity between it and feature T to obtain the feature similarity between each object in the previous frame and each point in the current frame; after that, we keep the point with the smallest distance. Feature similarity can be used as a measure of the degree of matching among embeddings. Additionally, it is regarded as the possible position of the object where the object of the previous frame may exist in the current frame. In this study, we mainly associate objects of two adjacent frames by calculating the similarity of the ReID feature, and the corresponding calculation process is shown in Equation (1). By calculating the similarity between the objects in the two frames, the possible position of each object in the previous frame in the current frame is obtained, and the feature fusion can be performed with the corresponding objects in the current frame. In this study, we also tried to achieve a similar effect through the point multiplication calculation of embeddings, which can improve the training and inference speed of the model. After the position information is obtained, feature fusion is performed, and the fusion method selected in this step is to add the corresponding feature matrix to the average. In the inference stage of the model, since there is no label information, the heatmap of the previous frame obtained by the model is used as auxiliary information. The number of possible targets in the previous frame is obtained from the heatmap.

$$similarity(x, y) = \frac{x \cdot y}{|x| \cdot |y|} = \frac{\sum_{i=1}^{n} x_i \cdot y_i}{\sqrt{\sum_{i=1}^{n} x_i^2} \sqrt{\sum_{i=1}^{n} y_i^2}} \tag{1}$$

where n represents the length of embeddings in the branches, and x and y represent the features of different objects between two frames.

The ReID embedding of the corresponding positions of these targets is input into the feature module as the feature of frame T-1. Since there is a situation in the inference phase where the object that appeared in the previous frame may disappear in the current frame, for this situation, a distance constraint needs to be added at this time. A threshold value needs to be set; namely, if the center point of the previous frame is far away from the center point matched in the current frame and exceeds the threshold, the matched point is considered unreliable. It should be ignored, and only a matched point with high reliability is retained. Distance constraints can be applied to the inference stage of the model to filter the predicted object positions. For the scaled size of the dataset images in this study, we set that if the distance of two predicted points is more than 50 pixels, the association can be considered unreliable matching, and we set it as the filtering threshold. The feature fusion method consistent with the training phase is performed to generate the final output of the ReID branch.

2.2.2. The ReID Branch with Structure B

The way that the structure makes use of temporal information is shown in Figure 4A. The information used in the structure includes three blocks, namely the heatmap of the previous frame and the feature t-1 and feature t of the two adjacent frames obtained by the decoder B. After the channel is spliced, the multi-layer convolution of the ReID branch is used to generate the final output on the branch. In the model's training phase, the previous frame's heatmap is also provided with label information. In the inference phase, the heatmap obtained from the model detection part is used as one of the inputs of the ReID branch.

2.3. The Post-Processing Part

The post-processing part functions mainly through adjusting the process of SORT [41] and DeepSORT [42] to complete the data association. Compared with single-category MOT, this research changes the post-processing part on multiple categories. Unlike the training stage, the post-processing stage does not assign an ID to each category; objects of various categories are assigned IDs together in sequence. The inference part uses DeepSORT as the main process framework, and there is a round for every three frames. The first frame normalizes the heatmap and standardizes ReID features obtained by the model, and performs non-maximum suppression processing on the heatmap according to the threshold to filter out possible objects. We assign an ID to objects in the first frame.

The second frame repeats the operation of the first frame; after getting the possible objects, it matches the object by the IoU value of the bounding boxes in the first frame, retains the expected detection, assigns the same ID, and includes those that are not matched. In the third frame, the ReID feature is added to the second frame, the cosine distance of the ReID feature is calculated on the detection target of the two adjacent frames, and the Kalman filter [47] is used for motion prediction. The appearance and motion characteristics are combined for data association. After that, the unmatched objects in the third frame and the objects in the second frame are subjected to IoU calculation. If it is smaller than a fixed threshold, it is regarded as a new target, and a unique ID is assigned. Finally, repeat the above steps for each subsequent frame to complete the post-processing steps of video MOT.

2.4. Training Strategy

In the model's training, a combination of multiple loss functions is used for training, and different training strategies are used primarily according to other branches of the model. The output of the heatmap of the target detection part is mainly used to train the model through focal loss. The corresponding label of the heatmap uses the label information to provide the center point position of every object. The corresponding response of the heatmap is obtained through Gaussian distribution processing, which is used as the training label of the heatmap, and the loss function is shown as follows. The information of offset and width and height branches are trained using the loss of L1 regression [48]. On the branch of ReID, the primary training method is classification type. After normalizing the target features obtained by the ReID head, classification training is carried out through classifiers. Since it is a multi-category MOT, multiple classifiers are also set at this time to train the model. The number of categories is the tota

$$L_H = -\frac{1}{n}\sum_m \begin{cases} (1-x)^a \log(x), & 1 \\ (1-x)^b (\hat{x})^a \log(1-\hat{x}) & \text{otherwise} \end{cases} \tag{2}$$

where m represents the number of points in the heatmap, $\hat{x}$ represents the model's predicted values, x represents the label's value, and a and b represent weights of focal loss.

$$L_B = \sum_{i=1}^{n} \|off^i - of\hat{f}^i\|_1 + \|wh^i - w\hat{h}^i\|_1 \tag{3}$$

$$L_{ID} = -\sum_{i=1}^{n}\sum_{j=1}^{J} A^i(j)\log(p(j)), \tag{4}$$

where off^i represents the values of the label, $of\hat{f}^i$ and $w\hat{h}^i$ represent the predicted outputs from the heads of the model, $A^i(j)$ represents the class label, and $p(j)$ is a class distribution obtained by the model.

$$L_T = \frac{1}{2}\left(\frac{1}{e^{w1}}(L_H + \frac{1}{2}(L_{B1} + L_{B2}) + \frac{1}{e^{w2}}L_{ID}\right) \tag{5}$$

where $w1$ and $w2$ represent the weights for different losses, L_{B1} represents the loss of Frame T, and L_{B2} represents the loss of Frame T-1.

The experiment of this study used pre-trained parameters, that is, the pre-trained parameters of CenterNet on the coco dataset [49]. Since CenterNet does not contain ReID parameters, only the parameters of the target detection part need to be used. After the parameter migration, training is performed on the training set of visdrone2019, and the learning rate is performed in an attenuated method, which will be reduced after a particular epoch, and we use Adam [50] as the optimizing process. In the testing phase, models are tested on the validation set with a single RTX 3090ti to compare the performance of the models.

3. Experimental Results

3.1. Data Introduction and Processing

This research dataset uses the UAV video sequences of visdrone2019 [51]. The dataset includes a training set and validation set. The training set contains 56 UAV video sequences. The training set has a total of 24,201 frames, and the validation set contains 7 video sequences, which have a total of 2819 frames. Each video is an optical image containing different scenes and targets. Each frame in the same video has the same size and format, and other video sequences have different image sizes and shooting methods. There are 10 categories in total. The main target categories of this dataset are pedestrian, person, car, van, bus, truck, motor, bicycle, awning-tricycle, and tricycle. The corresponding label format of the dataset used this time is the coco format. Table 1 is the distribution of the VisDrone2019 dataset in this study. Compared with the target detection dataset, the label of this dataset has ID information. Therefore, the information provided by the original data of the label mainly includes the serial number of the frame number, the target ID, the coordinates of the top-left vertex of the bounding box, the width and height of the bounding box, and whether the target is occluded and whether it needs to be ignored.

Table 1. The distribution of the VisDrone2019 dataset in this study.

	Training Set	Validation Set
Video Sequences	56	7
Number of Frames	24,201	2819
Category	10	10

The number of frames is the total frames in the sequences, and the category is the number of an object class in the video dataset.

Compared with the single-category MOT tasks, multi-category MOT requires additional processing on the dataset. This research is set to a total of 10 categories for the original label, so the training data construction needs to deal with the ID of multi-category. The primary way is to count the ID according to the category; that is, the ID of each category starts counting from 0. Compared with single-category tasks, it is necessary to count the number of target IDs of each category that appears in the entire video sequence and use them as the number of classes for the ReID branch during the model training. The model was trained with data augmentation, and each image imported into the model was processed by rotation and scaling, which improved the training effect of the model.

Evaluation Index

The evaluation indicators involved in this research mainly contain CLEAR metrics [52], including MOTA, MOTP, MT, ML, and IDF1 [53]. The corresponding calculation equations are as follows. MOTA represents the comprehensive performance of the model in MOT, and MOTP represents the average degree of overlaps of all tracked targets; MT and ML, respectively, represent the number of trajectories that are successfully tracked and the number of courses that have failed to follow. FN represents the number of detected objects missing in the target detection, FP represents the number of erroneously detected objects in the detection, and IDS represents the number of ID switches for the same objects.

$$\text{MOTA} = 1 - \frac{\sum_t (N_t + P_t + ids_t)}{\sum_t gt} \tag{6}$$

$$\text{MOTP} = \frac{\sum_{i,t} T_t^i}{\sum_t d_t} \tag{7}$$

where N_t represents the number of missed objects in frame T, P_t represents the number of the wrong predictions in frame T, idst represents the number of the ID switch in frame T, T_t represents the number of matches in frame T between objects and the hypothesis, and dt represents the distance between targets and the hypothesis.

MOTP has two different calculation and evaluation criteria in this paper. The criteria are that if the tracked match is perfect, MOTP is 100%, and if it deviates completely, it is 0. The larger the MOTA result, the better the overall performance of the model on multi-target tracking tasks, and the maximum value of MOTA is 100%. FN and FP represent the error of target detection, and IDS represents the number of ID switches in the tracking task. The larger the value of these indicators, the worse the MOT effect. The indicators used in this study are mainly from clear mot metrics. By measuring the differences between the indicators, the detection and tracking performance of the models can be evaluated well.

3.2. Experimental Results

This section mainly compares the performance of two feature enhancement structures constructed with temporal information. We add the two structures to the model for experiments, guarantee the same training and testing environment, and compare the performance of the two structures. The comparison method of the experiment in this section is mainly carried out by evaluating the difference of each indicator and comparing the performance shown by the visualization effect of each model.

The detailed information and structures of the modules are shown in Figure 4. Structure A combines the temporal information of two frames, including the embedding and heatmap obtained from the last frame. Structure B uses the features of two frames from the decoder part. In this part, the heatmap T-1 can be provided by the label information during the training phases. During the testing phase, the heatmap information can be obtained from the model in the last frame.

Figures 5 and 6 show the visualization effects of two different structures in the video sequences of the testing set. Figure 6 selects a part of the results of Figure 5 to enlarge the display, which can make the visualization effect of the model in the MOT task clearer. Table 2 shows the statistics of each indicator value on the testing set after adding two structures to the proposed model. The results in Table 2 show that the feature enhancement of structures A and B do not have much difference in the performance of the model's detection function. Structure A has a smaller total number of FN and FP than structure B. There is a difference in the function of ReID. It can be seen from the statistical table that the two models with different structures in the experiment have a specific difference in the IDS problem. When the FN is not notably different, structure A has a smaller IDS value than B, and structure A improves MOTA by 1.9% and MOTP by 0.3%, so it has a better improvement in ReID. The smaller number of IDS means fewer ReID errors in the testing phase, which provides better visual performance.

(A)

(B)

Figure 5. The visual results of MOT obtained by the two structures added to the ReID head, which displays two video sequences, including (**A**,**B**), in the validation set. The first row shows the visual results of structure B, and the second row shows the visual results of structure A. Every detected object is positioned by a bounding box of different sizes and is assigned an ID. In the three frames above, the same object is assigned the same ID in every frame. There are ten categories of objects detected in the video dataset at this time. The visualization of this multi-target tracking task only shows specific ID information, location and bounding boxes of every detected object.

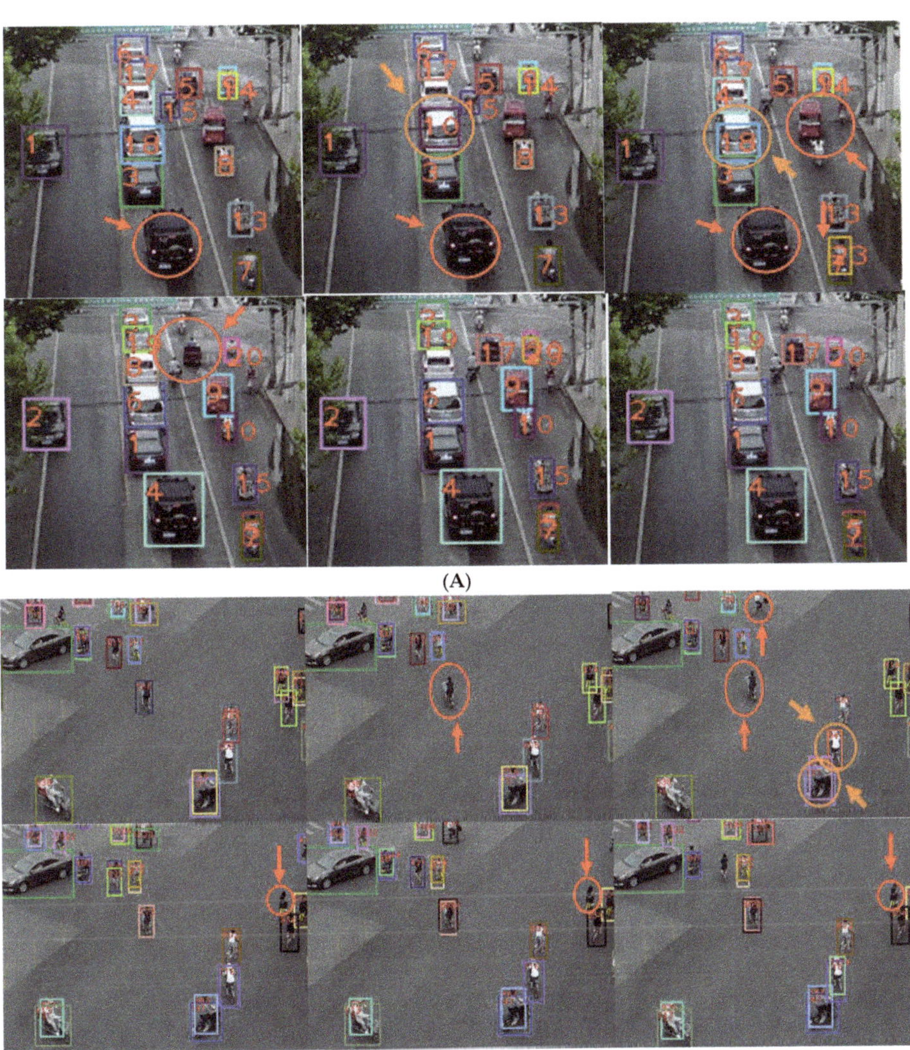

(A)

(B)

Figure 6. (**A**,**B**) are partially enlarged results of (A,B) in Figure 5. The arrangement of the images is the same as in Figure 5, which displays two video sequences in the validation set. The first row shows the visual results of structure B, and the second row shows the visual results of structure A. In every image, the red circle and red arrow mean that the object to be detected is missed in this image, and the orange circle and arrows mean that the object has an ID switch between two adjacent frames.

Table 2. Quantitative comparison of two structures used in the ReID head of the proposed model.

	MOTA	MOTP	IDF1	MT	ML	FN	FP	IDS
Structure A	34.7	74.5	45.2	164	265	57,818	14,005	2049
Structure B	32.8	74.2	**45.9**	**175**	257	57,939	16,106	2706

Structures A and B are used for feature enhancement added to the ReID branch, and the details are explained in Figure 4. The bold values in the table mean the best results.

As shown in Figures 5 and 6, the Figures illustrate the visual results of two structures in the two video sequences. The visualization results show that the two structures have similar effects on target detection. Most targets with different classes can be detected in the following samples, which means that the models in this experiment can achieve the object detection task. Only a part of the targets is not well detected. Compared with the larger and closer targets, some distant targets and occluded objects are more challenging to be accurately identified.

As for another aspect, there also are some IDS issues in the visualization results. From Figure 6, we can see that the IDS problem always appears when some objects are occluded in the last frame, and these objects appear in the next frame. If some objects are occluded, the embeddings of the object may be changed or not precious for the same object, which will influence the process of ReID and the MOT performance of the model. In the visualization results, compared with structure B in the task of ReID, structure A has a better ReID effect, and there are fewer IDS issues.

In order to verify whether different model structures have an impact on the performance of MOT, in this section, we have added corresponding ablation experiments. Table 3 counts the numerical results of the ablation experiment. Methods of this ablation experiment include: (1) FairMOT, which is the baseline for this experiment; (2) adjusted FairMOT, which only splits the detection and ReID branches; (3) the model changes the outputs from a single frame to the two adjacent frames during the training phase; (4) based on (3), the model is improved by adding the heatmap information of the previous frame and adding feature enhancement structure A.

Table 3. Quantitative comparison of the ablation experiment.

	MOTA	MOTP	IDF1	MT	ML	FN	FP
baseline	29.8	73.3	**46.1**	**183**	279	58,657	17,683
+split structure	32.6	73.3	44.9	164	278	59,865	14,855
+two-frame output	32.5	74.2	45.3	167	**260**	58,695	15,376
+centermap, attention	**34.7**	**74.5**	45.2	164	265	**57,848**	**14,385**

The structures of different models in the table are explained in the following paragraph. The bold values in the table mean the best results.

The result shows that compared with the original baseline, the final model significantly improved target detection and tracking performance. In the target detection part, the number of missed and wrong detections are reduced, the total number of FN and FP has dropped, and the effect has increased by 4.9% and 1.2% on MOTA and MOTP, respectively, compared with the baseline. Model (2) splits the detection and ReID heads, which improves the efficiency of multi-object detection and tracking tasks during training and can obtain more representative features for the MOT. From the result, we can see that the detection performance has been improved compared with the baseline. As for Model (3), the output part has been changed from a single frame to two adjacent frames. Compared with Model (2), Model (3) improved the MOTP by 0.9%, though there is no noticeable improvement effect in object detection from the numbers of FN and FP. Figures 7 and 8 show the visual comparison of the performance of the proposed model compared with the FairMOT under this validation dataset. The visualization effect shows that the proposed model of this paper has a certain degree of improved performance in multi-category MOT compared with the baseline.

The results show that in the MOT task based on UAV video, the baseline model did not detect the most targets. As for detecting some small objects and objects that may be occluded, the model may miss these objects. On the other hand, it can be seen from the visualization results that the ID switch has appeared in the samples in the function of ReID of the baseline model. The problem of ID switch mainly occurs when two objects meet, which may change the feature representation of the object between two frames. Therefore, the use of temporal information can improve this problem.

The numerical results of the quantitative comparison experiment are presented in Table 4, which includes an MOT evaluation matrix. Compared with the models that complete the MOT task on this dataset, the proposed model increases MOTA by 1.8% and IDF1 by 2.9% compared with HDHNet [54] and increases MOTA by 4.9% and MOTP by 1.2% compared with FairMOT. The proposed model has a smaller number of the sum of FN and FP than the other. The proposed model has a better object detection and ReID effect in the visualization results, which shows that the model can complete the MOT task. Thus, from the numerical and visualization results, the proposed model has improved the performance of the multi-category MOT task in this dataset.

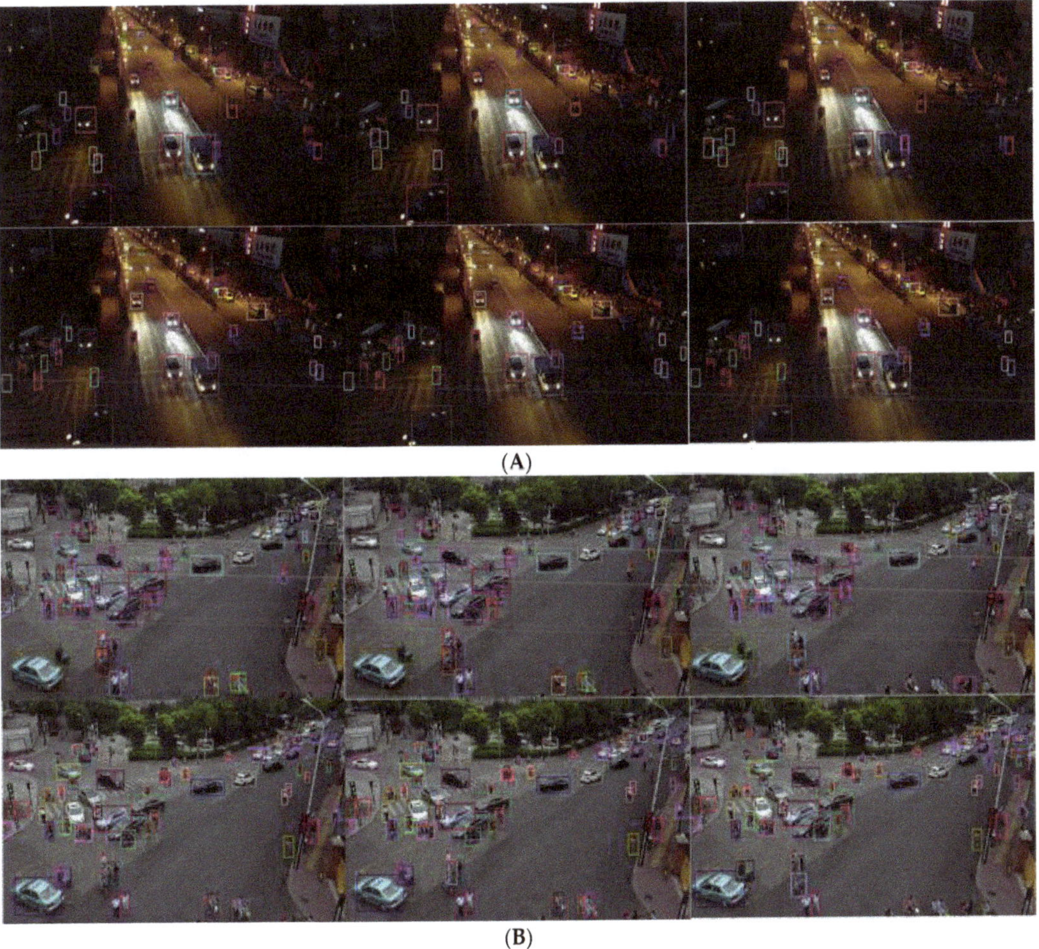

(A)

(B)

Figure 7. The visual results of MOT obtained by the two models, including the baseline and the proposed model, which displays two video sequences in the validation set. (**A**,**B**) are two examples in the two video sequences. The first row shows the visual results of FairMOT, and the second row shows the visual results of the proposed model. Every detected object is positioned by a bounding box of different sizes and is assigned an ID. In the three frames above, the same object is assigned the same ID in every frame.

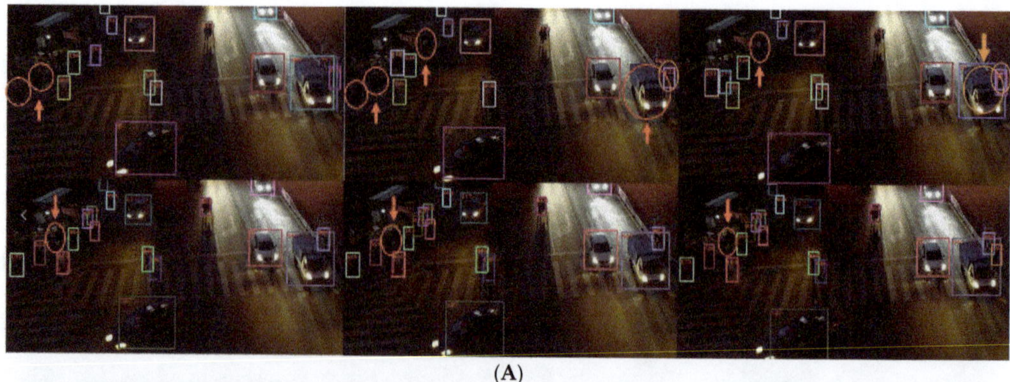

(A)

(B)

Figure 8. (A,B) are partially enlarged results of (A,B) in Figure 7. The arrangement of the images is the same as Figure 7, which displays two video sequences in the validation set. The first row shows the visual results of structure B, and the second row shows the visual results of structure A. In every image, the red circle and red arrow mean that the object to be detected is missed in this image, and the orange circle and arrows mean that the object has an ID switch between two adjacent frames.

Table 4. Quantitative comparison of the different models in the VisDrone2019 dataset.

	MOTA	MOTP	IDF1	MT	ML	FN	FP
GGDTRACK [55]	23.4	/	**48.1**	/	/	42,917	12,630
SORT	18.1	65.1	32.2	/	/	78,467	104,453
HDHNet	32.9	**76.9**	42.3	/	/	35,686	80,454
FairMOT	29.8	73.3	46.1	**183**	279	58,657	17,683
Proposed Model	**34.7**	74.5	45.2	164	**265**	**57,848**	**14,385**

Some indicators of the models mentioned in this table are not provided in the corresponding references. The bold values in the table mean the best results.

4. Discussion

In this research, the multi-category multi-object tracking task based on UAV video sequences is realized using the proposed model in this paper. It can be seen from the results of the first part of the ablation experiment that making good use of the temporal information and adjusting the corresponding network structure can improve the model's performance in detecting and tracking MOT tasks. Compared with the baseline structure, the structure that separates the detection and ReID branches improves the detection ability and reduces missed and false detection problems.

Different function parts can be separated to improve the structure and train the model on detection and ReID tasks. The original design integrates detection and ReID modules, and there will be some training conflicts during the training phase; that is, detection minimizes the distance between objects of the same category, and ReID maximizes the distance of objects within the same categories. In this study, separating the detection and ReID branches of the model in the encoder part can make the training of the two tasks more independent. From the overall structure of the model, the model uses the information of two adjacent frames in the input, and in the detection part, it realizes more independent training and inference based on the use of historical features, which weakens the influence of the ReID structure. In the output part of the model's detection branches, it changes the output format from single-frame output to a double-frame format, which can make the output of adjacent frames perform loss calculation at the same time, improves the training efficiency of the model, and uses the temporal information in the detection part during training. The model with the output structure of adjacent frames has an improvement of about 2.7% in MOTA performance compared with the baseline.

The proper way to use temporal information can help improve the ReID performance of the model. According to the results of the ablation experiment, by adding feature enhancement of structure A, the MOT performance of the model has been improved to a certain extent. In contrast, the feature enhancement structure B did not improve the model's performance in the ReID branch. After analyzing the structure, structure B needs to use the heatmap of the previous frame as auxiliary information for consideration, which is directly input into the network for calculation. At this time, the label information is used. However, the training efficiency of the model can be maintained during training; during the testing phase, the model will use the heatmap provided in the previous frame as the input of the branch, and the accuracy of the heatmap offered by the model is not as accurate as the label information, which calculates it directly as features will cause a specific deviation in the result. In subsequent experiments, other solutions were also implemented, such as now using the heatmap generated by the model as input during training. Firstly, using label information as training input after a particular round of iteration, then replacing it with the model's output. After comparing the two schemes, the effect has not been improved. As for structure A, the temporal information of the videos is also used. The object's center position tracked in the previous frame is input into the prediction as auxiliary information for the current frame. The ReID feature between the two frames is matched by the similarity of the embedding quality. From the result, it can be seen that structure A is better than structure B in the data association performance of MOT.

Adding temporal data to the model can also improve detection ability on the MOT task. From the data association process of DeepSORT, the matched target first needs to be detected. If an object in the video sequence is occluded or the size has changed, the detection result will change, and the target may be classified as the background. There is a case of missing detection or a change in the ReID embedding, which makes the model match the wrong object during the inference phase. This module expands the range of feature matching, merges embedding with the feature with the most significant similarity on the feature map, and combines the apparent elements of the previous frame, making the ReID features obtained during the training and testing phases of the model more sequential, which can improve the performance of the ReID task to a certain extent. Compared with the structure of the heatmap branch of the baseline, the model in this paper can improve the response to the position of the object by generating the centered map of the previous frame and adding it to the feature map of the current frame, which is used as auxiliary information in the target detection part of the model. It can be seen from the structure of the model that, compared with the original structure, the model in this paper has been adjusted in the input and calculation process. The performance of the multi-target tracking of the model is improved, and the speed of training and inference of the model itself is slightly reduced compared with the original model. For example, in the post-processing process, with a single 3090ti, the processing speed of the video streams can be maintained at about 15 FPS, which is lower than the original structure. However, the real-time performance of the model for MOT tasks of video sequences can still be guaranteed, and the performance of multi-target tracking can be improved better at the expense of a little calculation speed compared with the original model. We will also try to improve the model algorithm and post-processing flow in the future to explore methods to enhance speed and accuracy.

5. Conclusions

Our study proposes an improved MOT model based on FairMOT, which can realize end-to-end detection and tracking of multi-category objects in UAV video sequences. As for the original structure, the target detection and ReID tasks may have some conflicts during training. We separate the detection and ReID branches to make the two parts more independent and improve the detection accuracy. Additionally, the model in this paper uses temporal information in target detection and the ReID head, combines the central point features of historical frames, and includes a feature enhancement structure to improve the tracking performance of the model on UAV video sequences. Finally, compared with other MOT models on this study's drone video dataset, the use of the proposed model can achieve better multi-category and multi-object tracking performance. Although making good use of temporal information can improve the tracking performance of the model, there are still some scenes where objects have similar appearance characteristics, which will affect the results of data association during the process of history frame association. Therefore, it is necessary to focus on the association of similar objects in historical frames in subsequent research. Furthermore, the temporal information contained in adjacent frames is still limited. In the follow-up research, we will also try to explore the use of multi-frame and long-term information and apply it to tracking tasks to improve long-span tracking tasks while ensuring the accuracy and real-time performance of the model.

Author Contributions: Conceptualization, Y.L. (Yong Liu) and Y.L. (Yeneng Lin); methodology, Y.L. (Yeneng Lin); software, Y.L. (Yeneng Lin); validation, Y.L. (Yeneng Lin); formal analysis, Y.L. (Yeneng Lin); writing—original draft preparation, Y.L. (Yeneng Lin); writing—review and editing, M.W., W.C., L.L., Y.L. (Yong Liu) and W.G.; visualization, Y.L. (Yeneng Lin); supervision, M.W. and W.C.; project administration, Y.L. (Yong Liu); All authors have read and agreed to the published version of the manuscript.

Funding: The research was funded by the Key Research and Development Project of Zhejiang Province under Grant 2021C01035.

Data Availability Statement: Publicly available dataset was analyzed in this study. VisDrone 2019 dataset can be found here: http://aiskyeye.com/. The detailed indication of the dataset presented in this study is available in reference [51].

Conflicts of Interest: The authors declare no conflict of interest.

References

1. Voulodimos, A.; Doulamis, N.; Doulamis, A.; Protopapadakis, E. Deep Learning for Computer Vision: A Brief Review. *Comput. Intell. Neurosci.* **2018**, *2018*, 7068349. [CrossRef] [PubMed]
2. Patrício, D.I.; Rieder, R. Computer vision and artificial intelligence in precision agriculture for grain crops: A systematic review. *Comput. Electron. Agric.* **2018**, *153*, 69–81. [CrossRef]
3. Janai, J.; Güney, F.; Behl, A.; Geiger, A. Computer Vision for Autonomous Vehicles: Problems, Datasets and State of the Art. *Found. Trends Comput. Graph. Vis.* **2020**, *12*, 1–308. [CrossRef]
4. Ballesteros, R.; Intrigliolo, D.S.; Ortega, J.F.; Ramírez-Cuesta, J.M.; Buesa, I.; Moreno, M.A. Vineyard yield estimation by combining remote sensing, computer vision and artificial neural network techniques. *Precis. Agric.* **2020**, *21*, 1242–1262. [CrossRef]
5. Cheng, G.; Han, J. A survey on object detection in optical remote sensing images. *ISPRS J. Photogramm. Remote Sens.* **2016**, *117*, 11–28. [CrossRef]

6. Xu, Z.; Xu, X.; Wang, L.; Yang, R.; Pu, F. Deformable ConvNet with Aspect Ratio Constrained NMS for Object Detection in Remote Sensing Imagery. *Remote Sens.* **2017**, *9*, 1312. [CrossRef]
7. Kampffmeyer, M.; Salberg, A.-B.; Jenssen, R. Semantic Segmentation of Small Objects and Modeling of Uncertainty in Urban Remote Sensing Images Using Deep Convolutional Neural Networks. In Proceedings of the 2016 IEEE Conference on Computer Vision and Pattern Recognition Workshops (CVPRW), Las Vegas, NV, USA, 26 June–1 July 2016; pp. 1–9.
8. Ahmed, I.; Ahmad, M.; Ahmad, A.; Jeon, G. Top view multiple people tracking by detection using deep SORT and YOLOv3 with transfer learning: Within 5G infrastructure. *Int. J. Mach. Learn. Cybern.* **2020**, *12*, 3053–3067. [CrossRef]
9. Wang, Z.; Miao, D.; Zhao, C.; Luo, S.; Wei, Z. A Robust Long-Term Pedestrian Tracking-by-Detection Algorithm Based on Three-Way Decision. In Proceedings of the International Joint Conference on Rough Sets (IJCRS), Debrecen, Hungary, 17–21 June 2019; pp. 522–533.
10. Xie, Y.; Huang, Y.; Song, T.L. Iterative joint integrated probabilistic data association filter for multiple-detection multiple-target tracking. *Digit. Signal Process.* **2018**, *72*, 232–243. [CrossRef]
11. Wang, Z.; Zheng, L.; Liu, Y.; Li, Y.; Wang, S. Towards Real-Time Multi-Object Tracking. In *Lecture Notes in Computer Science (Including subseries Lecture Notes in Artificial Intelligence and Lecture Notes in Bioin-Formatics)*; Springer: Berlin/Heidelberg, Germany, 2020; Volume 12356. [CrossRef]
12. Munjal, B.; Aftab, A.R.; Amin, S.; Brandlmaier, M.D.; Tombari, F.; Galasso, F. Joint detection and tracking in videos with identification features. *Image Vis. Comput.* **2020**, *100*, 103932. [CrossRef]
13. Sun, P.; Cao, J.; Jiang, Y.; Zhang, R.; Xie, E.; Yuan, Z.; Wang, C.; Luo, P. TransTrack: Multiple-Object Tracking with Transformer. *arXiv* **2012**, arXiv:2012.15460.
14. Lin, X.; Guo, Y.; Wang, J. Global Correlation Network: End-to-End Joint Multi-Object Detection and Tracking. *arXiv* **2021**, arXiv:2103.12511.
15. Redmon, J.; Divvala, S.; Girshick, R.; Farhadi, A. You only look once: Unified, real-time object detection. In Proceedings of the IEEE Conference on Computer Vision and Pattern Recognition, Seattle, WA, USA, 27–30 June 2016; pp. 779–788.
16. Redmon, J.; Farhadi, A. YOLO9000: Better, faster, stronger. In Proceedings of the 30th IEEE Conference on Computer Vision and Pattern Recognition (CVPR), Honolulu, HI, USA, 21–26 July 2017; pp. 6517–6525.
17. Redmon, J.; Farhadi, A. YOLOv3: An Incremental Improvement. *arXiv* **2018**, arXiv:1804.02767.
18. Bochkovskiy, A.; Wang, C.Y.; Liao, H.Y.M. YOLOv4: Optimal Speed and Accuracy of Object Detection. *arXiv* **2020**, arXiv:2004.10934.
19. Wang, X.; Shrivastava, A.; Gupta, A. A-Fast-RCNN: Hard positive generation via adversary for object detection. In Proceedings of the 30th IEEE Conference on Computer Vision and Pattern Recognition, Honolulu, HI, USA, 21–26 July 2017; pp. 3039–3048.
20. Girshick, R.; Donahue, J.; Darrell, T.; Malik, J. Rich feature hierarchies for accurate object detection and semantic segmentation. In Proceedings of the 27th IEEE Conference on Computer Vision and Pattern Recognition, CVPR, Columbus, OH, USA, 24–27 June 2014; pp. 580–587.
21. Ren, S.; He, K.; Girshick, R.; Sun, J. Faster R-CNN: Towards Real-Time Object Detection with Region Proposal Networks. *IEEE Trans. Pattern Anal. Mach. Intell.* **2017**, *39*, 1137–1149. [CrossRef] [PubMed]
22. Liu, W.; Anguelov, D.; Erhan, D.; Szegedy, C.; Reed, S.; Fu, C.-Y.; Berg, A.C. SSD: Single shot multibox detector. In *Lecture Notes in Computer Science (Including Subseries Lecture Notes in Artificial Intelligence and Lecture Notes in Bioinformatics)*; Springer: Berlin/Heidelberg, Germany, 2016; Volume 9905. [CrossRef]
23. Carion, N.; Massa, F.; Synnaeve, G.; Usunier, N.; Kirillov, A.; Zagoruyko, S. End-to-End Object Detection with Transformers. In *Lecture Notes in Computer Science (Including Subseries Lecture Notes in Artificial Intelligence and Lecture Notes in Bioinformatics)*; Springer: Berlin/Heidelberg, Germany, 2020; Volume 12346. [CrossRef]
24. Lin, T.Y.; Goyal, P.; Girshick, R.; He, K.; Dollar, P. Focal loss for dense object detection. *IEEE Trans. Pattern Anal. Mach. Intell.* **2020**, *42*, 318–327. [CrossRef] [PubMed]
25. Lin, T.Y.; Dollár, P.; Girshick, R.; He, K.; Hariharan, B.; Belongie, S. Feature pyramid networks for object detection. In Proceedings of the 30th IEEE Conference on Computer Vision and Pattern Recognition, CVPR, Honolulu, HI, USA, 21–26 July 2017.
26. Dai, J.; Li, Y.; He, K.; Sun, J. R-FCN: Object detection via region-based fully convolutional networks. In Proceedings of the 30th Conference on Neural Information Processing Systems (NIPS), Barcelona, Spain, 5–10 December 2016.
27. Voigtlaender, P.; Krause, M.; Osep, A.; Luiten, J.; Sekar BB, G.; Geiger, A.; Leibe, B. Mots: Multi-object tracking and segmentation. In Proceedings of the 32nd IEEE/CVF Conference on Computer Vision and Pattern Recognition (CVPR), Long Beach, CA, USA, 16–20 June 2019; pp. 7934–7943.
28. Hu, Y.T.; Huang, J.b.; Schwing, A.G. MaskRNN: Instance level video object segmentation. In Proceedings of the 31st Annual Conference on Neural Information Processing Systems (NIPS), Long Beach, CA, USA, 4–9 December 2017.
29. Zhou, X.; Austin, U.T.; Wang, D.; Berkeley, U.C.; Austin, U.T. Object as Point. In Proceedings of the 32nd IEEE/CVF Conference on Computer Vision and Pattern Recognition (CVPR), Long Beach, CA, USA, 16–20 June 2019.
30. Tian, Z.; Shen, C.; Chen, H.; He, T. FCOS: Fully convolutional one-stage object detection. In Proceedings of the IEEE International Conference on Computer Vision, ICCV, Seoul, Korea, 27 October–2 November 2019; pp. 9626–9635.
31. Yin, T.; Zhou, X.; Krahenbuhl, P. Center-based 3D Object Detection and Tracking. In Proceedings of the 32nd IEEE/CVF Conference on Computer Vision and Pattern Recognition, CVPR, Nashville, TN, USA, 20–25 June 2021; pp. 11779–11788.

32. Lu, Z.; Rathod, V.; Votel, R.; Huang, J. RetinaTrack: Online single stage joint detection and tracking. In Proceedings of the IEEE/CVF Conference on Computer Vision and Pattern Recognition (CVPR), Seattle, WA, USA, 13–19 June 2020; pp. 14656–14666.

33. Zhou, X.; Koltun, V.; Krähenbühl, P. Tracking Objects as Points. *arXiv* **2020**, arXiv:2004.01177.

34. Wu, J.; Cao, J.; Song, L.; Wang, Y.; Yang, M.; Yuan, J. Track to Detect and Segment: An Online Multi-Object Tracker. *arXiv* **2021**, arXiv:2103.08808.

35. Liang, C.; Zhang, Z.; Zhou, X.; Li, B.; Lu, Y.; Hu, W. One More Check: Making "Fake Background" Be Tracked Again. *arXiv* **2021**, arXiv:2104.0944. [CrossRef]

36. Wu, J.; Su, X.; Yuan, Q.; Shen, H.; Zhang, L. Multi-Vehicle Object Tracking in Satellite Video Enhanced by Slow Features and Motion Features. *IEEE Trans. Geosci. Remote Sens.* **2021**, *60*, 1–26. [CrossRef]

37. Xuan, S.; Li, S.; Han, M.; Wan, X.; Xia, G.-S. Object Tracking in Satellite Videos by Improved Correlation Filters With Motion Estimations. *IEEE Trans. Geosci. Remote Sens.* **2019**, *58*, 1074–1086. [CrossRef]

38. Yang, X.; Wang, Y.; Wang, N.; Gao, X. An Enhanced SiamMask Network for Coastal Ship Tracking. *IEEE Trans. Geosci. Remote Sens.* **2021**, *60*, 1–11. [CrossRef]

39. Shao, J.; Du, B.; Wu, C.; Zhang, L. Tracking Objects From Satellite Videos: A Velocity Feature Based Correlation Filter. *IEEE Trans. Geosci. Remote Sens.* **2019**, *57*, 7860–7871. [CrossRef]

40. Lei, L.; Guo, D. Multitarget Detection and Tracking Method in Remote Sensing Satellite Video. *Comput. Intell. Neurosci.* **2021**, *2021*, 7381909. [CrossRef]

41. Bewley, A.; Ge, Z.; Ott, L.; Ramos, F.; Upcroft, B. Simple online and realtime tracking. In Proceedings of the International Conference on Image Processing, ICIP, Phoenix, AZ, USA, 25–28 September 2016; pp. 3464–3468.

42. Wojke, N.; Bewley, A.; Paulus, D. Simple online and realtime tracking with a deep association metric. In Proceedings of the International Conference on Image Processing, ICIP, Beijing, China, 17–20 September 2017; pp. 3645–3649.

43. Wu, J.; Cao, C.; Zhou, Y.; Zeng, X.; Feng, Z.; Wu, Q.; Huang, Z. Multiple Ship Tracking in Remote Sensing Images Using Deep Learning. *Remote Sens.* **2021**, *13*, 3601. [CrossRef]

44. Kraus, M.; Azimi, S.M.; Ercelik, E.; Bahmanyar, R.; Reinartz, P.; Knoll, A. AerialMPTNet: Multi-Pedestrian Tracking in Aerial Imagery Using Temporal and Graphical Features. In Proceedings of the 25th International Conference on Pattern Recognition (ICPR), Milan, Italy, 10–15 January 2021; pp. 2454–2461.

45. Zhang, Y.; Wang, C.; Wang, X.; Zeng, W.; Liu, W. FairMOT: On the Fairness of Detection and Re-identification in Multiple Object Tracking. *Int. J. Comput. Vis.* **2021**, *129*, 3069–3087. [CrossRef]

46. Yu, F.; Wang, D.; Shelhamer, E.; Darrell, T. Deep Layer Aggregation. In Proceedings of the 31st IEEE Conference on Computer Vision and Pattern Recognition, Salt Lake City, UT, USA, 18–23 June 2018; pp. 2403–2412.

47. Welch, G.; Bishop, G. An introduction to the Kalman filter. *Course Notes ACM SIGGRAPH* **1995**, *8*, 127–132.

48. Zhao, H.; Gallo, O.; Frosio, I.; Kautz, J. Loss Functions for Image Restoration With Neural Networks. *IEEE Trans. Comput. Imaging* **2017**, *3*, 47–57. [CrossRef]

49. Lin, T.; Maire, M.; Belongie, S.; Hays, J.; Perona, P.; Ramanan, D.; Dollár, P.; Zitnick, C.L. Microsoft COCO: Common objects in context. In *European Conference on Computer Vision (ECCV)*; Springer: Berlin/Heidelberg, Germany, 2014; Volume 8693, pp. 740–755.

50. Kingma, D.P.; Ba, J.L. Adam: A method for stochastic optimization. *arXiv* **2014**, arXiv:1412.6980.

51. Wen, L.; Zhu, P.; Du, D.; Bian, X.; Ling, H.; Hu, Q.; Zheng, J.; Peng, T.; Wang, X.; Zhang, Y.; et al. VisDrone-MOT2019: The vision meets drone multiple object tracking challenge results. In Proceedings of the IEEE/CVF International Conference on Computer Vision Workshops, Seoul, Korea, 27–28 October 2019; pp. 189–198.

52. Bernardin, K.; Stiefelhagen, R. Evaluating Multiple Object Tracking Performance: The CLEAR MOT Metrics. *EURASIP J. Image Video Process.* **2008**, *2008*, 246309. [CrossRef]

53. Ristani, E.; Solera, F.; Zou, R.; Cucchiara, R.; Tomasi, C. Performance measures and a data set for multi-target, multi-camera tracking. In Proceedings of the 14th European Conference on Computer Vision (ECCV), Amsterdam, The Netherlands, 11–14 October 2016; Volume 9914, pp. 17–35.

54. Huang, W.; Zhou, X.; Dong, M.; Xu, H. Multiple objects tracking in the UAV system based on hierarchical deep high-resolution network. *Multimed. Tools Appl.* **2021**, *80*, 13911–13929. [CrossRef]

55. Ardo, H.; Nilsson, M. Multi-target tracking from drones by learning from generalized graph differences. In Proceedings of the IEEE International Conference on Computer Vision, ICCV, Seoul, Korea, 27 October–2 November 2019; pp. 46–54.

Article

A Method for Designated Target Anti-Interference Tracking Combining YOLOv5 and SiamRPN for UAV Tracking and Landing Control

Dong Wu [1,2], Hang Zhu [1,2,*] and Yubin Lan [3]

1 School of Mechanical and Aerospace Engineering, Jilin University, Changchun 130025, China; wudong20@mails.jlu.edu.cn
2 Chongqing Research Institute, Jilin University, Chongqing 404100, China
3 School of Agricultural Engineering and Food Science, Shandong University of Technology, Zibo 255000, China; ylan@sdut.edu.cn
* Correspondence: hangzhu@jlu.edu.cn; Tel.: +86-1808-866-5997

Citation: Wu, D.; Zhu, H.; Lan, Y. A Method for Designated Target Anti-Interference Tracking Combining YOLOv5 and SiamRPN for UAV Tracking and Landing Control. *Remote Sens.* **2022**, *14*, 2825. https://doi.org/10.3390/rs14122825

Academic Editors: Wenjiang Huang, Giovanni Laneve, Yingying Dong and Chenghai Yang

Received: 29 April 2022
Accepted: 10 June 2022
Published: 12 June 2022

Publisher's Note: MDPI stays neutral with regard to jurisdictional claims in published maps and institutional affiliations.

Abstract: With the rapid development in the field of computer vision, the vision-based approach to unmanned aerial vehicle (UAV) tracking and landing technology in weak global positioning system (GPS) or GPS-free environments has become prominent in military and civilian missions. However, this technique still suffers from problems such as interference by similar targets in the environment, low tracking accuracy, slow processing speed, and poor stability. To solve these problems, we propose the designated target anti-interference tracking (DTAT) method, which integrates YOLOv5 and SiamRPN, and built a system to achieve UAV tracking and the landing of a designated target in an environment with multiple interference targets. The system consists of the following parts: first, an image is acquired by a monocular camera to obtain the pixel position information of the designated target. Next, the position of the UAV relative to the target is estimated based on the pixel location information of the target and the known target size information. Finally, the discrete proportion integration differentiation (PID) control law is used to complete the target tracking and landing task of the UAV. To test the system performance, we deployed it on a robot operating system (ROS) platform, conducted many simulation experiments, and observed the real-time trajectories of the UAV and the target through Gazebo software. The results show that the relative distance between the UAV and the target during the tracking process when the target was moving at 0.6 m/s does not exceed 0.8 m, and the landing error of the UAV during the landing process after the target is stationary does not exceed 0.01 m. The results validate the effectiveness and robustness of the system and lay a foundation for subsequent research.

Keywords: UAV tracking and landing; visual anti-interference; deep learning

1. Introduction

The UAV industry has grown rapidly in recent decades and the market is becoming larger [1]. The tracking and landing of moving targets has been an active research area during the development of UAV systems [2]. There are a wide range of application scenarios, including agricultural remote sensing [3], marine transportation [4], police search-and-rescue [5], etc. The production method of UAV-supervised agricultural machinery has become an important part of unmanned agricultural remote sensing. To build a UAV tracking and landing system, more technologies are being developed and applied, including multi-sensor fusion [6], visual navigation [7], communication between UAV and ground targets [8,9], target search [10], suitable landing area search [11], etc. However, there are still many challenges to be faced in refining the system, such as target localization in similar target interference environments, real-time target tracking, and robust flight control.

To solve the above problems, many researchers have used traditional vision methods to build autonomous UAV tracking and landing systems. Compared with acquiring the target position by radar [12,13], GPS [14], or SBAS [15,16], the vision-based method is more resistant to interference, faster, more accurate, and suitable for many scenarios. For example, Le et al. [17] proposed an autonomous tracking and landing system based on the Aruco code as the target with a high accuracy, but the target is limited and influenced by the light factor. Phang et al. [18] proposed an autonomous tracking system based on an infrared camera; it can help the UAV to land and still detect targets at night in low-light conditions. However, the accuracy is low and the tracking effect is poor.

In recent years, the field of computer vision has developed rapidly, especially in the area of target detection and tracking [19,20]. Based on deep learning, target tracking and detection methods improve target localization accuracy, and also have significantly improved computing speed [21], which is suitable for deployment in embedded systems such as UAV. Based on this, many studies have been carried out. Chen et al. [22] summarized the target tracking methods applied to UAV, and reported that computer vision tracking performs better than traditional tracking methods in terms of target occlusion, deformation, and similar background interference, but still suffers from environmental interference and difficulty in distinguishing between designated targets. Yang et al. [23] used the YOLOv3 object detection method and incorporated depth camera-based state estimation to successfully track and land on a target in a GPS-free environment, but the processing speed was slow and the target was easily lost.

The problems, such as interference from similar targets, slow processing speed, and inaccurate tracking, still exist in the above tracking and landing algorithms. To solve these problems, this paper proposes the DTAT method, which integrates YOLOv5 and SiamRPN, and uses it to build a system to achieve UAV tracking and landing of the designated target in an environment with multiple interference targets. The proposed method is proved to be effective in solving the above problems through experiments. The contributions of this work are as follows:

- First, we propose the DTAT method to obtain the pixel position information of the target. It effectively solves the problem of tracking a designated target in an environment with interference from similar targets. The method integrates target detection and tracking algorithms to provide a solution for subsequent research.
- Second, we design a method to obtain the relative positions of the UAV and the target, which is acquired from the pixel coordinates of the vertices of the target frame in the image through a coordinate system transformation. Compared with the traditional method, the estimation accuracy of the relative position is effectively improved.
- Finally, we employ a strategy based on the discrete PID control law to manage the UAV flight state and integrate the above methods to complete the closed-loop control system for UAV tracking and landing. After extensive simulation experiments, the system was verified to be significantly better than the general system in terms of tracking and landing accuracy, which has practical application significance.

The rest of the paper is organized as follows. In Section 2, we present the framework for the tracking and landing system and the details of the DTAT method. In Section 3, we describe the simulation experiments of the methods and the overall system. In Section 4, we analyze the experimental results and discuss the design process of the method, as well as its advantages and disadvantages, and finally analyze the shortcomings of the system. Finally, we conclude the paper in Section 5.

2. Autonomous UAV Tracking and Landing System

In this section, the authors introduce the UAV autonomous tracking and landing system in detail. The system hardware includes the UAV and the monocular camera. For the UAV, we selected the P450-NX produced by AMOVLAB as the experimental model, which has open-source control system code and easily completes algorithm deployment. The system uses a DF500-1944P industrial camera (Jieruiweitong Inc., Shenzhen, China), which is small in size and convenient to carry. The system software framework is shown in Figure 1. It consists of two main parts: the designated target anti-interference tracking system and the UAV flight control system. For the former, we designed a DTAT method to enable the UAV to accurately track the target under conditions of visual interference. The system flow is shown in the panel on the left. The camera image is passed through the YOLOv5 algorithm to get all potential targets in the field of view and we select the tracking target to initialize the DTAT. When the next image frame arrives, the DTAT calculates and outputs the pixel position of the designated target frame. The flow of the flight control system is shown in the panel on the right. We use the output pixel coordinates to estimate the relative position of the UAV to the target and obtain the coordinates of the target under the inertial system of the airframe. With these coordinates used as the error term, a PID control law is used to control the flight speed of the UAV to achieve tracking and landing. The implementation process is described in detail below.

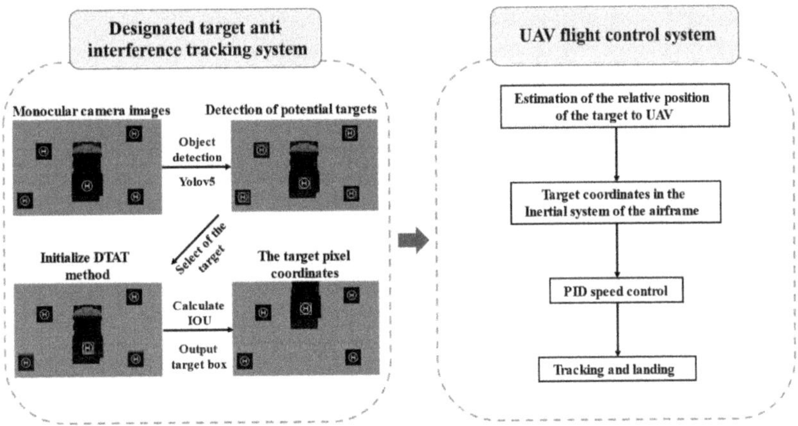

Figure 1. UAV autonomous tracking and landing system.

2.1. Specifying the Acquisition of Target Pixel Coordinates

The position of the designated target under the pixel coordinate system is required for the subsequent control of the UAV flight speed. For the speed and accuracy requirements of the algorithm during system deployment in the embedded environment, this paper proposes the DTAT method, integrating the YOLOv5 target detection algorithm, which has excellent accuracy performance, with the SiamRPN target tracking algorithm, which has fast speed and good embedded portability, to obtain the location of the designated target. The YOLOv5 algorithm can get the candidate frame position and size of all potential targets in the image with high detection accuracy but cannot locate the specified target. The SiamRPN algorithm uses a Siamese network to extract the given target information and is capable of tracking a single target, but the tracking accuracy is low and the target position is inaccurate. Therefore, the authors combined the two to construct the DTAT method, which ensures target tracking and improves accuracy at the same time. The implementation process of the algorithm is described in detail below.

2.1.1. Yolov5 and SiamRPN

Among target-detection algorithms, the YOLO series [24,25], a single-stage detection algorithm, performs well in many official datasets. YOLOv5 (Ultralytics LLC, Washington, DC, USA) is an improved version of YOLOv4 with better detection speed and accuracy that can be deployed in embedded systems. Its network model is shown in Figure 2; the detection results of the target are represented by the pixel box center coordinates and size.

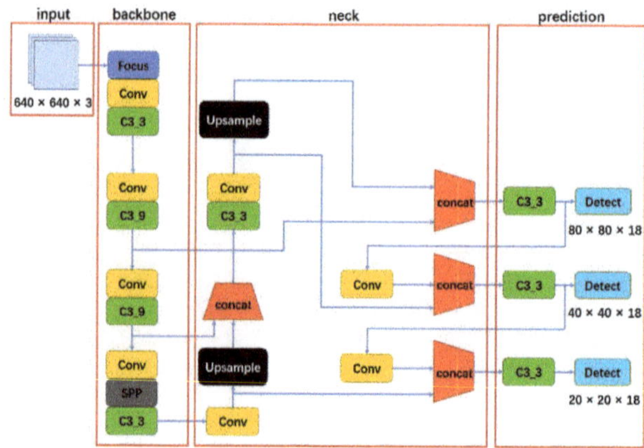

Figure 2. Yolov5 model structure.

Since the UAV processor uses the Jetson Xavier NX (NVIDIA Corporation, Los Angeles, CA, USA) embedded platform, to improve the algorithm inference speed, we use the open-source TensorRT high-performance optimized inference engine from NVIDIA. It works in three stages: network model parsing, engine optimization, and inference execution. In practical applications, TensorRT does not require any deep learning framework support to achieve the inference acceleration of the existing network models, which improves the real-time performance of the system.

SiamRPN, a single-target tracking algorithm that was proposed by Li et al. [26] at a computer vision and pattern recognition conference (CVPR) in 2018, is used for large-scale end-to-end image training and online detection of targets in an offline environment. It addresses the problem of target interference by lighting, distortion, and occlusion in visual tracking and the time-consumption problem of traditional tracking algorithms. Due to its good real-time performance, it can be applied to airborne embedded platforms. The SiamRPN algorithm consists of a Siamese network and a region proposal network (RPN), and its network structure is shown in Figure 3. The tracking result is determined based on the score of the classification and the offset of the regression, which means the target location and size are obtained.

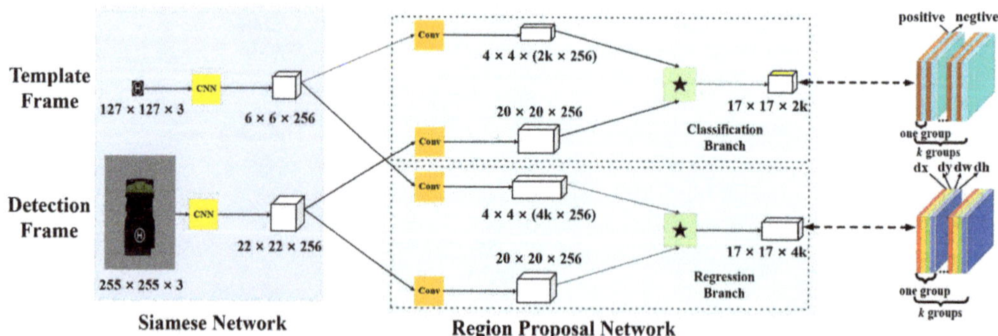

Figure 3. SiamRPN model structure. The star symbol represents Depthwise convolution.

2.1.2. Designation of Target Design

To build a target dataset to train the network model, the target shape and size need to be designed. In the calculation for the processing of relative position estimation, we assumed that the XY plane of the UAV body coordinate system would remain parallel to the horizontal plane of the target. For the accuracy of estimation, we designed the external contour of the target as a circular frame, making the length and width of the detection frame as equal as possible. Its outermost border was 50 cm in length and 50 cm in width, the same size as a typical landing target. To make the network learn the general features of the target and improve the recall rate when the target is occluded, we experimented with various internal markers, such as "A", "W", and "H". The "H" target worked best. The overall width and height of the internal marker was 30 cm. To resist the mosaic effect, the target and background plate were patterned in white and black. The final target effect is shown in Figure 4.

Figure 4. Designation of the target.

2.1.3. Designated Target Anti-Interference Tracking Method

As mentioned earlier, YOLOv5 was trained with a large-scale target detection dataset and had high model capacity and good algorithm accuracy. It can effectively detect all the target locations when the background changes, but lacks the specified target information. SiamRPN can be used for target tracking alone, but it has lower accuracy than YOLOv5 and is prone to tracking failure when the environment changes, resulting in target loss. Both algorithms are suitable for embedded platforms because of their relative lightness. To obtain the pixel coordinates of the target and improve the tracking accuracy, this paper proposes a DTAT method that integrates YOLOv5 and SiamRPN. The structure of DTAT is shown in Figure 5.

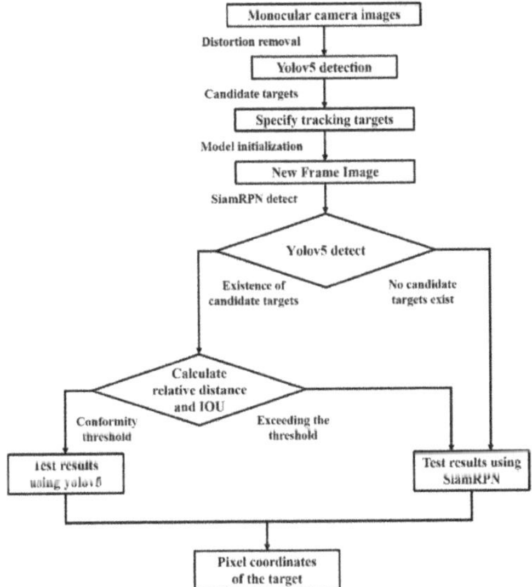

Figure 5. DTAT method structure.

The DTAT method first uses the YOLOv5 algorithm to detect the de-distorted images transmitted by the UAV's on-board camera to obtain all potential target candidate frames. The tracking target is artificially specified and its YOLOv5 detection frame is used as a sampling sample to initialize the SiamRPN model. When the new image frame arrives, it is simultaneously fed into the YOLOv5 and SiamRPN networks to obtain two detection results. When there is no YOLOv5 detection result in the image, the detection result of SiamRPN is used, while the sampling sample is not updated. Intersection over union (IoU) is an evaluation metric used to measure the accuracy of an object detector on a particular dataset. When the YOLOv5 target detection results are present, the IoU and centroid distance between all target detection frames and SiamRPN detection frames are judged. When the IOU value is greater and the centroid distance is less than the set threshold, the YOLOv5 detection frame with the largest IOU value is used as the target-tracking result, and the pixel coordinates of the target frame are output; then the target frame is used as the SiamRPN sampling sample for the new image frame. If there is no target detection frame within the threshold, the SiamRPN detection result is used and the sampling sample is not updated. In summary, according to the DTAT method, the final output of the target frame information with center pixel coordinates in the image is prepared for the UAV speed control. The above IoU and distance thresholds are designed empirically and are related to the camera parameters.

2.2. Target and UAV Relative Position Estimation

To control the flight speed of the UAV, its relative position to the target has to be calculated in real time. It is assumed that the camera is rigidly connected to and mounted directly below the UAV with the lens pointing vertically downward. Due to the compact structure of the UAV, the camera origin and the UAV body origin are considered to be the same point. The right-handed coordinate system is used as the reference to establish the coordinate system of the relative position of the UAV and the target, as shown in Figure 6. The figure includes five coordinate systems at different levels: A, pixel coordinate system $O(u, v)$; B, image plane coordinate system $O_s(x, y)$; C, camera coordinate system $O_c(X_c, Y_c, Z_c)$; D, target body inertia system $O_w(X_w, Y_w, Z_w)$; and E, UAV body coordinate system $O_b(X_b, Y_b, Z_b)$. The following describes the conversion relationship between coordinates and the relative position estimation method.

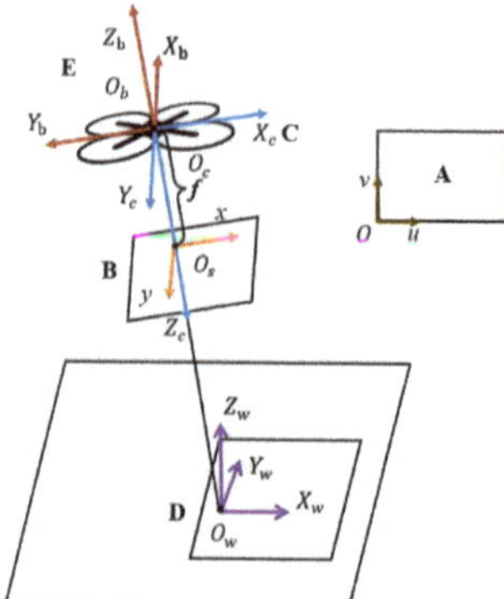

Figure 6. Relative position estimation coordinate system.

2.2.1. Coordinate SYSTEM Conversion Relationship

The image plane coordinate system is the coordinate system of the camera's light-sensitive imaging element. If the light-sensitive size of each pixel is d_x and d_y and the coordinates of the image plane coordinate system origin O_s in the pixel coordinate system are (u_0, v_0), the conversion relationship between the two coordinate systems is the following equation:

$$\begin{cases} u = \frac{x}{d_x} + u_0 \\ v = \frac{y}{d_y} + v_0 \end{cases} \tag{1}$$

Based on the pinhole imaging principle, we set the camera focal length as f and the distance from the optical center to the object as U, and the following basic relationship exists:

$$\frac{X}{U} = \frac{x}{f} \tag{2}$$

The relationship between the image plane coordinates (x, y) of the image point p and the camera coordinates (X_c, Y_c, Z_c) of the object point P is as follow:

$$\begin{cases} \frac{x}{f} = \frac{X_c}{Z_c} \\ \frac{y}{f} = \frac{Y_c}{Z_c} \end{cases}, \tag{3}$$

The relationship between the pixel and camera coordinate systems can be obtained by substituting Equation (1) into Equation (3) as follows:

$$\begin{cases} u = \frac{X_c}{Z_c} f_u + u_0 \\ v = \frac{Y_c}{Z_c} f_v + v_0 \end{cases}, \tag{4}$$

In the formula, $f_u = \frac{f}{d_u}, f_v = \frac{f}{d_v}$. Rewriting the above equation into matrix form, we get:

$$Z_c \begin{bmatrix} u \\ v \\ 1 \end{bmatrix} = \begin{bmatrix} f_u & 0 & u_0 \\ 0 & f_v & v_0 \\ 0 & 0 & 1 \end{bmatrix} \begin{bmatrix} X_c \\ Y_c \\ Z_c \end{bmatrix} = KP, \tag{5}$$

In the above equation, the matrix composed of intermediate quantities is called the internal reference matrix of camera K. In summary, the conversion relationship between pixel and camera coordinates is obtained, and the method for finding the relative position is introduced below.

2.2.2. Relative Position Estimation Method

To solve the relative position, first we solve the coordinates of the target within the camera coordinate system. Knowing u, v, and the internal reference matrix K, to solve for X_c, Y_c, and Z_c, the Z_c value needs to be found first. The length of the target is the same as the width, and the actual length is assumed to be known. We use the two diagonal points of the border in the pixel coordinate system to estimate the Z_c value. Let the two points be (u_1, v_1) and (u_2, v_2). Their corresponding camera coordinates are (X_{c1}, Y_{c1}, Z_{c1}) and (X_{c2}, Y_{c2}, Z_{c2}). The relationship can be derived as the following equation:

$$\begin{cases} u_1 = \frac{X_{c1}}{Z_{c1}} f_u + u_0 \\ v_1 = \frac{Y_{c1}}{Z_{c1}} f_v + v_0 \end{cases} \qquad \begin{cases} u_2 = \frac{X_{c2}}{Z_{c2}} f_u + u_0 \\ v_2 = \frac{Y_{c2}}{Z_{c2}} f_v + v_0 \end{cases} \tag{6}$$

The two equations are subtracted to obtain the following equation:

$$\begin{cases} u_1 - u_2 = \frac{X_{c1}}{Z_{c1}} f_u - \frac{X_{c2}}{Z_{c2}} f_u \\ v_1 - v_2 = \frac{Y_{c1}}{Z_{c1}} f_v - \frac{Y_{c2}}{Z_{c2}} f_v \end{cases}, \tag{7}$$

Assuming that the UAV is parallel to the target plane, $Z_{c1} = Z_{c2}$, the following equation is obtained:

$$\begin{cases} u_1 - u_2 = f_u * \left(\frac{X_{c1} - X_{c2}}{Z_c} \right) \\ v_1 - v_2 = f_v * \left(\frac{Y_{c1} - Y_{c2}}{Z_c} \right) \end{cases}, \tag{8}$$

This leads to the following equation:

$$\begin{cases} Z_c = f_u * \left(\frac{X_{c1} - X_{c2}}{u_1 - u_2} \right) \\ Z_c = f_v * \left(\frac{Y_{c1} - Y_{c2}}{v_1 - v_2} \right) \end{cases}, \tag{9}$$

Depth Z_c can be calculated from the target width and height. During the experiment, to address the situation where only a part of the image is in the field of view when the target is at the edge, the longer edge of the target detection is selected to calculate depth Z_c, and then the X_c and Y_c values are calculated to obtain the coordinates X_c, Y_c, Z_c of the target in the camera coordinate system.

From the above definition and as shown in Figure 6, the origins of the camera and of the body coordinate systems are the same, dual x-, y-direction and opposite z-direction, and the matrix is expressed as follows:

$$\begin{bmatrix} X_b \\ Y_b \\ Z_b \end{bmatrix} = \begin{bmatrix} 0 & -1 & 0 \\ -1 & 0 & 0 \\ 0 & 0 & -1 \end{bmatrix} \begin{bmatrix} X_c \\ Y_c \\ Z_c \end{bmatrix}, \tag{10}$$

Based on the three-axis rotation matrix, the equation for converting from the UAV body coordinate system to the body inertial system is as follows:

$$\begin{bmatrix} X_e \\ Y_e \\ Z_e \end{bmatrix} = R \begin{bmatrix} X_b \\ Y_b \\ Z_b \end{bmatrix}$$

$$R = \begin{bmatrix} \cos\theta\cos\psi & \cos\theta\sin\psi & -\sin\theta \\ \sin\phi\sin\theta\cos\psi - \cos\phi\sin\psi & \sin\phi\sin\theta\sin\psi + \cos\phi\cos\psi & \sin\phi\cos\theta \\ \cos\phi\sin\theta\cos\psi + \sin\phi\sin\psi & \cos\phi\sin\theta\sin\psi - \sin\phi\cos\psi & \cos\phi\cos\theta \end{bmatrix} \tag{11}$$

In the above equation, R is the rotation matrix, where θ, ψ, and ϕ are the pitch, yaw, and roll angles, respectively, and the magnitudes are calculated by the UAV flight control based on the IMU information. The coordinate values X_e, Y_e, and Z_e of the target within the body inertial system are obtained, which means that the position estimation of the UGV relative to the UAV is obtained, and then the UAV tracking and landing are controlled according to the estimated coordinate values.

2.3. Drone Tracking and Landing Control

Based on the relative position information obtained above, the three-axis velocities v_{xe}, v_{ye}, and v_{ze} of the UAV under the body inertia system need to be calculated. Since the UAV height can be controlled individually, the main focus is on the two-axis velocities v_{xe} and v_{ye}. The tracking speed is controlled by the discrete PID method during UAV tracking. Using the above relative position information as the error term, the velocity control method is established as follows:

$$\begin{cases} v_{xe2} = K_p \times x_{2e} + K_d(x_{2e} - x_{1e}) \\ v_{ye2} = K_p \times y_{2e} + K_d(y_{2e} - y_{1e}) \end{cases}, \tag{12}$$

In the equation, $v_{xe2}, v_{xe1}, v_{ye2}$, and v_{ye1} refer to the velocity of the UAV on the x- and y-axes of the body inertia system for the current and previous frame, respectively. x_{2e}, x_{1e} are the position of the target's current and previous frame on the x-axis under the body inertia system, respectively. y_{2e}, y_{1e} are the position of the target's current and previous frame on the y-axis under the body inertia system, respectively. K_p is the scale factor and K_d is the differential factor. Due to the uncertainty of the target, the UAV tracking process changes dynamically, so the integration term is not set.

For the scale factor K_p, when K_d is 0, the UAV control state changes to linear control; assuming the target speed is 1 m/s and the value of K_p is 0.5, there will be a steady-state error of 2 m between the UAV and the target, which in turn leads to tracking failure. Based on the consideration of stability and oscillation reduction, $K_p = 1.5$ is set, and the target may be lost when it moves too fast. To avoid oscillation, the trend of the target is predicted by increasing the differential term and $K_d = 0.6$ is set.

During the UAV landing, the drone continuously tracks the target. When the detected object is located in a circular area with a radius of 0.03 m from the origin in the UAV body inertial system, the system instructs the UAV to land directly. Due to the existence of steady-state errors in the tracking process and practical considerations, the target needs to be stationary before the landing process to achieve a safe landing.

3. Results

To evaluate the performance of the autonomous tracking and landing system, a number of simulation experiments were conducted. First, to evaluate the image target tracking performance, a target tracking experiment was conducted. The experiment was divided into two parts: (1) YOLOv5 target detection and (2) DTAT target tracking. Second, we conducted experiments to estimate the relative position of the UAV to evaluate the accuracy. Finally, to evaluate the system's effectiveness, we conducted overall UAV tracking and landing experiments.

The simulation environment uses the Jetson Xavier NX processor, which is the same as the actual UAV flight environment. The simulation environment is built based on the Gazebo function package under the ROS framework. We added an unmanned vehicle with landing landmarks and a UAV with a monocular camera to imitate the real environment of UAV flight, and the experimental procedure and results are as follows.

3.1. Experiments and Results of Target Tracking in Images

3.1.1. Yolov5 Target Detection Experiment

First, the target dataset was established. Since the UAV tracking and landing experiment was conducted in the Gazebo simulation environment, the dataset was collected by intercepting the screen in that environment. The dataset consisted of 240 target images, including five top view angles of 90°, 80°, 70°, 60°, and 50° of the Z_c axis relative to the horizontal plane of the target, eight yaw angles of the target around the Z_c axis, and six size occupancies of the target in the field of view. Figure 7 shows some examples of the original and labeled target images at a top view angle of 90°, a yaw angle of 90°, and a large field-of-view occupancy.

Based on the above dataset, the YOLOv5 network model was trained and pretrained models were used. The hyperparameters were the same as the official recommendations, and the learning rate was trained with 100 epochs using cosine annealing. To qualitatively evaluate the model performance, we detected the target in a complex scene and the results are shown in Figure 8. The YOLOv5 target detection performance was good when the camera view was tilted, the target was occluded, and the target was small (fewer than 36 pixels in the field of view). For quantitative analysis of model performance, the three indices of mean average precision (mAP)@0.5, mAP@0.5:0.95, and frames per second (FPS) of YOLOv5 and YOLOv5 + TensorRT models were compared and analyzed based on this dataset and the results are shown in Table 1. We can see that for the YOLOv5 model, mAP@0.5 detection accuracy is 99.55%, mAP@0.5:0.95 is 87.16%, and FPS is 51.25 detection frames. YOLOv5 combined with TensorRT improved FPS by 32.2, although there was a decrease in detection accuracy, which better meets the real-time target detection task.

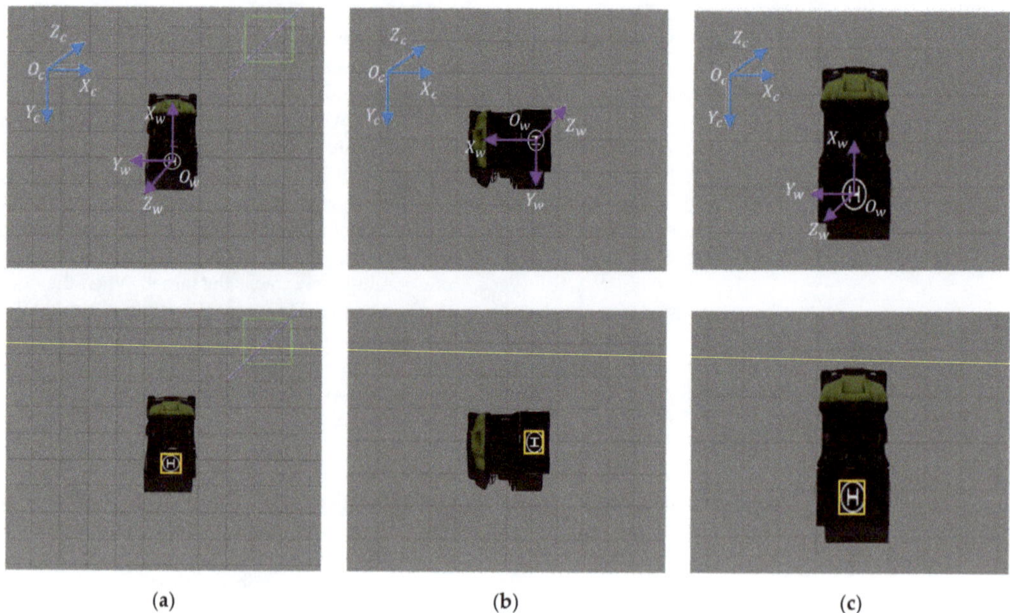

Figure 7. Examples of target images with labels: (**a**) top view angle of 90°; (**b**) yaw angle of 90°; and (**c**) large field-of-view occupancy.

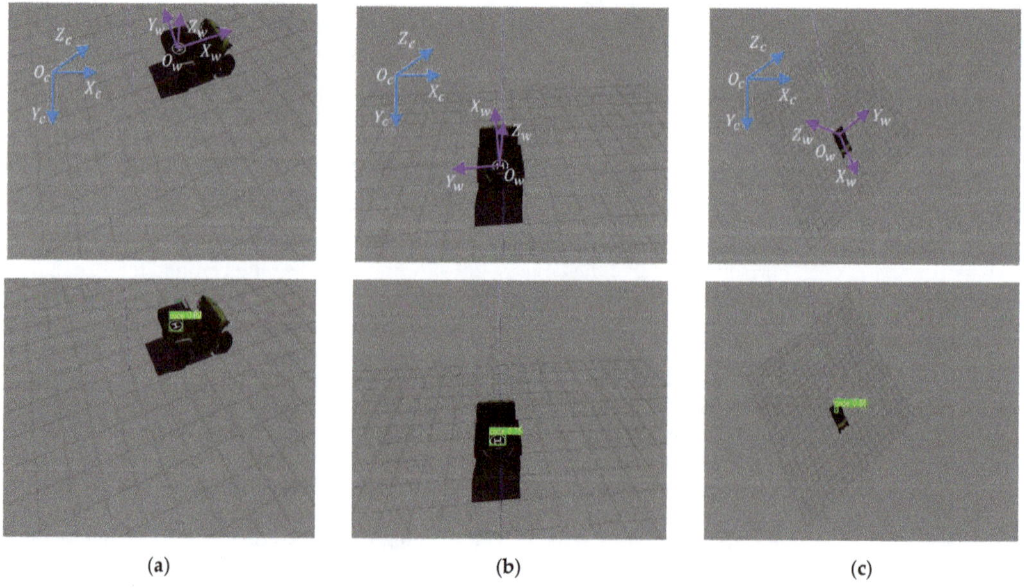

Figure 8. Examples of target images with labels: (**a**) camera tilt view; (**b**) target occlusion; and (**c**) small target.

Table 1. Model Performance Comparison.

Model	mAP@0.5(%)	mAP@0.5:0.95(%)	FPS
YOLOv5	99.55	87.16	51.25
YOLOv5+TensorRT	99.46	86.23	83.45

3.1.2. Designated Target Anti-Interference Tracking Experiment

To effectively evaluate the target tracking performance of the DTAT method, we built a multi-target disturbance scenario based on the Gazebo platform, including a UAV platform with a moving target to be tracked and one stationary disturbance target. The tracking experiments were based on video of the moving target captured by the UAV. The target moved forward while the UAV was hovering and the airframe carried a camera to collect video information. The 200 frames where the target was in the UAV's field of view were intercepted as the target tracking evaluation dataset, and DarkLabel software was used to construct the label information of the target for model evaluation.

One-pass evaluation (OPE) [27] was carried out, using the position of the first ground-truth frame of the target to initialize different tracking algorithms, and the recognition results of the algorithms for each frame were compared with the true value to obtain the precision rate (P) and success rate (S) to quantitatively analyze the tracking model. The results of accuracy rate and success rate of the SiamRPN and DTAT algorithms are shown in Figure 9a,b, respectively. Due to the small dataset, both models exhibited some degree of overfitting in both the accuracy and success rate plots. To effectively differentiate the effect, we changed the accuracy rate metric to the integrated area of the curve with respect to the x-axis, as shown in the figure legend. On the precision rate index, the DTAT method performed better at a low pixel threshold with high tracking accuracy. On the success rate index, the DTAT method could still recall the target under a high IOU threshold, and the tracking regression rate was high. In summary, compared with the single SiamRPN algorithm, the DTAT method performed better.

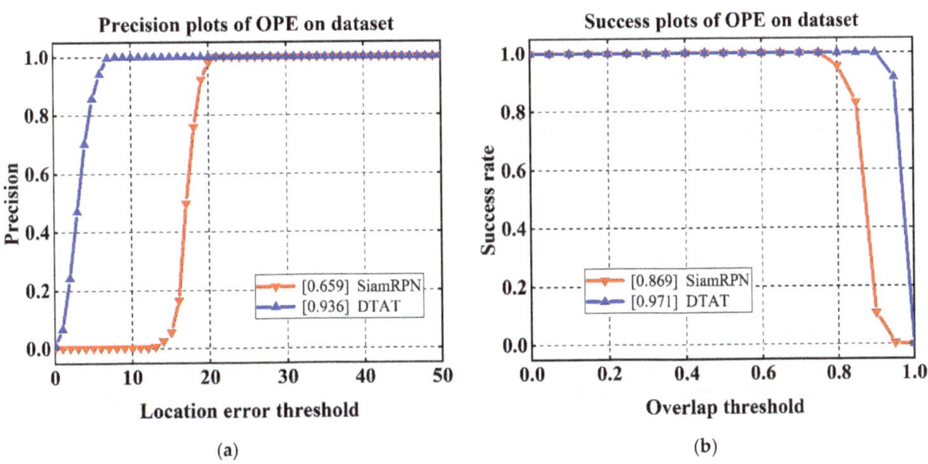

Figure 9. Performance comparison of tracking algorithms: (**a**) OPE accuracy; and (**b**) OPE success rate.

We selected frames 20 and 167 to show the tracking results, as shown in Figure 10. The yellow dotted and solid boxes shown in the figure represent the interference target and designated target obtained by the DTAT method. The red box represents the tracking of the designated target obtained by the SiamRPN algorithm. Both Figure 10a,b showed that the DTAT algorithm was able to distinguish between the specified target and the interfering target, and also had higher tracking accuracy for the designated target compared with the SiamRPN algorithm.

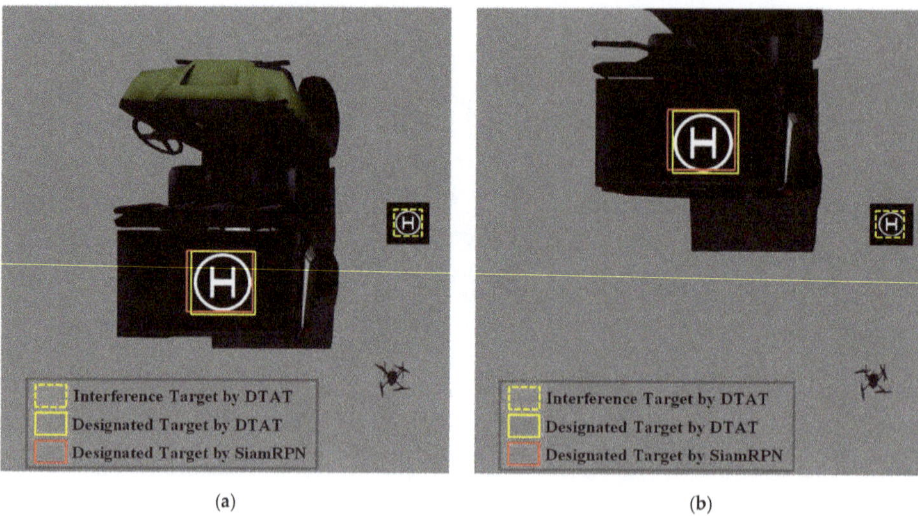

(a)   (b)

Figure 10. Multi-frame target tracking result. (**a**) Frame 20. (**b**) Frame 167.

3.2. Simulation Experiments and Results of Relative Position Estimation

Based on the pixel coordinates obtained by DTAT, we estimated the relative positions of the UAV and the ground target. To verify the effectiveness of relative position estimation, we estimated the relative position of the UAV to the target in the following four cases. As shown in Figure 11, the target was placed on the ground and the drone was in hovering mode. The yaw angle of the body rotation relative to the UAV system Z_b was $0°$, $90°$, $180°$, and $270°$. To eliminate the occasional factor, the experiment was repeated four times at each yaw angle.

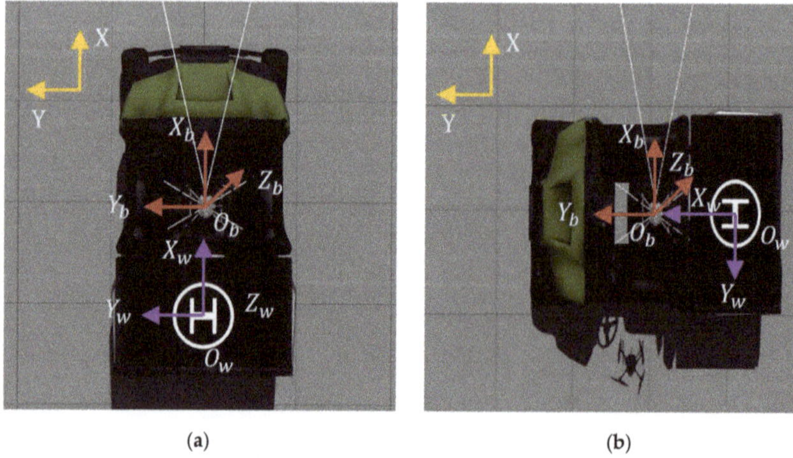

(a)   (b)

Figure 11. *Cont.*

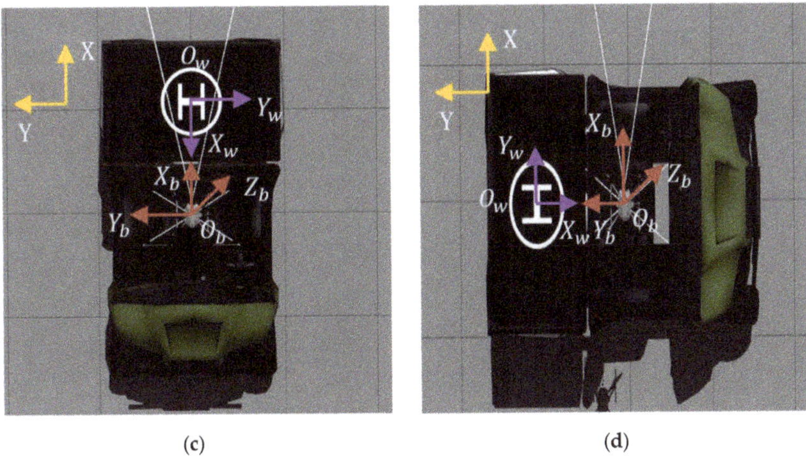

(c)   (d)

Figure 11. Stationary target with different yaw angles. (**a**) Yaw = 0°. (**b**) Yaw = 90°. (**c**) Yaw = 180°. (**d**) Yaw = 270°.

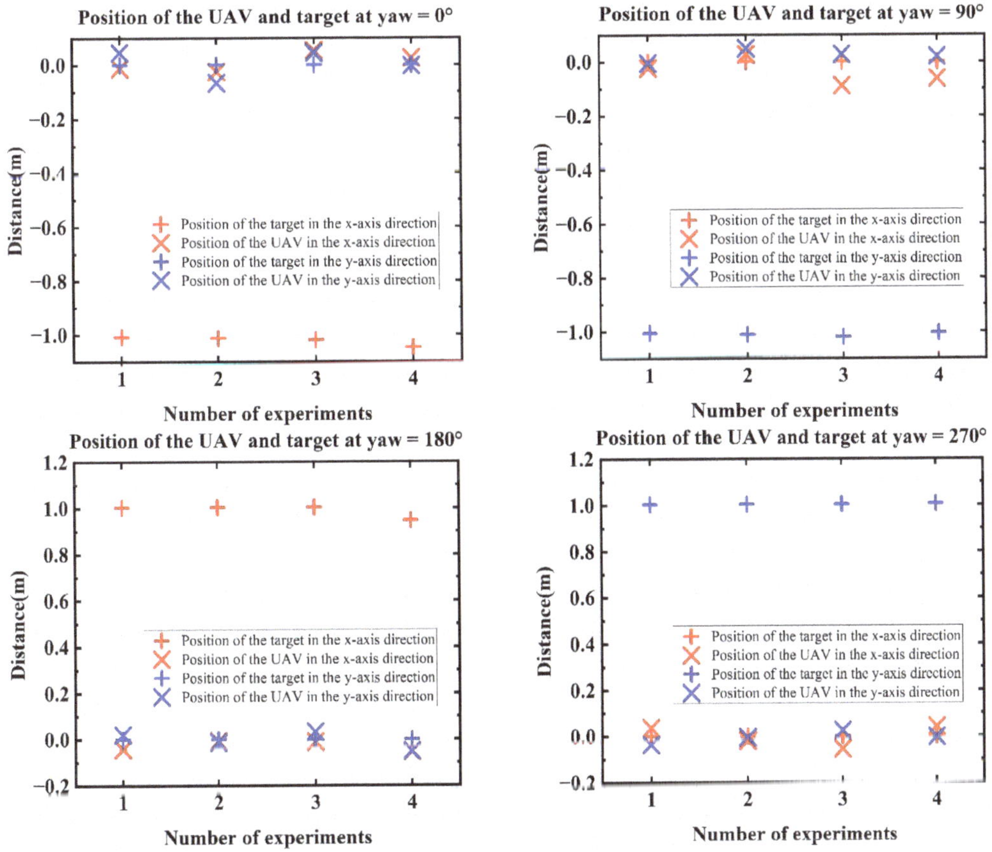

(a)

Figure 12. *Cont.*

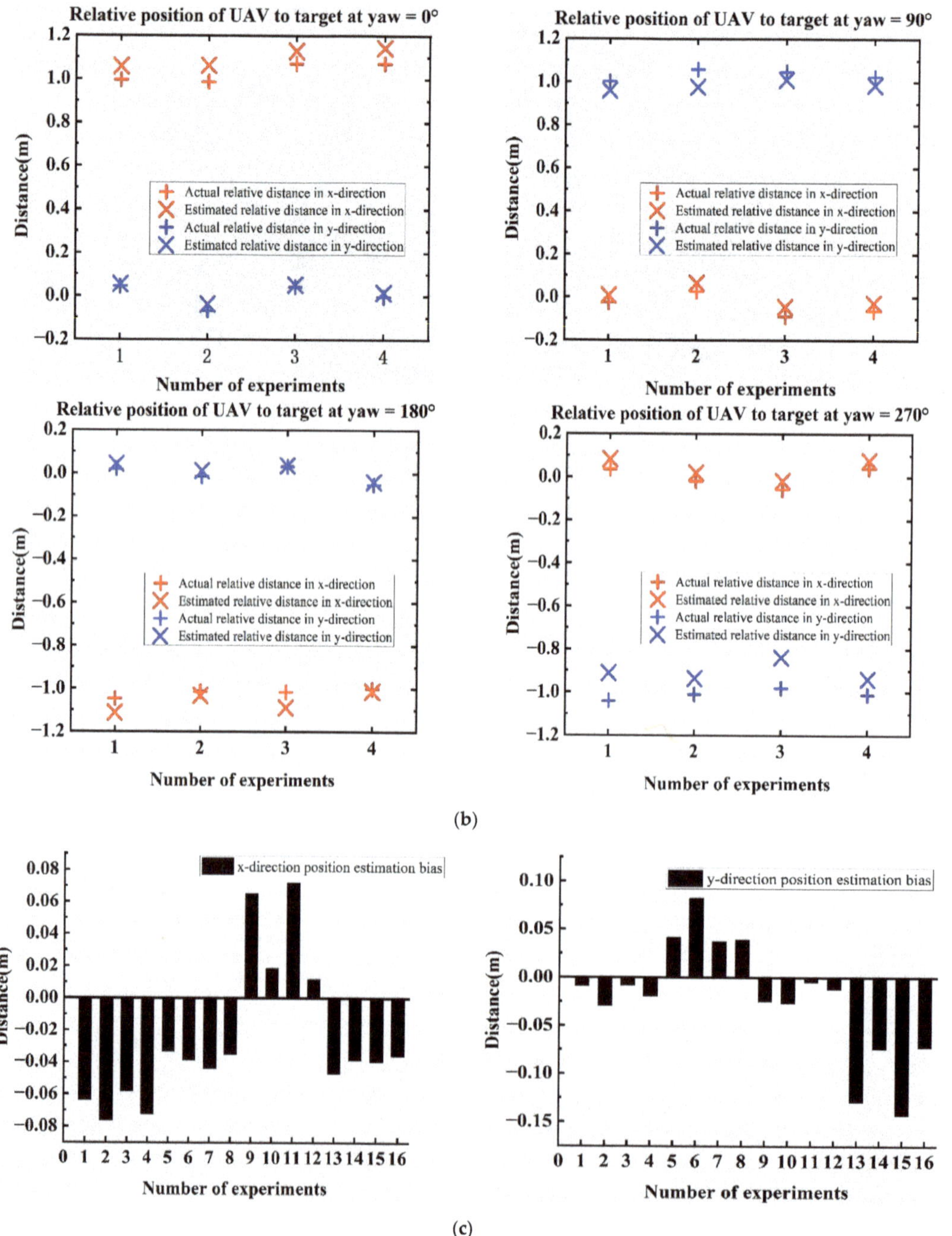

Figure 12. (**a**) True position of drone and target; (**b**) relative difference between real and estimated distance; and (**c**) bias in relative position estimation algorithm.

Based on the monocular camera information from the UAV, the pixel coordinates of the target were detected and the relative position was estimated. The estimation bias was obtained by subtracting the estimated value from the true value provided by Gazebo to verify the estimation effect, and the experimental results are shown in Figure 12. Figure 12a shows the real positions of the target and the UAV provided by Gazebo, that is, the coordinate positions of the x- and y-axis. The target and UAV coordinates were subtracted to obtain the true value of the relative position. Figure 12b shows the estimated value obtained by the relative position algorithm and the true value obtained by the true position calculation. Finally, the results of four replicate experiments for each yaw were summarized to obtain the errors of relative position estimation in the x- and y-directions in 16 experiments, as shown in Figure 12c.

As shown in Figure 12c, the error range of the relative position estimation algorithm with a static target was no more than 8 cm maximum error in the x-direction and 15 cm in the y-direction. The experimental results indicate that the relative position estimation algorithm met the UAV tracking and landing accuracy requirements.

3.3. Target Tracking and Landing Simulation Experiments and Results

To effectively evaluate the performance of the system, we designed a simulation experiment of UAV tracking and landing. The target was set to move along different paths (reciprocal straight line, circle, and rectangle), and the UAV tracked the target after a period of time and landed when the target was stationary. We chose the reciprocal straight-line trajectory for quantitative analysis, and the experimental results are shown in Figure 13. It should be noted that the effective estimated height range of the airborne monocular camera was about 0.3–6 m. When the flight height of the UAV was higher than the effective range of the camera, the position estimation would be inaccurate. Therefore, the flight height of the UAV was controlled within the effective range.

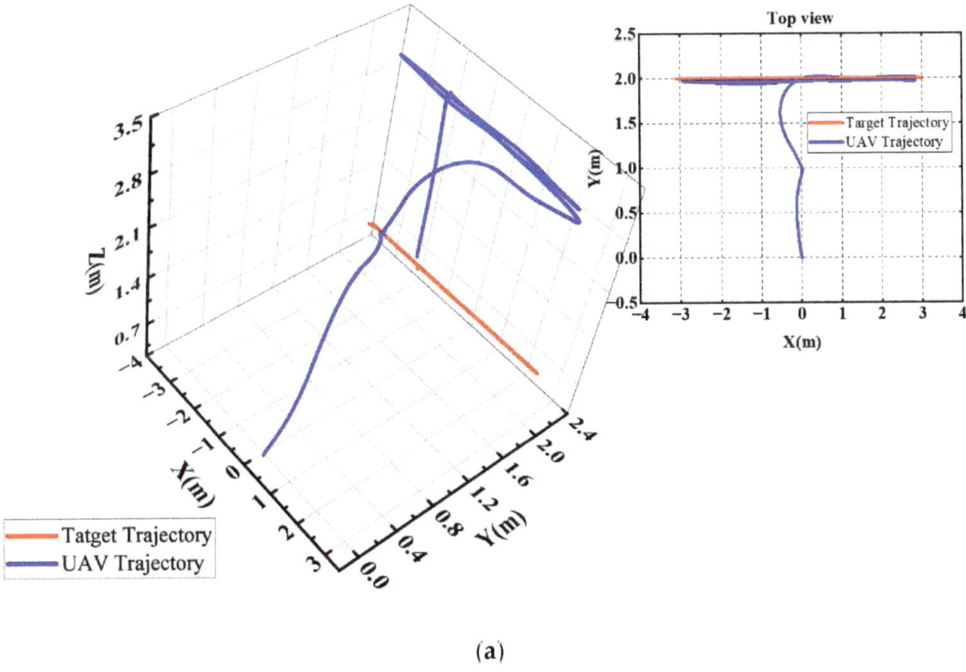

(a)

Figure 13. *Cont.*

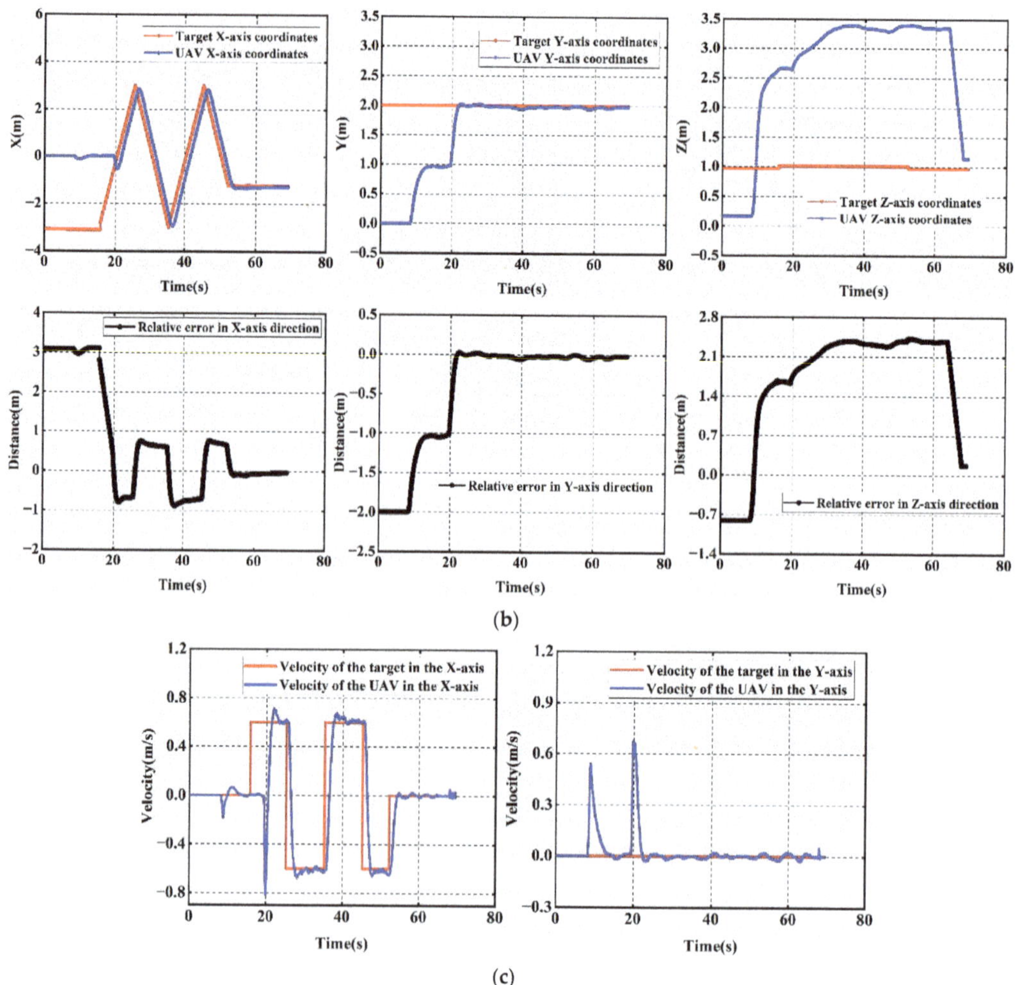

Figure 13. (a) 3D trajectory of drone and target; (b) tracking results and errors; and (c) speed comparison between drone and target.

Figure 13a shows the process of target tracking and landing of the UAV. The red and blue paths represent the real position trajectory curves of the target and the UAV, respectively. First, the UAV took off to the designated point, while the target made a circular reciprocal motion along the x-axis at 0.6 m/s. It can be seen from the Figure 13a that the UAV constantly followed the moving ground target, conformed to the target motion curve in the x- and y-directions, and maintained a relatively fixed height difference with the target in the z-direction. Figure 13b shows the tracking results of the UAV in the three axis directions with errors. As shown in the figure, the errors ranged from -0.8 to 0.8 (m) and -0.2 to 0.2 (m) in the x- and y-directions, respectively. During the motion, the relative height of the z-axis was set to 2 m, and the real-time error was up to 0.5 m. When the target was stationary and the UAV started to land, the landing error of both x- and y-axis was 0.01 m, indicating high landing accuracy. Figure 13c shows the real-time velocity of the UAV and the target, indicated by two colors. From the figure, it can be seen that the UAV's operation speed tended to be close to the real speed of the target, and when the target speed changed, the UAV could react faster and track the target continuously.

Figure 14 showed the UAV's following trajectory results when the target was moving in a circular and rectangular shape. When the target was moving at 0.4 m/s in rectangular and circular trajectories, respectively, the UAV's tracking trajectory matched the target's running trajectory and the system still worked well, although it could not overlap the target in real time.

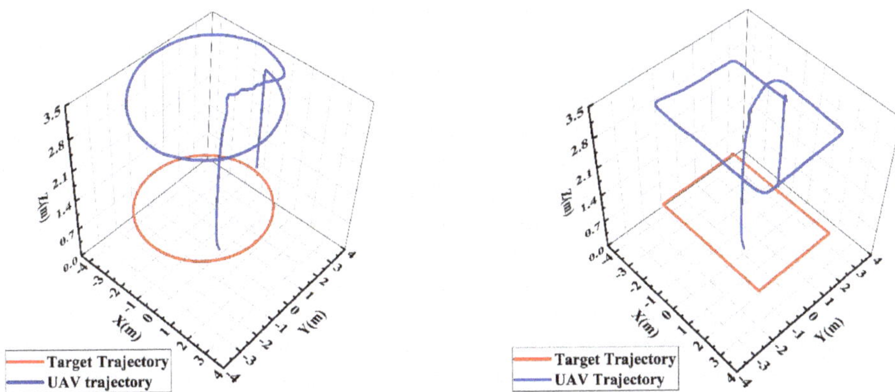

Figure 14. Drones tracking targets with different trajectories.

4. Discussion

With the above experimental results, we can eliminate the influence of interfering targets in the tracking process. The tracking system based on the DTAT method has more advantages of anti-interference and tracking accuracy compared with other tracking systems mentioned before. In this section, we further discuss the factors that affected the performance of the UAV tracking and landing system. First, for the DTAT method, we analyze the reasons for the method design and discuss the experimental results to propose the factors that affect the performance of the method and possible improvements. Second, for the relative position estimation experiments, the reasons affecting the experimental accuracy are analyzed. Finally, the overall situation of the system is discussed, and the limitations and areas for improvement are analyzed.

4.1. DTAT Method Design and Performance Analysis

The DTAT method combines YOLOv5 and SiamRPN, two algorithms used for target detection and tracking. The system was not designed using only a tracking algorithm based on the following considerations: the SiamRPN algorithm trains the target based on video frames, which has smaller data volume and fewer scenes. At the same time, the model capacity is smaller and the tracking accuracy is lower compared to the target-detection algorithm. The DTAT method is based on the advantages of the target-detection method and uses a large-scale image dataset, and the model has a better detection effect, while incorporating the target-tracking algorithm, which can achieve tracking of a specified target. In addition, target-detection and -tracking algorithms are now more widely studied; this paper mainly considers the features of light weight and portability and selects two classical algorithms, YOLOv5 and SiamRPN; the use of other model combinations can be considered in subsequent research.

In the tracking experiments using the DTAT method with the SiamRPN algorithm for a specified target, it can be seen from Figure 9 that SiamRPN had very low or even zero accuracy at low pixel thresholds. The DTAT method can quickly improve the accuracy at low pixel thresholds, which shows that the YOLOv5 algorithm is more robust in the tracking process and SiamRPN is easily affected by the results of the previous frame and has poor tracking accuracy. DTAT performs better in success-rate tracking experiments compared to SiamRPN and easily tracks targets, presumably mainly because the large YOLOv5 dataset provides robustness to target background changes and improves target recall.

In the DTAT method, we place the YOLOv5 detection results at a higher confidence level, but because of the limitations of the training dataset, background environmental noise, and the scaling of the field of view caused by the different flight altitudes of the UAV, sometimes neither target detection nor tracking can obtain accurate pixel coordinates, which in turn affects the relative position estimation of the target and leads to UAV tracking and landing failure. To effectively estimate the pixel coordinate position of the target, subsequent experiments could add Kalman filtering. When the target is within a certain distance threshold, only the original target detection result is used instead of the Kalman filter to fuse the target detection and tracking results, with the YOLOv5 detection result as the measured value and the SiamRPN tracking result as the state value, so as to estimate the target pixel coordinates and obtain improved accuracy.

4.2. Performance Analysis of Relative Position Estimation Method

The UAV was fixed and the target was placed around the z-axis of the UAV body system at four angles, and four sets of static estimation experiments were performed for each angle to verify the effectiveness of the relative position estimation algorithm. The final estimation error is shown in Figure 12c. It can be seen that each group of experiments had a certain regularity. When the x-axis of the target was parallel to the global x-axis, the relative position estimation method had a larger deviation in the x-direction and a smaller deviation in the y-direction. When the x-axis of the target was parallel to the global y-axis, the relative position estimation method deviated more in the y-direction and less in the x-direction. These deviations are always present, independent of the target's position in the airframe. We can conclude that the relative position estimation method is inaccurate in estimating the position of the target in the x-direction, i.e., the direction of the two long edges of H, and relatively accurate in estimating the position of the target in the y-direction, i.e., the direction of the shorter middle edge of H.

Therefore, the factor affecting the relative position estimation may be the training accuracy of the YOLOv5 network model, which is more accurate when it is better trained for the short-edge direction. Poor training for the long-edge direction results in lower accuracy. In actual flight, the UAV will have a certain degree of offset, which does not meet the assumptions of the relative position estimation algorithm and will therefore also produce a certain degree of error. To improve the relative position estimation accuracy, a model with higher accuracy can be selected for tracking, while controlling the degree of UAV offset and reducing the violent jitter during flight.

4.3. Tracking and Landing System Analysis and Constraints

In this system, the UAV uses a Jetson Xavier NX processor for autonomous tracking and landing. To evaluate the execution efficiency of the system, we built the same experimental simulation environment based on the ROS system framework. The environment consists of four nodes: the gazebo simulation node, the pix4 feedback node, the target detection and tracking node, and the nonlinear control node. The environment was deployed in the Jetson Xavier NX processor, and the processing time of each module is shown in Table 2. The system took 30.44 ms from inputting a picture to the UAV performing flight control, generating roughly 33 Hz to meet the actual flight requirements.

Table 2. Node test time.

Content	Jetson Xavier NX Test Time (ms)
Gazebo simulation node	3.24
Pix4 feedback node	2.73
Target detection and tracking node	19.72
Nonlinear control node	4.75
Total running time	30.44

Through qualitative and quantitative analysis, this paper verifies the tracking performance of a system for a designated target in the case of multi-target confusion; however, there are still some limitations of the system, mainly the following two aspects:

(1) The flight control algorithm takes the relative position of the airframe and the target as input variables and uses feedback PID control, so there is a certain steady-state error in the tracking process. In the process of following the target at low altitude, if the target moves too fast, it will rush out of the UAV's field of view and the UAV cannot respond effectively, which in turn leads to the failure of target tracking.

(2) The system needs to have real information about the target in advance. The YOLOv5 detection process requires datasets made by real targets, and the relative positional solution process relies on the real size information of the target. In the actual tracking process, it is difficult to track unfamiliar targets.

5. Conclusions

In this paper, we proposed a DTAT method and built an autonomous UAV tracking and landing system that enables UAV tracking and the landing of a designated target in an environment with multiple interference targets. The image information transmitted by the monocular camera was processed by DTAT and coordinate transformation to obtain the spatial coordinate information of the specified target; then, the discrete PID control method was adopted to realize tracking and the landing of the UAV on the designated target.

Extensive simulation experiments were conducted to evaluate the system performance. The effectiveness and reliability of the algorithm and system were analyzed qualitatively and quantitatively with regard to various aspects, such as the target pixel coordinates, the position of the UAV relative to the target, and the tracking and landing process. The results show that the relative distance between the UAV and the target during the tracking process when the target was moving at 0.6 m/s did not exceed 0.8 m. The landing error of the UAV during the landing process after the target was stationary did not exceed 0.01 m. In addition, the advantages and disadvantages of the system were discussed at the level of both processor hardware and algorithm design, and corresponding suggestions were made for different problems. In the future, we will use multi-sensor fusion, such as radar and a depth camera, to achieve target tracking and landing without a priori information, and explore more intelligent and stable tracking algorithms to make UAVs more adaptable to complex environments.

Author Contributions: Conceptualization, H.Z. and D.W.; methodology, D.W.; software, D.W.; validation, H.Z., D.W. and Y.L.; formal analysis, D.W.; investigation, D.W.; resources, H.Z.; data curation, D.W.; writing—original draft preparation, D.W.; writing—review and editing, D.W. and H.Z.; visualization, D.W.; supervision, H.Z.; project administration, H.Z.; funding acquisition, H.Z. All authors have read and agreed to the published version of the manuscript.

Funding: This research was funded by Jilin Province Development and Reform Commission (2020C 018-2, 2020). The vision-based navigation for autonomous UAV tracking and landing control technology project.

Data Availability Statement: The data and code presented in this study are openly available in Zenodo at https://doi.org/10.5281/zenodo.6500300.

Conflicts of Interest: The authors declare no conflict of interest.

References

1. Shakhatreh, H.; Sawalmeh, A.H.; Al-Fuqaha, A.; Dou, Z.; Almaita, E.; Khalil, I.; Othman, N.S.; Khreishah, A.; Guizani, M. Unmanned Aerial Vehicles (UAVs): A Survey on Civil Applications and Key Research Challenges. *IEEE Access* **2019**, *7*, 48572–48634. [CrossRef]

2. Fan, Y.; Ding, M.; Cao, Y. Vision algorithms for fixed-wing unmanned aerial vehicle landing system. *Sci. China-Technol. Sci.* **2017**, *60*, 434–443. [CrossRef]

3. Yao, H.; Qin, R.J.; Chen, X.Y. Unmanned Aerial Vehicle for Remote Sensing Applications-A Review. *Remote Sens.* **2019**, *11*, 1443. [CrossRef]

4. Niu, H.; Ji, Z.; Liguori, P.; Yin, H.; Carrasco, J. Design, Integration and Sea Trials of 3D Printed Unmanned Aerial Vehicle and Unmanned Surface Vehicle for Cooperative Missions. In Proceedings of the IEEE/SICE International Symposium on System Integration (SII), Electr Network, Iwaki, Japan, 11–14 January 2021; pp. 590–591.

5. Meng, Y.; Wang, W.; Han, H.; Ban, J. A visual/inertial integrated landing guidance method for UAV landing on the ship. *Aerosp. Sci. Technol.* **2019**, *85*, 474–480. [CrossRef]

6. Lv, M.; Li, Y.; Hu, J.; Zhao, C.; Hou, X.; Xu, Z.; Pan, Q.; Jia, C. Multi-sensor Data Fusion for UAV Landing Guidance Based on Bayes Estimation. In Proceedings of the 3rd International Conference on Unmanned Systems (ICUS), Harbin, China, 27–28 November 2020; pp. 721–726.

7. Wu, Z.; Han, P.; Yao, R.; Qiao, L.; Zhang, W.; Shen, T.; Sun, M.; Zhu, Y.; Liu, M.; Fan, R.; et al. Autonomous UAV Landing System Based on Visual Navigation. In Proceedings of the IEEE International Conference on Imaging Systems and Techniques (IST). IEEE International School on Imaging, Abu Dhabi, United Arab Emirates, 8–10 December 2019.

8. Chang, C.-W.; Lo, L.-Y.; Cheung, H.C.; Feng, Y.; Yang, A.-S.; Wen, C.-Y.; Zhou, W. Proactive Guidance for Accurate UAV Landing on a Dynamic Platform: A Visual-Inertial Approach. *Sensors* **2022**, *22*, 404. [CrossRef] [PubMed]

9. Jiang, J.; Qi, G.; Huang, G.; IOP. Design for Guidance Method of UAV Autonomous Landing on Mobile Platform Based on Prediction of Intersection Points. In Proceedings of the 9th Asia Conference on Mechanical and Aerospace Engineering (ACMAE), Singapore, 29–31 December 2018.

10. Santos, N.P.; Lobo, V.; Bernardino, A. A Ground-Based Vision System for UAV Tracking. In Proceedings of the Oceans 2015 Genova, Ctr Congressi Genova, Genova, Italy, 18–21 May 2015.

11. Cheng, H.-W.; Chen, T.-L.; Tien, C.-H. Motion Estimation by Hybrid Optical Flow Technology for UAV Landing in an Unvisited Area. *Sensors* **2019**, *19*, 1380. [CrossRef] [PubMed]

12. Pavlenko, T.; Schuetz, M.; Vossiek, M.; Walter, T.; Montenegro, S. Wireless Local Positioning System for Controlled UAV Landing in GNSS-Denied Environment. In Proceedings of the IEEE International Workshop on Metrology for AeroSpace (MetroAeroSpace), Torino, Italy, 19–21 June 2019; pp. 171–175.

13. Milani, I.; Bongioanni, C.; Colone, F.; Lombardo, P. Fusing Measurements from Wi-Fi Emission-Based and Passive Radar Sensors for Short-Range Surveillance. *Remote Sens.* **2021**, *13*, 3556. [CrossRef]

14. Lee, J.Y.; Chung, A.Y.; Shim, H.; Joe, C.; Park, S.; Kim, H. UAV Flight and Landing Guidance System for Emergency Situations. *Sensors* **2019**, *19*, 4468. [CrossRef] [PubMed]

15. Krasuski, K.; Wierzbicki, D.; Bakula, M. Improvement of UAV Positioning Performance Based on EGNOS + SDCM Solution. *Remote Sens.* **2021**, *13*, 2597. [CrossRef]

16. Vezinet, J.; Escher, A.C.; Guillet, A.; Macabiau, C.; ION. State of the Art of Image-aided Navigation Techniques for Aircraft Approach and Landing. In Proceedings of the International Technical Meeting of the Institute-of-Navigation, San Diego, CA, USA, 27–29 January 2013; pp. 473–485.

17. Lebedev, I.; Erashov, A.; Shabanova, A. Accurate Autonomous UAV Landing Using Vision-Based Detection of ArUco-Marker. In Proceedings of the 5th International Conference on Interactive Collaborative Robotics (ICR), Electr Network, St. Petersburg, Russia, 7–9 October 2020; pp. 179–188.

18. Phang, S.K.; Chen, X. Autonomous Tracking And Landing On Moving Ground Vehicle With Multi-Rotor UAV. *J. Eng. Sci. Technol.* **2021**, *16*, 2795–2815.

19. Dergachov, K.; Bahinskii, S.; Piavka, I. The Algorithm of UAV Automatic Landing System Using Computer Vision. In Proceedings of the IEEE 11th International Conference on Dependable Systems, Services and Technologies (DESSERT)-IoT, Big Data and AI for a Safe & Secure World and Industry 4.0, Kyiv, Ukraine, 14–18 May 2020; pp. 247–252.

20. Jia, C.; Zhen, Z.; Ma, K.; Yang, L. Target Tracking for Rotor UAV Based on Multi-Scale Compressive Sensing. In Proceedings of the IEEE Chinese Guidance, Navigation and Control Conference (CGNCC), Nanjing, China, 12–14 August 2016; pp. 1220–1225.

21. Hao, J.; Zhou, Y.; Zhang, G.; Lv, Q.; Wu, Q. A Review of Target Tracking Algorithm Based on UAV. In Proceedings of the IEEE International Conference on Cyborg and Bionic Systems (CBS), Chinese Acad Sci, Shenzhen Inst Adv Technol, Shenzhen, China, 25–27 October 2018; pp. 328–333.

22. Chen, P.; Zhou, Y. The Review of Target Tracking for UAV. In Proceedings of the 14th IEEE Conference on Industrial Electronics and Applications (ICIEA), Xi'an, China, 19–21 June 2019; pp. 1800–1805.

23. Yang, T.; Ren, Q.; Zhang, F.; Xie, B.; Ren, H.; Li, J.; Zhang, Y. Hybrid Camera Array-Based UAV Auto-Landing on Moving UGV in GPS-Denied Environment. *Remote Sens.* **2018**, *10*, 1829. [CrossRef]

24. Bochkovskiy, A.; Wang, C.Y.; Liao, H.Y.M. YOLOv4: Optimal Speed and Accuracy of Object Detection. *arXiv* **2020**, arXiv:2004.10934.

25. Redmon, J.; Farhadi, A. YOLOv3: An Incremental Improvement. *arXiv* **2018**, arXiv:1804.02767.

26. Li, B.; Yan, J.; Wu, W.; Zhu, Z.; Hu, X. High Performance Visual Tracking with Siamese Region Proposal Network. In Proceedings of the 31st IEEE/CVF Conference on Computer Vision and Pattern Recognition (CVPR), Salt Lake City, UT, USA, 18–23 June 2018; pp. 8971–8980.

27. Henriques, J.F.; Caseiro, R.; Martins, P.; Batista, J. High-speed tracking with kernelized correlation filters. *IEEE Trans. Pattern Anal. Mach. Intell.* **2014**, *37*, 583–596. [CrossRef]

remote sensing

MDPI

Communication

Remote Sensing Monitoring of Winter Wheat Stripe Rust Based on mRMR-XGBoost Algorithm

Xia Jing [1,*], Qin Zou [1], Jumei Yan [1], Yingying Dong [2] and Bingyu Li [1]

[1] College of Geometrics, Xi'an University of Science and Technology, Xi'an 710054, China; 19210210046@stu.xust.edu.cn (Q.Z.); 19210061040@stu.xust.edu.cn (J.Y.); 20210061021@stu.xust.edu.cn (B.L.)
[2] Key Laboratory of Digital Earth Science, Aerospace Information Research Institute, Chinese Academy of Sciences, Beijing 100094, China; dongyy@aircas.ac.cn
[*] Correspondence: jingxia@xust.edu.cn

Abstract: For the problem of multi-dimensional feature redundancy in remote sensing detection of wheat stripe rust using reflectance spectrum and solar-induced chlorophyll fluorescence (SIF), a feature selection and disease index (DI) monitoring model combining mRMR and XGBoost algorithm was proposed in this study. Firstly, characteristic wavelengths selected by successive projections algorithm (SPA) were combined with the vegetation indices, trilateral parameters, and canopy SIF parameters to constitute the initial feature set. Then, the max-relevance and min-redundancy (mRMR) algorithm and correlation coefficient (CC) analysis were used to reduce the dimensionality of the initial feature set, respectively. Features selected by mRMR and CC were input as independent variables into the extreme gradient boosting regression (XGBoost) and gradient boosting regression tree (GBRT) to monitor the severity of stripe rust. The experimental results show that, compared with CC analysis, the monitoring accuracy of the features selected by mRMR in the XGBoost and GBRT models increased by 12% and 17% on average, respectively. Meanwhile, the mRMR-XGBoost model achieved the best monitoring accuracy (R^2 = 0.8894, RMSE = 0.1135). The R^2 between the measured DI and predicted DI of mRMR-XGBoost was improved by an average of 5%, 12%, and 22% compared with mRMR-GBRT, CC-XGBoost, and CC-GBRT models. These results suggested that XGBoost is more suitable for the remote sensing monitoring of wheat stripe rust, and mRMR has more advantages than the commonly used CC analysis in feature selection. Field survey data validation results also confirm that the mRMR-XGBoost algorithm has excellent monitoring applicability and scalability. The proposed model could provide a reference for data dimensionality reduction and crop disease index monitoring based on hyperspectral data.

Keywords: max-relevance and min-redundance; extreme gradient boosting regression; wheat stripe rust; feature selection; spectral index; solar-induced chlorophyll fluorescence; hyperspectral remote sensing

Citation: Jing, X.; Zou, Q.; Yan, J.; Dong, Y.; Li, B. Remote Sensing Monitoring of Winter Wheat Stripe Rust Based on mRMR-XGBoost Algorithm. *Remote Sens.* **2022**, *14*, 756. https://doi.org/10.3390/rs14030756

Academic Editor: Saeid Homayouni

Received: 15 December 2021
Accepted: 3 February 2022
Published: 6 February 2022

1. Introduction

Wheat stripe rust is a pandemic disease caused by *Puccinia striiformis f.* sp. *tritici* that can achieve cross-regional initial and re-infection through airflow. It is one of the most important disease types for wheat prevention and control in China [1]. Due to the lack of precise remote sensing monitoring technology for crop diseases, the production is generally based on undifferentiated regional control. This increases the cost of wheat planting and pesticide residues in the cultivated land. With the advancement of agricultural informatization, hyperspectral data is widely used in remote sensing monitoring of crop diseases. However, the high-dimensional and small sample data characteristics of hyperspectral data make their direct application ineffective. Extracting disease-sensitive features in hyperspectral data is the most potential method to solve feature redundancy.

In the VIS-NIR spectrum, various symptoms and physiological changes of diseases show specific responses in spectral reflectance [2]. The identification of crop diseases and prediction of disease severity can be realized by using the sensitive wavelengths of spectral response and the variation of the abnormal spectrum. On this basis, the spectral vegetation index constructed by sensitive spectrum combinations in hyperspectral data shows a clear relation to the physiological and biochemical processes of crops that are infected by pathogens. Several researches have shown that the spectral vegetation indices have additionally the potential to detect and differentiate plant diseases [3,4]. Meanwhile, specific form changes of the original hyperspectral data can enhance the difference in spectral characteristics [5]. These methods mainly extract and construct relevant spectral features by searching for hyperspectral bands that are more sensitive to disease severity. But a quantitative statement or the identification of a specific disease is impossible so far since these methods lack disease specificity. Therefore, choosing a suitable model construction method is the key to realizing the crop diseases remote sensing monitoring research in recent years [6,7]. The combination of hyperspectral feature selection and machine learning model has been applied to the research of some crop diseases identification and detection [8]. However, these researches are mostly based on reflectance spectrum data, which is greatly affected by background noise. Moreover, the reflectance spectrum mainly reflects the concentration information of biochemical components, which cannot directly reveal the photosynthetic physiological state of vegetation [9]. Solar-induced chlorophyll fluorescence (SIF) can non-destructively detect the photosynthetic physiology and stress status of plants [10]. More importantly, the SIF signal comes entirely from the measured crop, which is purer than the reflectance data. Comprehensive utilization of the advantages of reflectance spectroscopy in the detection of crop biochemical parameters and the advantages of SIF in the diagnosis of photosynthetic physiology, which can objectively reflect the real condition of crops under disease stress and improve the accuracy of remote sensing detection [11]. Jing et al. used the GA-SVR model to optimize the initial feature set and model parameters composed of reflectance features and SIF parameters, which achieved high-precision prediction of the severity of stripe rust [12]. However, as a random search method, genetic algorithm takes a long time to get a more accurate solution, and its efficiency still needs improvement. Compared with the current popular machine learning models, the extreme gradient boosting regression (XGBoost) has the appealing properties of limited sample learning, fast model training, strong mathematical explanation ability, and data feature invariance [13].

These researches only pay attention to the influence of the selected feature parameters as the input factors of the machine learning model on the prediction accuracy of crop disease severity and ignore the redundancy between the selected feature parameters. According to the above statement, the initial feature set of this study consisted of the selected reflectance indices and canopy SIF parameters. Then, the feature combination was selected from the initial set by the max-relevance and min-redundancy (mRMR) algorithm, which had the maximal relevance with the stripe rust disease index and the minimal redundancy among the selected features. These features were input as independent variables into the XGBoost model to construct a remote sensing monitoring model for the severity of wheat stripe rust disease.

2. Materials and Methods

2.1. Field Experimental Data Acquirement

The experiment was conducted at China Agricultural Science Experimental Station, Langfang City, Hebei Province (39°30′40″N, 116°36′20″E). The wheat cultivar in the study area is Mingxian 169, which is more sensitive to stripe rust. On 9 April 2018 (wheat rising period), a spore solution with a concentration of 0.09 mg/mL was used to inoculate wheat stripe rust by spraying. The study area was divided into healthy groups and infected groups. A 5-m isolation zone was set up between the healthy groups and the infected groups, and the healthy groups were sprayed with pesticides. Canopy spectrum data of wheat stripe rust in different severity were measured on May 18 (226 d after sowing), May 24 (232 d after sowing), and May 30 (232 d after sowing) by ASD Field Spec 4 surface spectrometer.

Measurements were carried out between 11:00 and 12:30 Beijing time to reduce the influence of observation angle and solar zenith angle. In addition, the canopy radiance data were corrected by the standard BsSO4 board before data collection. The disease index (DI) of wheat stripe rust was investigated using a 5-point sampling method [14]. On each inspection, the plants were grouped into one of nine classifications of disease incidence (0, 1%, 10%, 20%, 30%, 45%, 60%, 80%, and 100%). According to Equation (1), DI can be calculated based on the number of wheat leaves recorded at each severity level.

$$\mathrm{DI} = \frac{\sum(m \times f)}{n \times \sum f} \tag{1}$$

where, DI is the disease index, m is the value of each gradient, n is the highest gradient level value, and the f is the number of leaves in each gradient.

2.2. Extraction of Canopy SIF Parameters

Solar-induced chlorophyll fluorescence (SIF) has a filling effect at the Fraunhofer dark line. Based on this, scholars have proposed single-band SIF extraction methods such as FLD, 3FLD, and iFLD [15,16]. Existing studies have shown that 3FLD is more robust, which provides a more accurate estimation of SIF signal under different signal-to-noise ratios conditions [17]. Therefore, the 3FLD method was chosen to extract the canopy SIF radiation in the O2-A and O2-B bands, which can be estimated according to Equation (2). Moreover, to eliminate the influence of external factors at different time periods on canopy SIF, this study adopted relative SIF as the canopy SIF [18]. Its calculation is shown in Equation (3).

$$\overline{F}_{in} = \frac{\left(\omega_{left}I_{left} + \omega_{right}I_{right}\right)L_{in} - I_{in}\left(\omega_{left}L_{left} + \omega_{right}L_{right}\right)}{\left(\omega_{left}I_{left} + \omega_{right}I_{right}\right) - I_{in}} \tag{2}$$

$$F_{relative} = \overline{F}_{in} / I_{in} \tag{3}$$

where $\overline{F}_{in}$ is the canopy SIF radiation. ω_{left} and ω_{right} represent the weight of the left and right bands. L_{in}, L_{left}, and L_{right} represent the canopy reflectance radiance inside, left, and right of the absorption band. I_{in}, I_{left}, and I_{right} represent the solar irradiance inside, left, and right of the absorption band. In addition to calculating the canopy SIF directly by radiance, the reflectance band at 650–800 nm, which is greatly affected by chlorophyll fluorescence, can be used to obtain a reflectance index that can reflect the intensity of fluorescence as well. Therefore, this study also elects the reflectance ratio index [19] and the reflectance first derivative index [20] as the fluorescence feature input of the model. The definitions of selected canopy SIF parameters are listed in Table 1.

Table 1. Definition of canopy SIF parameters.

Type	Definition	Type	Definition	Type	Definition
$F_{relative}$ of O_2-A band	SIF-A		R_{740}/R_{720}		R_{685}/R_{655}
$F_{relative}$ of O_2-B band	SIF-B	Reflectance ratio index	R_{440}/R_{690}	Reflectance ratio index	R_{690}/R_{655}
Reflectance first derivative index	D_{705}/D_{722} D_{730}/D_{706}		R_{740}/R_{800} R_{750}/R_{800}		R_{690}/R_{600} $R_{675}{}^{*}R_{690}/(R_{683})^2$

2.3. Calculation of Hyperspectral Vegetation Indices

Varieties in physiological, biochemical characteristics and apparent morphology of crops under disease stress cause changes in spectral characteristics. Its spectral response could be seen as a function of changes in pigments, water, morphology, and structure [21]. Spectral index constructed by the sensitive bands can reflect the change of spectral response, so as to realize the monitoring of the disease from the perspective of the pathological mechanism. According to the above statement, this paper selects vegetation indices related to pigments such as GI, PRI, SIPI, PSRI, and MCARI [22–26], and water-related indices WI and NDWI [27,28], as well as the TVI and RTVI that characterizes the morphology and structure of the plant canopy and the HI that indicates whether the vegetation is healthy or not [3,29,30]. Differential spectrum has an advantage in eliminating or reducing the influence of background and noise spectrum. When the canopy coverage reaches more than 20%, the soil background has little effect on the first-order differential [31]. Combined with the existing research on wheat stripe rust monitoring with hyperspectral trilateral parameters [32], this study also selects Db, SDb, Dy, SDy, Dr, and SDr. And their definitions are listed in Table 2.

Table 2. Definition of vegetation indices and trilateral parameters.

Type	Index	Definition	Reference
Vegetation index	Greenness index (GI)	R_{554}/R_{677}	[22]
	Photochemical reflectance index (PRI)	$(R_{570} - R_{531})/(R_{570} + R_{531})$	[23]
	Structural independent pigment index (SIPI)	$(R_{800} - R_{445})/(R_{800} + R_{680})$	[24]
	Plant senescence reflectance index (PSRI)	$(R_{678} - R_{550})/R_{750}$	[25]
	Modified chlorophyll absorbtion reflectance index (MCARI)	$[(R_{700} - R_{670}) - 0.2{*}(R_{700} - R_{550})]{*}(R_{700}/R_{670})$	[26]
	Water index (WI)	R_{900}/R_{970}	[27]
	Normalized difference water index (NDWI)	$(R_{860} - R_{1240})/(R_{860} + R_{1240})$	[28]
	Triangular vegetation index (TVI)	$0.5{*}[120{*}(R_{750} - R_{550}) - 200{*}(R_{670} - R_{550})]$	[29]
	Ration triangular vegetation index (RTVI)	$[55{*}(R_{750} - R_{550}) - 90(R_{680} - R_{550})]/[90(R_{750} + R_{550})]$	[30]
	Healthy index (HI)	$(R_{534} - R_{698})/(R_{534} + R_{698}) - 0.5\,R_{704}$	[3]
Trilateral Parameters	Db	The maximum value of the 1st order differential in 490–539 nm	[32]
	SDb	The sum of 1st order differential in 490–539 nm	[32]
	Dy	The maximum value of the 1st order differential in 550–582 nm	[32]
	SDy	The sum of 1st order differential in 550–582 nm	[32]
	Dr	The maximum value of the 1st order differential in 670–737 nm	[32]
	SDr	The sum of 1st order differential in 670–737 nm	[32]

2.4. Extraction of Characteristic Band

As a deterministic search methodology, successive projections algorithm (SPA) has reproducible results. It is more reliable in the selection of the verification set. It can find the variable group with the lowest redundant information from the spectral information, while retaining most of the features of the original spectrum [33]. The selected waveband has a clear physical meaning, which can better explain the response of the spectral shape and intensity changes to crop diseases. Based on SPA, the characteristic bands were selected from the hyperspectral data after Savitzky-Golay smoothing. The pros and cons of the algorithm are measured by the root mean square error (RMSE). According to the internal cross-validation RMSE of the training set, nine characteristic bands (RMSE = 0.074) are obtained, and the selected wavelength positions are shown in Figure 1. Since SPA is greatly affected by the first band, this study only selects the first six important bands as the model feature input (R539, R513, R1086, R776, R713, R678). These wavelengths are located in the absorption peaks (R776, R1086), absorption valleys (R678, R513), and relatively large slopes (R539, R713) of the spectral curve, which are typical disease response intervals and can sensitively reflect stripe rust stress.

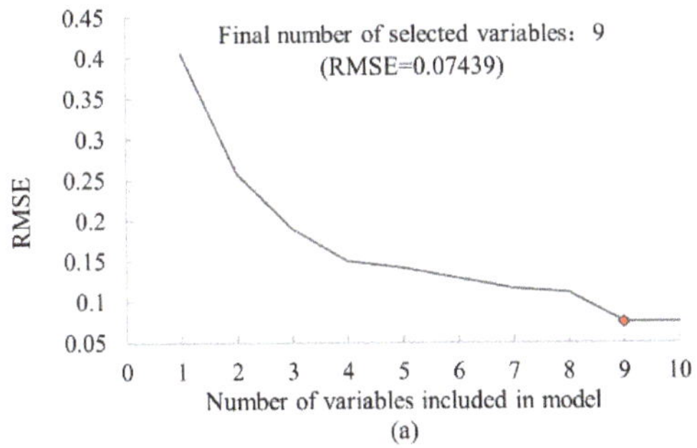
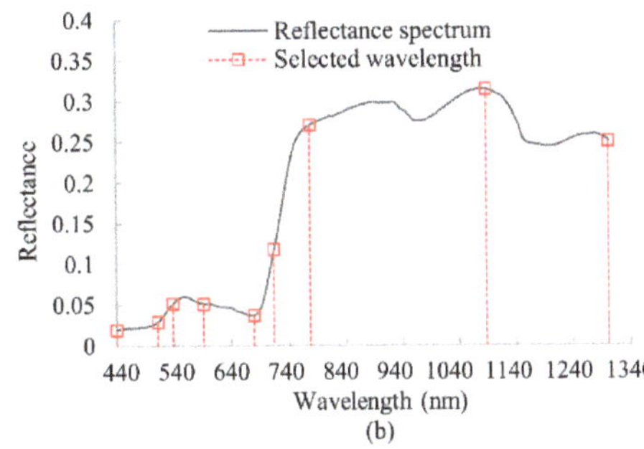

Figure 1. The number (**a**) and index (**b**) of characteristic wavelengths selected by SPA algorithm.

2.5. Methods

2.5.1. The Max-Relevance and Min-Redundancy Feature Selection

With mRMR, which is one of the feature selection methods, a subset is created in which the related properties of the data are retrieved and the unrelated features are discarded [34]. To measure the similarity among elements in the initial set X, the algorithm accepts each feature as a discrete variable and uses the mutual information. Suppose the probability density and joint probability density of feature a and b are $p(a)$, $p(b)$, and $p(a, b)$, respectively. The calculation formula of mutual information F between two features can be defined as Equation (4).

$$F(x, y) = \int\int_a^b p(a, b) \log\left(\frac{p(a, b)}{p(a)p(b)}\right) da\, db \tag{4}$$

Then, the mutual information between each feature and DI is calculated in turn according to Equation (5), and the feature with the maximum value is input into the subset S. In this way, a subset S with j features X_f can be found from the initial feature set X. That is, the features in the subset all have high relevance with DI.

$$maxR_1(S, DI),\ R_1 = \frac{1}{|S|}\sum_{X_f} F\left(X_f, DI\right) \tag{5}$$

Theoretically, the relevance between the first j features and the DI value is maximal. But the correlation between these features may also be large, which means that the redundancy is high. Therefore, the principle of least redundancy can be used to filter out redundant features.

$$minR_2(S),\ R_2 = \frac{1}{|S|^2}\sum_{X_f, X_l \in S} F\left(X_f, X_l\right) \tag{6}$$

where X_f and X_l are features in subset S. The combination of Equations (5) and (6) is the mutual information quotient (MIQ) criteria, the selected features have maximal relevance with the disease index and minimal redundancy with each other.

$$MIQ = max_{X_f \in \Omega_S}\left[F\left(X_f, DI\right) / \frac{1}{|S|}\sum_{X_f, X_l \in S} F\left(X_f, X_l\right)\right] \tag{7}$$

2.5.2. Extreme Gradient Boosting Regression

In this study, features selected by the mRMR algorithm were inputted as independent variables into XGBoost regression to construct a remote sensing monitoring model for stripe rust. XGBoost integrates weak classifiers into strong classifiers and iteratively generates new trees to fit the residuals of the previous trees [35]. As the number of iterations increases, the accuracy continues to improve. For a given training set consisting of n samples and m

feature groups, D = (x_i, y_i), where $|D| = n$, $x_i \in R^m$, $y_i \in R$. The mathematical model of the XGBoost algorithm can be regarded as an additive model composed of t regression trees. The predicted value of the model can be calculated by the following formula.

$$\hat{y}_i = \sum_{k=1}^{t} f_k(x_i) \tag{8}$$

where t is the number of trees, y_i and $\hat{y}_i$ are measured and predicted values, respectively, f_k is the function represented by the kth independent tree, and $f_k(x_i)$ is the space of the CART regression tree. The objective function of the XGBoost algorithm can be constructed as the Equation (9).

$$Obj(k) = \sum_{i=1}^{n} l\left[\left(y_i, \hat{y}^{(k-1)}\right) + f_k(x_i)\right] + \Omega(f_k) + cons \tag{9}$$

$$\Omega(f_k) = \gamma T_k + \frac{1}{2}\lambda||\omega_k||^2 \tag{10}$$

where $\Omega(f_k)$ is the regularization term, that is, the sum of the complexity of each tree. It can control the complexity of the model and prevent overfitting. γ and λ are the model's penalty coefficient and L2 regular term coefficient, T is the number of leaf nodes, and ω represents the leaf score. Compared with the traditional gradient boosting trees, the XGBoost regression incorporates initial derivative information into the optimization process, performs secondary Taylor expansion of the loss function, and adjusts the complexity of model fitting to prevent overfitting together with loss function regularization.

$$Obj(k) \approx \sum_{i=1}^{n} l\left[\left(y_i, \hat{y}^{(k-1)}\right) + g_i f_k(x_i) + \frac{1}{2}h_i f_k^2(x_i)\right] + \Omega(f_k) \tag{11}$$

where $g_i = \frac{\partial\left(l\left(y_i, \hat{y}^{(k-1)}\right)\right)}{\partial \hat{y}^{(k-1)}}$, $h_i = \frac{\partial^2\left(l\left(y_i, \hat{y}^{(k-1)}\right)\right)}{\partial \hat{y}^{(k-1)}}$, which are the first derivative and second derivative of the loss function to the current model, respectively. Therefore, the objective function can be expressed as Equation (12).

$$Obj(k) = \sum_{j=1}^{T} \left[\sum_{j\in I_j} g_j\omega_j + \frac{1}{2}\left(\sum_{j\in I_j} h_j + \lambda\right)\omega_j^2\right] + \gamma T \tag{12}$$

where I_j represents the set of samples on the leaf node whose sequence number is j. Find the optimal solution to equation (8) and bring it back to the equation to obtain the minimized objective function of the XGBoost model.

$$Obj(k) = -\frac{1}{2}\sum_{j=1}^{T} \frac{\left(\sum_{j\in I_j} g_j\right)^2}{\sum_{j\in I_j} h_j + \lambda} + +\gamma T \tag{13}$$

The smaller $Obj(k)$, the better the structure of the tree. During the training process, the XGBoost algorithm corrects the previous tree model through iterative residuals to optimize the specified loss function. XGBoost with the CART booster has more than 20 parameters, but the number of estimators, learning rate, minimum child weight, maximum tree depth, subsample ratio of training samples and alpha are the most important parameters. In this research, the optimal values of the above six parameters were obtained through the grid parameter optimization method. After obtaining the model parameters that best fit the train data, combining the mRMR algorithm and XGBoost regression could quickly achieve an accurate predictive disease index value with as few features as possible. The process framework of this research method is shown in the Figure 2.

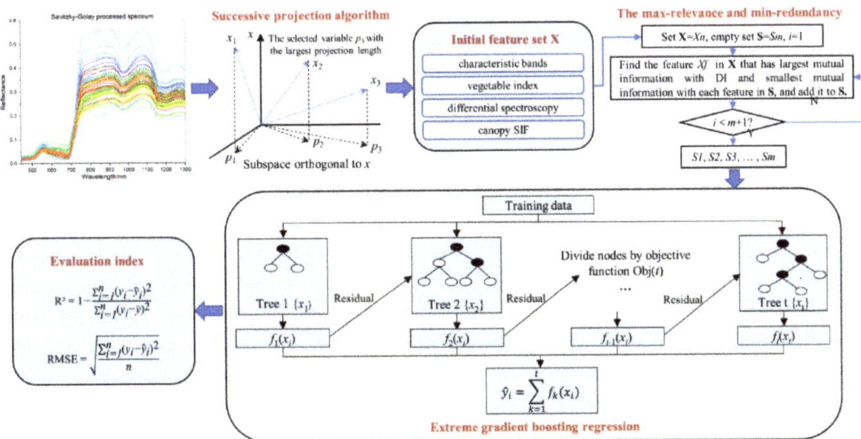

Figure 2. Schematic diagram of the algorithm flow in this study.

3. Results and Analysis

3.1. Features Selected by CC

The initial feature set is composed of 12 canopy SIF parameters, 10 vegetation indices, 6 trilateral parameters, and 6 characteristic bands. Parameters in the feature set are analysed with DI one by one, the correlation coefficient between each feature parameter and DI is shown in Figure 3. It can be seen from Figure 3 that the overall selected vegetation indices have high correlations with the DI. The winter wheat which has been affected by stripe rust has obvious responses in terms of pigment, water content, and canopy structure. In addition, canopy chlorophyll fluorescence parameters are sensitive to stripe rust stress. In the trilateral parameters, the area and amplitude of the red and yellow edges are more obvious in response to wheat stripe rust. Since SPA selects the wavelength combination with the least collinearity, and the combination contains the majority information of total spectral reflectance. Therefore, the correlation coefficient between each wavelength and the DI shows non-uniformity. According to Figure 3, this study selected 6 parameters that are extremely significantly related to DI ($p < 0.01$), including SIF-A, R440/R690, R740/R720, HI, GI and SDy.

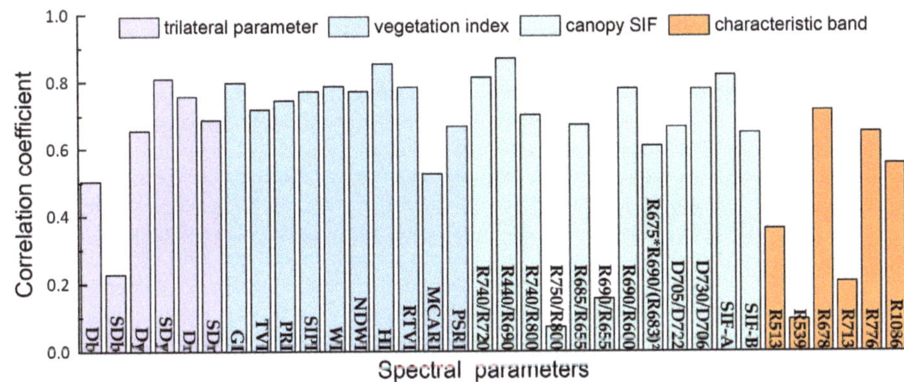

Figure 3. Correlation coefficient of each characteristic parameter and disease index.

3.2. Features Selected by mRMR

The mRMR feature selection algorithm is used to filter the initial feature set, so that related features are concentrated, and irrelevant features are minimized. The MIQ of each spectral parameter and DI is shown in Figure 4. The warmer the grid color, the higher the MIQ, that is, the feature in the grid has a high correlation with DI and low redundancy with other features. It can be seen from Figure 4, there are eight parameters (Dy, R740/R720, R440/R690, D705/D722, D730/D706, SIF-A, SIF-B, R1086, and R678) with higher MIQ values. Corresponding to the CC analysis, six features were selected according to grid color as input parameters for the remote sensing monitoring of the wheat stripe rust severity (R740/R720, SIF-A, Dy, R678, R1086, and R440/R690), including three SIF parameters, two characteristic bands, and one trilateral spectral index. Selected features had the maximum MIQ with each other compared to the others. In addition, although those vegetation indices had an obvious correlation with DI, their MIQ values were extremely low as seen in Figure 4. It was proved that vegetation indices have high redundancy among each other. Thus, mRMR feature selection did not choose any vegetation index.

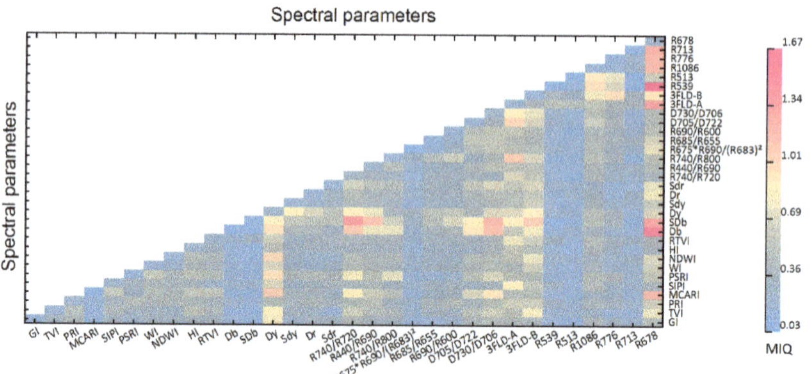

Figure 4. Mutual information quotient of each characteristic parameter and disease index.

3.3. Remote Sensing Monitoring Model of Wheat Stripe Rust

In order to ensure the stability and reliability of the evaluation results, improve the generalization ability of the model, and reduce the impact of a random grouping of sample data on the accuracy of the model. In this study, 52 samples (47 infected samples, 5 healthy samples) were grouped repeatedly for three times (A, B, C). Each group is randomly divided into training set and validation set according to a ratio of 3:1, including 39 samples in the training set and 13 samples in the validation set. The determination coefficient (R^2) and Root Mean Square Error (RMSE) between the predicted value and the measured value are selected as the model accuracy evaluation indicators.

Based on the given step size and optimization accuracy range, the parameters of the training set were continuously iteratively adjusted through grid parameter optimization. Figure 5 illustrates the RMSE values versus the combination of these four parameters based on train data. Since the n_estimators and learning_rate have the greatest impact on the model, these two parameters were optimized first. It is worth noting that when these two parameters were configured, the remaining parameters took the default values. On the basis of the optimal learning_rate and n_estimators, the subsample and reg_alpha were optimized. Repeat this process until the optimal values of the six parameters were found.

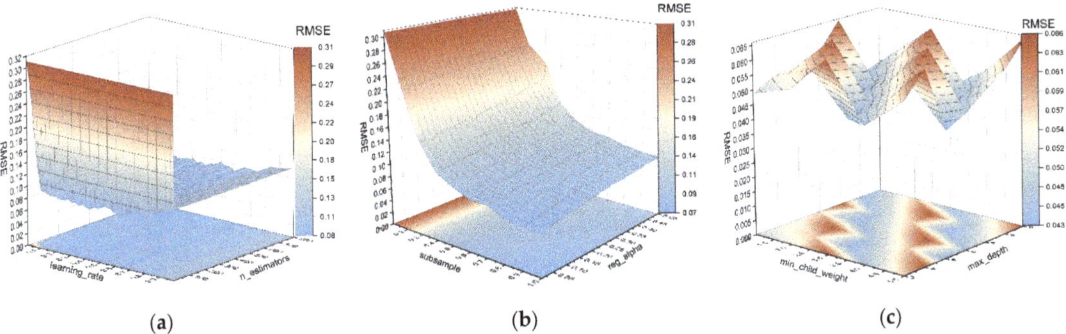

(a)   (b)   (c)

Figure 5. Root Mean Square Error (RMSE) values versus critical parameter combinations in grid parameters optimal process. (**a**) learning_rate with n_estimators; (**b**) subsample with reg_alpha; (**c**) min_child_weight with max_depth.

According to Figure 5, grid parameter optimization selects the two parameters corresponding to the grid at the minimum RMSE. It can be seen from Figure 5a that the learning rate has a stable RMSE value between 0.2 and 0.4. Beyond this interval, the algorithm becomes more conservative, and the RMSE increases with the increase of the learning rate. For the number of iterations, around 20 iterations can achieve a sufficiently low error, more iterations did not improve the model accuracy. Furthermore, according to the results of the sampling rate parameter, as shown in Figure 5b, the model error decreases as the sampling ratio increases. To prevent possible overfitting, especially in the case of limited training samples, the optimal range of sampling ratios should be between 0.5 and 1. In addition, changes in the L1 regularization term have no obvious effect on the accuracy. Tree depth and minimum child weights always exhibit a synergistic change (see Figure 3c), as the function of both parameters is to prevent the model from overfitting. In terms of the above optimization process and criteria, the best parameter combination is selected. The optimal values of the six parameters are listed in Table 3.

Table 3. The parameter settings of the XGBoost regression.

Parameter Type	Parameter	Adjustment Range	Step	Optimal Value
	learning_rate	[0, 1]	0.01	0.21
Booster	max_depth	[3, 10]	1	3
parameters	min_split_weight	[1, 6]	1	5
	subsample	[0, 1]	0.1	0.5
	reg_alpha	[0, 0.5]	0.01	0.02
Learning task parameters	n_estimators	[0, 800]	1	11

3.4. Model Evaluation

Based on optimal values of the above parameters, the characteristic parameters selected by the CC analysis and the mRMR algorithm in the test set samples are input as independent variables into the XGBoost regression model and the GBRT regression model. Four stripe rust disease index prediction models were constructed, namely mRMR-XGBoost, mRMR-GBRT, CC-XGBoost, and CC-GBRT models, and their prediction accuracy is shown in Figure 6. Regardless of whether the features selected by the mRMR algorithm or the CC analysis are used as model-independent variables to estimate the severity of wheat stripe rust disease, the prediction accuracy of the XGBoost model is better than that of the GBRT model. This is because XGBoost performs a second-order Taylor expansion on the objective function, which can achieve faster and more accurate gradient descent. In addition, iterative parameter adjustment of L2 regularization items was carried out during model construction, and the decision tree structure was constrained to prevent the model from overfitting. The model prediction accuracy of wheat stripe rust test samples was improved in this way. As a consequence, XGBoost has a better performance than GBRT.

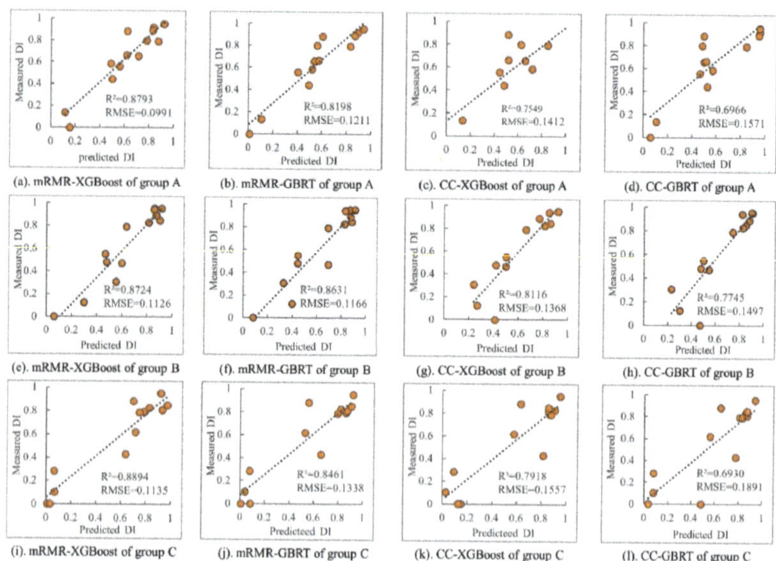

Figure 6. Accuracy comparison of validation set models based on mRMR-XGBoost, mRMR-GBRT, CC-XGBoost and CC-GBRT algorithms. (**a–d**), (**e–h**) and (**i–l**) represent the prediction results of the three groups, respectively.

Comparing the prediction accuracy of the two feature selection algorithms in the same model, it is found that in the three sets of XGBoost prediction models, the prediction accuracy based on mRMR-optimized features is improved by an average of 12% compared with the CC analysis. Correspondingly, the prediction accuracy in the GBRT model is improved by an average of 17%. The features selected by the mRMR algorithm have high relevance and low redundancy, which maximizes the information contained in the features under the condition that the number of features is equal to that of the CC analysis.

In addition, the features selected by the two methods both contain canopy SIF parameters. It shows that SIF parameters are sensitive to the changes of photosynthetic physiology and spectral reflectance of winter wheat caused by stripe rust stress. SIF has great potential application in crop disease monitoring. In three groups of random experiments, the mRMR-XGBoost model achieved the best prediction accuracy. Compared with the mRMR-GBRT, CC-XGBoost, and CC-GBRT models, the average R^2 between the predicted DI and the measured DI is increased by an average of 5%, 12%, and 22%, and the RMSE is reduced by an average of 14%, 33%, and 52%.

3.5. Field Survey Data Validation

In order to verify the applicability of the mRMR-XGBoost model in the field, this study carried out further research on the stripe rust survey data obtained in the field planting area of Ningqiang County, Hanzhong City, Shaanxi Province on 12 May 2018. The severity of stripe rust was monitored by the four models constructed above based on the obtained 34 field survey samples, and the accuracy evaluations were summarized in Table 4. From the verification accuracy of the three random sample groups in Table 4, compared with mRMR-GBRT, CC-XGBoost, and CC-GBRT, the R^2 between the predicted DI and the measured DI value in the mRMR-XGBoost model were improved by 44%, 32%, and 82% on average. It shows the highest monitoring accuracy among the four models in this study, which is consistent with the above result, indicating that the mRMR-XGBoost algorithm has excellent monitoring universality and scalability.

Table 4. Validation results of field survey data (m = 34).

Sample Group	mRMR-XGBoost		mRMR-GBRT		CC-XGBoost		CC-GBRT	
	R^2	RMSE	R^2	RMSE	R^2	RMSE	R^2	RMSE
A	0.915	0.181	0.721	0.157	0.890	0.161	0.769	0.166
B	0.830	0.201	0.346	0.125	0.695	0.165	0.359	0.127
C	0.676	0.131	0.608	0.119	0.245	0.137	0.200	0.162

4. Discussion

In some research, XGBooost is also used to select features [36]. In this study, we also tried to use the XGBoost to perform feature selection on the initial feature set, and the result is shown in Figure 7. It can be seen from the picture that GI has the highest importance, and the importance of SIF-A, TVI, and PRI exceed 20. Wheat stripe rust mainly occurs on the leaves. Small chlorotic spots (variegated spots) are formed at the affected parts, and then yellow or bright yellow piles of *Puccinia striiformis* appear quickly. GI and PRI, which characterize plant pigments, can capture the slight changes in leaf pigments and realize stripe rust detection. When wheat plants are continuously parasitized and infected by *Puccinia striiformis*, their cell viability and biochemical components will change, which will further cause changes in leaf morphology, leaf inclination distribution, and canopy structure. Therefore, the index TVI, which characterizes canopy structure, has a clear response to stripe rust. In addition, a previous study has demonstrated that the canopy structure is the dominant factor responsible for variation in far-red fluorescence under the saturation conditions [37]. The canopy-leaving broadband (641–800 nm) SIF variability is determined mainly by leaf optical properties and canopy structural variables [38]. As a consequence, SIF can reflect the severity of wheat stripe rust according to the changes in leaf and canopy structure.

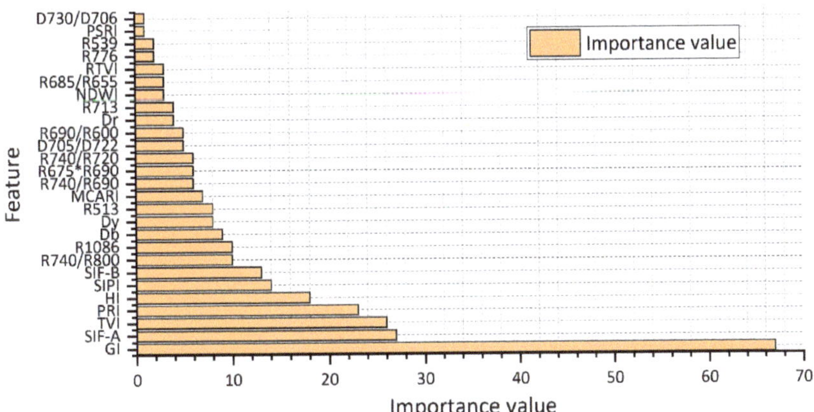

Figure 7. Ranking of features calculated by XGBoost.

Take the sorted features as independent variables and input them into the XGBoost model and GBRT model. In each modeling, add the most important feature to the feature combination, and the stripe rust remote sensing monitoring model is constructed based on the test samples. Model accuracy is checked by R^2 between predicted DI and measured DI, and the accuracy of the XGBoost and GBRT models are listed in Table 5. When the number of features is less than 6, both models showed low accuracy. After exceeding 6, the accuracies of the two models continued to improve and stabilize as the feature increased.

But they are still lower than the mRMR-XGBoost model. The reason may be that the features selected by the XGBoost algorithm were overfitted in train samples, so after transferred them into test samples, the model accuracy decreased.

Table 5. The effect of the features' number selected based on the XGBoost algorithm on the accuracy.

Numbers	Feature Combination	R^2 of XGBoost	R^2 of GBRT
1	GI	0.12	0.16
2	GI, SIF-A	0.17	0.28
3	GI, SIF-A, TVI	0.25	0.18
4	GI, SIF-A, TVI, PRI	0.23	0.27
5	GI, SIF-A, TVI, PRI, HI	0.23	0.29
6	GI, SIF-A, TVI, PRI, HI, SIPI	0.64	0.62
7	GI, SIF-A, TVI, PRI, HI, SIPI, SIF-B	0.78	0.74
8	GI, SIF-A, TVI, PRI, HI, SIPI, SIF-B, R_{740}/R_{800}	0.83	0.87
9	GI, SIF-A, TVI, PRI, HI, SIPI, SIF-B, R_{740}/R_{800}, R_{1086}	0.83	0.84
10	GI, SIF-A, TVI, PRI, HI, SIPI, SIF-B, R_{740}/R_{800}, R_{1086}, D_b	0.84	0.88
11	GI, SIF-A, TVI, PRI, HI, SIPI, SIF-B, R_{740}/R_{800}, R_{1086}, D_b, D_y	0.81	0.81

Combining the feature selection results of CC analysis and the mRMR algorithm, it is found that SIF-A is selected for all three methods. SIF-A is an indicative factor in monitoring wheat stripe rust. Chlorophyll fluorescence is closely related to plant photosynthetic physiology and participates in the energy distribution of crops. When crops are under stress such as diseases, chlorophyll fluorescence changes before chlorophyll content. Therefore, chlorophyll fluorescence has distinctive advantages in crop disease monitoring [39]. In this study, the mRMR algorithm selected three SIF parameters, indicating that chlorophyll fluorescence can be applied in wheat stripe rust monitoring. Our results were consistent with previously reported conclusions [40,41]. At the same time, the cooperative characteristic wavelength and trilateral parameter information can effectively avoid the omission of spectral information. The prediction accuracy based on the mRMR-XGBoost model provided better performance than the other three models, with a prediction accuracy of 87.2–88.9%. The reason may be that the mRMR algorithm selects feature factors by mutual information, which ensures the maximum relevance between features and DI and minimum redundancy among features and effectively reduces feature dimensionality. In addition, the XGBoost algorithm contains an L2 regularization term that can avoid overfitting, making it more suitable for the prediction of small sample data.

Although this study provided satisfactory results in predicting wheat stripe rust, some limitations must be addressed in future studies. First, in this study, although the XGBoost regression was successfully applied to monitor the stripe rust, the parameters of the XGBoost algorithm need to be further optimized. In the next research, the XGBoost model will be updated based on the spectrum and habitat parameters, to integrate information on the disease mechanism and model, as well as to improve the accuracy of the prediction model. Then, the main content of this paper is the compression of near-ground hyperspectral data and the extraction of stripe rust sensitive factors. Due to weather and manpower constraints, the obtained samples were few and the coverage was limited. Therefore, this study did not involve large-scale disease monitoring and early warning. In the following work, the meteorological information (e.g., temperature, precipitation, humidity, etc.) and remote sensing data will be integrated. After the meteorological factors and index features are jointly screened based on the mRMR algorithm, weights are assigned to the selected features through XGBoost. A remote sensing prediction model of stripe rust can be constructed by taking the weighted features as independent variables, so as to realize disease monitoring and prediction in large areas.

5. Conclusions

This study presents an optimized method for predicting the disease index of wheat stripe rust by using mRMR-XGBoost. The method not only reduces the dimension of characteristic parameters that are used to detect the disease index of winter wheat stripe rust by remote sensing but also improves the regression speed and prediction accuracy of the disease index of wheat stripe rust. For the two feature selection algorithms in this study, the features selected by mRMR contain more information about the severity of stripe rust disease under the circumstance of equal number than that of CC analysis. What's more, compared with mRMR-GBRT, CC-XGBoost, and CC-GBRT, the mRMR-XGBoost severity prediction model has better R^2 and RMSE performance parameter values. The R^2 between prediction DI and measured DI of the three random groups in test samples are all above 0.87, and the RMSE is reduced by an average of 14%, 33%, and 52%. The field survey data validation experiment also confirmed the applicability of the mRMR-XGBoost algorithm. The high accuracy and regional accurate monitoring value justified the feasibility of using the mRMR-XGBoost model for monitoring wheat stripe rust, which is promising for this technology to be applied in practical wheat production management.

Author Contributions: Conceptualization, X.J. and Q.Z.; methodology, X.J. and Q.Z.; software, Q.Z.; validation, X.J. and J.Y.; formal analysis, Q.Z. and B.L.; investigation, X.J., Q.Z, J.Y., Y.D. and B.L.; resources, X.J. and Y.D.; data curation, X.J., Y.D., Q.Z., J.Y. and B.L.; writing—original draft preparation, Q.Z.; writing—review and editing, X.J. and J.Y.; visualization, Q.Z.; supervision, Y.D.; project administration, X.J.; funding acquisition, X.J. All authors have read and agreed to the published version of the manuscript.

Funding: This research was funded by the National Natural Science Foundation of China under [Grant NO.42171394] and [Grant NO.41601467], in part by the Natural Science Foundation of Tibet Autonomous Region under [Grant NO.XZ202101ZR0085G].

Institutional Review Board Statement: Not applicable.

Informed Consent Statement: Not applicable.

Data Availability Statement: Due to the nature of this research, participants of this study did not agree for their data to be shared publicly, so supporting data is not available.

Acknowledgments: This research was funded by the National Science Foundation of China under Grant NO.42171394 and Grant NO.41601467 and Natural Science Foundation of Tibet Autonomous Region under Grant NO.XZ202101ZR0085G. The financial support is highly appreciated. We are grateful for the experimental site provided by the Langfang Research Pilot Base of the Chinese Academy of Agricultural Sciences. We are also thankful for the insightful comments raised by anonymous reviewers and the associate editor, who helped to improve this paper.

Conflicts of Interest: The authors declare no conflict of interest.

References

1. Lihua, C.; Shichang, X.; Ruiming, L.; Taiguo, L.; Wanquan, C. Early molecular diagnosis and detection of *puccinia striiformis* f. sp. *tritici* in China. *Lett. Appl. Microbiol.* **2008**, *46*, 501–506. [CrossRef] [PubMed]
2. Zhang, J.; Huang, Y.; Pu, R.; Gonzalez-Moreno, P.; Yuan, L.; Wu, K.; Huang, W. Monitoring plant diseases and pests through remote sensing technology: A review. *Comput. Electron. Agric.* **2019**, *165*, 104943. [CrossRef]
3. Mahlein, A.-K.; Rumpf, T.; Welke, P.; Dehne, H.W.; Plümer, L.; Steiner, U.; Oerke, E.C. Development of spectral indices for detecting and identifying plant diseases. *Remote Sens. Environ.* **2013**, *128*, 21–30. [CrossRef]
4. Graeff, S.; Link, J.; Claupein, W. Identification of powdery mildew (*Erysiphe graminis* sp. tritici) and take-all disease (*Gaeumannomyces graminis* sp. tritici) in wheat (*Triticum aestivum* L.) by means of leaf reflectance measurements. *Cent. Eur. J. Biol.* **2006**, *1*, 275–288. [CrossRef]
5. Liu, Z.Y.; Wu, H.F.; Huang, J.F. Application of neural networks to discriminate fungal infection levels in rice panicles using hyperspectral reflectance and principal components analysis. *Comput. Electron. Agric.* **2010**, *72*, 99–106. [CrossRef]
6. Li, X.; Yang, C.; Huang, W.; Tang, J.; Tian, Y.; Zhang, Q. Identification of cotton root rot by multifeature selection from Sentinel-2 images using Random Forest. *Remote Sens.* **2020**, *21*, 3054. [CrossRef]
7. Huang, L.; Wu, Z.; Huang, W.; Ma, H.; Zhao, J. Identification of Fusarium Head Blight in winter wheat ears based on Fisher's Linear Discriminant Analysis and a Support Vector Machine. *Appl. Sci.* **2019**, *18*, 3894. [CrossRef]

8.   Yuan, D.; Jiang, J.; Qi, X.; Xie, Z.; Zhang, G. Selecting key wavelengths of hyperspectral imagine for nondestructive classification of moldy peanuts using ensemble classifier. *Infrared Phys. Technol.* **2020**, *111*, 103518. [CrossRef]

9.   Davoud, A.; Mohammad, M.; Alfredo, H. Developing two spectral disease indices for detection of wheat leaf rust (*pucciniatriticina*). *Remote Sens.* **2014**, *6*, 4723–4740.

10.   Poblete, T.; Navas-Cortes, J.A.; Camino, C.; Calderon, R.; Hornero, A.; Gonzalez-Dugo, V.; Landa, B.B.; Zarco-Tejada, P.J. Discriminating xylella fastidiosa from verticillium dahliae infections in olive trees using thermal- and hyperspectral-based plant traits. *ISPRS J. Photogramm.* **2021**, *179*, 133–144. [CrossRef]

11.   Sankaran, S.; Mishra, A.; Ehsani, R.; Davis, C. A review of advanced techniques for detecting plant diseases. *Comput. Electron. Agric.* **2010**, *72*, 1–13. [CrossRef]

12.   Jing, X.; Zhang, T.; Bai, Z.; Huang, W. Feature Selection and Model Construction of Wheat Stripe Rust Based on GA and SVR Algorithm. *Trans. Chin. Soc. Agric. Mach.* **2020**, *51*, 253–263.

13.   Samat, A.; Li, E.; Wang, W.; Liu, S.; Lin, C.; Abuduwaili, J. Meta-XGBoost for hyperspectral image classification using extended MSER-guided morphological profiles. *Remote Sens.* **2020**, *12*, 1973. [CrossRef]

14.   Wenjiang, H.; David, W.L.; Zheng, N.; Yongjiang, Z.; Liangyun, L.; Jihua, W. Identification of yellow rust in wheat using in-situ spectral reflectance measurements and airborne hyperspectral imaging. *Precis. Agric.* **2007**, *8*, 187–197.

15.   Plascyk, J.A.; Gabriel, F.C. Fraunhofer line discriminator MK II—airborne instrument for precise and standardized ecological luminescence measurement. *IEEE Trans. Instrum. Meas.* **1975**, *24*, 306–313. [CrossRef]

16.   Maier, S.W.; Günther, K.P.; Stellmes, M. Sun-induced fluorescence: A new tool for precision farming. In *Digital Imaging and Spectral Techniques: Applications to Precision Agriculture and Crop Physiology*; Mcdonald, M., Schepers, J., Tartly, L., Eds.; American Society of Agronomy Special Publication: Madison, WI, USA, 2003; pp. 209–222.

17.   Damm, A.; Erler, A.; Hillen, W.; Meroni, M.; Schaepman, M.E.; Verhoef, W.; Rascher, U. Modeling the impact of spectral sensor configurations on the FLD retrieval accuracy of sun-induced chlorophyll fluorescence. *Remote Sen. Environ.* **2011**, *115*, 1882–1892. [CrossRef]

18.   Liu, L.; Cheng, Z. Detection of vegetation light-use efficiency based on solar-induced chlorophyll fluorescence separated from canopy radiance spectrum. *IEEE J. Sel. Top. Appl. Earth Obs. Remote Sens.* **2010**, *3*, 306–312. [CrossRef]

19.   Dobrowski, S.Z.; Pushnik, J.C.; Zarco-Tejada, P.J.; Ustin, S.L. Simple reflectance indices track heat and water stress-induced changes in steady-state chlorophyll fluorescence at the canopy scale. *Remote Sens. Environ.* **2005**, *97*, 403–414. [CrossRef]

20.   Zarco-Tejada, P.J.; Pushnik, J.C.; Dobrowski, S.; Ustin, S.L. Steady-state chlorophyll a fluorescence detection from canopy derivative reflectance and double-peak red-edge effects. *Remote Sens. Environ.* **2003**, *84*, 283–294. [CrossRef]

21.   Zhang, J.; Yuan, L.; Wang, J.; Luo, J.; Du, S.; Huang, W. Research progress of crop diseases and pests monitoring based on remote sensing. *Trans. Chin. Soc. Agric. Eng.* **2012**, *28*, 1–11.

22.   Zarco-Tejada, P.J.; Berjón, A.; López-Lozano, R.; Miller, J.R.; Martín, P.; Cachorro, V.; González, M.R.; de Frutos, A. Assessing vineyard condition with hyperspectral indices: Leaf and canopy reflectance simulation in a row-structured discontinuous canopy. *Remote Sens. Environ.* **2005**, *99*, 271–287. [CrossRef]

23.   Gamon, J.A.; Peñuelas, J.; Field, C.B. A narrow-waveband spectral index that tracks diurnal changes in photosynthetic efficiency. *Remote Sens. Environ.* **1992**, *41*, 35–44. [CrossRef]

24.   Peñuelas, J.; Baret, F.; Filella, I. Semi-empirical indices to assess carotenoids/chlorophyll a ratio from leaf spectral reflectance. *Photosynthetica* **1995**, *31*, 221–230.

25.   Merzlyak, M.N.; Gitelson, A.A.; Chivkunova, O.B.; Rakitin, V.Y. Non-destructive optical detection of pigment changes during leaf senescence and fruit ripening. *Physiol. Plant.* **1999**, *106*, 135–141. [CrossRef]

26.   Daughtry, C.S.T.; Walthall, C.L.; Kim, M.S.; Brown de Colstoun, E.; McMurtrey, J.E., III. Estimating Corn Leaf Chlorophyll Concentration from Leaf and Canopy Reflectance. *Remote Sens. Environ.* **2000**, *74*, 229–239. [CrossRef]

27.   Peñuelas, J.; Pinol, J.; Ogaya, R.; Filella, I. Estimation of plant water concentration by the reflectance water index wi (r900/r970). *Int. J. Remote Sens.* **1997**, *18*, 2869–2875. [CrossRef]

28.   Gao, B.C. NDWI—A normalized difference water index for remote sensing of vegetation liquid water from space. *Remote Sens. Environ.* **1996**, *58*, 257–266. [CrossRef]

29.   Broge, N.H.; Leblanc, E. Comparing prediction power and stability of broadband and hyperspectral vegetation indices for estimation of green leaf area index and canopy chlorophyll density. *Remote Sens. Environ.* **2001**, *76*, 156–172. [CrossRef]

30.   Verstraete, M.M.; Pinty, B.; Myneni, R.B. Potential and limitations of information extraction on the terrestrial biosphere from satellite remote sensing. *Remote Sens. Environ.* **1996**, *58*, 201–214. [CrossRef]

31.   Smith, K.L.; Steven, M.D.; Colls, J.J. Use of hyperspectral derivative ratios in the red-edge region to identify plant stress responses to gas leaks. *Remote Sens. Environ.* **2004**, *92*, 207–217. [CrossRef]

32.   Jiang, J.B.; Chen, Y.H.; Huang, W.J. Using hyperspectral derivative indices to diagnose severity of winter wheat stripe rust. *Opt. Tech.* **2007**, *4*, 620–623.

33.   Wu, D.; Nie, P.; He, Y.; Bao, Y. Determination of calcium content in powdered milk using near and mid-infrared spectroscopy with variable selection and chemometrics. *Food Bioprocess Technol.* **2012**, *5*, 1402–1410. [CrossRef]

34.   Peng, H.C.; Long, F.H.; Ding, C. Feature selection based on mutual information: Criteria of max-dependency, max-relevance, and min-redundancy. *IEEE Trans. Pattern Anal. Mach. Intell.* **2005**, *27*, 31507–31518.

35. Chen, T.; Guestrin, C. XGBoost: A Scalable Tree Boosting System. In Proceedings of the 22nd ACM SIGKDD International Conference on Knowledge Discovery and Data Mining, San Francisco, CA, USA, 13–17 August 2016.
36. Bhagat, S.K.; Tiyasha, T.; Awadh, S.M.; Tung, T.M.; Jawad, A.H.; Yaseen, Z.M. Prediction of sediment heavy metal at the Australian Bays using newly developed hybrid artificial intelligence models. *Environ. Pollut.* **2021**, *268*, 115663. [CrossRef]
37. Knyazikhin, Y.; Schull, M.A.; Stenberg, P.; Moettus, M.; Rautiainen, M.; Yang, Y.; Marshak, A.; Carmona, P.L.; Kaufmann, R.K.; Lewis, P. Hyperspectral remote sensing of foliar nitrogen content. *Proc. Natl. Acad. Sci. USA* **2012**, *110*, E185–E192. [CrossRef] [PubMed]
38. Verrelst, J.; Rivera, J.P.; Van der Tol, C.; Magnani, F.; Mohammed, G.; Moreno, J. Global sensitivity analysis of the SCOPE model: What drives simulated canopy-leaving sun-induced fluorescence? *Remote Sens. Environ.* **2015**, *166*, 8–21. [CrossRef]
39. Liu, L.; Zhang, Y.; Jiao, Q.; Peng, D. Assessing photosynthetic light-use efficiency using a solar-induced chlorophyll fluorescence and photochemical reflectance index. *Int. J. Remote Sens.* **2013**, *34*, 4264–4280. [CrossRef]
40. Atta, B.M.; Saleem, M.; Ali, H.; Bilal, M.; Fayyaz, M. Application of fluorescence spectroscopy in wheat crop: Early disease detection and associated molecular changes. *J. Fluoresc.* **2020**, *30*, 801–810. [CrossRef]
41. Jing, X.; Bai, Z.F.; Gao, Y.; Liu, L.Y. Wheat stripe rust monitoring by random forest algorithm combined with SIF and reflectance spectrum. *Trans. Chin. Soc. Agric. Mach.* **2019**, *35*, 154–161.

remote sensing

Article

Combining Disease Mechanism and Machine Learning to Predict Wheat Fusarium Head Blight

Lu Li [1], Yingying Dong [2,3,*], Yingxin Xiao [2], Linyi Liu [2], Xing Zhao [1] and Wenjiang Huang [2,3,4]

1   School of Mathematical Sciences, Capital Normal University, Beijing 100048, China; 2200502178@cnu.edu.cn (L.L.); 5273@mail.cnu.edu.cn (X.Z.)
2   Key Laboratory of Digital Earth Science, Aerospace Information Research Institute, Chinese Academy of Sciences, Beijing 100094, China; xiaoyingxin20@mails.ucas.ac.cn (Y.X.); liuly35@radi.ac.cn (L.L.); huangwj@aircas.ac.cn (W.H.)
3   University of Chinese Academy of Sciences, Beijing 100049, China
4   Key Laboratory of Earth Observation of Hainan Province, Hainan Research Institute, Aerospace Information Research Institute, Chinese Academy of Sciences, Sanya 572029, China
*   Correspondence: dongyy@aircas.ac.cn; Tel.: +86-10-82178178

Abstract: Wheat Fusarium head blight (FHB) can be effectively controlled through prediction. To address the low accuracy and poor stability of model predictions of wheat FHB, a prediction method of wheat FHB that couples a logistic regression mechanism-based model and k-nearest neighbours (KNN) model is proposed in this paper. First, we selected predictive factors, including remote sensing-based and meteorological factors. Then, we quantitatively expressed the factor weights of the disease occurrence and development mechanisms in the disease prediction model by using a logistic model. Subsequently, we integrated the obtained factor weights into the predictive factors and input the predictive factors with weights into the KNN model to predict the incidence of wheat FHB. Finally, the accuracy and generalizability of the models were evaluated. Wheat fields in Changfeng, Dingyuan, Fengyuan, and Feidong counties, Anhui Province, where wheat FHB often occurs, were used as the study area. The incidences of wheat FHB on 29 April and 10 May 2021 were predicted. Compared with a model that did not consider disease mechanism, the accuracy of our model increased by approximately 13%. The overall accuracies of the models for the two dates were 0.88 and 0.92, and the F1 index was 0.86 and 0.94, respectively. The results show that the predictions made with the logistic-KNN model had higher accuracy and better stability than those made with the KNN model, thus achieving remote sensing-based high-precision prediction of wheat FHB.

Keywords: wheat; fusarium head blight; mechanism; remote sensing; machine learning techniques

Citation: Li, L.; Dong, Y.; Xiao, Y.; Liu, L.; Zhao, X.; Huang, W. Combining Disease Mechanism and Machine Learning to Predict Wheat Fusarium Head Blight. *Remote Sens.* **2022**, *14*, 2732. https://doi.org/10.3390/rs14122732

Academic Editor: Saeid Homayouni

Received: 29 April 2022
Accepted: 3 June 2022
Published: 7 June 2022

Publisher's Note: MDPI stays neutral with regard to jurisdictional claims in published maps and institutional affiliations.

1. Introduction

Fusarium head blight (FHB) is a major wheat disease caused by *Fusarium graminearum* (*Gibberella zeae*) [1]. In recent years, due to the large-scale promotion of wheat and maize rotation and returning crop straw to the field, as well as the influence of climate change and other factors, wheat FHB in China has gradually spread from the south to the north, and the prevalence and severity of this disease have increased significantly [2]. In addition, wheat varieties planted in most areas in China lack resistance to FHB. After wheat is infected by FHB, toxic substances such as deoxynivalenol (DON), a metabolite of the pathogen, seriously endanger human and animal safety [3,4]. According to previous reports, in epidemic years, FHB can generally cause 5~15% yield loss, with losses of up to 40% in severe cases, which directly affects food security [5,6]. Because of the frequent occurrence and rapid development of wheat FHB, once the disease occurs, the damage to the wheat field is irreversible. Therefore, high-precision prediction plays an important role in the accurate formulation of prevention and control plans.

Selecting and determining important host and habitat factors are necessary for remote sensing-based prediction of crop diseases. At present, scholars worldwide mainly carry out disease prediction based on meteorological information [7,8]. By studying the quantitative relationship between meteorological factors, such as temperature, precipitation, rainy days and wind direction, and disease prevalence, and using these factors as input variables to build prediction models, regional disease prediction can be realized [9]. In recent years, the rapid development of Earth observation technology has provided a new opportunity for crop disease prediction [10]. Remote sensing technology can be used to obtain continuous spatial information of farmland. Many scholars try to monitor and predict crop diseases based on remote sensing. Li Xingrong et al. identified cotton root rot based on a remote sensing spectral vegetation index and achieved high accuracy. Bajwa S G et al. used a remote sensing vegetation index to monitor soybean diseases and achieved good results [11,12]. However, in the field of disease prediction, most studies are based on only meteorological information or remote sensing indexes that cover a single timeframe, and the accuracy and spatial resolution of these predictions are not high [13–15]. In this paper, by combining the multiscale meteorological factors of the integrated habitat with temporally varying remote sensing factors, according to the different sensitivities of factors to wheat FHB in different growth stages, factors were selected for timeseries prediction.

In recent studies, machine learning techniques have been extensively explored for plant disease studies due to their rapid and precise measurement capacities. Many scholars have tried to predict diseases based on machine learning models. Elham Khalili et al. predicted soybean carbon rot based on the linear regression of L1 and L2 regularization terms (LR-L1 and LR-L2), Multi-layer Perceptron (MLP), Random Forest (RF), Gradient Boosting Tree (GBT) and support vector machine (SVM) models. The accuracy of these models reached higher than 90%. Among the models, the sensitivity and specificity of GBT were the best, reaching 96.25% and 97.33%, respectively [16]. Diego Bedin Marin et al. predicted coffee leaf rust based on the Logical Model Tree (LMT), Random Tree (RT) and RF models. The results showed that the LMT method contributed the most to the accurate prediction of early and several infection categories [17]. Some scholars have also established mechanistic models to simulate the occurrence and development processes of diseases and have achieved good results [18]. Sarinya Kirtpaiboona et al. used a Susceptible Exposed Infectious Recovered (SEIR) model to simulate the occurrence and development of rice blasts. Then, they predicted the severity of rice blast and achieved good results [19]. Henderson et al. used the logistic regression mechanism model to predict potato late blight. Their results showed that the logistic model could accurately predict the occurrence of late potato blight, with sensitivity and specificity values of 75% and 62.5%, respectively [20]. However, although the prediction accuracy based on the machine learning model was high, it is a black box model [21] that fails to consider the biological mechanism of disease occurrence and development and has poor generalization and stability. The SEIR model requires many parameters, and its accuracy and stability depend on whether the input information is accurate and complete [19,22]. The structure of the logistic mechanism-based model is more suitable for local simulation [18,23]. To address the above-mentioned problems, this paper first quantitatively expressed the factor weight of the disease occurrence and development mechanism in the disease prediction model by using a logistic mechanism-based model, and then, factor weights were fused into the disease prediction factors and input into the KNN model. By inputting the prediction factor of the fusion weight, the prediction accuracy and stability of the KNN model for wheat FHB can be improved, and high-precision prediction of wheat FHB can be realized.

In this paper, a logistic mechanism-based model and KNN model were coupled to study the high-precision prediction of winter wheat FHB. First, we obtained satellite images of the study area with high temporal and spatial resolutions and information about the study area. Then, we extracted the prediction factors by integrating remote sensing and meteorology. Factors were selected according to different wheat growth stages, and the factor weights of disease occurrence and development mechanism in the disease prediction model were quantitatively expressed according to the logistic mechanism-based model. Then, the factor weights were combined with the predictive factors and input into the prediction model to predict wheat Fusarium head blight. Remote sensing-based high-precision prediction of wheat FHB was realized.

2. Materials and Methods

2.1. Study Area and Data

The study area addressed in this paper is located in Changfeng county, Dingyuan county, Fengyuang county, and Feidong county (29°24′N~34°39′N, 114°53′E~119°39′E), Anhui Province (Figure 1). The area is located in the lower reaches of the Yangtze River. The planting area of winter wheat is approximately 2450 square kilometres, accounting for 30% of the total area. The wheat seeding time in Anhui, including our study area, was in October 2020. The main wheat variety here is Yangmai 25, which is susceptible to FHB. The region has a subtropical humid continental monsoon climate, with an annual precipitation of approximately 1000 mm and an average annual temperature of 15 °C [24,25]. Due to the influence of monsoons, the amount of precipitation is highest in spring and summer. In addition, due to the high planting density of wheat, wheat varieties susceptible to FHB, sufficient bacterial sources, suitable precipitation and temperature, and other conditions, the occurrence of winter wheat FHB is frequent in this area [26].

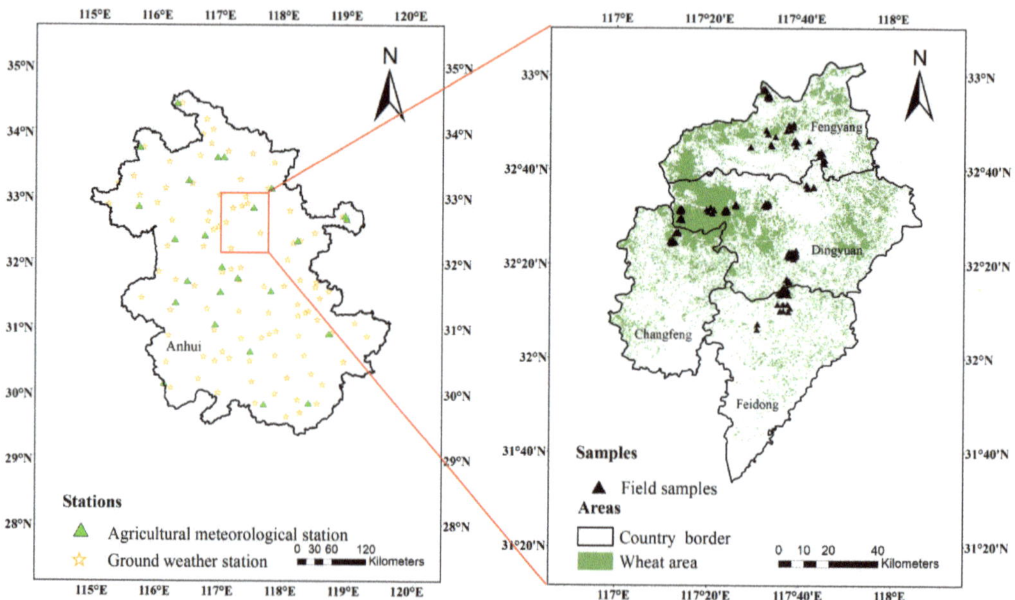

Figure 1. Study area and sampling sites.

We obtained ground survey data, satellite images, and meteorological data for the abovementioned study area. To collect reference data for wheat FHB predictions, we conducted two field investigations on 29 April and 10 May 2021. We selected 154 homogenous wheat areas with width, length, and the distance to each one all larger than 30 m were selected. A GNSS receiver was utilized to obtain the longitude and latitude of the centre of the area. In each area, five 1×1 m^2 plots were randomly selected out to investigate the number of wheat plants and diseased plants. Subsequently, the average disease incidence of the five plots was calculated to be the disease incidence in the area. Satellite images include Sentinel-2 reflectance products, MODIS reflectance products and Sentinel-3 reflectance products. The Sentinel-2 products have a spatial resolution of 20 m, MODIS products have a spatial resolution of 250 m and Fractional Vegetation Cover (FVC) product from the Copernicus Global Land Service (CCLS) have a spatial resolution of 250 m. The meteorological data include daily temperature, relative humidity and precipitation from ground meteorological stations. Spatial kriging interpolation of the meteorological data was carried out based on data from 47 meteorological stations in and around the study area. The spatial resolution of the interpolated meteorological data is 250 m.

2.2. Construction of a Remote Sensing Prediction Model of Wheat FHB

To solve the problems of low prediction accuracy and poor model stability caused by less consideration of the contribution of disease occurrence and development mechanisms in wheat FHB prediction models and achieve high-precision predictions of wheat FHB, in this paper, a coupled logistic mechanism-based model and KNN machine learning model was proposed to predict wheat FHB. The basic steps of the model are as follows. First, due to the different sensitivities of wheat FHB to disease prediction factors at different stages of wheat growth, we selected potential disease prediction factors and divided them into test and training sets at different stages of growth [13,27]. Second, the factor weights of disease occurrence and development mechanism in the disease prediction model were quantitatively expressed by using a logistic regression mechanism-based model, and the factor weight was fused with the prediction factor. Finally, the prediction factors with weights were input into the KNN model to create high-precision predictions of wheat FHB, and the model was evaluated (Figure 2).

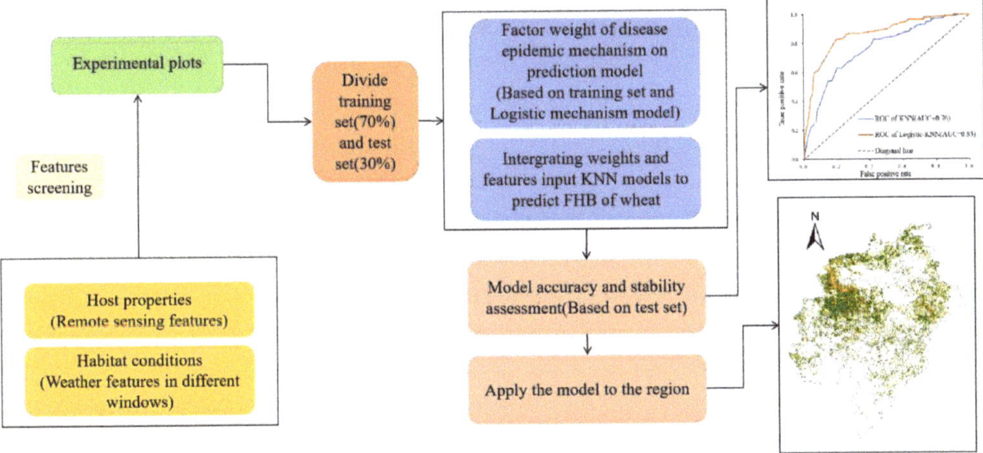

Figure 2. Flow chart of remote sensing-based prediction of wheat FHB area and samplings.

2.2.1. Selection of Disease Prediction Factors

According to the characteristics of wheat FHB, we preliminarily selected the remote sensing factors and meteorological factors. Wheat FHB usually causes changes in leaf morphology, colour, and other aspects of plant morphology, resulting in changes in optical properties [15]. Therefore, when selecting the remote sensing features used to describe the crop growth environment, we mainly considered the features that can reflect crop nutrition and disease stress, for example, the Fractional Vegetation Cover (FVC), Plant Senescence Reflection Index (PSRI), Red-edge Head Blight Index (REHBI), Modified Chlorophyll Absorption Ratio Index (MCARI), Differential Vegetation Index (DVI), Triangular Vegetation Index (TVI), etc. [28–32]. Among these indexes, the PSRI reflects stress in the crop canopy well [28,29], and the TVI describes the radiant energy absorbed by pigments [33].

For meteorological factors, we first investigated the heading stage and flowering stage of wheat in the study area. The heading date of wheat in our study area was

20 April 2021, and the flowering date was 24 April 2021. Then, 3, 5 and 7 days before and after flowering and heading, we calculated the average temperature (TAVG), humidity (RHAVG) and precipitation (PAVG) respectively. We also calculated the number of days with humidity greater than 60%, 70% and 80% (RHT60, RHT70, and RHT80, respectively) and temperature greater than 15 °C and less than 30 °C and the number of days with precipitation (PDAY) [34].

To improve the accuracy of the prediction model and prevent redundancy caused by the inclusion of too many features, at the two time points (29 April and 10 May), we adopted the method of Logistic + L1 regularization and Fisher Score to select the factor library containing 303 remote sensing factors and meteorological factors.

2.2.2. The Factor Weight of the Disease Epidemic Mechanism in the Prediction Model Was Expressed Quantitatively

The logistic regression model is a mechanism model used to describe the progress of wheat FHB [35], and its application and interpretation have been reviewed in detail (Campbell and Madden, 1990). The logistic regression expression is:

$$y = \frac{1}{1 + e^{-\beta^T * X}} \tag{1}$$

The vector $X = (x_0, x_1, x_2, ...x_n)$ is the prediction vector and is composed of each disease prediction factor, including both remote sensing-based factors and meteorological factors. $\beta = (\beta_0, \beta_1, \beta_2, ..., \beta_n)$ are the coefficients of each factor, and y is the disease probability of wheat FHB [18,36]. When there was no collinearity among the factors, the logistic regression model that fit with the training set quantitatively expressed the factor weights of disease occurrence and development mechanism in the disease prediction model [37]. The factor weights were expressed quantitatively after the square of the standard regression coefficient of each disease prediction factor was normalized.

There was usually collinearity among disease prediction factors, which affected the standard regression coefficient fitted by the logistic regression; therefore, it was necessary to reduce collinearity among the disease prediction factors [38]. First, the orthogonal representation of the original factors was found by using singular value decomposition to remove the correlation between the disease prediction factors [39], and it was assumed that X was a full rank matrix of disease prediction factors of size N*J, where N is the number of rows, J is the number of columns, and N > J. Then, singular value decomposition was performed on X:

$$X = P\Delta Q^T \tag{2}$$

where P is the feature matrix of XX^T, Q is the feature matrix of X^TX, and the matrix Δ is composed of the square root of the eigenvalues of X^TX. Therefore, the least-squares orthogonal standard of factor X was approximate:

$$Z = PQ^T \tag{3}$$

Since there was no collinearity between the various characteristic components of Z, we first used Z as the training set. The standard regression coefficient for disease prediction factor Z was fitted using the logistic regression model, and then the standard regression coefficient was converted to a standard regression for disease prediction factor X using the relationship between the original disease prediction factors X and Z [37].

Because the outcome variable of the logistic regression is usually the category, the method of calculating the standardized regression coefficient by multiple linear regression was not applicable to the logistic regression. Menard (2004) proposed a method for calculating the standard regression coefficient with Z as the disease prediction factor, and the expression is:

$$\beta_M^* = (b)(s_Z)(R_0) / (s_{y^*}) \tag{4}$$

where b is the nonstandardized logistic regression coefficient, s_Z is the standard deviation of the disease prediction factor Z, R_0 is the square root of the goodness of fit of the logistic regression, and s_{y^*} is the standard deviation of the predicted value y^* in the logistic regression [40]. Since the disease prediction factors in Z have no collinearity, β_M^{*2} can represent the relative importance of the disease prediction factors. Below, we convert β_M^{*2} into the relative importance coefficient when X is the disease prediction factor. First, we performed a linear regression for each component in X about Z and obtained the linear regression coefficient, the expression of which is as follows:

$$\omega^* = \left(Z^T Z\right)^{-1} Z^T X \tag{5}$$

Using Equations (4) and (5), we obtained the relative importance coefficient ε^{*2} of disease prediction factor X [37], expressed as follows:

$$\varepsilon^{*2} = \omega^{*2} \beta_M^{*2} \tag{6}$$

Finally, we standardized the value to obtain the factor weight:

$$\omega_i = \frac{\varepsilon^{*2}{}_i}{\sum_{i=1}^{m} \varepsilon^{*2}{}_i} \tag{7}$$

2.2.3. Prediction of Wheat FHB with KNN Coupled with the Logistic Mechanism-Based Model

The selected disease prediction factors from 29 April 2021 and 10 May 2021 were divided into a training set and test set at 7:3. We quantitatively expressed the factor weights of disease occurrence and development mechanism in the prediction model by using a logistic regression model and joined them with the prediction factors. The disease prediction factors with weights were input into the KNN model to predict the incidence of wheat FHB in the study area on 29 April 2021 and 10 May 2021. The prediction accuracy was calculated, and the model was evaluated using the F1 index as well as receiver operating characteristic (ROC) curves.

Although KNN is simple, it can achieve good classification results in many scenarios [41–44]. However, due to its lack of focus on mechanisms, the KNN model has a relatively general performance in predicting wheat disease [45]. Therefore, we used the logistic mechanism-based model to quantitatively express the factor weights of disease occurrence and development mechanisms in the KNN model, to integrate the weights with the prediction factors, and to input the prediction factors with the weights into the KNN model to obtain a KNN wheat FHB prediction model coupled with a logistic mechanism-based model (logistic-KNN). The distance formula for the logistic-KNN model is as follows:

$$d_{sj} = \sqrt{\sum_{i=1}^{m} \left(a_{si} - b_{ji}\right)^2 * \omega_i} \; i = 1, 2..., m \tag{8}$$

where d_{sj} is the Euclidean distance between the sth training sample and the jth test sample, a_{si} is the ith component of the sth training sample, b_{ji} is the ith component of the jth test sample, ω_i is the factor weight of the disease development mechanism in the prediction model, and m is the number of disease prediction factors. In the process of prediction, we randomly divided the set of prediction factors into a training set and a test set at a certain proportion. To ensure the best prediction effect from the model, we used cross-validation to obtain the best k value [46]. Then, to evaluate the performance of the KNN model with the best k value coupled with the logistic mechanism-based model, we tested the predictive ability of the model with the test dataset. We calculated the overall accuracy to evaluate the accuracy of the classification. Then, considering the precision and recall, we calculated the F1 score. The F1 score is defined as follows:

$$F_1 = \frac{2 * P * R}{P + R} \tag{9}$$

where P is the precision value, and R is the recall value. In addition, we used the ROC curve to evaluate the specificity and sensitivity of the model.

Due to the coupling of the logistic mechanism-based model and KNN model compared with traditional KNN models, the model had higher accuracy and better stability. It was more suitable for the high-precision prediction of wheat FHB.

3. Results

3.1. Selection of Remote Sensing-Based Disease Prediction Factors of Wheat FHB

Based on the host and habitat conditions, we created 303 prediction factors, including meteorological factors and remote sensing-based factors. According to the sensitivities of wheat FHB factors at different growth stages, we selected factors at the two time points of 29 April 2021 and 10 May 2021 by using the combination of logistic and L1 regularization and Fisher scores. On 29 April 2021, we obtained eight prediction factors (Table 1), including six meteorological factors and two remote sensing-based factors, and on 10 May 2021, we obtained seven prediction factors (Table 2).

Table 1. Feature selection results based on logistic and L1 regularization and Fisher scores (29 April 2021).

Index	Definition
F_RH70_mid_-11	The number of days with RH greater than 70% in the 11 days before and after flowering
H_RH70_-7	The number of days with RH greater than 70% in the 7 days before heading
H_RH_+3	The average RH in the 3 days after heading
F_P_+11	The average precipitation of the 11 days after flowering
H_T1530RH80_mid_-5	The number of days with temperature between 15 and 30 °C degrees and RH greater than 80% in the 5 days before and after heading
F_P_-3	The average precipitation during the first three days of flowering
0404_REHBI	Red-edge head blight index on 4 April 2021
0404_MCARI	Modified chlorophyll absorption ratio on 4 April 2021

Table 2. Feature selection results based on logistic and L1 regularization and Fisher scores (10 May 2021).

Index	Definition
H_T1530RH70_+5	The number of days with temperatures between 15 and 30 °C degrees and RH greater than 70% in the 5 days after heading
H_RH70_-7	The number of days with RH greater than 70% in the 7 days before heading
F_T1530RH80_mid_-7	The number of days with temperatures between 15 and 30 °C degrees and RH greater than 80% in the 7 days before and after flowering
H_P_-15	The average precipitation of the 15 days before heading
H_P_-7	The average precipitation of the 7 days before heading
0501_PSRI	Plant senescence absorption ratio on 1 May 2021
0320_FVC	Fractional Vegetation Cover on 20 March 2021

3.2. Quantitative Expression of the Weight of the Prediction Model Based on the Logistic Mechanism-Based Model

The selected disease prediction factors for 29 April and 10 May were divided into a test set and training set at a 3:7 ratio, and then, the disease prediction factors in the training set were decomposed by singular value decomposition to obtain linearly independent disease prediction factors. Then, we obtained the weights of the factors using the linearly independent disease prediction factors (Figure 3).

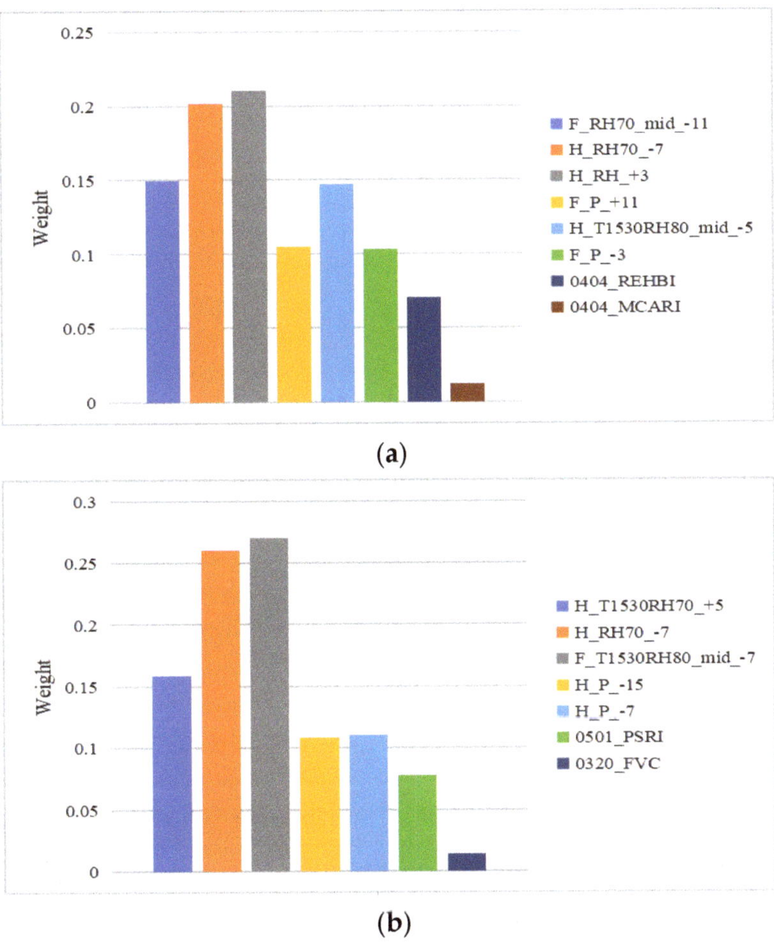

Figure 3. Weighting results of factors on 29 April 2021 (**a**) and 10 May 2021 (**b**).

We found that the weights of meteorological factors were generally larger than those of the remote sensing-based factors on 29 April 2021 and 10 May 2021, and the weights of the meteorological factors related to humidity were larger than those of the remaining meteorological factors. Thus, the weight of disease prediction factors expresses the mechanism of disease occurrence.

3.3. Remote Sensing-Based Prediction of Wheat FHB

Because remote sensing images vary with time, remote sensing factors will change over time. The meteorological factors were based on data acquired at meteorological stations rather than prediction data; thus, the meteorological factors associated with the samples did not change. That is, the factors associated with each sample include remote sensing-based factors and meteorological factors; the remote sensing-based factors change with the acquisition of satellite images, and the meteorological factors are fixed. Based on the characteristics of samples in different periods, the incidence of wheat FHB on 29 April 2021 and 10 May 2021 was predicted. The 154 sample points collected on these two dates were randomly divided into a training set and test set at a 7:3 ratio. The training set was used to determine the parameter values of the logistic-KNN model, and a prediction model was constructed for these two dates to calculate the accuracy of wheat disease predictions on each date. The prediction results were displayed on a map. Then, the total accuracy, classification performance, and generalizability of the models were evaluated with the corresponding test datasets. For comparison, we constructed and evaluated a KNN model in the same way.

First, to optimize the parameter value k, which is the number of nearest neighbours selected in each prediction, in the logistic-KNN model, we conducted a five-fold cross-validation with the training datasets from different dates [46,47]. Taking the prediction for 29 April 2021 as an example, Table 1 shows the accuracy of the coupled logistic-KNN model for different k values. For the prediction based on meteorological and remote sensing-based characteristics on 29 April 2021, the average prediction accuracy results (Table 3) indicate that the accuracy rate was highest when k = 3. The k value of the coupled logistic-KNN model for 10 May 2021 was determined in the same way.

Table 3. The mean overall accuracy according to fivefold cross-validation for the 29 April 2021 model when the threshold of probability was 0.

Prediction Model	29 April 2021	
	k	Mean Overall Accuracy
Logistic-KNN	3	0.865
Logistic-KNN	5	0.781
Logistic-KNN	7	0.709

We used the logistic-KNN model with the best k value to predict the occurrence of wheat FHB on 29 April 2021 and 10 May 2021. Moreover, we compared the KNN model with the logistic-KNN model. The prediction results were divided into two states: infected and not infected. As shown in Figure 4, the occurrence of wheat FHB is generally severe in Dingyuan county. The distribution of diseased wheat was approximately the same overall, and the trends were similar between the two models. In general, the prevalence of wheat FHB gradually increased over time. The comparison of the prediction results of the KNN model and logistic-KNN model showed that the KNN model indicated more severe infection on 29 April 2021 and less severe infection on 10 May 2021 than the logistic-KNN model.

To evaluate the advantages of the logistic-KNN model in predicting wheat FHB, we compared the performance of the KNN model with that of the logistic-KNN model. Taking 0 as the threshold of the disease incidence, wheat plots with a disease incidence value of 0 indicated absence of diseased wheat, and wheat plots with a disease incidence greater than 0 indicated diseased wheat. The overall accuracy and F1 index values of the two models on different dates were calculated (Tables 4 and 5). The results indicate that the logistic-KNN model was better than the KNN model in terms of the overall accuracy and F1 index value. The logistic-KNN model had more advantages in predicting wheat FHB.

In addition, we randomly divided the ground survey points from 29 April 2021 and 10 May 2021 into training and testing sets at a 7:3 ratio. We made 10 predictions, obtained 10 ROC curves, obtained the average ROC curve, and calculated the AUC (Figure 5a,b). The comparison results show that the logistic-KNN model had stronger specificity and sensitivity than the KNN model, and the AUC was greater than 0.5, indicating that the two models were better than a random guess for predicting wheat FHB and thus have reference value. In addition, the logistic-KNN model had a stronger prediction ability and generalizability than the KNN model. As the prediction date neared the onset of wheat FHB, the AUC index value and prediction accuracy increased.

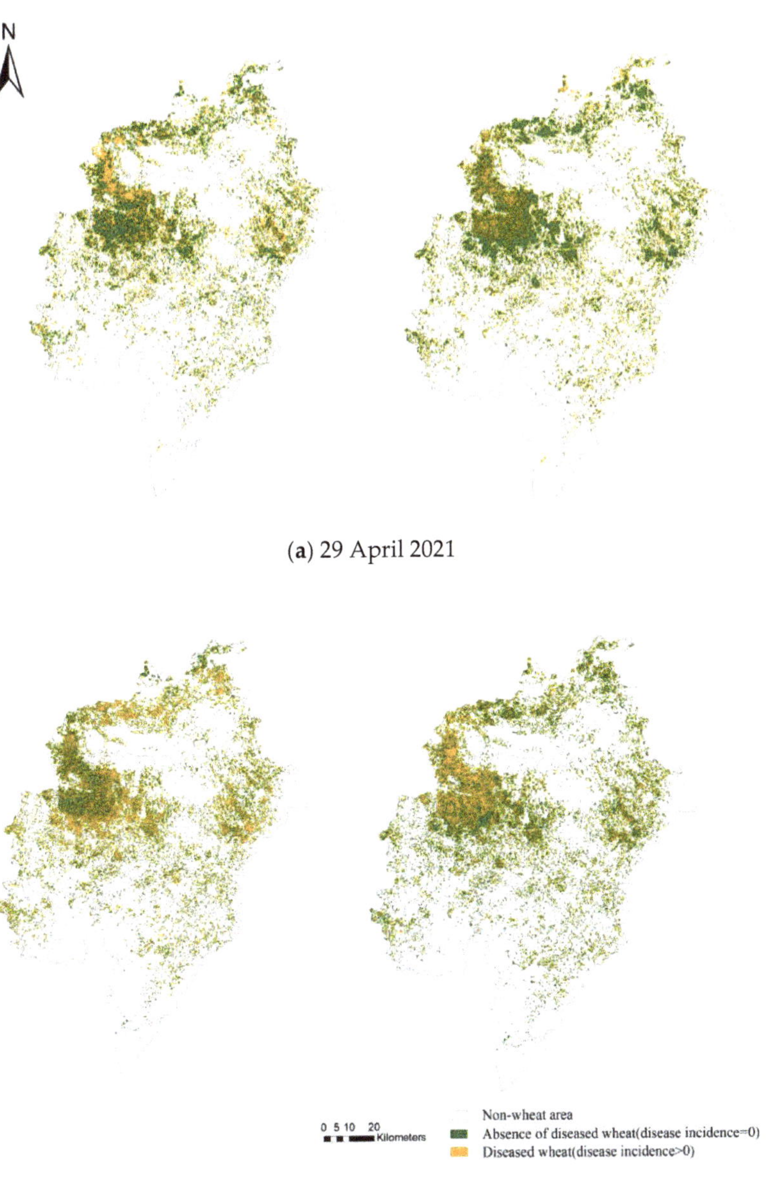

(**a**) 29 April 2021

(**b**) 10 May 2021

Figure 4. Predictions of FHB severity (**a**,**b**). The results of the logistic-KNN model are shown on the left, and those of the KNN model are shown on the right.

Table 4. Overall accuracies and F1 scores of the two models for 29 April 2021.

Prediction Model	29 April 2021	
	Accuracy	F1 Score
Logistic-KNN	0.88	0.86
KNN	0.75	0.68

Table 5. Overall accuracies and F1 scores of the two models for 10 May 2021.

Prediction Model	10 May 2021	
	Accuracy	F1 Score
Logistic-KNN	0.92	0.94
KNN	0.79	0.80

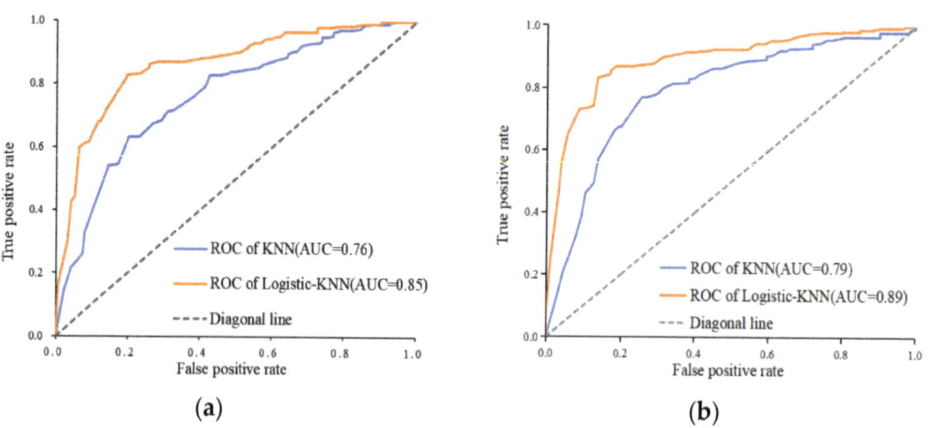

Figure 5. Average ROC curves and their corresponding AUC values: (**a**) 29 April 2021; (**b**) 10 May 2021.

4. Discussion

Researchers have used various methods to predict wheat FHB [1,48,49]. Compared with previous methods, the main advantage of the method proposed in this paper is the use of a logistic mechanism-based model to quantitatively express the factor weights of disease occurrence and development mechanism in disease prediction models. Considering a comprehensive set of meteorological and remote sensing factors, the disease prediction factors of wheat FHB for 29 April 2021 and 10 May 2021 were selected according to the incidence characteristics of wheat FHB in different growth stages. Finally, REHBI and MCARI were selected as the remote sensing-based factors for the 29 April 2021 models, and FVC and PSRI were selected as the remote sensing-based factors for the 10 May 2021 models, reflecting the growth status of wheat. In addition, five meteorological factors before and after heading or flowering were included in the prediction factors selected. Based on the selected disease prediction factors, we predicted the occurrence of wheat FHB on 29 April 2021 and 10 May 2021. In addition, we compared the logistic-KNN model with the KNN model and found that the former had higher classification accuracy and stability for wheat FHB.

Previously, researchers used machine learning models to predict wheat FHB and achieved good results, which proved the feasibility of applying machine learning models in the prediction of wheat FHB [50–52]. However, the interpretability of machine learning models is poor; thus, it is difficult to achieve stable predictions of wheat FHB [53]. In this paper, a high-precision model for predicting wheat FHB was constructed by coupling a logistic mechanism-based model with a KNN model. Compared with the KNN model alone, the generalizability and stability were improved in the logistic-KNN model. Although this study has achieved satisfactory results in the prediction of wheat FHB, there are still some deficiencies that need to be improved upon in future research. First, apart from these key meteorological factors that were considered in our methods, numerous other meteorological factors influence the occurrence of FHB infection, and the effect cannot be ignored under certain circumstances. Thus, in future studies, we will try to add predictors such as wind speed to predict wheat FHB. Then, the logistic-KNN model needs to include all the points in the training set when searching for the nearest neighbour; this will result in a large number of calculations in the prediction [54–57]. Many methods have been used to reduce the number of calculations associated with finding nearest neighbour points, such as constructing a k-d tree [58,59]. To construct a more extensive and effective prediction model of wheat FHB, these methods can be used to reduce the number of calculations in future research. Second, the method proposed in this paper produces binary classifications based on multiple features. To achieve more accurate predictions of wheat FHB, we will try to classify wheat FHB according to the disease grade in future research.

Our method results in a more effective model for predicting wheat FHB. Remote sensing-based and meteorological factors are included in the prediction models as input. As the remote sensing-based factors were more closely related than the meteorological factors to the heading and flowering periods, the performance of the model was better when remote sensing-based factors were included, with accuracies reaching more than 90% and the F1 index reaching approximately 0.9. We evaluated the KNN model and logistic-KNN model, and the results indicated that the former had better model accuracy, stability, and generalizability than the latter, and the prediction accuracy improved by approximately 13% when the logistic-KNN model was used.

5. Conclusions

In this study, a remote sensing-based prediction method for wheat FHB was established. The mechanism of disease occurrence and development plays a key role in the prediction of wheat FHB. By inputting disease prediction factors that integrate host and habitat conditions into the logistic mechanism-based model, the factor weights of disease occurrence and development mechanism in the disease prediction model were quantitatively expressed. The disease prediction factors joined with factor weights, which were input into the KNN model to predict the incidence of wheat FHB in the study area. The results show that the logistic-KNN model can improve the prediction accuracy and stability of machine learning models.

In future research, we will consider some important problems. For example, we will try to use the k-d tree to reduce the computational complexity of the model and make the model more widely used, and we will try to classify the severity of wheat FHB and achieve a more accurate prediction of wheat FHB.

Author Contributions: Data curation, L.L. (Linyi Liu); formal analysis, Y.D., Y.X. and L.L. (Lu Li); funding acquisition, Y.D. and W.H.; investigation, L.L. (Linyi Liu); methodology, L.L. (Lu Li), Y.D., Y.X. and X.Z.; resources, W.H.; writing—original draft, L.L. (Lu Li). All authors have read and agreed to the published version of the manuscript.

Funding: This work was supported by the National Key R&D Program of China (2021YFE0194800), National Natural Science Foundation of China 42071423, Alliance of International Science Organizations (grant no. ANSO-CR-KP-2021-06), Beijing Nova Program of Science and Technology (Z191100001119089), and the Program of Bureau of International Cooperation, Chinese Academy of Sciences (183611KYSB20200080).

References

1. Dweba, C.C.; Figlan, S.; Shimelis, H.A.; Motaung, T.E.; Sydenham, S.; Mwadzingeni, L.; Tsilo, T.J. Fusarium head blight of wheat: Pathogenesis and control strategies. *Crop Prot.* **2017**, *91*, 114–122. [CrossRef]
2. Jia, H.; Zhou, J.; Xue, S.; Li, G.; Yan, H.; Ran, C.; Zhang, Y.; Shi, J.; Jia, L.; Wang, X.; et al. A journey to understand wheat Fusarium head blight resistance in the Chinese wheat landrace Wangshuibai. *Crop J.* **2018**, *6*, 48–59. [CrossRef]
3. Wegulo, S.N.; Baenziger, P.S.; Nopsa, J.H.; Bockus, W.W.; Hallen-Adams, H. Management of Fusarium head blight of wheat and barley. *Crop Prot.* **2015**, *73*, 100–107. [CrossRef]
4. Ma, H.; Zhang, X.; Yao, J.; Cheng, S. Breeding for the resistance to Fusarium head blight of wheat in China. Front. *Agric. Sci. Eng.* **2019**, *6*, 251–264.

5. Shah, L.; Ali, A.; Yahya, M.; Zhu, Y.; Wang, S.; Si, H.; Rahman, H.; Ma, C. Integrated control of fusarium head blight and deoxynivalenol mycotoxin in wheat. *Plant Pathol.* **2018**, *67*, 532–548. [CrossRef]

6. Chen, Y.; Zhang, A.F.; Gao, T.C.; Zhang, Y.; Wang, W.X.; Ding, K.J.; Chen, L.; Sun, Z.; Fang, X.Z.; Zhou, M.G. Integrated use of pyraclostrobin and epoxiconazole for the control of Fusarium head blight of wheat in Anhui Province of China. *Plant Dis.* **2012**, *96*, 1495–1500. [CrossRef]

7. Guo, X.; Wang, M.T.; Zhang, G.Z. Prediction model of meteorological grade of wheat stripe rust in winter-reproductive area, Sichuan Basin, China. *Ying Yong Sheng Tai Xue Bao = J. Appl. Ecol.* **2017**, *28*, 3994–4000.

8. Rodríguez-Moreno, V.M.; Jiménez-Lagunes, A.; Estrada-Avalos, J.; Mauricio-Ruvalcaba, J.E.; Padilla-Ramírez, J.S. Weather-data-based model: An approach for forecasting leaf and stripe rust on winter wheat. *Meteorol. Appl.* **2020**, *27*, e1896. [CrossRef]

9. El Jarroudi, M.; Kouadio, L.; Bock, C.H.; El Jarroudi, M.; Junk, J.; Pasquali, M.; Maraite, H.; Delfosse, P. A threshold-based weather model for predicting stripe rust infection in winter wheat. *Plant Dis.* **2017**, *101*, 693–703. [CrossRef]

10. Zhang, J.; Yuan, L.; Nie, C.; Wei, L.; Yang, G. Forecasting of powdery mildew disease with multi-sources of remote sensing information. In Proceedings of the 2014 The Third International Conference on Agro-Geoinformatics, Beijing, China, 11–14 August 2014; pp. 1–5.

11. Li, X.; Yang, C.; Huang, W.; Tang, J.; Tian, Y.; Zhang, Q. Identification of cotton root rot by multifeature selection from sentinel-2 images using random forest. *Remote Sens.* **2020**, *12*, 3504. [CrossRef]

12. Bajwa, S.G.; Rupe, J.C.; Mason, J. Soybean disease monitoring with leaf reflectance. *Remote Sens.* **2017**, *9*, 127. [CrossRef]

13. Xiao, Y.; Dong, Y.; Huang, W.; Liu, L.; Ma, H.; Ye, H.; Wang, K. Dynamic remote sensing prediction for wheat fusarium head blight by combining host and habitat conditions. *Remote Sens.* **2020**, *12*, 3046. [CrossRef]

14. Li, W.; Huang, W.; Dong, Y.; Chen, H.; Wang, J.; Shan, J. Estimation on winter wheat scab based on combination of temperature, humidity and remote sensing vegetation index. *Trans. Chin. Soc. Agric. Eng.* **2017**, *33*, 203–210.

15. Li, W.; Liu, Y.; Chen, H.; Zhang, C.C. Estimation model of winter wheat disease based on meteorological factors and spectral information. *Food Prod. Process. Nutr.* **2020**, *2*, 5. [CrossRef]

16. Khalili, E.; Kouchaki, S.; Ramazi, S.; Ghanati, F. Machine learning techniques for soybean charcoal rot disease prediction. *Front. Plant Sci.* **2020**, *11*, 2009. [CrossRef] [PubMed]

17. Marin, D.B.; Santana, L.S.; Barbosa, B.D.S.; Barata, R.A.P.; Osco, L.P.; Ramos, A.P.M.; Guimarães, P.H.S. Detecting coffee leaf rust with UAV-based vegetation indices and decision tree machine learning models. *Comput. Electron. Agric.* **2021**, *190*, 106476. [CrossRef]

18. Shah, D.A.; Molineros, J.E.; Paul, P.A.; Willyerd, K.T.; Madden, L.V.; De Wolf, E.D. Predicting Fusarium head blight epidemics with weather-driven pre-and post-anthesis logistic regression models. *Phytopathology* **2013**, *103*, 906–919. [CrossRef] [PubMed]

19. Kirtphaiboon, S.; Humphries, U.; Khan, A.; Yusuf, A. Model of rice blast disease under tropical climate conditions. *Chaos Solitons Fractals* **2021**, *143*, 110530. [CrossRef]

20. Henderson, D.; Williams, C.J.; Miller, J.S. Forecasting late blight in potato crops of southern Idaho using logistic regression analysis. *Plant Dis.* **2007**, *91*, 951–956. [CrossRef]

21. Rudin, C. Stop explaining black box machine learning models for high stakes decisions and use interpretable models instead. *Nat. Mach. Intell.* **2019**, *1*, 206–215. [CrossRef] [PubMed]

22. Papastamati, K.; van den Bosch, F. The sensitivity of the epidemic growth rate to weather variables, with an application to yellow rust on wheat. *Phytopathology* **2007**, *97*, 202–210. [CrossRef]

23. HarDIan, J.M. A logistic model simulating environmental changes associated with the growth of populations of rice weevils, Sitophilus oryzae, reared in small cells of wheat. *J. Appl. Ecol.* **1978**, *15*, 65–87. [CrossRef]

24. Shan, C.; Wang, W.; Liu, C.; Sun, Y.; Hu, Q.; Xu, X.; Tian, Y.; Zhang, H.; Morino, L.; Griffith, D.W.T.; et al. Regional CO emission estimated from ground-based remote sensing at Hefei site, China. *Atmos. Res.* **2019**, *222*, 25–35. [CrossRef]

25. Liu, L.; Dong, Y.; Huang, W.; Du, X.; Ren, B.; Huang, L.; Zheng, Q.; Ma, H. A disease index for efficiently detecting wheat fusarium head blight using sentinel-2 multispectral imagery. *IEEE Access* **2020**, *8*, 52181–52191. [CrossRef]

26. Chen, Y.; Yang, X.; Gu, C.Y.; Zhang, A.F.; Gao, T.C.; Zhou, M.G. Genotypes and phenotypic characterization of field Fusarium asiaticum isolates resistant to carbendazim in Anhui Province of China. *Plant Dis.* **2015**, *99*, 342–346. [CrossRef]

27. Huang, L.; Li, T.; Ding, C.; Zhao, J.; Zhang, D.; Yang, G. Diagnosis of the severity of Fusarium head blight of wheat ears on the basis of image and spectral feature fusion. *Siensors* **2020**, *20*, 2887. [CrossRef]

28. Ren, S.; Chen, X.; An, S. Assessing plant senescence reflectance index-retrieved vegetation phenology and its spatiotemporal response to climate change in the Inner Mongolian Grassland. *Int. J. Biometeorol.* **2017**, *61*, 601–612. [CrossRef]

29. Hatfield, J.L.; Prueger, J.H. Value of using different vegetative indices to quantify agricultural crop characteristics at different growth stages under varying management practices. *Remote Sens.* **2010**, *2*, 562–578. [CrossRef]

30. Zhang, Z.; Liu, M.; Liu, X.; Zhou, G. A new vegetation index based on multitemporal Sentinel-2 images for discriminating heavy metal stress levels in rice. *Sensors* **2018**, *18*, 2172. [CrossRef]

31. Guo, A.; Huang, W.; Dong, Y.; Ye, H.; Ma, H.; Liu, B.; Wu, W.; Ren, Y.; Ruan, C.; Geng, Y. Wheat yellow rust detection using UAV-based hyperspectral technology. *Remote Sens.* **2021**, *13*, 123. [CrossRef]

32. Wu, C.; Niu, Z.; Tang, Q.; Huang, W. Estimating chlorophyll content from hyperspectral vegetation indices: Modeling and validation. *Agric. For. Meteorol.* **2008**, *148*, 1230–1241. [CrossRef]

33. Ren, Y.; Huang, W.; Ye, H.; Zhou, X.; Ma, H.; Dong, Y.; Shi, Y.; Geng, Y.; Huang, Y.; Jiao, Q. Quantitative identification of yellow rust in winter wheat with a new spectral index: Development and validation using simulated and experimental data. *Int. J. Appl. Earth Obs. Geoinf.* **2021**, *102*, 102384. [CrossRef]
34. Gilbert, J.; Haber, S. Overview of some recent research developments in Fusarium head blight of wheat. *Can. J. Plant Pathol.* **2013**, *35*, 149–174. [CrossRef]
35. Xiangxiang, W.; Quanjiu, W.; Jun, F.; Lijun, S.; Xinlei, S. Logistic model analysis of winter wheat growth on China's Loess Plateau. *Can. J. Plant Sci.* **2014**, *94*, 1471–1479. [CrossRef]
36. King, E.N.; Ryan, T.P. A preliminary investigation of maximum likelihood logistic regression versus exact logistic regression. *Am. Stat.* **2002**, *56*, 163–170. [CrossRef]
37. Tonidandel, S.; LeBreton, J.M. Determining the relative importance of disease prediction factors in logistic regression: An extension of relative weight analysis. *Organ. Res. Methods* **2010**, *13*, 767–781. [CrossRef]
38. Owen, A.B.; Roediger, P.A. The sign of the logistic regression coefficient. *Am. Stat.* **2014**, *68*, 297–301. [CrossRef]
39. de Souza, J.C.S.; Assis, T.M.L.; Pal, B.C. Data compression in smart distribution systems via singular value decomposition. *IEEE Trans. Smart Grid* **2015**, *8*, 275–284. [CrossRef]
40. Menard, S. Six approaches to calculating standardized logistic regression coefficients. *Am. Stat.* **2004**, *58*, 218–223. [CrossRef]
41. Uddin, S.; Haque, I.; Lu, H.; Moni, M.A.; Gide, E. Comparative performance analysis of K-nearest neighbour (KNN) algorithm and its different variants for disease prediction. *Sci. Rep.* **2022**, *12*, 6256. [CrossRef]
42. Noola, D.A.; Basavaraju, D.R. Corn leaf image classification based on machine learning techniques for accurate leaf disease detection. *Int. J. Electr. Comput. Eng.* **2022**, *12*, 2088–8708. [CrossRef]
43. Huang, Y.; Zhang, J.; Zhang, J.; Yuan, L.; Zhou, X.; Xu, X.; Yang, G. Forecasting Alternaria Leaf Spot in Apple with Spatial-Temporal Meteorological and Mobile Internet-Based Disease Survey Data. *Agronomy* **2022**, *12*, 679. [CrossRef]
44. Mendigoria, C.H.; Concepcion, R.; Bandala, A.; Alajas, O.J.; Aquino, H.; Dadios, E. OryzaNet: Leaf Quality Assessment of Oryza sativa Using Hybrid Machine Learning and Deep Neural Network. In Proceedings of the 2021 IEEE 13th International Conference on Humanoid, Nanotechnology, Information Technology, Communication and Control, Environment, and Management (HNICEM), Manila, Philippines, 28–30 November 2021; pp. 1–6.
45. Ruan, C.; Dong, Y.; Huang, W.; Huang, L.; Ye, H.; Ma, H.; Guo, A.; Ren, Y. Prediction of Wheat Stripe Rust Occurrence with Time Series Sentinel-2 Images. *Agriculture* **2021**, *11*, 1079. [CrossRef]
46. Kaabneh, K.; Tarawneh, H. Dynamic Tomato Leaves Disease Detection using Histogram-based K-means Clustering Algorithm with Back-Propagation Neural Network. In Proceedings of the 2021 22nd International Arab Conference on Information Technology (ACIT), Muscat, Oman, 21–23 December 2021; pp. 1–5.
47. Arlot, S.; Celisse, A. A survey of cross-validation procedures for model selection. *Stat. Surv.* **2010**, *4*, 40–79. [CrossRef]
48. Weng, S.; Han, K.; Chu, Z.; Zhu, G.; Liu, C.; Zhu, Z.; Zhang, Z.; Zheng, L.; Huang, L. Reflectance images of effective wavelengths from hyperspectral imaging for identification of Fusarium head blight-infected wheat kernels combined with a residual attention convolution neural network. *Comput. Electron. Agric.* **2021**, *190*, 106483. [CrossRef]
49. Prandini, A.; Sigolo, S.; Filippi, L.; Battilani, P.; Piva, G. Review of predictive models for Fusarium head blight and related mycotoxin contamination in wheat. *Food Chem. Toxicol.* **2009**, *47*, 927–931. [CrossRef] [PubMed]
50. Zhu, Z.; Chen, L.; Zhang, W.; Yang, L.; Zhu, W.; Li, J.; Liu, Y.; Tong, H.; Fu, L.; Liu, J.; et al. Genome-wide association analysis of Fusarium head blight resistance in Chinese elite wheat lines. *Front. Plant Sci.* **2020**, *11*, 206. [CrossRef]
51. Ma, H.; Huang, W.; Dong, Y.; Liu, L.; Guo, A. Using UAV-Based Hyperspectral Imagery to Detect Winter Wheat Fusarium Head Blight. *Remote Sens.* **2021**, *13*, 3024. [CrossRef]
52. Qiu, M.; Zheng, S.; Tang, L.; Hu, X.; Xu, Q.; Zheng, L.; Weng, S. Raman Spectroscopy and Improved Inception Network for Determination of FHB-Infected Wheat Kernels. *Foods* **2022**, *11*, 578. [CrossRef]
53. Borrellas, P.; Unceta, I. The Challenges of Machine Learning and Their Economic Implications. *Entropy* **2021**, *23*, 275. [CrossRef]
54. Gao, X.; Li, G. A KNN model based on manhattan distance to identify the SNARE proteins. *IEEE Access* **2020**, *8*, 112922–112931. [CrossRef]
55. Ma, C.; Du, X.; Cao, L. Improved KNN Algorithm for Fine-Grained Classification of Encrypted Network Flow. *Electronics* **2020**, *9*, 324. [CrossRef]
56. Zhang, S.; Li, X.; Zong, M.; Zhu, X.; Wang, R. Efficient kNN classification with different numbers of nearest neighbors. *IEEE Trans. Neural Netw. Learn. Syst.* **2017**, *29*, 1774–1785. [CrossRef] [PubMed]
57. Pujari, M.; Awati, C.; Kharade, S. Efficient Classification with an Improved Nearest Neighbor Algorithm. In Proceedings of the 2018 Fourth International Conference on Computing Communication Control and Automation (ICCUBEA), Pune, India, 16–18 August 2018; pp. 1–5.
58. Shan, Y.; Li, S.; Li, F.; Cui, Y.; Li, S.; Chen, M.; He, X. A density peaks clustering algorithm with sparse search and Kd tree. *arXiv* **2022**, arXiv:2203.00973.
59. Chen, Y.; Zhou, L.; Tang, Y.; Singh, J.P.; Bouguila, N.; Wang, C.; Wang, C.; Du, J. Fast neighbor search by using revised kd tree. *Inf. Sci.* **2019**, *472*, 145–162. [CrossRef]

MDPI AG
Grosspeteranlage 5
4052 Basel
Switzerland
Tel.: +41 61 683 77 34

Remote Sensing Editorial Office
E-mail: remotesensing@mdpi.com
www.mdpi.com/journal/remotesensing